Contributors:

I am extremely grateful to the following individuals who made major contributions to this book. They are listed in alphabetical order. In particular, Ron Bosch and Ted Frankiewicz were involved in the initial conception and planning of the book. But also many others provided valuable input who are not listed here, particularly those people affiliated with the Produced Water Society and the facilities discipline of the Society of Petroleum Engineers.

Munib Ahmad: Munib and I worked together in Oman. We shared a keen interest in using science to solve practical problems while at the same time gathering as much data we could on water chemistry and equipment performance in the field. Munib contributed very strongly to the quality of Chapter 12 on Flotation.

Kris Bansal: Shortly after Kris retired from ConocoPhillips, he accepted my offer to co-teach the SPE course. We worked on the course from the CETCO offices. During those months, he reviewed several chapters and gave me excellent feedback on the book.

Ron Bosch: Ron, Ted and I developed the outline for the book in the form of a detailed Table of Contents in November 2013. While we recognized the enormous ambition of the project, none of us could part with any subject in the outline. In the end, I am grateful to these two gentlemen for laying out an ambitious plan. In the several years that followed, Ron made intellectual contributions and provided ideas for Chapter 5 on Chemical Treatment.

Ted Frankiewicz: As mentioned in Ron's acknowledgement above, Ted, Ron and I developed the outline for the book. In addition, Ted provided copy editing for several chapters. Much of the Chapter on Primary Separation was written by Ted.

Morris Hoagland: Morris wrote a few sections and provided feedback on a number of subjects. The section on activated carbon in Chapter 13 is almost entirely his material. He is one of the world experts on that subject.

Dan Shannon: Dan and Ron Bosch provided the chemist perspective that was needed for Chapter 5 on production chemicals. Both gentlemen provided insightful material.

Ramesh Sharma: Ramesh and I worked together in New Mexico where we studied the high concentration of very fine particles associated with shale produced water. Ramesh provided significant intellectual insight on that subject, and on many other subjects of produced water.

Greg Simpson: Greg wrote the chapter on microbial biology and biological control. Also, Greg mentored me on the process of self-publishing. He is author of a multivolume series of texts on chlorine dioxide which he self-published. Greg is a true scholar. He wrote his chapter with minimal input from me and delivered it with an extensive set of references.

Colin Tyrie: There is a good chance that there would not have been a Produced Water Society if it were not for the efforts of Colin Tyrie and Dan Caudle. If there had not been a Produced Water Society, I and many others might not have become produced water specialists. Thus, much of the credit for the existence of this book goes to Colin. Also, I am grateful to him for helping me in my career, as he did for so many others.

Ming Yang: Ming helped write the chapter on sampling and analysis. It should also be mentioned that Gary Bartman provided a detailed review of that chapter. Between the two of these gentlemen, there is no one who knows the subject better.

PRODUCED WATER

Volume 1:

Fundamentals
Water Chemistry
Emulsions
Chemical Treatment

John M. Walsh

Produced Water

Volume 1:
   Fundamentals
   Water Chemistry
   Emulsions
   Chemical Treatment

Volume 2:
   Equipment
   Process Configuration
   Applications

Authored by: John M. Walsh

Published By: Petro Water Technology, LLC

First Edition, Volume 1 ISBN: 978-1-7322736-0-3
Library of Congress Control Number:

First Edition, Volume 2 ISBN: 978-1-7322736-1-0
Library of Congress Control Number:

Table of Contents

Preface:

The material for this book originated in a course that I started teaching in Oman in 2007. I was an expatriate there working for a multinational oil company. Water treatment is an important issue in Oman since over 95 % of the fluid produced is water. As an expatriate, one of my duties was to mentor and coach the local staff so I began teaching a course in water treatment.

In 2010, I became the instructor for the SPE course on water treatment. I taught the course three or four times a year, across the globe, for more than ten years. Also, I chaired a series of SPE workshops each of which was held in an important area of produced water around the world. The SPE course and workshops provided a wealth of information.

For much of my career I spent a lot of time in the field both onshore and offshore. In addition to that practical experience, another valuable input for this book comes from the scientific literature. Eventually I amassed a collection of a couple thousand papers. One of the more gratifying experiences in writing this book has been the tying together of scientific literature with field observations. Understanding the chemical and physical fundamental mechanisms involved in produced water treatment provides a powerful tool in design and problem solving.

In 2013, Ron Bosch and Ted Frankiewicz had the idea to take the course material and write a book. Together we developed an inspiring table of contents. Kindly, John Occhipinti, my boss at CETCO gave me time to write. My experience at CETCO had a profound effect on the material in the book. The range and complexity of problems, the variety of systems and equipment, the ability to conduct lab studies expediently, and the team atmosphere made my time there a wonderful experience which is hopefully captured in the book.

In November 2018 I joined Jacobs, one of the largest and highly rated engineering firms in the world. The legacy Ch2m was still intact to a great extent within Jacobs. The contents of this book benefitted greatly by my interaction with this talented group of industrial water specialists including Ken Martins, Bruce Thomas-Benke, Michael Dunkel, and Derek Evans.

Looking back, I have been extremely fortunate to have made so many friendships around the subject of produced water. Produced water is a by-product of oil and gas production. As such it tends to be overlooked in the industry and is often considered of secondary importance. But it is also extremely complex, challenging and humbling. People who work in this area enjoy challenges and believe that we are doing something good for the environment. This book is my way of giving something to the many people who work in this area.

John
March 2019

Acknowledgement

This book would not have come into existence without the support and encouragement of my wife, Bawani. She helped to solve many publishing problems. I am extremely grateful for her help.

I would also like to thank my parents, Maryann and James for their constant encouragement.

CHAPTER ONE

Introduction

Chapter One Table of Contents

1.0 Introduction

Scope of this Book: The subject of this book is the treatment of produced and flow back water in the upstream oil and gas industry. The chemistry of produced and flow back water is discussed in detail since there is a strong correlation between water chemistry and viable treatment options. A wide range of treatment technologies are covered in order to address the many different treatment, disposal and reuse options for these fluids. Technologies include coagulation and flocculation chemical treatment and various mechanical technologies such as separators, hydrocyclones, flotation (conventional water treatment technologies), as well as tertiary technologies (filtration, coalescing media), and some technologies that would be considered to be niche technologies for the oil and gas industry including oxidation, biotreatment, clarifiers, and electrocoagulation. Seawater treatment for offshore water flood is also discussed briefly but is not considered in detail. Treatment locations include onshore (conventional and unconventional), offshore, and deep water. Disposal options include overboard discharge, underground injection disposal, water flood, recycle, beneficial use, and enhanced oil recovery. Finally troubleshooting case studies and applications are presented.

This introductory chapter defines the scope of the book and how it is organized. In this section (1.0), the objectives of the book are stated. Section 1.1 defines what is meant by "produced" and "flow back water." The Overview section (Section 1.5 below) gives perspective and background on the subject. It provides a distinction between the different types of produced and flow back water and it helps to explain why certain technologies predominate in certain regions of the world and not in others.

It is always helpful to understand other perspectives on a subject. For those readers who wish to have short introductory material, there are a number of monographs available [1 – 4]. For greater depth, the book by Charles Patton was first published in 1986, and has been updated by Alan Foster in 2007 [5]. The books by Arnold and Stewart provide an understanding of the entire production system including gas and oil conditioning and water treatment [6].

Objectives of this Book: There are multiple objectives for this book. First, it is intended to provide an introduction to the subject for people who may never design or operate a water treatment system but who require an awareness of water properties, the equipment, and processes involved. There are many Facilities, Process and Project engineers who are not water treatment specialists but who need such awareness in order to manage facilities operations and upgrades. For this purpose, each chapter starts with background and basic information. The first half of most chapters is devoted to practical information that can be put to use in everyday applications.

For those scientists and engineers who need to master the subject, the later part of each chapter provides rigorous detail and references to the scientific and engineering literature. That material is intended for specialists who are or will be involved in design, testing, or troubleshooting of water treatment in the oil and gas industry and require advanced knowledge. Each chapter is constructed to take the reader through the four levels of learning: Awareness, Knowledge, Skill and Mastery. Of course, mastery of this subject also requires at least a few years of field experience, conference attendance, and good mentors.

Another goal of the book is to provide a resource for troubleshooting. In order to carry out a competent troubleshooting project, a specialist must have a deep understanding of the *chemical and physical mechanisms* involved in water treatment. A guiding principle in writing this book was to provide enough detail to explain field observations from a mechanistic standpoint. In some cases, a simple basic explanation is sufficient. In other cases, considerable background and depth is required. The chapters on chemistry and emulsion were guided by the need to explain why certain produced and flow back waters are inexpensive to treat, and others require far more cost and effort. In this regard, it is important to explain why some suspensions or emulsions readily settle into component phases while others can be stable for several weeks. For this purpose, considerable effort is spent on the chemistry of produced fluids and their interfacial properties. Having a mechanistic understanding of oil and water chemistry, emulsions, production chemicals, and separation technologies greatly enhances the ability to troubleshoot and solve problems.

The final objective of this book is to provide a rigorous explanation of produced and flow back water treatment for water specialists outside of the oil and gas industry. In the author's experience, there is a large group of water specialists that have something to offer the oil and gas industry. Many of these people want to understand the technology needs of the industry, or are looking for entry points for technology that they have developed. Some of these people have worked in water treatment outside the oil and gas industry for decades but they find it difficult to understand the drivers for technology selection within the oil and gas industry. They generally require a basic understanding of how water is involved in the various hydrocarbon recovery processes, the cost elements of the processes, the variability of fluid characteristics, and the relative importance of space and weight in facilities design. Also, it is intended that the contents of this book will help improve the readiness of those technologies that are intended for technology transfer from outside the oil and gas industry.

Sources of Information: In general, the author has incorporated material from several different disciplines and industries. Almost all aspects of water treatment in the oil and gas industry have their counterparts in other industries such as food and beverage, pulp and paper, minerals processing, and municipal utilities, to name a few. Water treatment in these other industries is, in many respects, as challenging as that in the oil and gas industry. Much of the mechanistic understanding and design correlations of equipment used for water treatment in oil and gas comes from these other industries. For example, there are less than a half dozen papers on Walnut Shell Filters (WSF) within the oil and gas literature. Among them, Hank Rawlins's paper is an excellent example [7]. But that is only one paper. On the other hand, WSF are a type of deep bed filtration for which there are thousands of papers and several book chapters available in the industrial water treatment literature. Combing these sources of information provides a comprehensive understanding of WSF.

As discussed below, the material in this book can be classified into five principle areas: chemistry of produced water, production chemistry, equipment, process configuration, and applications / troubleshooting. Depending on which area is discussed, different sources of information were used. For example, the section on produced water chemistry draws heavily on general chemistry, surface chemistry, emulsion and colloid science, etc.. These fundamental subjects are covered in great detail in the scientific literature. While source material is relatively abundant, effort was required to identify material relevant to the upstream oil and gas industry and to translate the jargon. In doing so, anecdotal field observations for which our industry is famous are explained in fundamental and mechanistic detail.

Data and information from both field studies and laboratory studies has been used. These two sources of information are quite different. In the development of most technologies there are four stages: Discovery, Development, Demonstration, and Deployment. In the early stages of the Development stage, most of the work is carried out in the lab. In the Demonstration and Deployment stages, testing in the field must be performed. In the study of produced water, an important distinction must always be made between laboratory and field studies due to the many changes that can occur in the chemistry of produced water when it is sampled and shipped from the field to the lab. The practical tradeoffs between lab and field studies are discussed in Chapter 16 (Characterization). Suffice for now to say that both laboratory and field data are used as sources of information with consideration of the benefits and drawbacks of both.

Sources for information on applications and troubleshooting were drawn almost exclusively from the field experience of the authors. Troubleshooting activities was another very rich area for information. When a water treatment system is not performing adequately provides an excellent opportunity to collect data. Troubleshooting projects always occurs, by definition, when there are problems and there is pressure to fix the problem as soon as possible. On the other hand, troubleshooting can be gratifying in the sense that the recommendations that are made, if made properly, will result in action. If the analysis is done well, and the data presented clearly, money will be spent, modifications will be made, and a demonstrated improvement will occur in a short period of time. It is particularly rewarding when the recommendations are correct, the predictions are proven accurate and the results do indeed provide a solution to the problem. However, troubleshooting can be humbling. It is generally true that one will "get it wrong" before "getting it right." A troubleshooting program that is well organized and thought-through will minimize the time and cost and number of false steps in developing a final solution.

1.1 Produced & Flow Back Water – Definition and Characteristics

According to the American Petroleum Institute (API), as of the year 1995 roughly 49 billion barrels of produced water were generated each day by U.S. onshore oil and gas operations [8]. According to Khatib and Verbeek [9], as of 1999 roughly 210 million barrels of produced water were generated each day worldwide. At the time this represented a worldwide water to oil ratio of roughly 3. In the U.S. this ratio is closer to 7, as of 2002 [10]. As oilfields mature, they generally produce more water. Thus, the water/oil ratio of the U.S. is characteristic of mature oilfield development.

To begin the discussion, a definition of produced water and flow back water is provided:

Produced water is any water stream that flows from an oil or gas producing reservoir.

Flow back water is any water stream that flows from an oil and gas producing well and was the result of a stimulation treatment or other drilling and completion operation.

There are two reasons why it is necessary to define what is meant by "produced water." The first reason is that there is considerable variation in terminology in the industry. Practitioners are not consistent, nor precise in what is meant by produced water. This leads to confusion. To avoid this confusion, a definition is given here and applied consistently throughout this book.

The second reason to define what is meant by produced water is that there are several types of produced water, each with unique characteristics. These different types include produced water from primary sandstone or carbonate reservoirs, produced water from shale reservoirs, secondary recovery, and tertiary recovery including produced water from steam flood, miscible gas floor, or chemical EOR (Enhanced Oil Recovery). These are most of the different types of produced water. In subsequent chapters of this book the similarities and differences of these produced water types will be discussed.

Whether one is designing a new facility, designing a Brownfield expansion, planning a PWRI (Produced Water Re-Injection) project, or carrying out a troubleshooting project, it is essential to recognize the different types of produced water. This is particularly true in troubleshooting where a transition over the life of a facility from one type of produced water to another can cause a number of problems.

Produced versus Flow Back Water: Flow back water originates from drilling and completions operations and includes water-based drilling fluids, brines, as well as stimulation, hydraulic fracturing fluids, and proppant slurry. In most producing basins, the volumes of flow back water are generally small when compared with produced water. This is even true for unconventional basins such as shale.

In the majority of onshore cases, flow back water does not undergo any special treatment and is injected into a disposal well. In arid regions, and in regions where disposal wells are scarce, a fraction of the flow back fluid is recycled. In the majority of offshore cases, temporary rental equipment is used to segregate the flow back fluid from the normal produced water and treat it for overboard discharge. Co-mingling flow back fluid with produced water often leads to system upsets that require weeks to completely resolve. All things considered, flow back fluids are quite different from most produced water and deserve a separate category.

Water Flood Produced Water: An important third category of water consists of so-called flood water. In a typical water flood, water from a surface source such as a river, lake, or the sea is injected into a hydrocarbon bearing formation to maintain the pressure of the reservoir or to sweep or push out oil. This third category is warranted because the properties are quite different from both produced water and flow back water, and the specifications required can be quite different as well. Deoxygenation and biological control are major treatment objectives. Usually the hydrocarbon reservoir pore diameters are smaller than disposal well and water for flooding must be filtered to a greater degree in order to maintain injectivity. There are a few variations of the typical water flood such as Enhanced Oil Recovery (EOR) where chemicals are added to the water for improved oil recovery. In some cases produced water is used for water flood. This can improve, though not guarantee compatibility with the reservoir water since the CO_2 content will have shifted. In any case, most water used for flooding is not somewhat different from produced water. This is an important category of water but it is not considered in great detail in this book in order to keep the length of the book and the time required to write it to a manageable extent.

The table below gives a description of several different types of water, the category that shall be used in this book, and a sub-category that will be necessary to discuss important characteristics or specifications.

Table 1.1. Different categories of water referred to in this book. The characteristics, requirements, and specifications for these categories are described below in the text.

Description	General Category	Sub-Category
water produced from a conventional hydrocarbon formation	produced water	conventional produced water
water associated with dewatering a coal seam	produced water	coal seam water
water that condenses from gas in a well or well head to facility pipeline	produced water	condensed water
water produced from a hydrocarbon reservoir that is subsequently injected into a hydrocarbon reservoir (PWRI< IOR, EOR)	produced water	polymer breakthrough water
		ASP breakthrough water
		seawater breakthrough water
water-based drilling and completion fluids	flow back water	drilling & completion fluids
water-based stimulation fluids used onshore or offshore, for conventional or unconventional (coal seam, tight gas, shale gas or liquids)	flow back water	acid stimulation fluid
		conventional HF water
		unconventional HF water

Some of these water types may seem a bit outside of the expected scope of produced water. However, they are included here for practical purposes. From a practical standpoint, oil and gas companies are responsible for the environmental impact of any and all water (and hydrocarbon for that matter) that is produced as a result of their operations. Part of the purpose of this book is to aid in that endeavor. Thus, from an industry standpoint, produced water is any water that is produced through the well head, and which enters the facility, at any time of field development or operation.

Produced versus Formation Water: It will be necessary in certain discussions to distinguish between formation versus produced water. They are not the same. Formation water is literally the water in the hydrocarbon bearing formation (rock). Interstitial water is a type of formation water. It is water that is found in the pore spaces of the formation rock. Connate water is water that was trapped in the pore spaces of the rock as the rock was being formed. It has not been produced. Once the formation water passes into the production system (e.g. well head, flow line, separator, etc), then it becomes produced water. Important changes take place in the water properties as it is produced. Temperature and pressure are two obvious properties that change. Along with those changes, the composition and volume of the gas/oil/water phases change. This results in a change in pH, which causes organic acids to redistribute between the oil and water phases. Lower pH makes organic acids more oil soluble than at higher pH. Importantly, the mixing of oil and water in the presence of acids and other surface active compounds causes the oil/water interface to undergo dramatic changes in composition and properties. Much more will be discussed about this in Chapters 1 (Chemistry) and 3 (Emulsions).

Types of Produced Water: There is a wide variety of types of produced water in the industry. The dispersed phase can range from small solids particles, to droplets of oil dispersed in water, to a foam-like structure of large water droplets separated by thin films of oil. When these foam-like emulsions form they migrate to the oil / water interface and can be difficult to resolve. Dispersed contaminants can include oil, tar-like materials such as asphaltenes, waxes, formation fines, precipitated minerals, iron solids, polymer added as a hydraulic fracturing additive, various production chemicals, and so on. The dissolved components of produced water can range in salinity from a few parts per million to two or three hundred parts per million. A variety of minerals, metals, and heavy metals can be found in the dissolved phase. In addition to these contaminated waters, condensed water is also considered in the category of produced water since it fits the definition of any water stream that flows from a hydrocarbon producing reservoir. Condensed water is liquid water that existed as gaseous water in the formation. Upon production, this water has condensed. In doing so, it dilutes any saline produced water that may be present. It is discussed in Section 2.1.3 (Gas Production – Interstitial and Condensed Produced Water).

Refinery Water: The treatment of refinery waste water is not covered in this book. Nevertheless, it is worthwhile to compare and contrast produced water with refinery waste water since it helps to illuminate what produced water is. While produced water and oily waste water in a refinery have some features in common, they are by and large quite different. They both contain oil droplets that are dispersed in salty brine. In a refinery, various waste water streams are often combined in one or more waste water treatment plants. This gives rise to variability in the waste water composition from one location to another within a refinery. This can also be true in an upstream facility where there is often variability in the produced water composition and flow rate from one well to another, and particularly from one reservoir to another.

However, there are three major differences between upstream (produced water) and downstream (refinery) waste water challenges. First, produced water is subjected to many sources of shear, many of which are immediately upstream of the oil/water separation facility. Fluids are initially subjected to shear in the reservoir, near the well bore, where pressure drop, and the narrowing radius of the flow creates an acceleration of fluid. Lifting techniques, if they are used, such as electrical submersible pumps create turbulence. The well head choke is another, sometimes intense, source of shear. The list goes on and includes the control valves and pumps used in the facility itself. This results in the generation of small oil droplets suspended in the produced water. While a refinery may have some sources of shear, such as centrifugal pumps, the general trend is that upstream facilities have higher shear.

Refinery waste water is typically exposed to air. The general trend in produced water is to prevent contact with air. This is done for two reasons. The first reason is to prevent corrosion and so that inexpensive steels can be used. The second reason is to prevent the precipitation of oxides such as iron oxide. Iron oxide flocs can pose a significant challenge in upstream water treatment. Whereas in downstream water treatment, air contact will ultimately occur in an API separator for example or in an aerobic biotreater. Thus, there is no point to prevent air contact. Further, iron oxide flocs can be used to form a sludge blanket that actually improves the removal of other contaminants as well as the iron oxide. Equipment that can do this, such as flocculation clarifiers are too large for offshore upstream installation. They are not used onshore again because of the need to prevent corrosion of inexpensive steel in the transportation flowlines.

The third major difference between upstream and downstream is water volume and retention time. If the upstream water treatment processes are working effectively, most of the produced water does not reach the refinery. As discussed above, the ratio of water to hydrocarbon as a global average is 3/1. The amount of water entering with the raw crude oil is between 0.5 to 2 % water. Thus, refineries deal with much smaller amounts of water on average. This fact, combined with the fact that refineries are strictly onshore, results in a completely different strategy toward water treatment in the upstream versus the downstream industry. Upstream tends to employ equipment that has much smaller flux and throughput and larger residence time. This is one of the reasons why Dissolved Gas Flotation is so prevalent in downstream and relatively rare in upstream. Another reason is lower temperature. But, we are getting ahead of ourselves. Suffice to say for now that there are few similarities between produced water treatment and refinery oily waste treatment, and they have evolved into completely different markets for water treatment equipment, with different specialists and different solutions.

Drilling and Completion Water: The term "drilling and completion" refers to all drilling and chemical treatments of the well prior to hydrocarbon production. The fluids used in drilling are obviously drilling mud, and completion brines. The completion brines are various fluids that counterbalance the formation fluid pressure to prevent hydrocarbon production until the surface facilities are ready to handle the produced water and hydrocarbons.

Prior to production, various chemical treatments may be used to clean out the well and nearby formation. These treatments may include stimulation fluids which are designed to increase the permeability and/or surface area of the rock by dissolving and dispersing various residues, precipitates, and fine particles of reservoir rock material. Stimulation fluids have a wide range of composition which includes solvents, acids, and hydraulic fracturing fluids. These fluids are pumped into the well for the purpose of stimulating hydrocarbon production. The residence time of these fluids in the well varies depending on the purpose and chemistry of the fluid. Once they have been in the well for the intended amount of time, the fluids are allowed to flow back out of the well. The reservoir pressure provides the motive force. The fluids that flow back after a stimulation job are referred to as flow back fluids.

Hydraulic fracturing is a form of stimulation and as such is considered to be a part of drilling and completion. Thus, the hydraulic fracturing flow back fluids are a form of produced water that comes from the drilling and completion operation. The fluids that flow back after a hydraulic fracturing operation may contain the injected fluids, usually at smaller volume, and will likely also contain some formation water, formation fines, and minerals dissolved from the formation.

Importance of Water Treatment in Upstream Oil & Gas: Water treatment in the oil and gas industry has become an important economic and environmental issue. Serious water treatment problems can develop within weeks. Through the addition of new wells, or due to water flood breakthrough, a smoothly operating facility can be suddenly faced with a water treating problem that results in millions of dollars per day in deferred oil production. Once oil production becomes deferred due to a water treatment problem, the importance of water treatment becomes evident. This is part of the reason by asset and production managers in the oil and gas industry typically have a strong appreciation for the importance of water treatment.

The acquisition of large amounts of water for drilling and completions (including hydraulic fracturing) in an arid location may involve costly acquisition and transportation steps. Reuse of produced water can be an economically viable alternative. Implementation of sulfate removal technology on a deep water offshore platform for water flood requires a relatively sophisticated knowledge of filtration, hydraulics, bacteria control, and process control. As a final example, water treatment for polymer flood involves water treatment on both the injection side and the production side of the system in order to minimize the amount of polymer required and to dispose of or recycle the produced water which contains polymer.

Competent treating and handling of water can be an enabler for improved and enhanced recovery of hydrocarbons. The economical hydraulic fracturing of shales is highly dependent on economical water management. As the existing oil fields of the world become more mature, there is a greater need to increase ultimate recovery. Most, but not all, of the important IOR / EOR methods involve injection of water in one form or another. In many locations around the world there is competition for water resources. In those locations, treating and reuse of produced water is required. Desalination of such water is sometimes necessary and this can require oil removal to low or even sub-ppm levels in order to avoid fouling of the desalination equipment.

Facilities for steam or polymer flood require large volumes of high quality water. Most oilfield locations are challenged in terms of water supply. This often means that a high salinity aquifer source is used initially, followed by recycling of the produced water. Such water treating systems can more than double the cost of the facilities. In such projects, operation of the water treating facility is necessary for hydrocarbon recovery and they can be, in a real sense, the tail that wags the dog.

From an engineering standpoint, there is no doubt that the subject of produced water treating is broad and complex. Water that is co-produced with hydrocarbons contains many different substances. It contains formation solids (clay, sand, silt), dissolved mineral solids, natural acids, natural surfactants, dissolved gases, and of course oil that is dispersed into droplets and stabilized by various chemical and interfacial mechanisms.

The systems and processes used to recover hydrocarbons from produced fluids in a reliable, safe, and cost effective manner often place significant demands on the water treating portion of those systems. For example, a water treating system on a deep water platform would have the following typical challenges:

- short residence time,
- multiphase flow with slugging,
- sampling complications,
- high rates of turbulence and shear,
- rapidly changing conditions,
- fluids that originate in the ground and contain a range of reservoir minerals,
- emulsions with important but seemingly mysterious properties,
- target qualities that are measured in the parts per million range (ppmv, mg/L, etc).

The goal of any water treating system is to achieve a required level of water quality. There are a large number of parameters that characterize quality. These include dispersed and dissolved oil concentration, dissolved and suspended solids concentration, toxicity, trace element concentrations, pH, temperature, dissolved gases, etc.. The limits or acceptable values of these parameters vary widely from one location to another across the world and depend on the details of the discharge point, whether it is on land, to surface waters, subsurface, or to open sea water.

Upstream Water Treatment as an Engineering Discipline: In the last ten or fifteen years water treatment in the oil and gas industry has advanced significantly. Today, it is much more of a science and engineering discipline. There have been great improvements in the quantification, modeling, and understanding of produced water characteristics, and how to treat different types of produced water. There is much more knowledge today than there was just ten years ago. This is not to say that all water treating studies are of high quality – some are and some are not. But the improvement in knowledge across the industry has been significant.

From an individual perspective, water treating expertise was mostly gained by spending hundreds of days in the field either with experienced operators, or, in the best case, with one of the gurus of the industry. In the past thirty or so years there have been only a small number of such gurus. Thus, gaining knowledge in oilfield water treating was a long and slow process. Many water specialists, the author included, learned the subject through extensive field experience and years of problem solving. While such experience provides practical knowledge, it also can lead to large gaps in knowledge and misunderstanding based on anecdotal observations without firm scientific or engineering principles behind it. Over many years, the author supplemented field experience with literature studies in order to close those gaps. This has the important advantage of being able to see trends in equipment performance from one location to another, or seeing common elements in fluid type that might otherwise be considered as unique. This has practical benefit in solving problems.

1.2 Organization of the Book

All subjects relevant to water treatment in the upstream oil and gas industry are covered, including conventional water treating associated with drilling and completion, primary, secondary or tertiary recovery; reuse and recycling of flow back and produced water; and treatment of water from various sources including brackish or saline aquifers, or seawater for water flood. Operating locations are discussed both in terms of geographical location, and in terms of the system category such as land-based, offshore, deep water and subsea. The aim is to provide a comprehensive handling of the subject for the upstream oil and gas industry.

Handbook versus Textbook: The material in this book is organized and written such that it can be used as both a handbook, and as a textbook. As a handbook, information can be quickly and easily located and extracted for solving problems. At the beginning of each chapter there is a detailed Table of Contents. There is a detailed Index at the back. There are many bold font sub-titles that help the reader scan the material to locate sections of interest. The writing style is intended to address Frequently Asked Questions. One of the guiding principles in organizing this book was to address a need to access reference materials quickly. In order to serve the oil and gas engineering community, this is a basic requirement.

The book is also intended to be read and understood in the same way that a textbook would be used. As a text book, the presentation starts with basic background and progresses logically to more detailed and precise concepts. The first few chapters are more fundamental and scientifically based. They lay the foundation in chemistry, fluid mechanics, and surface and colloid chemistry. The middle material is mostly about equipment and processes. Application chapters are at the end where the mode and manner of the application of the various technologies are discussed.

Industry Jargon: A certain amount of industry jargon is required since there are many commonly accepted names for equipment and processes. Jargon is explained in this book. Also, a good source for understanding jargon is the Schlumberger Oilfield Glossary (http://www.glossary.oilfield.slb.com/Terms/.aspx).

Units of Measure: A consistent set of units is used throughout the book. It is a hybrid of the International System of Units (SI), the cgs (centimeter, gram, second), and oilfield units. The hybrid nature of the units used here is a consequence of the nature of the subject. Oil droplet diameters are measured in microns, their concentration is expressed as milligrams/liter or parts per million, their rise or settling rate is expressed in centimeters per minute, and they are separated in vessels that are designed to handle barrels of water per day. Thus, it is most convenient to use a mixture of units. Most calculations are carried out in SI units with a final conversion to oilfield units if necessary. As will be discussed, anyone who spends time in the field must be able to make rough calculations quickly without a calculator (unless it is intrinsically safe). For this purpose, the SI system of units is most convenient. An explanation of the units used in this book, as well as the tricks of the trade for doing calculations rapidly and easily are given in Appendix A.

Major Themes or Subjects: The chapters are organized into five major themes that correspond to the major categories in a well-planned troubleshooting program. A well-planned troubleshooting project is thorough and will ensure that all possible causes of a problem are considered. No potential cause of a problem will be overlooked. This approach to organizing the book was developed at an early stage of writing and did turn out to be a very effective way of organizing the material.

This does not mean that the entire book is about troubleshooting. The book provides material that is directly applicable to design, sampling and analysis, equipment specification, chemical selection, technology development, pilot testing, and other aspects of water treatment. However, in an effort to ensure that all relevant material is covered, the major themes of the book were selected on the basis of a troubleshooting program that the author developed over nearly three decades of work [11, 12].

The five main themes of the book are:

- Produced fluid characterization: chemistry, fluid dynamics, and emulsions

- Chemical treatment

- Equipment types, selection and performance

- Process design, configuration and recycling

- Operations

Conceptually, the organization of the book can be thought of in the following figure. Fluid Characterization is the first section. The characteristics of the incoming fluids, together with the target effluent quality define the water treating challenge. Between the incoming fluid and the effluent are the four elements that provide design specifics for the water treating system: Equipment, Process Configuration, Chemical Treatment and Operations. These must all contribute to the transformation of a given feed water to the desired effluent quality. The elements fit together like a jig saw puzzle, closely overlapping with each other to form the complete picture.

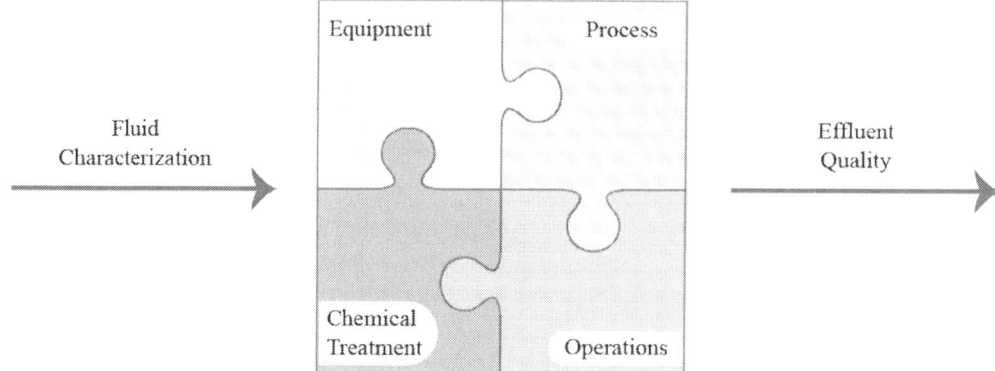

Figure 1.1 The major themes by which technical aspects of produced water treatment can be organized.

The themes of the book are briefly described here. This description of content is intended to give the reader a sense of the high level organization of the book.

<u>Produced Water Properties & Chemistry:</u> This theme is covered in Chapters 2 through 7. It includes most of the scientific background for the book. It includes the fundamental aspects of oil in water emulsions as they pertain to produced water treating. Subjects such as drop size analysis, shearing of fluids and chemical effects from both naturally occurring oil components as well as from production chemicals are dealt with from a colloid and surface science perspective. This theme includes most of the sampling and analysis techniques.

<u>Chemical Treating:</u> This theme is covered in Chapter 5, which is also part of the fundamental material. It includes the application of chemicals in water treating and how chemicals not directly related to water treatment can impact the equipment performance. Besides the chemistry itself, application concepts such as the selection and design of injection points, chemical selection, dosage, monitoring performance, etc. are included.

<u>Equipment:</u> This theme is covered in Chapters 8 through 14. It includes the most common equipment used to separate oil and water. This equipment starts with the primary separators (gas/oil/water), and includes gravity based equipment (separators, tanks, hydrocyclones, centrifuges), flotation equipment, filtration and media based equipment, coalescing elements and systems, as well as more specialized equipment. Again, fundamental scientific principles are presented as needed to describe the principles of separation.

<u>Process Design, Engineering and Integration:</u> This theme is covered in Chapter 14. It covers process engineering topics. These include integration of individual pieces of equipment into a coherent process with rejects handling and avoidance of recycle loops. Also, process control instrumentation and strategies are discussed. On-line measurement of oil concentration and drop size is also discussed.

<u>Operating Practices:</u> This theme does not have a specific chapter. It is covered throughout the book particularly in the chapters on equipment (8 - 13). Also, Chapter 15 on Troubleshooting has material related to operations and management of water treatment systems. Chapter 19 discusses the best practices (i.e., lessons learned) for operating a produced water treating system, as well as the subtle practices that can lead to hidden problems. This section is mostly based on examples from the field.

Troubleshooting & Problem Solving: This theme is covered in Chapter 15. It presents a methodology for troubleshooting water treating problems and several examples from actual troubleshooting projects. The troubleshooting methodology is based on fact based problem solving. Several examples are given which both demonstrate the methodology and utilize the scientific and technology principles discussed in the previous chapters.

Applications: Chapters 16 – 21 bring together most of the concepts of the book to show how complex water treating problems have been solved in practice. Examples are given from the US, Brazil, Middle East, Asia, Europe, the North Sea, Russia, and Australia. Each application starts with a characterization of the fluids to be treated. The equipment selection, process configuration, chemical treating system and operating practices are presented and discussed. The range of applications spans from onshore to offshore, simple treating for injection disposal, and more complex examples where produced water is treated for reinjection as water flood, steam, or polymer flood. The range of fluids include gas condensate, medium and heavy oil, sweet and sour hydrocarbons.

1.3 Historical Development of Water Treatment in the Upstream Oil & Gas Industry

The following is a brief and high level overview of historical developments in water treatment technology for upstream oil and gas [6, 13].

1930's: Limited regulation.

- Disposal wells were the predominant PWT technology in use. Essentially all oil and gas production is carried out onshore.

- Settling tanks and some horizontal API separators are the dominant technologies for removing oil and solids.

1940's: First offshore oil and gas production.

- Federal Water Pollution Control Act - started to drive water treatment technology development for offshore applications.

1950's: Corrugated Plate Interceptors and desanders were introduced and widely applied.

1960's: Flotation units brought to oil field from minerals and mining industry. Initial application was onshore in California. Application spread relatively quickly to offshore application in the Gulf of Mexico.

1970's: Flotation units become widespread, added downstream of CPIs, and installed near shore in the Gulf of Mexico.

- Desanders evolve into solids cyclones.

- The Clean Water Act (1972), and the US Safe Drinking Water Act (1974) sets not only US, but industry standard for surface discharge of produced water onshore and offshore.

- OPEC price shocks drive the initial development of North Sea EP industry. North Sea facilities built with water flood and produced water treatment in mind. North Sea development provides greater focus on improved environmental performance both in the North Sea and other regions of the world.

1980's: Solids cyclones evolve into deoiling hydrocyclones which spread rapidly throughout the industry.

- Offshore water flooding becomes more common (early 1980's) particularly in the North Sea.

- Automation greatly improves operability of water treating equipment.

- With the development of horizontal drilling, the implementation of Steam Assisted Gravity Drainage becomes significant, although the price collapse of the mid-1980's shows this down. It so happens that most SAGD fields are in regions of the world where fresh water sources are inadequate. Thus, recycling the produced water to make steam becomes significant with the application of softening, silica removal, and eventually mechanical evaporation.

1990's: Tightening of offshore discharge OiW limits based on evaluations of Best Available Technology.

- North Sea water floods begin to experience water breakthrough.

- Toward the end of the 1990's, on-Line oil concentration analyzers begin to be installed offshore.

- Treatment of hydraulic fracturing flow back fluid offshore becomes important due to the incompatibility of these fluids with produced water particularly in deepwater where residence times are short and transportation of these fluids to shore by boat is expensive.

2000's: Hydraulic fracturing of shales, and the enormous quantities of water involved, encourages new technology development and the transfer of technology from industrial water treatment to oil and gas. Initial applications do not live up to promises.

- Oil companies scramble to understand water treatment and management, while spending obscene amounts of money on trucking flow back water to disposal wells.

- Water departments are set up in oil and gas companies that previously had no water expertise. There is an influx of specialists from industrial and municipal water companies. However, the unique aspects of this application are only slowly appreciated.

- Electrocoagulation was introduced to the oil and gas industry with moderate success until the price collapse of 2015 makes it essentially uneconomical. Initial EC designs fouled excessively but these problems are eventually worked out.

- Mobile chemical treatment technology for hydraulic fracturing flow back finds success.

- Constructed Wetlands for produced water treatment find a niche in the industry. CWL had been in use in refineries for decades but upstream oil and gas installation in Oman (Nimr field), and in California provide industry experience and an understanding of the technical requirements of CWL for treatment of produced water..

2010's: Effects of price collapse of 2015 are felt deep and wide.

- Application of water technology for hydraulic fracturing of shales begins to become rational and the search for the "Silver Bullet" technology gives way to sensible and cost effective application of chemical, filtration, and flotation for recycling the water.

- By far, the biggest innovation is been the development and application of machinery to lay down and connect lay-flat hose very rapidly and at low cost, as well as the erection of temporary ponds and tanks. While the operators and the service companies finally agree on practical solutions, price pressure continues to disrupt the application of technology in some regions. Those operators that had failed to install infrastructure continue to pay enormous costs for trucking which has a significant impact on the costs.

1.4 Water Treatment System Design Strategy

At this point it is worthwhile to discuss the manner in which water treatment systems for upstream oil and gas application are selected and designed. This section is an introduction to the next section (Section 1.5) which gives a summary of produced water treatment systems around the world. The considerations and factors that go into design are discussed here. The resulting systems from different regions are then presented in the next section.

The manner in which produced water treatment systems are designed is different from the approach used in industrial and municipal water applications. In the oil and gas industry, the design process depends more strongly on identifying analogous systems and less on water quality data than in industrial and municipal industries. There are good reasons for this such as lack of samples for analysis in the design phase, variability in water properties, and the need to select equipment that operators are familiar with already, to name a few. The use of analogous systems for design is also practiced as well in industrial and municipal applications, but more so in oil and gas.

Industrial Design Process: Industrial water treatment systems are generally designed on the basis of the following four factors:

- quantitative analysis of the feed water contaminants

- quantitative specification of the effluent water quality requirements

- selection of technology that will remove the necessary feed contaminants.

- equipment design and process configuration based on the Basis of Design.

Given a flow rate, generally expected operating conditions, and an understanding of what contaminants need to be removed, a water specialist will design the water treatment system. In doing so, the selection of technology is based on known performance of the candidate technologies and design correlations for the separation efficiency for specific contaminants. The design correlations are used to determine the required residence time and thus size of the equipment. This approach is often used in industrial water treatment applications such as pulp and paper, food and beverage, and chemical processing. It is described in several books such as [14, 15].

In the Table below, for a given feed contaminant that must be removed, the applicable technologies are listed. For example, as shown in the table below, if the feed contains a relatively high concentration of Total Organic Carbon, the most likely technology that will reduce the TOC is pre-treatment to remove the disposed oil followed by some form of biotreatment (biotreatment pond, MBR – membrane bioreactor, MBBR – moving bed bioreactor). If the TOC contains some components that are not readily biodegradable, then pre-treatment with an oxidant will likely help the biotreatment process. Heavy metals can be removed by coagulation / flocculation, electrocoagulation, ion exchange, or some specific chelating agents. The point is that in many industrial water treatment applications the equipment selection process and the design of the overall system are based on an understanding of the difference between the feed and effluent characteristics and the capabilities of the available technologies.

Table 1.2 relation between contaminants and selection of water treatment equipment in industrial water treatment.

Contaminant	Technology
salinity	desalination (membranes or thermal)
hydrocarbons (TOG)	flotation, filtration
heavy metals	ion exchange, chelating agent, coagulation / floc-culation, electrocoagulation
alkalinity	softening
TOC, COD, BOD	pre-treatment to remove disposed oil followed by biotreatment
refractory BOD (high TOC/BOD ratio)	oxidation followed by biotreatment
phenols	oxidation and/or biotreatment
dissolved organics	activated carbon
BTEX	gas stripping

This is not to say that industrial water treatment systems are simple or that their design is easy. They, like essentially all chemical engineering processes have tight design constraints and targets in terms of reliability, performance, capital cost, size, and operating costs. Numerous problems must be solved in order to deliver a process that achieves all of its objectives. The design process however does tend to be based on information and facts from which a logical design will be developed. Also, analogous systems are usually identified. Analogous systems are those installations that have similar feed contaminants and effluent requirements, and for which performance data and operating experience are available. As logical as this approach may be, it is often not suitable for the upstream oil and gas industry.

Water treatment systems in the oil and gas industry are built for process reliability, not necessarily performance. By reliability, we mean that operators can neglect the equipment without significant performance deterioration, and without major or frequent breakdown. The term, robust is often used to describe the requirement. Reliability or robustness is the most important criterion because the economics of an oil and gas facility depends on the flow of oil, not water. Water only becomes a serious problem when it impacts oil and gas production. There are much more elegant technologies available for application in upstream water treating. But reliability is the most important criterion.

Design Process for Water Treatment in Oil & Gas: The design process described above presumes knowledge of the feed contaminants, and the discharge requirements. Unfortunately, these are not so well known in the design stage for many upstream oil and gas applications. For many upstream projects, little is known about the characteristics of the water that will be produced before the project is put online. This can be overcome in some cases by simply setting aside space and weight for water treatment equipment, waiting until water is produced, sampling it, and designing the facility accordingly. This can be done when initial production is dry, as was done in the early days of deepwater platforms in the Gulf of Mexico.

Also, the produced water flow rate, and characteristics are likely to vary considerably over the life of the field. Further, drilling and completion options may generate waste water streams that are handled through the existing produced water system. This will be an intense operation in the early stages, but will also continue through the life of the field. The variability and uncertainty associated with feed water characteristics are often justifiable reasons for why the industrial design approach is not utilized by the oil and gas industry.

In the development of an oil or gas field, the water disposal options may or may not be well known in advance. If there is an opportunity for overboard discharge into the sea, then that disposal option will almost always be selected as it is generally the least expensive option. But for onshore applications, where a disposal well is not available, then other options must be considered.

In designing a water treatment facility for an oil and gas field, a large number of factors must be considered. A good oil and gas engineer tend to be an outside-the-box thinker, constantly looking at the big picture and looking for alternative development schemes with less expensive options. An engineer may find another disposal option that is much less expensive than the original concept. This would change the treatment design.

Basis of Design Questions: In designing a water treatment system for an upstream oil and gas project there is a short list of questions that must be answered. The answers to these questions drive the design.

1. **What is the required treated water quality (is the water needed for oil production and will it be reused)?** The first and most important question is whether or not the produced or hydraulic flow back water will be reused, recycled, or used for enhancing the oil production. The term reuse includes water flood, chemical EOR, steam flood, hydraulic fracturing and so on. Obviously the water will need to be treated in order to be suitable for most of these applications but the degree and complexity of treatment required will depend on the application.

   a. If the water will be reused, then subsequent questions are related to the specifics of the required water quality, flow rates, etc.. Reuse of produced water for steam flood requires evaporation in either a Once Through Steam Generator (OTSG), or possibly in an evaporative desalination system such as a Mechanical Vapor Recompression (MVR) unit.

   b. Beneficial reuse generally involves treatment for agriculture or surface discharge. In most reuse applications to date, the treated water is not directly consumed by humans or animals. Most reuse options are directed to fiber crops, crops for building materials, fuel and vegetable oils. Due to the high salinity of produced and HF flow backwater, desalination is usually required for beneficial use. Desalination is challenging due to the fouling tendency of the water.

   c. If the water will not be reused, then it will be disposed of or discharged. The typical options include well injection, surface discharge (rarely due to environmental considerations), and overboard discharge. The water quality required for disposal well injection depends on the permeability of the rock and the viscosity of the oil. Viscous oil has a much greater tendency to plug pores. For the overboard discharge option, the water quality depends on environmental regulations. Subsequent questions relate to disposal options and the water quality required for those options. Water intended for disposal is almost never desalinated.

2. **What is the produced water quality (how is it generated)?** As discussed in the previous paragraphs, the question of produced water quality is always one of the drivers for design. It naturally follows the previous question about where is the water going. The important parameters of produced water quality can usually be estimated on the basis of the hydrocarbon extraction technique (primary, water flood, steam flood, etc). For example, water from a water flood will usually contain some H_2S, and a significant concentration of sulfate. This will require some attention to metallurgy in order to prevent stress corrosion cracking. Also, if barium is present in the formation water, then some form of anti-scaling treatment will be required. On the other hand, if the water is from a steam flood, then it is likely to

contain dissolved as well as dispersed silica, and dissolved organics. Both of which require consideration in the design.

3. **When is the water treatment system needed (from day-one versus later in the field)?** In hydraulic fracturing water is required in large quantities in the drilling and completion stage of the field. Thus, decisions about transportation, disposal, and reuse must be made weeks if not months in advance of initial field development. On the other hand, many facilities in the Gulf of Mexico were designed and commissioned without a water treatment system. Instead, space and weight were made available for the subsequent installation of a water treatment system. If any wells started producing water, they were shut in. This allowed for less upfront time and effort to be spent on designing a water treatment system and allowed the eventual design to be developed on the basis of known water properties.

4. **What are the volumes?** If the water is needed for hydrocarbon production, then there must be reconciliation between quantity of water produced and the quantity of water needed. If the water is not needed, then there must be reconciliation between disposal or discharge options and water production rates over the life of the field. In hydraulic fracturing operations, this may require temporary storage of produced water. Or it may require one option for drilling and completion reuse and another option for produced water disposal.

5. **Where is the facility located (onshore, offshore, or deepwater)?** There are often significant differences between water facilities designed for offshore installation which depend on the geographic location of the producing field. For example, water treatment systems designed for the deep water Gulf of Mexico differ from those designed for use in the North Sea. Facilities located near shore are often similar to onshore facilities since problematic water streams can be transported to shore and treated where facilities weight is not a factor, and space is far less constrained.

These are the essential questions that need to be answered in designing a water treatment facility in the upstream oil and gas industry. The answer to these questions can be used to identify analogous existing facilities that are already in operation. In the next section, an overview is given of the various water treatment systems for different hydrocarbon recovery methods, around the world. Each of these systems is discussed with the above five questions in mind.

1.5 Oveview of Water Treatment in the Upstream Oil & Gas Industry

This introduction provides a framework and high level overview of the subject of water treatment in the upstream oil and gas industry. It explains why certain water treatment equipment and systems predominate in certain regions of the world, and for certain applications. This knowledge is required for specialists in large multi-regional oil and gas companies. Equipment providers will find it useful in order to identify the water treatment problems that need to be solved. It will also help in understanding which emerging technologies are likely to be used in various parts of the world and for which types of problems. It is intended to show the similarities and differences in contaminated water types, discharge requirements, and in the equipment and systems that have been designed, built and operated to achieve the required task.

General Trends: Water treatment in the oil and gas industry is a very broad and diverse subject. Some specialists believe that the produced water characteristics of each oil and gas producing facility is unique. It is true that there are often differences in the chemistry from one location to another. But there are also broad similarities within certain categories. For example, a water treatment system for produced water from a shale formation has similarities with other such facilities, regardless of location. They, as a group, will be different from those systems designed for produced water from a conventional reservoir undergoing primary recovery. Also, for example, an onshore treatment facility for reuse will have common features in different locations while they will be significantly different from those for a salt water disposal well. Produced water facilities for a steam flood in California have similarities and differences with those in Alberta. It is at this level of analysis that interesting similarities and differences can be pointed out. Quite a bit can be learned by comparing the facilities for the major types of produced water which include shale, primary conventional, water flood, chemical EOR, steam flood, etc.. The obvious drawback of such generalizations is of course the tendency to over-simplify and to make generalizations where distinctions should be made instead. Nevertheless, it seems appropriate to make some generalizations in the Introduction since the rest of the book will identify the unique features and differences.

In both design and troubleshooting it is helpful to understand why water treating equipment and their process configuration differ depending on whether the water will be reused, the geographic location of the facility, hydrocarbon extraction strategy, and other variables. Understanding what the design differences are, and why they occur provides a deeper insight into water treatment in general.

Categories of Water Treatment Facilities: It is useful to have a framework by which water treatment systems can be categorized. The following table is intended to capture the factors that contribute to water treatment design. It is intended for both produced water and flow back water. There is some redundancy in the factors but that should not deter from the objective of the framework which is to categorize different types of water treatment in the upstream oil and gas industry.

A. **Hydrocarbon Recovery Strategy:** The hydrocarbon recovery strategy, also known as the development strategy, will determine many characteristics about the contaminated effluent, its characteristics, flow rate and the treated effluent quality required. As an obvious example, if the development strategy is a steam flood, then steam water quality will be required and the produced water will have high concentrations of those reservoir minerals that dissolve at high temperature. Depending on the location, most of the produced water will be treated for recycle back into steam.

B. **Location:** Location has another important impact which has been manifested lately in the development of shales. If the shale well is remote from infrastructure, and from supplies of spare parts, then the water treatment technology must be mobile, and very robust. It cannot break down or require specialist parts or repairs.

C. **Fluid Characteristics:** Another set of parameters, again from the standpoint of the oil and gas industry, is the petrochemistry of the hydrocarbons. For example, are there high concentrations of H_2S or CO_2. These so-called sour gases will have an effect on the pH of the contaminated water, as well as the content of iron in the water and the possible presence of a corrosion inhibitor. These factors can be expected to have a detrimental impact on water quality. In most oil and gas projects, this information is readily available from knowledge of the composition of the gas that is to be or is being produced.

D. **Regulations and Disposal or Reuse Options:** Finally, the water quality requirement for the treated effluent is critical. That is the target or goal of the water treatment system. But again, rather than try to specify the contaminant concentrations, it is far more practical to consider the disposal options. The options themselves will provide a rough guide to treated water effluent quality.

A classification scheme for water treatment in the oil and gas industry is shown in Table 1.3 below. There is some overlap in the factors but that should not deter from the objective of the framework which is to provide a simple means to categorize different types of water treatment in the upstream oil and gas industry.

Table 1.3 An oil industry classification scheme for water treatment systems.

Parameter	Range of Values
Hydrocarbon Recovery Strategy	Conventional oil/gas production; Gas Facility; Enhanced Recovery: water flood; Steam flood, CSS, SAGD; Chemical EOR; Well Stimulation Flow Back.
Field Location	Onshore / near shore / offshore; Stranded / accessible / constrained wt & space.
Fluid Characteristics	API gravity; H_2S, CO_2; Total Acid Number; Sulfur
Regulations & Disposal Options	Overboard discharge; Disposal Well; Produced Water Re-Injection (PWRI); Water flood; Recycle: Steam flood / Huff-and-Puff / SAGD; Surface discharge for agriculture, crops, fiber production; Reuse for stimulation (hydraulic fracturing)

In the Table below, the above classification scheme is applied to a number of different specific cases. The following table is intended to capture the factors that contribute to water treatment design. It is intended for both produced water and flow back water.

Table 1.4 Summary of the water treatment challenges faced in each type of water treatment system.

Hydrocarbon Recovery Strategy	Field Location	Fluid Characteristics	Disposal Options & Regulations	Challenges
Primary	onshore	moderate gravity	disposal well	large solids, oily solids
Primary	offshore	wide range	overboard	TOG, toxicity
Primary	near shore	high GOR	reuse; surface discharge	TOC, COD, BOD
Water flood	onshore	moderate gravity	flood	solids, oily solids; iron compounds
Water flood	onshore	low gravity	flood	oily solids are a major challenge
Water flood	offshore	not relevant – sea water is typically used	flood	solids, oxygen, perhaps H2S control (desulfation, nitrate addition)
Steam flood	onshore	heavy oil, bitumen	recycle	Silica, hardness, TOC
Chemical EOR	onshore	various	polymer makeup	TSS (polymer), TDS
Shale	onshore	gas, light oil	disposal well	Sourcing water, transportation and storage
Shale	onshore	gas, light oil	reuse / recycle	TSS, TDS in some cases
Coal Bed Methane	onshore	gas, light oil	evaporation, surface discharge	May require desalination for surface discharge

1.5.1 Hydrocarbon Recovery Strategy

Early in a project, the strategy for development is selected. Listed here are the major strategies that give rise to different water treatment systems:

1. Primary recovery

2. Water flood

3. Steam flood (SAGD, CSS)

4. Chemical Enhanced Oil Recovery

5. Shale

The upstream oil and gas industry employs a number of different strategies for extracting oil and gas. The strategy that is employed has an enormous impact on the equipment selected, and the process design.

Decades ago, most hydrocarbon production was categorized in terms of three simple stages of recovery. The thinking was that most hydrocarbon fields would be developed sequentially starting with primary and moving to secondary and in some cases to tertiary recovery. Roughly speaking, primary production is the result of drilling into the hydrocarbon bearing formation and allowing the natural pressure of the hydrocarbons to force them to the surface. Over time, pressure would diminish and secondary recovery (water injection) would be required to continue production. Finally, and only in some cases, a modified water or other fluid would be injected such as steam flood or chemical enhanced oil recovery which includes polymer flood and alkali surfactant polymer flood and variations. Each of the recovery strategies has a profound effect on the produced water characteristics.

For example, in Steam Assisted Gravity Drainage (SAGD) steam is injected into the ground to raise the temperature of the oil and reduce its viscosity so that it will flow to the surface. The produced water will likely be recycled for generating steam. Silica and carbonate scale control, salinity management, as well as corrosion control will be important. Sludge generation, and a system for dewatering, controlling organic acids, and sludge disposal will be needed. Some of the details such as the use of Once-Through Steam Generators (OTSG) versus drum boilers will need to be determined. There is no doubt that they will have an effect on the water treatment system, as well as many other details. Much about the water treatment system is simply determined from the development strategy together with geographic location.

1.5.2 Field Location

There are four types of location in the industry:

1. Onshore Accessible (within a few miles of infrastructure)

2. Onshore Stranded (remote)

3. Offshore Relatively Unconstrained (accessible, lower cost of weight and space)

4. Offshore Tightly Constrained (e.g. Deep Water, floating facility, expensive deck space)

Onshore Accessible Facilities: The onshore accessible facilities are near enough to infrastructure that operations, engineering and maintenance staff do not have to make a long commute. Spare parts, and repair specialists can readily get to the facility in a few hours or a day. Roads and rail lines are adequate for hauling large pieces of equipment to a location near the site. Another characteristic of importance to water treatment is the availability of laboratories. Examples of this type of location would be the NAM facilities in the North of the Netherlands. Some of the facilities in the corridor between Alberta and Calgary are similarly situated, although many facilities in Alberta are remote. Bakersfield California is similar, at least in the last decade or two. Certainly when Bakersfield and Calgary/Alberta were under development, they were considered to be somewhat remote. From a water treatment perspective, water clarifier and skim tanks with simple internal devices having hours of residence time are the most common technology used. Also, open settling tanks, such as the API separator were originally (but are no longer) common. Since these technologies are simple and inexpensive, the cost of processing produced water is also relatively low. This allows economic production of fluids with as high as 95 to 98 % water cut. Other factors come into play such as the need for artificial lift and numerous disposal or waterflood wells along with high capacity and in some cases high pressure injection pumps. In fact, the simplicity of the water treatment technology can be a trap. In such systems corrosion prevention and control of bacteria must be rigorously pursued in order to maintain the economic viability of the field.

Figure 1.1 The Shell Mars deep water Tension Leg Platform (TLP), located about 150 miles south of New Orleans, in the Gulf of Mexico. This picture was taken before the wells were connected to the platform.

Onshore Remote Facility: The onshore stranded or remote category is more typical of onshore oil and gas facilities around the world. These facilities are characterized by having essentially the opposite of what is described in the above paragraph. Training of local staff, patience and the application of simple solutions to problems are necessary for the successful operation of these facilities. From a water treating perspective, the simpler the technology the better. Also, since qualified laboratories are not available, simple onsite diagnostic tests and an experienced eye are critical for monitoring and maintaining the performance of water treatment systems.

Figure 1.2. The Shell Fluminense deep water Floating Production Storage Operation (FPSO)
producing from the Bijupira-Salema fields deep water located about 100 miles East of Rio de Janeiro.
This picture was taken during one of the crude offloading operations.

Another example of remote oil and gas operation is that of shale development. In the early days of a shale development, isolated and remote well locations were often brought on line in order to retain a production lease, or satisfy a procution contract with a mineral rights holder. Operators in the US and elsewhere would develop a single or a small number of wells on a lease in order to fulfill such obligations. Later, when infrastructure could be installed, a more comprehensive and less isolated development could occur. Being far from infrastructure, the conveyance of water to or from existing infrastructure would be prohibitively expensive. Thus, water treatment technology would be brought to the site on a periodic, as-needed basis. This required the development of effective and robust mobile technology. In this case, it is interesting to see which water treatment technologies were successful.

Offshore Near Shore Accessible Facilities: Offshore facilities built near the shore or in shallow water generally have lower costs associated with weight and space than deep water facilities. These locations include geographic areas where the water is not excessively deep, and areas where large converted tankers can be used to house production facilities. The water depth in the North Sea is roughly 150 meters (500 ft). This allows the economic development of fixed leg platforms.

Offshore Remote or Deepwater Facilities: Deep water offshore platforms were, in the early days (1995 – 1998), rather inaccessible. Few companies had developed the services and equipment infrastructure to support the deep water platforms let alone understand what is required to do so. In today's environment, there are many companies that have developed such expertise and have developed a well organized supply infrastructure to support deep water operations. In deep water Gulf of Mexico the water depth is 1,000 m or more. Similar depths are found in offshore Brazil, Angola and Nigeria. In deep water Gulf of Mexico the floating structures must be able to withstand some of the most powerful storms on earth. The combination of water depth and strong hurricanes results in significant constraints on deck space and weight. In other areas, such as offshore Nigeria, Angola, and Brazil, strong storms are rare. In those locations, large Floating Production and Storage Operation (FPSO) vessels can be deployed. The cost of deck space is much less on these vessels although still somewhat more expensive than that in the shallow water regions. From a water treatment perspective, residence times can be somewhat longer than in deep water, though shorter than onshore.

Figure 1.3. The Shell Draugen platform in the North Sea

The difference between onshore and offshore facilities has largely to do with the cost of the supporting facility. If that cost is high, then the equipment must be as compact as possible. If the cost is moderate, then the constraint on size and weight is less. Although, there is always some degree of constraint.

1.5.3 Fluid Characteristics

The characteristics of the oil, gas and water have a significant impact on the concentration of contaminants and on their composition in the produced water. The term Petrochemistry is sometimes used to describe this characteristic. It is intended to refer to the aspects of hydrocarbon composition that have an effect on the development strategy and equipment required to produce the fluids.

The most general distinction that can be made is:

1. gas / gas condensate

2. oil

3. heavy oil & bitumen

Granted there is a range of Gas to Oil Ratios (GOR). But at a high level it has been found that produced water from a predominantly gas producing field is distinctly different from that of a field producing a moderate or a heavy oil. Gas fields tend to have much less water production than do oil fields. While there is a wide range of water production from oil fields, the statement is nevertheless generally true. Gas fields tend to have a high concentration of dissolved components in the produced water compared to oil fields. The dissolved components include short chain organic acids, and aromatic components such as BTEX (benzene, toluene, ethyl benzene and xylene).

Besides these parameters, another important distinction is the H_2S and CO_2 content of the produced or associated gas. If the H_2S concentration is greater than say 0.1 %v (1000 ppmv) then significant measures are required to produce the fluids and avoid corrosion, as well as injury to operating staff. Likewise, the presence of a high concentration of CO_2 will result in lower pH for the produced water and an aggressive fluid from a corrosion standpoint. When gas is produced at high pressure, then as the pressure of the produced water is reduced during processing, the CO_2 and H_2S concentration of the evolved gas will increase due to the higher solubility of these acid gases in water compared to the solubility of, for example, CH_4. These changes in gas composition and the changes in mineral solubility which take place with even small changes in the pH of the water must be considered when designing or troubleshooting a water treatment facility.

One final broad distinction that goes hand-in-hand with the development strategy, and location, is the presence of injection fluids in the production wells. If the field is composed of heavy oil, then some form of enhanced recovery will be required, such as steam flood, Steam Assisted Gravity Drainage (SAGD), or Cyclic Steam Stimulation (CSS). The source and quality of water for steam generation and the extent of recycling required for steam production will have an enormous impact on the water treatment system and on the prevention of silica and carbonate scale formation.

Figure 1.4. The Kashagan onshore facility in Kazakhstan. This facility is designed to process sour hydrocarbons. The main processing train is nearly one mile long.

1.5.4 Regulations and Disposal or Reuse Options

The U.S. EPA has subdivided produced water disposal into categories each of which has specific restrictions on contaminant concentration, composition, and toxicity:

- Onshore
- Stripper well
- Beneficial use
- Coastal
- Territorial seas
- Outer Continental Shelf (OCS)

In order to consider a broader range of locations, besides just the U.S., and to put the disposal options into economic context, a broader range of options must be considered.

There are many other options for disposal, recycle and beneficial use as indicated in the list and text below:

1. Overboard discharge (surface water discharge)
2. Underground injection disposal (SWD – Saltwater disposal well)
3. Onshore surface discharge for disposal

4. Onshore surface discharge for beneficial use

5. Recycle for IOR (waterflood) or EOR (Enhanced Oil Recovery)

6. Recycle for steam flood

7. Recycle for hydraulic fracturing

Each of these is discussed below within the context of the regulations, and when they would be economically viable. Most offshore produced water is treated and discharged overboard into the sea. Most onshore produced water is treated and disposed of by injection into a permeable rock formation.

Literature on Regulations in the U.S.: Caudle [16] gives a review of the historical development of regulations regarding produced water in the U.S. Veil [8] describes the current legislative strategy. Davies and Scott [17] provide a good summary for the U.S. and a few other countries.

Brief Summary of U.S. Regulations: The 1972 Federal Water Pollution Control Act (Clean Water Act, 33 USC 1251 et seq.) had the goal of making all rivers and streams within the U.S. fishable and swimmable. It established the goals and authority for the federal regulatory agencies, and it appointed the U.S. Environmental Protection Agency (EPA) to administer the act. It also established a number of federal, state, and local bodies to provide various functions such as research, monitoring, reporting, and permit granting and administration. States can apply for Primacy (primary enforcement authority, rather than federal authority). To do so, the state must demonstrate to the EPA that they do indeed have control over injection activities in the state.

Originally (early 1970's) it was thought that environmental protection could be achieved by eliminating all discharge of pollutants to rivers, lakes and the sea. This was obviously not a workable plan. Subsequently the National Pollutant Discharge Elimination System (NPDES) was put into place, as mandated by the Clean Water Act Amendments of 1977. This legislation established a permitting system to authorize and regulate discharges to surface water. The NPDES permitting system is described in Section 402 of the Clean Water Act. It contains effluent limitations for both point and non-point sources that discharge pollutants into surface waters of the United States. It also provides penalties for non-compliance which include prison sentences for negligent and knowing violations.

Onshore produced water may not be discharged to navigable waters (rivers, streams, lakes) unless it is being used for agriculture or wildlife, and it meets a maximum daily limitation for oil and grease of 35 mg/L, and that there is no visible oil sheen [17].

Deepwater facilities in the Gulf of Mexico are outside of the state controlled waters which extend only to 3 nautical miles from shore. The deepwater platforms are located in the Outer Continental Shelf area of the EPA Region 4 and 6. Region 6 includes the federal waters beyond the Texas and Louisiana coasts, among other on-shore areas and states.

To administer the regulations, the Region 6 office of the EPA works closely with the Minerals Management Service (MMS) of the Department of Interior. MMS inspectors perform most of the NPDES offshore platform monitoring and compliance inspections for EPA. Under the CWA, the EPA may authorize individual states to implement all or various parts of the federal program. Certain states have strong environmental administrations, while others rely on the federal government. The federal EPA has the ultimate authority to decide if the programs of a state are sufficient or are in conflict with the federal programs.

U.S. Deepwater: When produced water is disposed to surface waters (streams, rivers, lakes, seas and oceans) there are strict water quality regulations. In the U.S., the Clean Water Act (CWA) requires that all such water must be authorized by a permit issued under the National Pollutant Discharge Elimination System (NPDES) program [8], which typically include TOG (Total Oil and Grease) and toxicity. In the case where the water will be discharged to surface waters salinity is usually required to be below a certain limit. In some regions of the world, discharge of produced water to onshore surface water or land requires the removal of organic contaminants that contribute to TOC (Total Organic Carbon), COD (Chemical Oxygen Demand) and BOD (Biological Oxygen Demand). This is particularly true if surface disposal could impact the water quality of a nearby stream or lake.

Water-wet suspended solids are usually not a problem. Water-wet solids do not contain hydrocarbons and therefore do not contribute to TOG. Such solids can include sand, carbonate material, clay and various other minerals. They may originate in the water bearing zone of the reservoir, or may have formed (precipitated) at some point in the production process. Provided that they remain water-wet through the production facility, they can usually be discharged overboard from an offshore production facility. Suspended solids can become a problem if the water is injected into an underground formation.

The oil and grease content of the produced water discharge must be below a daily maximum of 42 mg/L and a monthly average of 29 mg/L. Samples for oil and grease monitoring shall be analyzed at least once per month. A produced water sample shall be collected, within 2 hours of when a sheen is observed, and analyzed for oil and grease. The regulatory procedure is US EPA-1664 [18].

The 7-day average minimum and monthly average minimum No Observable Effect Concentration (NOEC) must be equal to or greater than the critical dilution (percent effluent) concentration specified in Appendix D, Table 1 of the permit. The critical dilution concentration calculated from Table 1 depends on the discharge rate, discharge pipe diameter, and the water depth between the discharge pipe and the bottom. The NOEC is defined as the greatest effluent dilution which does not result in lethality, compared to the control sample.

U.S. Onshore Disposal Wells: In 1974, the U.S. congress passed the so-called Safe Drinking Water Act. It was subsequently amended in 1988. This law, and its amendments, established the Underground Injection Control (UIC) program which provides the guidelines for injection disposal wells including hydraulic fracturing flow back water, and produced water. For oilfield waste fluids going into disposal wells, there are no radioactivity restrictions.

To obtain a permit for a new Class II well, the operator must file an application with the UIC Director containing information listed in 40 CFR 146, or in the applicable State requirements.

Key requirements include:

- well site layout relative to USDW locations
- structural integrity of the well
- operations of the well (injection pressure curve)
- status of nearby wells
- proposed monitoring of the facility

Most onshore oil and gas activities fall under the Class II description which states: Salt water disposal wells are those disposing of fluids brought to the surface that may be co-mingled with various waste streams … as long as the fluid is not hazardous at the point of injection.

While the economics of onshore versus offshore produced water injection disposal are different, the water treatment processes required are largely similar, as discussed presently. The main distinction that must be made is between injection into a disposal formation versus injection into a hydrocarbon bearing formation. One of the important properties of water that is destined for injection is the tendency of the water to plug the formation. If the formation has relatively small pores, low permeability, and if the injected water has a relatively high concentration of contaminants, the tendency to plug will be relatively high. In that case, more pressure will be required to overcome the plugging and the formation will have a greater tendency to fracture. Thus, the reservoir engineer will try to select a permeable reservoir with large pores. This will allow dirtier water to be injected thus reducing the cost of water treatment. The details of water treatment are of course discussed throughout this book. If such a reservoir is not available nearby, then the water may require transportation, which can be expensive. These are the tradeoffs that must be made.

The term waterflood refers to the injection of water into the hydrocarbon bearing formation. Injecting produced water into the hydrocarbon bearing reservoir is different from injecting produced water into a disposal formation. In the former case, the formation properties (pore size and permeability) are given and cannot be selected. In the later case, there may be several formations available nearby from which to choose. When this is the case, there is a probability that a permeable disposal reservoir with large pores can be found nearby. In general, this leads to a sweeping generalization that water quality for disposal injection does not have to be as clean as that for waterflood. This is of course a generalization that has many exceptions.

U.S. RCRA: In 1976 the U.S. Resource Conservation and Recovery Act (RCRA) was passed. It governs the disposal of hazardous and non-hazardous waste material in the U.S.. In 1988, the US EPA determined that disposal of E&P wastes shall not be controlled under the hazardous waste restrictions of the RCRA. In other words, E&P waste, within a specific definition of that material, was determined to be exempt. These wastes include drill cuttings, drill muds, and production brines that are uniquely associated with E&P activities. They are sometimes referred to as Normal Oilfield Waste (NOW) in order to emphasize that they are exempted from the RCRA hazardous waste regulations. The 1988 determination also exempts residual salts derived from evaporation and demineralization of produced waters. However state regulations and permitting programs that are more stringent than the RCRA have been adopted. For example, California law does not exempt E&P wastes from its hazardous waste program. State and local radioactivity restrictions for solid waste disposal would apply. Solid waste is required to pass the Paint Filter Liquids Test (SW-846 Method 9095).

U.S. Recycle for IOR and EOR: The water quality requirements for recycling vary significantly depending on the particular recycle employed. Recycle options include waterflood, discussed briefly above, and various forms of Enhanced Oil Recovery (EOR) including polymer flood, alkaline surfactant polymer flood, and hydraulic fracturing which is employed in a wide range of formations including sandstone, coal bed, tight gas, and shale.

As expected, the presence of certain contaminants in the water stream will have an impact on the water treatment system design. One of the most obvious examples would be the difference between light oils (condensates) and moderate to heavy oils. In some cases, gas condensates have a high concentration of dissolved organics. These dissolved organics, unlike most dissolved organics contribute to the TOG content of the water. Such compounds as BTEX, acids, etc. make up this fraction. In some cases, a very aggressive water treatment system is required with activated carbon on the end of the system.

In other cases, phenols must be removed. Total Suspended Solids are common in high rate wells, especially when the fluid velocity near the wellbore exceeds the "critical velocity" for fines migration within the reservoir rock. Producing from unconsolidated sandstone can also be a challenge since the produced solids can rapidly accumulate in vessels and tanks. Di- and multivalent cations are a problem when the water must be desalinated since they typically contribute to scaling.

If Normally Occurring radioactive Materials (NORM) concentrations are high then scaling must be avoided. NORM-containing scale will render the separator, vessel, or pipe radioactive and necessitate special disposal methods.

Regulations regarding water quality and waste discharge vary significantly around the globe both in terms of the values of certain parameters and in terms of what are measured and how it is measured. In order to facilitate discussion in the subsequent chapters of this book, a number of parameters are defined here. Also, some notable effluent quality regulations are discussed.

Effluent quality can be measured by a number of different parameters such as Total Oil and Grease (TOG), Total Organic Carbon (TOC), Chemical Oxygen Demand (COD), and Biological Oxygen Demand (BOD).

The TOG is intended to measure the mass of dispersed oil in a water sample. TOG analytical procedures are typically measured by liquid/liquid extraction of the sample using a method-specified hydrocarbon extraction solvent. The solvent is then either evaporated and the residue weighed (gravimetric procedure) or analyzed by a specified gas chromatographic procedure. The EPA-1664 analytical procedure is an example of a gravimetric TOG measurement procedure. In Region 6 of the EPA, the maximum monthly average TOG in samples analyzed by a certified laboratory must be below 29 mg/liter with a sample maximum of 42 mg/L. Other regions of the world, such as the North Sea, use the gas chromatographic onshore method to determine TOG as mandated by OSPAR Convention.

U.S. Onshore Surface Discharge for Beneficial Use: Onshore surface discharge is used for agriculture, and fiber production in some locations. In selecting a water treatment system for onshore surface discharge, the concentration of chloride ion must be examined first. If the chlorides (or Total Dissolved Solids) of the produced water must be reduced in order to allow surface discharge, then some form of desalination is necessary. Desalination of produced water is generally a challenge because of the dissolved and dispersed organics which have a strong tendency to foul desalination equipment. This is particularly true of membrane filters. In the case of evaporators, scaling and fouling tendency of the fluids must also be controlled. If desalination is not required, then the removal of dissolved and dispersed organics is usually the main water quality criteria.

The strategy toward environmental protection and the regulations regarding the measurement, transportation, storage, recycling, and disposal of produced water vary from state to state, and country to country. It would be impractical to provide a summary of this topic from around the globe. Instead a summary is given here for the U.S., and a few other regions, with comparisons to the approach taken in other parts of world, and comments as to how the regulations impact water treatment strategies.

A critical issue that is immediately faced in legislation and regulation of E&P waste is that the contaminant limits (e.g. oil-in-water concentration), depend significantly on the sampling and measurement techniques. Thus, understanding how to comply with the regulations requires an understanding of the regulatory limits, the required sampling and measurement methods, and how the chemical and process technology perform in reducing contaminant concentrations to acceptable levels. Unfortunately, this is a complex situation. The important discussion of sampling and analysis methods is deferred until Chapter 7 (Sampling and Analysis of Produced Water) since several other topics must be covered first before sampling and analysis can be adequately introduced.

North Sea OSPAR (Oslo-Paris Convention): OSPAR generally refers to The Convention for the Protection of the Marine Environment of the North-East Atlantic. Several conventions were held over a period of several years. The early conventions were held in Oslo and Paris. As part of the 1992 convention held in Paris [19] a set of documents were put forward for member nations to ratify. Representatives from sixteen contracting parties were involved including 14 countries. The net result at that time was a set of documents that provide definitions, general obligations, decisions, recommendations, and statements of intent to protect the environment. The document is referred to as the OSPAR Convention. It sets up a management structure, means to share information, and intervals for monitoring and review. The North-East Atlantic region was split into five Maritime Areas and Regions including Region II: the Greater North Sea. The documents define the difference between decisions and recommendations. The former is binding for all signatory parties. The later having no binding force. The OSPAR Convention does not provide specific sampling and measurement methods, nor discharge limits, etc.. Recognizing that technology will improve with time, it defines in general terms what is meant by Best Available Techniques and Best Environmental Practice without going into specific detail, and provides a means for continuous improvement in its approach to environmental protection. The current scope includes overboard disposal of drilling fluids, produced water, sewage, onshore waste disposal, as well as emissions to the air, cooling water discharge, and decommissioning of offshore structures.

In addition, the OSPAR Convention established the OSPAR Commission [20] made up of representative members that meets regularly to supervise implementation, review results, and to address concerns or needs through a program of work. The OSPAR Commission provides administration of various testing, research and development programs. Depending on particular issues, tasks, white papers, and reports that are under discussion or development, various working groups, committees, and correspondence groups are formed. Out of these programs, various decisions and recommendations are made for specific sampling, measurement and reporting techniques, and limits that would apply to them. Depending on the nature of the recommendations, they will be adopted by only some of the contracting parties. One of the mandates for the operators in member counties was a reduction, by 15 %, in the total discharge tonnage of oil compared to levels permitted in the year 2000.

One of the major developments in the OSPAR recommendations over the years is the adoption of a risk-based approach. It was recognized at an early stage that the use of BAT (Best Available Technology) and BEP (Best Available Practices) have strengths and weaknesses. The main strength was that it forced operators to apply at least a minimum standard of design. One of the main drawbacks was the handling of particular substances such as heavy metals, aromatic hydrocarbons, and alkyl phenols which vary significantly from one region to another. In order to minimize the discharge of these components it became obvious that some platforms need to do more than others. Therefore in 2012, OSPAR adopted a Recommendation for a Risk-Based Approach to the Management of Produced Water Discharges from Offshore Installations (RBA Recommendation). The RBA is a method for prioritizing certain discharges and requiring mitigation for certain substances that pose the greatest risk to the environment.

EIF (Environmental Impact Factor): In addition to the OSPAR programs, there are other environmental protection programs in operation in the North Sea. For example, in the Norwegian sector, the EIF (Environmental Impact Factor) and the DREAM (Dose Related Risk and Effect Assessment Model) approach are used. The origin of this approach goes back to a white paper issued by the Norwegian Ministry of the Environment in 1998 (White Paper No. 58) which required the oil and gas companies operating in the Norwegian sector of the North Sea to demonstrate "zero environmental harmful discharges" by the year 2005. In response, the Norwegian Oil Industry Association (OLF) developed the EIF as a quantitative method for assessing environmental impact. This approach is not adopted by other OSPAR members and is essentially a self-imposed approach to environmental protection on the part of the Norwegian government and the Norwegian oil and gas companies.

The EIF is based on the Predicted Environmental Concentration (PEC), the Predicted No-Effect Concentration (PNEC), and the amount of contaminant discharge [21 – 25, 27]. The lower the PEC/

PNEC ratio, the smaller the impact. The DREAM program (Dose Related Risk and Effect Assessment Model) [25] is used to calculate the volume of sea water for which the ratio is greater than 1.0. Presumably, this volume is a measure of the environmental impact. The larger the volume, the greater the impact. This method can be applied to individual platforms on a continuous basis and is used to demonstrate continuous reduction in environmental impact due to overboard discharge [21]. Oil companies take steps in order to continuous reduce the EIF such as water shutoff, migration to green production chemicals, and implementation of advanced water treatment technologies. Due to the large volumes of produced water that are involved on some Norwegian platforms, even a small reduction in contaminant concentration can have a significant effect on EIF. This is why so many new water treatment technologies are constantly being developed and tested in Norway. One of the drawbacks to this approach is the difficulty in obtaining PNEC data on produced water constituents. Due to this difficulty, the organic acids as a class of compounds are usually not considered in EIF calculations. The presence of organic acids in produced water is discussed in Section 2.63 (Organics in Produced Water).

Saudi Arabia: In Saudi Arabia, environmental protection is controlled under the following ministries and departments: the Meteorology and Environmental Protection Administration (MEPA), the Ministry of Agriculture and Water, the Ministry of Health, and the National Commission for Wildlife Conservation and Development. A summary of the environmental laws can be found in reference [26].

One of the ongoing concerns in the Kingdom is oil spills along the Red Sea and Arabian Gulf coasts. These spills are a consequence of heavy traffic from oil tankers, spillage from barges, tankers, and pipelines, and the result of conflict in the region. Regarding E&P related waste, the leadership of Saudi Arabia has tried to balance the need to protect wildlife and fisheries along the coast with the need for economic development.

One of the interesting features of the Saudi approach to environmental protection is the active involvement of Saudi Aramco in research, development, and sponsoring of third party organizations such as the UK-based Oil Spill Services Centre, and the Gulf Area Oil Companies Mutual Aid Organization. A good summary of the issues and the active involvement of Saudi Aramco in environmental stewardship can be found in reference [26].

1.6 Concluding Remarks

The following subjects are covered in this book:

1. A wide range of locations are considered including onshore (conventional and unconventional), offshore, and deep water, including producing regions as the North Sea, Asia-Pacific, onshore U.S. and Canada, the Middle East and North Africa, and deepwater Brazil, West Africa, and the Gulf of Mexico.

2. Disposal options include overboard discharge, underground injection disposal, water flood, recycle, beneficial use, and enhanced oil recovery.

3. Technologies include separators, hydrocyclones, flotation (conventional water treatment technologies), as well as tertiary technologies (filtration, coalescing media), and some technologies that would be considered to be niche technologies for the oil and gas industry including oxidation, biotreatment, clarifiers, and electrocoagulation.

4. Seawater treatment for offshore water flood is discussed briefly.

5. The chemistry of produced and hydraulic fracturing flow back water, and production chemicals are discussed in detail since there is a strong correlation between water chemistry and treatment requirements.

6. Troubleshooting case studies and applications are presented.

The main goals and objectives of the book are:

1. Provide an introduction to the subject for people who may never design or operate a water treatment system but who require an awareness of water properties, the equipment, and processes involved.

2. Provide detail and references for those scientists and engineers who want to master the subject.

3. Explain the chemical and physical mechanisms involved in water treatment. This knowledge is critical for troubleshooting and problem solving, as well as for design.

4. Provide a ready reference to formulas and convenient units of measure.

5. Provide a rigorous explanation for water specialists outside of the oil and gas industry so that they can gain rapid entry and success for their products and services.

The book is organized into seven themes:

1. Properties & Chemistry

2. Equipment

3. Process Design, Engineering and Integration

4. Chemical Treating

5. Operating Practices

6. Troubleshooting & Problem Solving

7. Applications

An overview of water treatment in the oil and gas is given in which four major categories are applied:

1. Hydrocarbon Recovery Strategy (conventional, unconventional shale, coal seam gas, waterflood, chemical floor and steam).

2. Location (onshore, offshore and deepwater).

3. Fluid Characteristics (API, H2S, CO2, iron, presence of oxygen, production chemicals particularly corrosion inhibitor).

4. Regulations and Disposal Options (options such as disposal well, discharge to surface, reuse, recycle, targets and limitations on TOG, TOC, COD, BOD, etc.).

References to Chapter 1

1. J. Drewes, "An integrated framework for treatment and management of produced water. Technical assessment of produced water treatment technologies," 1st Edition, Report prepared by the Colorado School of Mines for RPSEA Project 07122-12 (2009).

2. J.D. Arthur, B.G. Langhus, C. Patel, "Technical summary of oil & gas produced water treatment technologies," ALL Consulting, paper produced in partial fulfillment of NETL project (2005).

3. S. Basu, "A review of the chemical characteristics of Frac / Flow Back / Produced water," paper presented at the Workshop on Water Management in the Marcellus Shale, Atlantic City, NJ (2011).

4. T. Hayes, J.D. Arthur, Overview of emerging produced water treatment technologies," paper presented at the 11th Annual International Petroleum Environmental Conference, Albuquerque (2004).

5. C. Patton, Applied Water Technology, First Edition 1986, Third Ed: C. Patton, A. Foster (2007).

6. K. Arnold, M. Stewart, Surface Production Operations, Gulf Publishing (2008).

7. C.H. Rawlins, A.E. Erickson, C. Ly, "Characterization of deep bed filter media for oil removal from produced water," paper presented at the Produced Water Society, Houston (2010).

8. J.A. Veil, M.G. Puder, D. Elcock, R.J. Redweik, "A white paper describing produced water from production of crude oil, natural gas, and coal bed methane," report written at the Argonne National Laboratory for the U.S. Department of Energy under contract W-31-109-Eng-38 (2004). C.E. Clark, J.A. Veil, "Produced water volumes and management practices in the United States," report prepared at the Argonne National Laboratory, for the U.S. DOE under contract no.: DE-AC02-06CH11357 (2009).

9. Z. Khatib, P. Verbeek, "Water to value – produced water management for sustainable field development of mature and green fields," J. Pet. Tech., v. 55, p. 26 (2003).

10. R. Lee, R. Seright, M. Hightower, A. Sattler, M. Cather, B. McPherson, L. Wrotenbery, D. Martin, M. Whitworth, "Strategies for produced water handling in New Mexico," paper presented at the 2002 Ground Water Protection Council Produced Water Conference, Colorado Springs, CO (Oct 2002).

11. J.M. Walsh, J. Fanta, W. Bryson, C. Toschi, "Troubleshooting produced water – methods and lessons learned. Part1 and Part 2" World Oil, p. 111 and p. 151, March (2007).

12. J.M. Walsh, W. Bryson, M. Stacy, C. Toschi, J. Petty, J. Langer, "Troubleshooting produced oil and water treating problems – methodology and a case study," paper presented at the Produced Water Society meeting, Houston January (2006).

13. A. Delgado, H.M. Lee, "The chronology of water-oil handling equipment," SPE-39878, paper presented at the SPE IPCE, Villahermosa, Mexico (1998).

14. G. Tchobanoglous, F. Burton, H.D. Stensel, R. Tsuchihashi, (Staff of Metcalf and Eddie, Inc.), <u>Wastewater Engineering Treatment and Resource Recovery</u>, McGraw-Hill, New York (2014).

15. W.W. Eckenfelder, D.L. Ford, A.J. Englande, <u>Industrial Water Quality</u>, McGraw Hill, New York (2009).

16. D.D. Caudle, "Produced water regulations in the United States: Then, now and in the future," SPE-77389, paper presented at the SPE Annual Technical Conference and Exhibition, San Antonio (2002).

17. M. Davies, P.J.B. Scott, Oilfield Water Technology, National Association of Corrosion Engineers (NACE) Press, Houston (2006).

18. United States Environmental Protection Agency, Method 1664, Revision A: n-Hexane extractable material and silica gel treated n-hexane extractable material by extraction and gravimetry, EPA-821-R-98-002 (1999).

19. Anonymous, <u>Convention for the protection of the marine environment of the North-East Atlantic</u>, also known as the <u>OSPAR Convention</u>, published by the OSPAR Commission (1992). Available on the web at: http://www.ospar.org/site/assets/files/1290/ospar_convention_e_updated_text_in_2007_no_revs.pdf

20. Information about the OSPAR Convention and the OSPAR Commission can be found at: http://www.ospar.org/.

21. S. Johnsen, T.K. Frost, M. Hjelsvold, T.R. Utvik, "The environmental impact factor – a proposed tool for produced water impact reduction, management and regulation," SPE – 61178, paper presented at the SPE International Conference on Health, Safety, and the Environment, Stavanger (2000).

22. T.I.R. Utvik, J.R. Hasle, "Recent Knowledge about produced water composition and the contribution from different chemicals to risk of harmful environmental effects," JPT December (2002).

23. International Association of Oil & Gas Producers, "Aromatics in Produced Water: Occurrence, Fate & Effects, and Treatment," Report No. 1.20/324 January (2002).

24. A. Descousse, K. Monig, K. Voldum, "Evaluation study of various produced-water treatment technologies to remove dissolved aromatic compounds," SPE – 90103, paper presented at the SPE Annual Conference and Exhibition, Houston (2004).

25. B.L. Kundsen, M. Hjelsvold, T.K. Frost, M.B.E. Svarstad, P.G. Grini, C.F. Willumsen, H. Torvik, "Meeting the zero discharge challenge for produced water," SPE – 86671, paper presented at the Seventh SPE International Conference on HSE, Alberta (2004).

26. F.K. Alturki, "Promoting sustainable development through environmental law: Prospect for Saudi Arabia," thesis presented to the Pace University School of Law, New York (2015).

27. P.G. Grini, M. Hjelsvold, S. Johnsen, "Choosing produced water treatment technologies based on environmental impact reduction," SPE – 74002, paper presented at the SPE International Conference on Health, Safety and Environment, Kuala Lampur (2002).

CHAPTER TWO

Chemistry of Produced and Flow Back Water

Chapter Two Table of Contents

2.0 Introduction

This chapter provides both a practical introduction and an in-depth discussion of the chemical composition of produced and flow back water in upstream oil and gas operations. Produced and flow back water are composed of ions, dissolved minerals and salts, thousands of dissolved and dispersed hydrocarbons and hetero-organic compounds along with dissolved gases, suspended minerals, polymers, and production chemicals. The chemical composition of these contaminants is important in selecting, designing, and operating water treatment equipment.

In order to understand the composition of produced water, all four phases of produced fluids (gas, oil, water and solids) must be characterized. Produced water is unique in the degree to which the composition, properties and character of the water can change within the system as a result of changes in process conditions (pressure, temperature, mixing of streams). Components such as CO_2, H_2S and dissolved iron compounds, for example, can have a significant effect on the ability of produced water treatment equipment to perform as required. Thus, the constituents of all four phases are discussed.

There are many reasons why the chemical composition of oil / water systems is important. For example, the following questions are relevant:

- Which production chemicals, oil components, and produced water components stabilize an emulsion?

- Which produced fluid components and production chemicals make the fluids sensitive to shear?

- How does one differentiate between problems caused by mechanical equipment versus chemical causes?

- How does one use chemicals to overcome mechanical performance limitations?

- Which flocculating agents are best for which types of produced fluids?

In this chapter, the chemical nature of the produced oil, water, solids, gas as well as the interfaces between these phases is discussed. Also discussed are the origins of these contaminants, and their geographical occurrence.

The chemical composition of produced water contributes to the:

- sensitivity of oil droplets to shear

- tendency to oil droplets to coalesce

- chemical treatment required to coagulate and flocculate (selection, dosage, injection point) contaminants

- precipitation and scaling tendency of dissolved minerals

- corrosivity of the water

- toxicity of the water

- selection and design of treatment equipment required (residence times, equipment types)

- water discharge or disposal options

Sources of Produced Water: There are several kinds and sources of produced water. These include primary production, water from water flood, steam flood, and chemical enhanced oil recovery. Produced water from a heavy oilfield is different than produced water from a gas condensate field. In design and troubleshooting it helps to know what kind of produced water is being processed. Once this is known, then experience from similar fields can be applied to resolve local challenges. Throughout this chapter the origins of chemical constituents will be discussed along with their chemistry and analysis.

Chemistry & Emulsions: Many of the contaminants that must be removed from produced and flow back water are in the form of emulsions. Emulsions must be resolved (or broken) in order to clean the water. Some of the chemical compounds that occur naturally in produced water are responsible for stabilizing oil-in-water (as well as water-in-oil) emulsions. Some of the man-made surfactants introduced as production chemicals also stabilize emulsions. There are several different emulsion structures (suspensions, liquid by-layers, high internal phase volume). But all of them are composed fundamentally of oil-water interfaces. The oil-water interface is stabilized by surface active molecules, ions, oil-wet solids, and other contaminants at the interface. When a problematic produced water is encountered, it is worthwhile to determine which constituents are responsible for the stability of the emulsion. This chapter provides the background chemistry for these naturally occurring and man-made chemical components. This chapter provides the fundamental chemistry background for the discussion of emulsions in Chapter 4 and for the discussion of chemical treatment in Chapter 5.

Chemistry & Design: There is a direct connection between the chemical composition of the produced fluids, the properties of the fluids, and the water treatment equipment, processes and chemicals that are required to treat the water. There is no one set of technologies that can be used for all types of produced water. Each water treatment process must be designed and operated taking into account the characteristics of the water being produced, and the target quality of the effluent water. The connection between water chemistry and treatment processes is made in the subsequent chapters of this book.

Chemistry & Regulatory Measurements: Another reason for this detailed discussion of produced water chemistry. First, this chapter sets the stage for a discussion of what is actually measured in various regulatory statutes. In order to achieve regulatory compliance, it is important to know what the regulatory tests are and are not measuring. This is discussed in greater detail in Chapter 4 (Sampling and Analysis of Produced Water), and from a more specialized standpoint in Chapter 21 (Applications – Water Soluble Organics).

Chemistry and its Impact on Sampling and Analysis: In order to achieve an understanding of produced water composition, a number of sophisticated analytical procedures have been developed and applied over the years. An illustrative selection of the analyses is presented in this chapter. However, it must be emphasized that many of these analytical procedures are not necessary for resolving the common problems related to water treatment process design and troubleshooting. While sophisticated analytical methods provide detailed descriptions of produced water and its contaminants, most problems can be solved by applying common industry test methods which are much simpler, less expensive, and for which many analytical labs are capable. The specifics of these tests are discussed in Chapter 17 where the subject of "Characterization" is discussed, and Chapter 7 (Sampling and Analysis). Chapters 7 and 17 not only provide the common test methods, they also explain how to interpret the results for equipment design and problem solving. There are of course some difficult and unique problems where sophisticated analysis is required.

Produced Water Composition Depends on Process Conditions: Throughout upstream separation, the chemical composition of produced water depends to some degree on the system's processing conditions (temperature, pressure, contact with the atmosphere – if any) as well as the dynamic conditions (shearing, coalescence, turbulence intensity, mixing energy, settling time), and the use of production chemicals. During the separation process, chemical constituents of produced water are precipitating, dissolving, and moving to the interfaces between oil/water/solids, etc. In other words, the composition of produced water is both a complex and a highly dynamic variable. This is one factor that makes produced water treatment unlike water treatment in other industries where the water composition is relatively fixed and the main dynamic variable is blending of different sources of water. Blending of various well streams is also an important factor in produced water chemistry and treatment.

2.1 Produced and Flow Back Water

The definitions of produced water and flow back water, that are used in this book, are given in Section 1.3. They are also summarized below. In this chapter, the chemical nature of produced and flow back water is discussed. Definitions and sources of produced and flow back water are summarized here:

- **Produced water is any water that flows from an oil or gas producing reservoir. This includes:**

  o **formation (reservoir) water from primary production, including interstitial water**

  o **condensed water**

  o **water from water flood, chemical enhanced oil recovery, steam flood, etc.**

- **Flow back water is any water that flows from an oil and gas producing well and was the result of a stimulation treatment or other drilling and completion operation. This includes:**

  o **drilling fluid**

  o **completion brine**

  o **squeeze chemicals**

  o **stimulation fluids such as solvent, acid, hydraulic fracturing fluid**

Produced water includes water that has been injected, has traveled through at least a portion of the reservoir, and is produced through a hydrocarbon producing well. In the case of IOR / EOR, the water that was injected and has at some stage of production started to be produced is sometimes called "break-through" water, or "back-flow" water. Though confusing, "back-flow" water is not the same as "flow-back water. As indicated in the definitions above, a distinction is made between back-flow and flow back water. The term "flow back" is generally reserved for stimulation fluids which are injected into and are subsequently produced from the same well. Since this water has not traveled through the reservoir, nor had significant contact with the hydrocarbons in the reservoir, it is generally not considered to be a type of produced water. This distinction in naming helps to distinguish one type of water from another and ultimately leads to a better understanding of the characteristics and history of the water.

Within the broad definition of produced water given above, several types of produced water can be distinguished. These are listed above in the sub-bullets. It is important in many cases to distinguish between the type of produced water since each type can have distinct chemical and physical properties.

2.1.1 Formation Water

The term "formation water" refers to any water that resides in a porous rock formation. The porous rock formation can be sandstone, silt, limestone, shale, coal, etc. Some of these materials have very low porosity. Some have high porosity. Some have their pore spaces filled with water, others have it filled with a mixture of gas, hydrocarbon and/or water. Be that as it may, if the rock contains water, we refer to this water as formation water.

Water residing in a porous rock matrix is in chemical and physical equilibrium with the rock minerals, and with other fluids which are present. The subject of the origin of this water will be discussed later in this chapter. By "origin" we mean, the likely geologic processes that resulted in the water residing in the formation rock. Knowledge of the type of rock, and the origin of the water, can provide clues about the composition of the water. For example, water that resides in a carbonate formation will be saturated with calcium carbonate. This is an obvious example. Other examples include the presence of substances such as mercury, arsenic, radium, metals such as iron and zinc, etc., and substances such as hydrogen sulfide.

A distinction needs to be made between "formation water" and "produced formation water". Formation water refers to the water that is still in the reservoir rock. Produced formation water is the water that has been produced from the formation. These two water types differ from each other because of the compositional changes that occur upon production. Obviously, production changes the fluid temperature, and pressure. It also shears the fluid and alters the thermodynamic balance between dissolved and precipitated solids. During and after production, the water's properties can dramatically change. For example, mineral salts, asphaltenes, and/or waxes may precipitate. Fine solids and other grains of rock may be dragged into the production fluid, and become coated with naturally occurring acid compounds. Production causes dissolved gases to vaporize. This can have a significant effect on pH. Oil droplets can become dispersed in the water, and vice versa. Polar organic compounds will migrate to the newly formed oil/water interface of the droplets. These are all important changes that affect the design and troubleshooting of produced water treatment systems.

There is also a physical aspect to the water properties. In some circumstances, the porous rock formation of interest supports the overlying rock. In this case, the fact that the rock has fluid in it has little effect on the pressure of that fluid. If a well were to be drilled, the fluids would flow out of the rock, into the well, and up the well bore. They would not, however, reach the surface. The fact that the rock supports the overburden means that the formation fluid is not compressed or pressurized by the overburden rock.

More commonly the formation rock is compressible and the pressurized fluid within the rock helps to support the overburden. In this case, the well would fill with the fluid all the way to the surface. As fluid is withdrawn, the compressible rock would indeed compress and subsidence of the surface may occur. Understanding the concept of formation pressure is important in designing water flood systems.

2.1.2 Interstitial and Connate Water

The term "connate water" is used with various meanings in the industry. The following definition is adopted here: Connate water is formation water that was trapped in the pores of sedimentary rock when that sedimentary rock was deposited. At some later time, the bulk water either flowed out of the formation or was substantially evaporated. However, some of the original water remained trapped in the pores as interstitial water. Thus, connate water refers to original water trapped in the pores and interstitial water refers to the water trapped in the pores at the time of production. This definition is consistent with a geological or hydrogeological definition. The interstitial water is of far greater concern to water treatment issues than the connate water.

Capillary pressure holds the interstitial water in the pores, interstices and small crevices between rock grains. As oil or gas is produced, the velocity of the escaping fluids can sweep some of this water out of the formation, particularly near the well bore where the fluid velocities are highest. Interstitial water typically has high salinity, particularly in gas formations, due to the evaporation of some of the interstitial water. If there is formation water in a reservoir below the gas, there is a tendency for coning to occur near the well bore which can drag this interstitial water into the production stream. The properties and significant of interstitial water will be discussed below in the next section.

2.1.3 Gas Production – Interstitial and Condensed Produced Water

Gas reservoirs typically contain some amount of formation water. Some water is dissolved in the gas as vapor; it may also be trapped as a liquid in the pores that are distributed throughout the reservoir; it may also be present as a continuous phase beneath the gas. The trapped water is held in place by capillary pressure and is referred to as interstitial water, which was just discussed. Interstitial water, as mentioned, can have high salinity. Depending on the reservoir temperature and pressure, the gas in the reservoir may or may not be saturated with water.

As the gas is produced, the temperature is reduced in part due to Joule-Thompson expansion and in part due to heat loss to the surroundings. The pressure is also reduced. As the temperature decreases, water will condense out of the gas phase. However, pressure reduction has the opposite effect. As pressure is reduced and the gas expands, liquid water will vaporize into the gas phase. Thus whether produced gas is wet or dry depends on the specific temperature and pressure changes that occur during production. In some circumstances, droplets of salty liquid water become entrained in the gas as it is produced from the reservoir. As the produced fluid makes its way up the wellbore, these droplets may then become vaporized (evaporated). The salt contained in the water droplets then precipitates and forms salt blockages in the production tubing or flow lines.

Condensed produced water has no dissolved ions, and no dissolved mineral content. Operators often assume that gas wells are only producing condensed water because the water volumes can be very low. But in most cases, analysis will show the produced water contains a proportion of formation water together with a portion of condensed water. Typically, small amounts of interstitial water (see above) are pulled out of the pores. This can be a source of scaling material. Many wells seem to 'dribble' low amounts of formation water for long periods. If in doubt, check the conductivity, and if it indicates more than 1000 mg/l TDS, then check the ionic composition - that should help distinguish between formation and injected waters (e.g., from drilling or completion fluids).

Consequences of Ignoring the Formation Water: It is almost never the case that produced water is composed entirely of condensed water. A design basis that makes this assumption is incorrect and will have disastrous consequences in terms of scaling. For example, in 2007 the Kauther gas field in Oman was put on production. It was assumed by the operator that the produced water was entirely condensed and contained no mineral salts. In reality, the produced water contained calcium carbonate which precipitated upon mixing with the alkaline hydrogen sulfide scavenger. The hydrogen sulfide scavenger was an amine chemical which raised the pH and caused the calcium carbonate to precipitate. The facility was shut down for 45 days to jack-hammer the calcium carbonate scale out of the piping. More recently a field in the North Sea ran a PLT/caliper in a well producing 'condensed water.' In this case, the operator discovered the buildup of calcium carbonate before a costly shutdown became necessary. As recently as 2010, a facility in the North Sea was shutdown when the MEG (methylene glycol) unit became scaled due to the presence of 1 m^3 of formation water / MM

m^3 of gas. The assumption had been made that all produced water was condensed water and that zero formation water would be produced. The facilities had no detection or monitoring methods for scale. Nor did it have any injection points for solvent wash, or scale inhibitor treatment.

One contributing factor that leads to mistakes of this kind is the narrow focus of the reservoir and facilities engineers. Each lacks an understanding of the other discipline. The reservoir engineers lack an understanding of facilities design. The facilities engineers do not typically understand the sampling, analytical, and data reporting methods and conventions used by the reservoir engineers. Both disciplines need to learn more about the other discipline.

From a practical standpoint, many people in the industry consider produced water with a TDS of less than 1,000 mg/L to be predominantly condensed water. But such water is a combination of formation water together with condensed water. This produced water typically contains low but detectable amounts of common ions (sodium, chloride etc). In addition it is common to see organic acids (formate, acetate etc) in the water. Generally speaking, the higher the reservoir temperature the higher the levels of organic acids in condensed water. HPHT (high pressure / high temperature) fields often have greater than 200 mg/l of short chain acids in the condensed water. Boron and bicarbonate have also occasionally been found in quite high levels.

Characteristics of Condensed Water: As discussed previously in this book, any water that comes from the reservoir will be referred to as produced water. Condensed water is initially dissolved in the gas in the reservoir. Upon production it condenses. Condensed produced water has no dissolved ions, and no dissolved mineral content. In some cases, condensed water is mistakenly viewed as having simple chemistry and characteristics. However, the lack of dissolved ions does not make this type of water easy to treat. Condensed water contains dissolved gases. The most important dissolved gases are carbon dioxide, hydrogen sulfide, the BTEX compounds (benzene, toluene, ethylbenzene, and xylene), and the short chain volatile acids (VFA) such as formic, acetic, propionic, butyric, etc. This cocktail of compounds, together with shearing that often occurs and the use of corrosion inhibitors and hydrate inhibitors creates a very challenging situation.

Consequences of Low pH: The water that condenses will contain dissolved CO_2 and H_2S, if they are present in the gas. Typically, there will be no or very little alkalinity (e.g. sodium bicarbonate) in the condensed water. Thus, pH buffering capacity will be low or essentially zero. If CO_2 or H_2S are present, the pH of the water can be as low as 4. The low pH, and low mineral content, makes this type of produced water quite corrosive. In some cases the operating company is forced to inject high dosages of corrosion inhibitor. The most common oilfield corrosion inhibitors have an amine head group and a hydrocarbon tail group. They have surfactant-like character and can stabilize an emulsion.

Surface activity is an important part of the mechanism by which corrosion inhibitors prevent or at least reduce the corrosion rate. A major consequence of this surface activity is that they also stabilize O/W and W/O emulsions. There are two mechanisms involved. One mechanism is the lowering of interfacial tension. Lower interfacial tension leads to greater susceptibility to forming small oil droplets when the fluid is subjected to shear, from pumps or valves. For a given shear intensity, lower interfacial tension creates smaller oil droplets. The other mechanism involved in stabilizing emulsions is the creation of an interfacial barrier to coalescence. The presence of any surface active compound at the oil / water interface will create such a barrier. Asphaltenes at the interface create a stiff and immobile barrier that prevents coalescence. Corrosion inhibitors also form a barrier. Thus the corrosion inhibitor promotes the formation of small droplets and prevents their growth by coalescence.

Most production streams also contain some produced formation solids such as sandstone, clay, silica, carbonates, etc.. At the low pH of the condensed water, some of these minerals will partially dissolve. Typically, these, and their counter ions amount to a few hundred or maybe a thousand mg/L of Total

Dissolved Solids depending on the pH and the amount of solid material produced. Most of these constituents have a relatively minor consequence to the production system however they do show up in a water analysis.

Effect of Short Chain Acids: The short chain acids are somewhat surface active and behave like corrosion inhibitor to stabilize emulsions. They migrate to the oil / water interface. This has two effects. First, they reduce the oil/water interfacial tension. The reduced interfacial tension makes the oil droplets sensitive to breakage due to shearing (see Chapter 3 – Fluid Mechanics). Thus, a typical or normal amount of shear when applied to these fluids results in very small droplets of oil in water. The second effect of their presence at the oil / water interface is that they create a charge barrier to droplet / droplet coalescence. Both mechanisms result in small, relatively stable droplets of hydrocarbon condensate dispersed in the produced water.

Norwegian Sector Experience: High pressure gas fields in the Norwegian sector of the North Sea suffered produced water problems due to a combination of high shear which created small drops of condensate, low salinity of produced water, and high concentrations of acids. The low salinity helped to stabilize the condensate droplets by electrostatic repulsion (large negative Zeta potential). The acids helped to reduce the interfacial tension which promoted the formation of small droplets under the shear conditions. The acids also helped to prevent coalescence by creating a rigid repulsive interface between droplets.

2.1.4 Water from Improved and Enhanced Oil Recovery (IOR and EOR)

Water is often used as the fluid of choice in attempts to improve or enhance hydrocarbon production. Water is relatively inexpensive and easier to handle than gas, hydrocarbon liquids or solvents, although these are sometimes used.

Water that is pumped into the ground for IOR (Improved Oil Recovery) or EOR (Enhanced Oil Recovery) purposes almost always contains chemical additives. Common additives include biocide, filter aids, oxygen scavengers for corrosion control, corrosion inhibitors, and scale inhibitors. Whether the water is referred to as a water flood, polymer flood, steam flood, hot water flood, or alkali-surfactant flood depends on how much and/or which type of chemicals have been injected. For example, there are many water floods that have a friction reducing agent injected at a few to a few tens of mg/L. The same friction reducing agent is often used in a polymer flood where the concentration may be a few hundred mg/L. Fluids containing polymer that have migrated from the injection well and are produced are referred to as back-produced or "flow back" fluids.

Over time injected water will sweep through the formation and be produced up the well bore. When this happens, the flood is said to have "broken through." Reservoir engineers run models which help to predict the "breakthrough" time. Once breakthrough occurs, the produced water properties start to change from the original formation water to a mixture of formation water and injected water. Over time, the percentage of "break-through water will go from zero to several tens of percent. Over sufficient time, the produced water may become indistinguishable from the injected water.

Compatibility of injected and formation water is a critically important issue. If the mixture of these waters causes any constituents to precipitate, the solids may clog the formation and lead to reduced production and to lower ultimate recovery. Thus injected water must be compatible with the formation water and the formation mineralogy. Low salinity water, having a low concentration of divalent cations can cause clay to swell. Clay swelling and fines migration can lead to reduced injection permeability and in severe cases to reservoir plugging.

2.1.5 Drilling and Completions Flow Back Water

The term "drilling and completion" refers to all drilling and chemical treatments prior to production startup. There are three classes of fluid:

- Drilling mud (water-based or oil-based)

- Completion brine

- Stimulation Fluid

Drilling Mud & Completion Brine: The fluids used in drilling are obviously drilling mud, and various fluids that counterbalance the formation fluid pressure to prevent hydrocarbon production until the well is ready to handle hydrocarbons. These counterbalancing fluids are also referred to as completion fluids or completion brines. They are typically brines that are formulated to whatever density is required to achieve a certain hydrostatic head. The following dissolved salts are commonly used in the formulation of completion fluids: $NaCl$, KCl, KBr, $CaCl_2$, ZnCl, ZnBr. The selection of which salt is used and its concentration is dictated by the required weight of the fluid and various other considerations such as compatibility with other completion fluids and with the formation fluids. Other additives are also typically used such as corrosion inhibitor and biocide.

Stimulation Fluids: Wells for either conventional or unconventional hydrocarbon fields may be stimulated, depending on the production profile and the economics. For example, in the deep water Gulf of Mexico essentially all wells are stimulated before production and most wells are re-stimulated several times during the production life while producing from the same reservoir. In order for the wells to be profitable, they must be productive. Due to the water and reservoir depth and the remoteness of the operation, drilling these wells is very expensive. Stimulation is a matter of economic necessity.. Stimulation commonly takes the form of an acid flush to clean out drilling debris and to remove carbonates and sandstones that may have accumulated near the well bore during the drilling operation. This is commonly followed by hydraulic fracturing with proppant injection. This is referred to as "frac-and-pack."

Stimulation fluids have a wide range of composition. They include:

- Acid (HF, HCl, formic, acetic, or combinations of these, usually with oxygen scavenger)

- Hydrocarbon solvent (heavy aromatic naphtha, xylenes, or toluene)

- Surfactants to stabilize emulsions of acid in a hydrocarbon solvent

- Warm or hot water (to melt wax or dissolve salt, usually with a biocide and oxygen scavenger)

- Hydraulic fracturing fluids (see below)

These fluids are pumped into the well and are then partially produced back.

Hydraulic Fracturing: Shale is an important resource and the water management issues during and after hydraulic fracturing of shale resources are economically significant and complex. Hydraulic fracturing (HF) is not new. Not counting the recent activity in shale, nearly 2.5 million conventional HF operations have been carried out in the world. Conventional wells, both onshore and offshore coal seam gas (or coal bed methane, as it is known in the United States) wells, and wells where some form of improved (IOR) or enhanced recovery method is used (EOR) have been hydraulically fractured.

HF of conventional oil and gas fields uses essentially the same fluids as for shale and coal seam gas. In one form or another, HF has been practiced for at least 50 years. The chemicals used and the water injection rates and volumes have evolved over time, and differ from one application or field to another. But in general, the same types of chemicals and proppants have been used. The historical development and current practice of hydraulic fracturing is discussed by Hlidek and Economides et al. [1].

Hydraulic fracturing fluids generally contain the following ingredients:

- Polymer – water soluble, high molecular weight friction reducer; may be used to carry proppant into the well

- Surfactant – used to promote hydrocarbon wetting of the fractured rock

- Biocide – used to prevent bacterial souring and bacterial blockage of the fractures

- Proppant – used to prop open the fractures

- Breaker – an oxidant or enzyme that will reduce the viscosity of the polymer to facilitate fluid flow back while leaving the proppant in place in the fractures

- Others – several other chemicals can be included for specific purposes

Once the fracturing operation is completed, the pressurized fluids in the reservoir push the hydraulic fracturing fluids out of the well. The fluids that flow back after a hydraulic fracturing operation will contain the injected fluids, usually with some precipitation and degradation of the polymer, and will contain some formation water, formation fines, and minerals dissolved from the formation. In many of the shales, the formation is dry, i.e. the shale does not contain water and thus retains some of the injected HF water.

2.2 The Geologic Processes that Impact Formation Water

One of the recurring themes of this section, indeed of the entire text, is that produced water varies in properties, composition and character from one location to another. In the journey to understand the differences in produced water, it is helpful to know what causes produced water to vary in the first place. In the preceding material, the different types of produced water were discussed. In this section, the origin of formation water is discussed. Carpenter gives an excellent though somewhat dated description [2] of the origin and evolution of brines.

One of the barriers to applying geology and hydrogeology concepts is a lack of accessibility in the literature. While they are fascinating subjects, they are specialized. People in the petroleum industry tend to either know the subject well, or not at all. The literature is crammed with jargon and loosely defined terminology. It is difficult for the non-specialist to understand. Many people say the same thing about the subject of produced water.

Without adequate guidance it is difficult for the water treating engineer to recognize the different types of produced water and to know how to account for these differences in the design, operation and troubleshooting of water treatment processes. The intent of this section is to provide enough background for a water treating specialist to derive useful information from discussions of petroleum geology and hydrogeology, without having to spend long hours studying the details. This chapter does not contain detailed information on any particular water. It does however provide an entry point to the literature.

There are many reasons why formation water at one location differs from that at another location. One of the contributing factors is the geologic history of the water. The geologic history includes such topics as:

1. How did the water become buried;

2. Once buried, how did it migrate from one location to another;

3. During its migration what temperatures and pressures was it subjected to;

4. What biological processes have occurred;

5. What other sources and bodies of water did it mix with;

6. What hydrocarbon liquids and gases did it come in contact with;

7. What minerals did it dissolve;

8. What minerals have precipitated out of the water;

This may seem like an implausible level of detail for formation water. However, such information is generated as a by-product of petroleum geology studies which are carried out in the hunt for oil. The geological history of water gives clues to the petroleum geologist about the history of the hydrocarbons, how to interpret seismic data, and where to find more hydrocarbons. Thus, wherever the geology of an area has been studied for hydrocarbon extraction, there is typically some information available on the geology of water in the same region.

How can this information be used? Knowing the difference, for example, between Devonian and Permian helps explain the differences between various salinities in shale plays. Also, knowing what the term "maturity" means can help explain whether sandstone is consolidated or unconsolidated, which gives a clue to the probability of sand production. Sand production, and the production of carbonate or shale fines are an important consideration for produced water treatment. These solids can become coated with oil or, if very fine, can concentrate at the oil/water interface and help stabilize an emulsion. Also, knowing something about "sea water evaporates" and the processes of migration, dehydration, and biological degradation can help provide insight into potential precipitating constituents, the stability of asphaltenes in the oil, the tendency of asphaltenes to migrate to the oil / water interface, why formation brines contain very little potassium (presence of potassium ion can be a telltale for completion fluid contamination), and the presence of naturally occurring surfactants.

It is difficult to find water treating specialists who know geology and hydrogeology well enough for it to be a useful subject in the art of water treating. Those few people who do know the subject are able to quickly assess certain general characteristics of a particular formation water. The kinds of conclusions that can be drawn include:

1. Expected alkalinity and pH;

2. Expected ratios of ions such as K/Cl, Ba/Cl, Ca/Mg, SO_4/CO_3 and HCO_3, and a few others;

3. Likelihood that the water will be saturated in certain components;

4. Likelihood that the water will contain naturally occurring radioactive material (NORM);

5. Whether an analysis of ionic composition is reliable or potentially inaccurate;

6. Whether a sample is contaminated with drilling mud or completion fluid.

In the reservoir, oil, water and rock are in chemical and physical equilibrium. Chemical equilibrium governs the concentrations of ions in water, organic chemicals in oil, constituents in the gas, and minerals in the rock. Physical equilibrium governs the temperature and pressure of all materials and constituents.

2.2.1 The Origin of Crude Oil

Since produced water is in chemical, mechanical, and thermodynamic equilibrium with formation oil, it is worthwhile to briefly discuss the chemical nature of oil. Oil is not a chemical. It is a large group of hydrocarbon and organic molecules that span a very broad range of physical (gravity, density, viscosity, etc), and chemical characteristics. To understand the essentially unlimited variety of organic compounds it is helpful to consider the origin of crude oil.

Crude oil is the product of plant and animal matter that has been subjected to millions of years of digestion at high temperature and pressure. During this long conversion process, the proportion of hydrocarbon molecules gradually increases. Since oil ultimately comes from dead plant and animal matter, it may contain some undigested "molecules of life" such as lipids, fatty acids, proteins, sterols, porphyrins, chitin, sugars, resins, lignins, etc.. This is particularly true if the process has been short (less than a few million years). This complex mixture of compounds has an appreciable percentage of surfactant-like compounds (e.g. naphthenic acids). The cell walls of plants and animals are composed of surfactant-like molecules. Older or more mature oil will contain the digestion products which tend to be hydrocarbon-like and less surfactant-like. However, biodegraded oil can contain very high concentrations of acids including the surfactant-like alkyl acids, depending on the conditions under which biodegradation occurred.

For our initial purposes, oil can be considered to have three basic chemical groups:

- Nonpolar compounds typically referred to as hydrocarbons

- Polar surfactant-like compounds that are predominantly oil soluble

- Polar surfactant-like compounds that are predominantly water soluble

All three of these chemical groups are of interest in water treatment since they can contribute to the concentration of contaminants and the stability of emulsions.

Fatty acids (short and long chain carboxylic acids) and cell membrane material of plants and animals, the so-called lipid bilayer, is present in the original material from which crude oil is generated. Lipids are surfactant-like molecules with a polar hydrophilic head and a hydrophobic hydrocarbon tail. In the early stages of hydrocarbon maturation, these acids are converted to hydrocarbons. Thus, a young source rock will contain a high acid content due to the limited conversion of original organic material.

Over time, and under the right conditions, biodegradation occurs as a result of bacteria metabolism. The result of this process is conversion of hydrocarbons to acids and eventually to CO_2, water and microbial cell mass. A number of factors can cause the bacteria to die off. When they do, they leave behind their own cell mass which is rich in organic acids, as was the original organic material.

2.2.2 The Water Cycle

In order to understand the geologic origins of formation water, it is helpful to understand the transport and migration patterns of water, as they exist in the present day. These processes, described below, are the same processes that have always occurred throughout geologic time. These processes have occurred continuously over millions and hundreds of millions of years.

Meteorological (weather-related) water falls to the ground and is absorbed or runs off directly into streams, lakes or rivers. Eventually these water bodies typically drain into the sea. If absorbed into the ground, the water flows underground and helps recharge the subsurface water table. Water from the water table also flows into lakes and streams, or into the sea, or into other aquifers. This is the normal hydrogeological cycle, as shown in Figure 2.1.

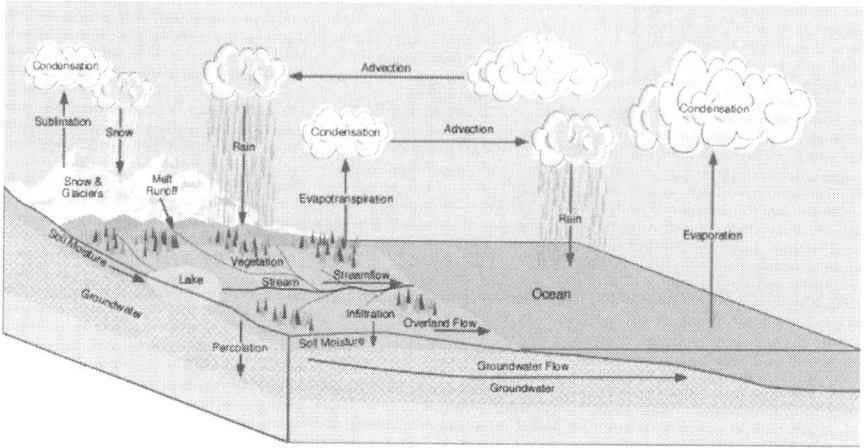

Figure 2.1. The normal hydrogeological cycle of surface and shallow subsurface waters.

The term surface water is often used to refer to water that participates in the near-surface hydrogeological cycle. Surface water is not necessarily on the surface. It can even refer to water in clouds. It is water that moves freely, without being trapped in the air, on the surface and underground within say a few hundred meters of the surface. The salinity of this water is typically less than a few hundred mg/L of Total Dissolved Solids. However, it can be somewhat higher. Salinity of surface water is a consideration from a municipal and agricultural standpoint. This is one of the principle studies of hydrogeology.

Surface Salinity: One of the important issues in hydrogeology is the accumulation of salts in the ground and in the surface water. Due to anthropomorphic (e.g. human) activity such as deforestation, agriculture, settlement, etc., salt that would otherwise be immobilized in the soil is released and enters into the hydrogeological water cycle. When this happens, the salinity of a region may become too high to sustain vegetation and high value agricultural crops. This subject will be discussed in considerable detail in relation to water treatment for coal seam gas (coal bed methane) operations (see Chapter xx).

Present Day Water Stranding Mechanisms: Over time, bodies of water can become "stranded." The term stranded water is used here to denote a body of water that is cut off from the normal water cycle described above. Examples of such processes are readily apparent in the present day. Of course, the processes themselves take place in geologic timescales and are thus not possible to observe. Nevertheless, various water bodies around the world are in various stages of becoming stranded, and eventually buried.

The Caspian Sea: The Caspian Sea is an example of an Endorheic or terminal lake. It was originally composed of sea water. It became landlocked about 5.5 million years ago due to tectonic uplift and a fall in sea level. During warm and dry periods, the landlocked sea started to dry up. During the evaporation process, various components reached their solubility limit and would have therefore precipitated. Carbonates would have precipitated first, followed by sulfates and eventually halite. These precipitated minerals were deposited in various locations around the receding lake. Some of the precipitated minerals were covered by wind-blown deposits.

With time, the global climate became less arid and moister. Fresh water drained into the basin refilling it. Some of the precipitated mineral has dissolved. But since the mineral was partially or completely buried, and distributed around the previous evaporating lake, the reformed lake (the present day Caspian Sea) does not have the same composition as the original body of sea water. Currently, the mean salinity of the Caspian is one third that of the ocean (approximately 12,000 mg/L). The mineralogy is different as well as shown in the table below.

Table 2.1. Composition of the Caspian Sea, as an example of the geological processes that can occur in a terminal lake

Salt Type	Ocean Water (%)	Caspian Sea Water (%)
NaCl	78.3	62.2
MgSO4	6.4	23.6
MgCl2, MgBr2	9.44	4.54
CaCO3	0.21	1.24
KCl	1.69	1.21
CaSO4	3.93	6.92

The example of the Caspian Sea demonstrates various mechanisms that are involved in the geologic history of a body of water. In this case the mechanisms include formation of a stranded body of water (terminal or Endorheic Lake), evaporation, precipitation, burial of precipitated minerals, and refilling. Many variations in this sequence have been deciphered by geologists.

Subterranean Burial: Water can not only migrate in the surface layer, but it can also become buried. Fluids flow through porous rock due to a gradient (variation) in hydraulic head. Hydraulic head is the sum of the isotropic pressure plus the hydrostatic head. Isotropic pressure is the usual pressure due to the presence of fluids. It is the same in all directions. Hydrostatic head is also referred to as elevation head. It is the pressure due to gravity. It has the same characteristics below ground as above. The weight of the rock above the reservoir (overburden) contributes to the isotropic pressure but not the hydrostatic pressure. The path of water migration is not always in perfect alignment with the shortest path to the least pressure. In most instances, there are fractures, conduits, and regions of higher permeability which direct the flow. While the flow may not be in the direction of least pressure, it will always be in a direction of less pressure and never in a direction of greater pressure. There is another factor that must be taken into account in describing the flow of fluids underground and that is the adsorption / desorption phenomena. Of course outright precipitation of components from the water will significantly reduce the rate of migration of those components. In any case, the movement of subsurface fluids over geological time periods can be complex and substantial.

The evolution of seawater composition: Two of the powerful observations in understanding oilfield brine origin are that seawater composition has varied only slightly over the past 500 million or so years, and that during this time, as well as during present day, the composition of seawater is relatively uniform across the globe.

To be more precise, modern seawater salinity ranges between 32 to 37 g/L across most of the globe. There are exceptions of course. In the Arabian Gulf (also known as the Persian Gulf) seawater salinity is in the range of 41 to 44 g/L. This is due to the hot and dry conditions in the region which promote evaporation, and which therefore has limited fresh water influx from rainfall, and the limited exchange of water through the narrow Strait of Hormuz. For similar reasons, the salinity of the Red Sea ranges from 36 to 41 g/L. Conversely, the salinity of surface currents near the North Pole and South Pole varies seasonally due to ice cap melting in the summer. These are however, exceptions, and relatively minor variations as far as the discussion of oilfield brines is concerned. The vast majority of modern seawater has a small variation in salinity across the globe.

There have been some compositional changes to seawater during this period. But there are some aspects of the composition that have not changed. For example, it is widely assumed that the Cl/Br ratio has been constant over this long time period. Thus, oilfield brines having the same Cl/Br ratio as modern seawater are likely to have originated from seawater [2]. Seawater evaporation brines (also referred to as seawater evaporates) have a Cl/Br ratio of roughly 290. If an oilfield water has a Cl/Br ratio less than 290, then it has most likely been subject to oversaturation of halite. Halite (NaCl) precipitation will reduce the chloride concentration relative to bromine, thus reducing the Cl/Br ratio as well.

If an oilfield brine has a Cl/Br ratio in the range of 600, then it was likely formed by the interaction of relatively fresh water (such as meteoric or ground water) with a deposit of halite. Massive deposits of halite are relatively common in some parts of the world. Ground water, or meteoric water or essentially any source of water can come in contact with such deposits and dissolve the halite. Such halite deposits have a relatively low, but typically non-zero concentration of Bromine. When this occurs the resulting water is referred to as a halite dissolution brine.

If a seawater evaporate comes in contact with a halite deposit, it too will dissolve halite. When this occurs the Cl/Br ratio of the resulting water will be between that of seawater (290) and that of a halite dissolution brine (600). Thus, an oilfield water with a Cl/Br ratio between 290 and 600 has contributions from both halite dissolution and seawater evaporation.

Origin of Metal-Rich Oilfield Brines: Some oilfield brines contain relatively high concentrations of lead, zinc (> 25 mg/L) [2, 3], or mercury. Other metals can be found as well. It is useful to know in general the origin of such components in order to determine if the water analysis is consistent with other factors. It is also useful to know the geographical distribution of such components around the globe so that the presence of these metals can be anticipated. Generally mercury is only found in oil/gas production if the reservoir temperatures are high, e.g., >350°F. At these temperatures ionic, organic, and sulfidic mercury species are not thermodynamically stable relative to elemental mercury. So most of the time mercury in oil/gas production is elemental in nature. In the presence of CO_2 and H_2S, however, ionic mercury species can be generated by oxidative reactions with elemental mercury

Field data suggest that metal-rich brines are associated with red beds. Red beds are sedimentary rocks that consist of sandstone, siltstone, and shale. They are red in color due to the presence of ferric oxides. Some of the better known red beds are the Permian and Triassic formations of the western United States, and the Devonian Old Red Sandstone formations of Europe.

Metal-rich oilfield brines are found in the Gulf Coast of the United States, Southeast Asia, Northern Alberta, the Netherlands, Kazakhstan, North Africa, and offshore West Africa. Metal-rich brines in the Lower Cretaceous reservoirs of the Mississippi salt basin have a Cl/Br ratio of 130. Frankiewicz et al. report the occurrence of arsenic (25 mg/L) and mercury (5 mg/L) in produced water from the Erawan PSO in the Gulf of Thailand [4]. They report good removal of these contaminants through the use of ultra-filtration and reverse osmosis.

2.2.3 Seawater Evaporites

Many oilfield brines have their origin from seawater that has been trapped in a shallow sea, and subsequently buried [2]. Once buried, the seawater can migrate and undergo a number of processes which change its composition. One of the most common processes is simple dehydration by evaporation. This evaporation process gives rise to the observation of a wide range of salinities in oilfield brines. The salinities of oilfield brines vary from a few thousand to over 250 k mg/L.

The evaporation process is most easily understood by analysis of a marker constituent. There are several candidates that could be used. Typically bromine is chosen. A good marker constituent has the property that it does not readily precipitate during evaporation. Thus, as the seawater evaporates, the mass of the marker constituent dissolved in the liquid remains constant. Thus, any change is concentration of the marker constituent is a direct indication of the extent of dehydration by evaporation. Zherebtsova and Volkova [5] showed experimentally that, when seawater evaporates, essentially all of the potassium, rubidium, lithium and bromine remain in solution up to a relatively high concentration. Eventually potash salts will precipitate. Once this occurs, the potassium ion concentration no longer reflects the extent of dehydration. However, even during potash precipitation, lithium and bromine stay in solution. Bromine is usually chosen as the marker constituent.

The evaporation process of seawater is represented in the Figure 2.2 below.

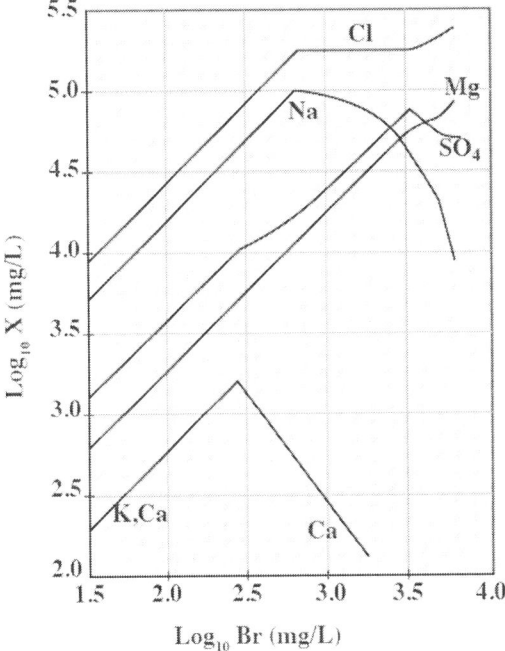

Figure 2.2. The concentration of ions plotted as a function of bromine concentration during the process of seawater evaporation. After Carpenter [2].

As shown in Figure 2.2, calcium precipitates at relatively low levels of seawater dehydration. Sodium and chloride show deviation from linear behavior at the same bromine concentration. This indicates that they precipitate as a pair, i.e. halide (NaCl). Sulfate precipitates in two stages indicating that different counter ions are involved.

2.2.4 Biodegradation and TAN

Biodegradation can occur at any time in the history of the reservoir depending on the location, temperature and the presence of water and nutrients [6]. The net result of biodegradation varies from one reservoir to another but there are certain trends that generally are followed. In a typical biodegradation process the main net result is that microbes consume straight chain hydrocarbon groups (-CH2-) and they generate organic acid molecules. This reduces the hydrocarbon content and increases the polarity of the crude oil. There tends to be a decrease in the API gravity of the oil, and a higher concentration of polar organics, organic acids, natural surfactants, naphthenic acids, and surface active components which stabilize oil in water emulsions and increase the dissolved organic concentration The net result is that water treatment is more difficult.

Calculations have shown that the water volumes needed for the necessary oxygen transport into biodegrading reservoirs are unrealistically high. Hence, in-reservoir biodegradation processes are probably anaerobic (without oxygen) with the most important processes being sulphate reduction and methanogenesis. Over the last 25 or so years, a number of anaerobic bacteria have been identified that have the ability to degrade hydrocarbons [7] without the presence of oxygen. Various correlations of reservoir temperature and biodegradation have shown that microbial life in petroleum reservoirs has an upper temperature limit of about 80°C [7].

Total Acid Number: The Total Acid Number (TAN) of the crude oil is a simple and often indicative test for the occurrence of biodegradation. However, TAN is not conclusive. There are other causes of high TAN besides biodegradation. Therefore if the objective is to confirm whether biodegradation has occurred, then other tests should be run as well. Organic acid content and composition are also simple indicators. More definitive tests include gas chromatographic hydrocarbon fingerprinting which reveals the presence of naphthene biomarkers. In the early stages of biodegradation there is often an abundance of straight chain alkyl groups. Most oil metabolizing microbes attack these functional groups first. With time, as the alkyl groups become depleted, the aromatic functional groups, resins, and asphaltenes can become biodegraded. The more aromatic, polynuclear, and higher molecular weight fractions are often the last to be biodegraded and tend to increase in percentage as the saturated hydrocarbons are depleted.

Note that Total Acid Number (TAN) is reported in units of mg KOH/g. Thus, a TAN value of 1.0 is equivalent to 1,000 mg of acid/kg of crude oil, or roughly 850 mg/L of a 35°API crude oil. As shown in the Figure 2.3 below, a TAN value of 2 is equivalent to roughly 6,000 mg/g (micro gram of acid per gram of oil). Assuming a density of oil of 800 kg/m3 (0.8 mg/L), 6,000 mg/g is equivalent to 4,800 mg/L (mg of acid per liter of crude oil). This is a high value.

The acid content of crude oil can vary from essentially zero to several tens of thousands of mg/L. Many of these acids are surface active. Paraffinic crude oils typically have low acid content, but there are exceptions. Acid concentrations are usually highest in the kerosene fraction and low in the gasoline and lube oil fractions.

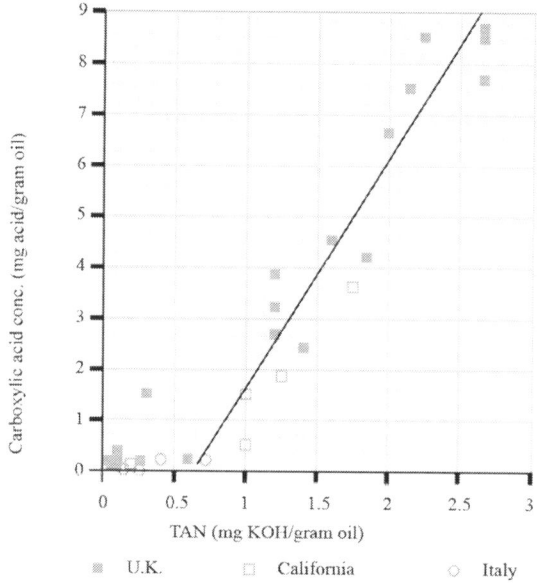

Figure 2.3. Concentration of carboxylic acids in oil (micro-grams of carboxylic acid/gram of oil as a function of the measured Total Acid Number (TAN) (mg KOH/gram of oil) [662].

Sources of Organic Acids: It must be kept in mind that the presence of carboxylic acids is not necessarily due to biodegradation. In order to understand the sources of carboxylic acids in produced water, it is necessary to understand their sources in the oil. There are three main sources [8]:

1. original plant and animal organic matter (particularly heavy crude oil from geologically young formations)

2. biodegradation

3. bacteria themselves (cell wall of bacteria).

The first source refers mostly to the fatty acids and cell membrane material of plants and animals, the so-called lipid bilayer. The second and third sources are related in that biodegradation occurs as a result of bacteria metabolism. The result of this process is conversion of hydrocarbons to acids and eventually to CO_2, water and cell mass. A number of factors can cause the bacteria to die off. When they do, they leave behind their own cell mass which is rich in organic acids, as was the original organic material. Crude oil containing a high concentration of carboxylic acid is either young (not sufficiently degraded), or has been biodegraded. Thus, biodegradation is of interest because of emulsion stabilizing tendency of such acids, and their potential contribution to Water Soluble Organics. Speight has classified crude oil into three categories regarding their acidity, as shown in the table below.

Table 2.2. Classification of Acidic Crude Oil [8].

Classification	Range of Acid Content (wt %)	Examples
Low	Nil to 0.1	Rangely CO; PA; Iraq
Medium	0.3 to 0.7	Lobitos, Peru; Balachany; Russia; Gulf Coast, USA
High	1.2 to 3.0	Lagunillas, Venezuela; Midway Sunset, CA; Romania

Mars Pink Reservoir - Example of Biodegradation: An excellent example of biodegradation is given in the following Table [9]. As shown, the Mars field had a number of hydrocarbon bearing reservoirs. They were stacked on top of each other with shale separating each reservoir. Over geologic time these reservoirs experienced different temperatures and pressures. However, it is likely that the bottom most reservoirs experienced higher temperatures than the top most reservoirs. This is generally the case for subterranean materials. Biodegradation is most likely for temperatures below 80°C.

The upper most reservoir (Pink) has been biodegraded. The other reservoirs either have not been biodegraded or the extent of biodegradation is very slight. As a result of the biodegradation, the saturates content of the Pink crude has decreased, and the asphaltene fraction is proportionately higher. The acid number is significantly higher in the Pink hydrocarbon sample. This is one of the more simple measurements that suggests that biodegradation has occurred.

Table 2.2. The properties of oil from different reservoirs in the Mars basin, deep water Gulf of Mexico. The Pink reservoir shows signs of biodegradation. Data from [9].

Saturates	Aromatics	Resins	Asphaltenes	Atomic S (wt%)	Total Acid Number (mg KOH/g oil)
21.0	61.6	14.0	3.4	2.7	4.35
40.9	47.6	9.5	2.1	1.8	0.34
41.3	37.5	13.9	7.3	2.6	
27.4	54.5	14.5	3.6	2.4	1.00
24.3	55.6	14.9	5.2	2.6	
28.1	54.1	13.5	4.3	2.6	0.61
25.0	59.3	12.6	3.1	2.7	1.20
30.8	51.8	15.1	22.0	2.2	
25.2	50.2	13.4	112.0	2.8	0.95

2.3 Composition of Produced Water – General Overview

The previous two Sections (2.1 and 2.2) discussed the general origins of produced water, where it comes from and what its general characteristics are. In the next several sections the constituents of produced water are discussed. The material below is focused on specific chemical compounds, the properties of the produced water containing those compounds, and brief comments about how to remove the compounds. The discussion about separation and removal of these constituents from the produced water is the subject of subsequent chapters of this book.

Each section includes sub-sections discussing which compounds are typically found dissolved in water and which compounds are typically dispersed. In many cases, the distinction between dissolved and dispersed compounds is difficult to make. Produced water contains a continuum of aggregate sizes from acid dimers, to a few molecules in small micelles, to many hundred molecules in large micelles, to large numbers of molecules suspended as drops. Much more will be said about these aggregates in the discussion of colloidal properties in Chapter 4 (Emulsions). Examples of produced water composition from various locations around the globe can be found in [10 - 16].

2.3.1 The Difference between Dissolved and Suspended Contaminants

A dissolved organic molecule is simply an individual organic molecule that is surrounded by water. Dispersed oil forms a distinct second phase, typically in the shape of drops that are suspended in the water phase. Due to buoyancy, the dispersed drops may cream (rise to the top) and form a free oil phase. Likewise solid particles suspended in the water may settle to the bottom of a vessel or tank.

But not all of the organic material in the water is quite so simple. Many acids form dimers in water. These dimers too would be regarded as dissolved. Other polar molecules form small aggregates composed of say a few to dozens of molecules. These might also be considered to be dissolved. Continuing the trend toward larger assemblies, produced water contains a wide range of concentrations and types of natural surfactant molecules. These can form micelles with polar groups at the oil/water interface and hydrocarbon molecules in the interior. There is a range of aggregate size for which the distinction between dissolved and dispersed organics becomes ambiguous.

While there is no universally accepted definition in colloid and surface chemistry, organic molecules in water and assemblies of organic molecules having a diameter of less than 5 nanometers are generally considered to be dissolved [17]. Droplets and particles having a diameter of greater than 0.5 microns are considered to be dispersed. These definitions leave a middle range of size that is not defined as either dissolved or dispersed. Assemblies of molecules having a diameter in between (5 nanometers to 0.5 micron) are considered to be colloidal. Colloidal material tends to be influenced by Brownian forces, which keep the particles suspended, and separation of these colloidal materials is dominated by surface chemistry, not simply settling or creaming.

Produced water contains an enormous number and types of chemical compounds that have surfactant-like character. This includes the medium and long chain acids, resins and asphaltene classes of constituents. It is likely that in the produced water associated with moderate or high acid crude oils, most of the dispersed material has these natural surfactant-like constituents residing at the organic/water interface. This detail is important in establishing the stability of the dispersed phase and whether or not it will form a free oil phase or remain dispersed. It is also important in establishing whether surfactant-like molecules reside in the dissolved phase or at the organic/water interface.

Methods to (somewhat arbitrarily) differentiate between dissolved and dispersed organics include the use of centrifugation or filtration. In either method, the result is a removal of dispersed organics with diameter in the range of 0.5 micron to 2 micron [18]. Any organics larger are generally considered to be dispersed. Those smaller are generally referred to as dissolved.

The Figure 2.4 below shows an outline of how different classes of organic entities are expected to be present in produced water. As mentioned above, "Free Oil" is assumed to have been fully separated prior to samples being taken for Oil-in-Water analysis.

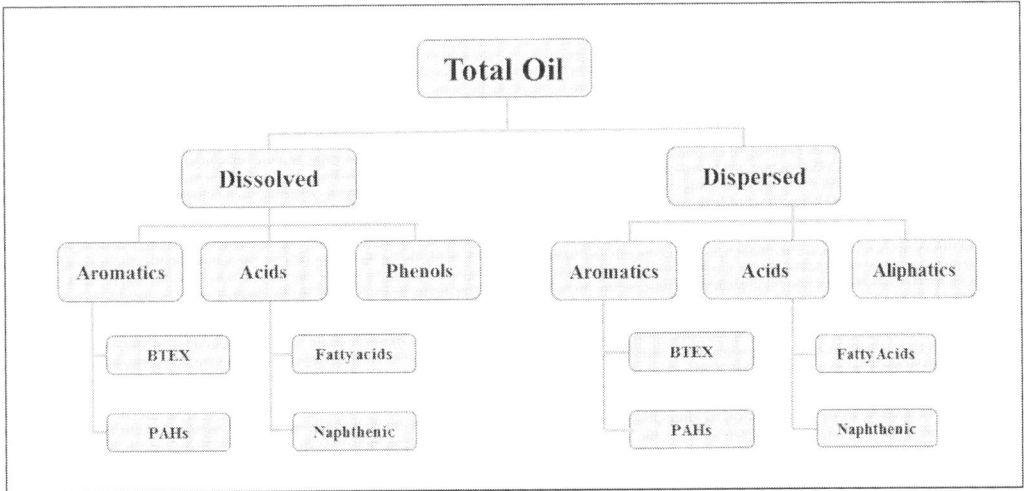

Figure 2.4 Components of crude oil typically found in produced water [19]

It is worth mentioning that the amount of dissolved and dispersed oil in the produced water can increase or decrease depending on the processing conditions such as temperature, pressure, flow rate, and also on which wells are producing at any given time. Some of these components may be present but might not contribute to the measured oil in water. The relative contribution that these components make to the oil in water content depends on the method used for analysis.

As mentioned above, the measured concentration of oil in produced water is method dependent. Each analytical method will measure a portion of dispersed and dissolved oil in produced water depending on the details of the methodology, the measurement principle, degree of acidification of the sample before extraction, type and quantity of solvent used for the extraction, how the sample is extracted, the use of florisil or silica gel for removing polar compounds, how the calibration is established, what calibration oil is used etc. For the discharge of produced water in the OSPAR region and USA Gulf of Mexico (GoM), oil in produced water is well defined by the regulatory procedures adopted in those regions. OSPAR and U.S. GoM are significantly different from each other, but they are both well-defined.

2.3.2 Aging and The Thermodynamic State of Produced Water

Water samples more than a few hours old can yield significantly different test results than testing on fresh samples. This is true for tests involving interfacial properties such as emulsion stability, interfacial tension measurement, and separation efficiency of various water treatment equipment. In bottle testing, as part of chemical program development, different results are obtained with "old" or "dead" samples versus live or fresh fluids. The lighter hydrocarbons (methane, ethane, etc) escape leaving a more polar hydrocarbon behind. Dissolved carbon dioxide can vaporize and significantly raise the pH which can cause various solids to precipitate such as iron hydroxide, and carbonates. A change in temperature can cause wax to precipitate. With time, oily solids can separate into a separate oil layer and a solids sediment.

Over longer shelf time, microbiological activity can change the pH of the water and significantly alter the surface characteristics of suspended oil droplets or particles. Bacteria will exude polymers that will attract/collect suspended matter. These polymers can even act as coagulants or flocculating agents. Bacteria can begin digesting certain portions of the organic contaminants further altering the chemical composition of the water. In addition, the metabolic byproducts will frequently be acidic causing a drop in pH.

Produced water is almost always in a thermodynamically reduced state. This means that there is essentially no oxygen present in the produced fluids. It also means that elements and ions that would be readily oxidized in the presence of air are in a reduced electronic configuration. For example, iron is typically found in the ferrous (Fe^{+2}) state rather than the ferric (Fe^{+3}) state. If produced water is exposed to air, various minerals, including iron, and organic functional groups will oxidize. Iron precipitation can cause significant problems particularly if the iron solids become oil-wet.

In addition to the minerals, the organic compounds are often found in various states of partial oxidation. Partial oxidation is the result of maturation and biodegradation processes which can occur over long timeframes. During maturation the concentration of asphaltenes and oxygen containing groups generally decreases. Whereas in biodegradation the concentration of asphaltene and oxygen containing groups increases. The asphaltene concentration increases due to the preference of microbes to metabolize straight and branch chain hydrocarbon groups. This leaves the asphaltenes behind and their concentration increases due to depletion of the saturate fraction. Oxygen containing groups, increase during biodegradation for various reasons explained below. In any case, it is true that further oxidation of the organic molecules will occur upon exposure to air. This has been studied in relation to the chemical composition of crude oils and their oil/water interfacial tension. Figure 2.5 below illustrates how oxidation of a crude sample reduces the interfacial tension after only a few minutes of exposure to air.

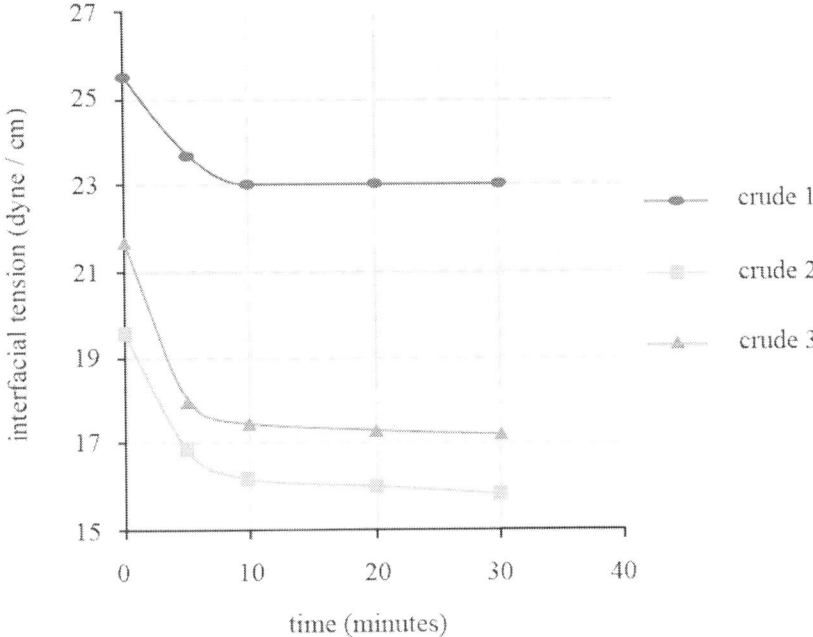

Figure 2.5. Effect of "aging" on interfacial tension of three crude oils. Each has slightly different SARA composition.

The size of oil drops formed during fluid shear depends on the viscosity of the crude, the turbulence intensity experienced by the crude, and on the interfacial tension. This is discussed in some detail in Chapter 3. A low interfacial tension allows many small drops to form, increasing the difficulty of water treatment.

Factors that contribute to low interfacial tension:

- production chemicals such as corrosion inhibitor or other surface active chemicals

- aging of fluids

- natural surfactants (volatile fatty acids, naphthenic acids)

Oil droplet coalescence can be hindered by solids such as:

- inorganic scales

- iron compounds

- clays

- asphaltenes or waxes

- charge stabilization in low TDS brines (i.e., relatively fresh water)

Sjoblom and co-workers studied the aging of crude oil [20] and asphaltene fractions [21] in air. In [20] it was found that interfacial tension decreases continuously with crude oil aging. In [21] they monitored the evolution of infrared peaks corresponding to double bonds, and carbonyl (C=O) groups, as well as other functional groups during aging in the presence of air. The most notable change was that the concentration of double bonds decreased significantly and simultaneously with the increase in concentration of carbonyl groups. This indicates that oxygen from air was being incorporated into the asphaltenes. The infrared method does not distinguish between carbonyls in the form of ketones, esters, or acids. The greatest changes occurred in the first 72 hours. After about four weeks essentially no other changes were observed.

In a study of asphalts, Lin et al. [22] exposed asphaltenes to air for extended periods of time (weeks) at elevated temperatures (88 to 99 C). The aging temperatures were comparable to hydrocarbon processing temperatures, while the aging times were much longer. On the other hand, the aging times were comparable to storage times in a laboratory, if unrealistically high in temperature for that application. The effect of aging on the oxygen content of the asphaltene and the elastic viscosity of the asphaltene were dramatic. The concentration of carbonyl groups increased dramatically. The elastic viscosity increased as well. For some of the asphaltene samples, the elastic viscosity rose by a factor of ten.

Another aging effect due to the presence of asphaltenes is that they tend to migrate to the interface and form a rigid elastic film. The combined process of migration and network formation can take between several minutes to a few hours. These processes may be too slow to be observed within the typical residence time of oil and water on an offshore platform. However, oil that is removed from produced water and recycled through the system may in fact have sufficient time to form a rigid interface. Also, the process of interface hardening may have an effect in onshore facilities where the oil and water retention times are much longer.

2.3.3 Water Quality Analysis (TDS, TOG, TOC, BOD, COD, TSS)

There are a number of standard tests that are used in the water treatment industry. These tests are discussed in detail in Chapter 7 (Sampling and Analysis) where proper sampling and analysis technique is explained. For the discussion in this chapter the main interest is what these tests tell us about the chemistry of produced water.

Total Dissolved Solids (TDS): TDS is the mass concentration of dissolved solids expressed as mg of solids/L of produced water sample. In order to measure TDS, a sample is filtered (typically at 0.45 microns) to remove suspended solids and dispersed oil. Once the sample is filtered, the water is evaporated. The residue left after evaporation is weighed and reported as TDS. TDS measures the concentration of ions, plus charged and uncharged organic compounds. The more volatile, usually low molecular weight, dissolved organic solids will have volatilized and lost during the water evaporation (drying) step. TDS for produced water varies from a few hundred for gas field condensed water to a few hundred thousand for highly saline produced water and for some shale produced water.

Total Oil & Grease (TOG): There are many different tests that are referred to as Total Oil and Grease. They each measure something different. In other words, if a few identical samples are each measured by a different TOG test, each method will detect and therefore measure different chemical compounds in the water and each will therefore give a different numerical result for TOG. In most cases, the TOG is intended to measure the mass of dispersed oil in a water sample.

There are three main types of method. They are gravimetric, spectrophotometric, and gas chromatographic. In the gravimetric methods a liquid/liquid extraction of an acidified water sample is carried out using a hydrocarbon extraction solvent. The solvent is then evaporated. What is left is reported as TOG.

The hydrocarbon solvent will extract most of the dispersed oil. It will also extract some of the dissolved organic compounds. It will not extract any of the ions. As a general rule, ions do not dissolve in hydrocarbon or even most organic solvents. There are, however, some organic compounds that are mostly dissolved but which have a finite tendency to partition between both the water and the hydrocarbon solvent. Examples of compounds that have this tendency are hexanoic and heptanoic acid, providing that the pH is slightly acidic. The lower molecular weight acids such as acetic acid, propionic acid are too polar to have appreciable partitioning into a hydrocarbon solvent. This is particularly true when the pH is neutral or slightly alkaline since the acid would in that case be ionized. Higher molecular weight acids such as naphthenoic acid partition predominantly into the hydrocarbon, again under slightly acidic conditions.

The EPA-1664 test [23] is an example of a TOG measurement procedure. In the US, the daily maximum value for the TOG (as measured by EPA 1664) of water being discharged into navigable waters must be below 42 mg/L. The monthly average must be below 29 mg/L. These values are codified in the NPDES (National Pollution Discharge Elimination System) permit required for all locations which discharge produced water into a navigable body of water. Other regions of the world, such as the North Sea use a solvent extraction and a gas chromatographic analytical procedure. This procedure is referred to as the "OSPAR Method" or Modified ISO 9377-2.

Total Organic Carbon (TOC): TOC is intended to measure the carbon content of dissolved organic material. It is reported as mg of carbon per liter of waste water. It is discussed in some detail in Section 7.5.2 (Total Organic Carbon). Compounds such as CO_2, H_2CO_3, and Na_2CO_3, for example, contain carbon but are not organic and therefore are not measured by the test. In order to exclude these compounds from the test, the sample is acidified and purged with nitrogen. This drives the carbonate equilibria to CO_2 which is then removed by the nitrogen. The remaining carbon content is sometimes referred to as Non-Purgable Organic Carbon (NPOC). This is basically synonymous with typical TOC. In order to measure only the dissolved material, the sample should be filtered before making the measurement of TOC, in order to remove any dispersed or suspended material.

The amount of TOC in a sample depends on the concentration of dissolved organic acids, esters, alcohol, phenols, and BTEX (benzene, toluene, ethyl benzene, xylene). Many of these compounds are dissolved in the water and do not fully extract into hydrocarbon solvents and so do not contribute significantly to typical measurements of TOG.

TOC measurement provides potentially important information. First, like all of the other measurements, a TOC measurement indicates to a water treatment engineer what equipment and system design would be needed to treat the water. In general, waste waters having a high concentration of TOC that must be removed usually require some form of biological treatment such as a bioreactor or biotreater. In some cases, partial oxidation is used upstream of the bioreactor to facilitate the bio-reaction. In other cases, activated carbon is used instead to remove the dissolved organics. Although it must be kept in mind that activated carbon will not remove ionized components. Depending on the pH, organic acids may or may not be in the ionized form. TOC is becoming more frequently used as a regulatory parameter for the oil and gas industry. In Trinidad for example produced water TOC is regulated.

TOC measures the organic carbon concentration. Other elements such as HNOS are not measured by the test. Thus, the TOC measurement cannot be used to determine the concentration of the organic molecules. The actual concentration of dissolved organic material cannot be determined by a TOC measurement alone. If on the other hand the elemental composition of the contaminant is known, and if most of the TOC is from one type or one class of organic compounds, then a correlation can be constructed between the TOC and the concentration of dissolved organic material.

Allen and Robinson [16] report that in produced water samples obtained from a survey of North Sea facilities TOC ranged from 100 to 1,000 mg/L. These samples were taken from the final overboard discharge. This is in agreement with the author's experience with produced water from the North Sea and Gulf of Mexico. As will be explained later, this corresponds to between 1,200 to 2,500 mg/L or dissolved organic material. Thus, produced water discharge samples from the Gulf of Mexico which meet the TOG requirement of less than 29 mg/L can contain a couple thousand mg/L of dissolved organics.

Biological Oxygen Demand (BOD): This parameter is intended to measure the mass of organic material that can be readily biodegraded. The test measures the amount of oxygen that is consumed by common bacteria as they digest the organic material in a water sample. The test is carried out over a five day period at 20 C. The BOD value is reported as ppm of oxygen consumed over the test period.

Chemical Oxygen Demand (COD): This parameter measures the amount of oxygen that is consumed by a sample when all of the organic carbon is oxidized to CO_2 and H_2O. The test is carried out using hot chromic acid. COD and TOC are related chemically. For many sources of produced water there will be a correlation between measured TOC and measured COD. Typically, the COD is greater than TOC. The ratio of TOC and COD indicates the molar ratio of oxygen in the organic material. During partial oxidation using hydrogen peroxide, or some other oxidant, the COD will decrease immediately. This is due to the incorporation of oxygen into the organic molecules forming acid, ether, ester and alcohol groups from CH_2 groups. The TOC will only decrease once the organic material has been converted to CO_2 and H_2O. This is discussed in some detail in Section 7.5.1 (Chemical Oxygen Demand). Like TOC, a COD sample must be filtered in order to first remove TOG. The COD represents both biodegradable material and non-biodegradable material.

Total Suspended Solids (TSS): A good starting point for the analysis of solids in produced water is the NACE Standard Test Method TM 0173 – 2005. Without going into details, solids are typically captured using filtration through a 0.45 micron filter with a constant 20 PSI ΔP being maintained across the filter membrane. Several sets of samples are often collected so that the samples can be subjected to various analyses which include gravimetric determination of the fraction of organic material, inorganic material, and the composition of the organic and inorganic constituents. Examples of results are given in Section 7.6 (Solids Content).

From an environmental standpoint, suspended solids may have a number of potential impacts, including:

- Some solids may be toxic themselves, contain toxic elements, or have radioactive constituents.

- Solids may trap or collect other contaminants (i.e. oil).

- Discharged solids may accumulate as mud or silt in the local environment.

- Discharged solids may result in turbidity in receiving environments with poor dispersion characteristics.

Suspended solids may have a significant impact on the performance of deoiling equipment. The turbidity in discharged water due to the presence of solids may have an environmental impact as well as being undesirable visual pollution.

2.3.4 Produced Water Characteristics and Crude Oil Types

It is helpful in design and troubleshooting to be able to classify or categorize produced water into a handful of types. The characteristics that can be used for classification include salinity, acid types and concentrations, presence of solids, type of solids (oil wet or water wet), presence of production chemicals, etc., as well as the type of crude oil produced with the water. The type of crude oil can have a profound effect on produced water characteristics, types of contaminants, and the difficulty of removing the contaminants. The idea of categorizing produced water is developed extensively in Chapter 5 (Characterization). In that chapter, produced water is characterized into several different types and the processes, equipment, and chemical treatment required to clean the water is discussed in detail. For now, in this chapter on produced water chemistry it is sufficient to draw a rough relationship between the oil type and the tendency to form oily solids, the size of oil droplets, the stability of emulsions, the concentration of oil-in-water, and the tendency (or lack thereof) of the oil droplets to coalesce into larger droplets. It is customary to categorize crude oil into three ranges of API. The characteristics of the oil and the produced water are given here. It should be noted that this is only a very rough categorization and there are numerous counter examples around the world.

1. **Light Crude Oil / High API:** any crude oil with API gravity greater than 31 degrees (density less than 0.87 gr/mL). This range of API gravity includes gas condensate liquids which are typically defined as API 45 and higher. This crude oil type is produced with a relatively small amount of formation water and some condensed water. The properties of the condensed water were discussed in Section 2.1.3 (Gas Production – Interstitial and Condensed Water). Many shale liquids are in this high API range. The crude oil can contain high concentrations of organic acids which will, if present, stabilize emulsions of oil droplets in water. The crude oil tends to have low concentrations of resins and asphaltenes which usually means that any solids present will not be oil wet.

2. **Medium Crude Oil / Moderate API:** roughly 20 to 31 API gravity (density between 0.87 and 0.93 gr/mL). This crude oil type can, in most cases, be produced using conventional methods. The viscosity is typically moderate. There is a wide range of resin, asphaltene and wax content (these components are discussed below in Section 2.6.4 Resins and Asphaltenes). The stability of the wax and asphaltene also has a range from very stable to unstable. The crude oil is moderately sticky (stickiness is related to wetting angle and viscosity). Within this crude oil type, there is a range of biodegradation and a wide range of acid content. As discussed elsewhere, high acid concentration will increase the shear sensitivity making the crude oil / produced water mixture become heavily contaminated with small droplets of oil. In the presence of high acid concentration or unstable asphaltenes, the oil droplets do not readily coalesce.

3. **Heavy Crude Oil / Low API:** any crude oil with API gravity less than 20 (density greater than 0.93 gr/mL). There are various types of heavy oil such as high sulfur, low sulfur, and acidic crude oil. This crude oil type is associated with immature oil or bitumen and is typically recovered by steam due to its high viscosity. The produced water and water treatment system required for steam flood is unique since most steam flood facilities must recycle the produced water to make steam. These oils tend to be very sticky with a high fouling tendency. They contain a high percentage of resins and asphaltenes. Those heavy oils that have a high asphaltene / resin ratio tend not to coalesce and facilities are usually designed to minimize shear. High concentrations of oily solids are common.

2.4 Dissolved Gasses (CO$_2$, H$_2$S, O$_2$)

Dissolved gases have an important effect on the properties of produced water. Carbon dioxide for example, is one of the controls on water pH which in turn affects the surfactant character of natural acids in the oil and water. When produced water is exposed to oxygen, dissolved iron, manganese, chrome and other metals can precipitate as oxides. These issues are discussed in this section.

2.4.1 Carbon Dioxide (CO₂)

Carbon dioxide is commonly found in both oil and gas reservoirs. Under reservoir conditions it is in chemical / phase equilibrium with the water, oil, gas and mineral phases. The most important mineral equilibria are with the carbonates ($FeCO_3$, $CaCO_3$, $MgCO_3$).

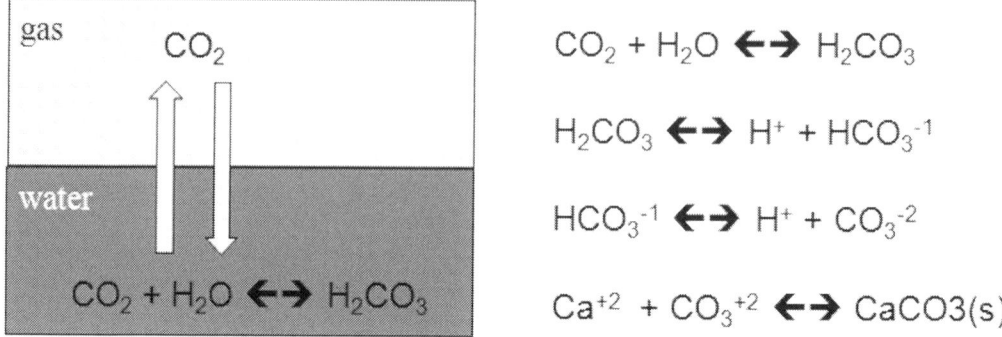

Figure 2.6 Chemical and phase equilibria of carbon dioxide in a gas and water mixture. Gaseous carbon dioxide dissolves in water and undergoes hydrolysis to form carbonic acid. Further equilibria occur, which are not shown here, to form bicarbonate and carbonate anions.

Carbon dioxide readily dissolves in water as shown in Figure 2.6. When it does so, some of it remains in its original molecular form (CO2). Some of it hydrolyzes to form carbonic acid (H_2CO_3). Carbonic acid is a weak organic acid that partially dissociates to form the bicarbonate anion (HCO_3^{-1}) and the carbonate divalent anion (CO_3^{-2}). The terminology "weak" says nothing about the acidity, only that the carbonic acid does not fully dissociate as say HCl would when dissolved in water. These reactions are written as follows:

$$CO_2(aq) + H_2O \quad \leftrightarrow \quad H_2CO_3 \qquad \text{Eqn (2.1)}$$

$$H_2CO_3 \quad \leftrightarrow \quad H^{+1} + HCO_3^{-1} \qquad \text{Eqn (2.2)}$$

$$HCO_3^{-1} \quad \leftrightarrow \quad H^{+1} + CO_3^{-2} \qquad \text{Eqn (2.3)}$$

In Figure 2.7 below, the equilibrium concentrations of carbonic acid, bicarbonate ion, and carbonate ion are shown as a function of pH. As the pH rises from 5 to 7, the predominate form of carbonate changes from carbonic acid to bicarbonate ion. As this happens, the corrosive nature of the water declines. Notice that carbonate ion concentrations do not become significant until the pH exceeds 9 – a situation not normally encountered with produced water (but will occur during some process operations such as lime or caustic softening).

As the pH increases above 9.6, hydroxyl alkalinity due to the presence of the hydroxide ion starts to occur. In Section 2.13 (Precipitation and Scaling Tendency) the scaling tendency is discussed. It is pointed out that the solubility of $CaCO_3$ changes with the partial pressure of CO_2 in the gas with which the water is saturated.

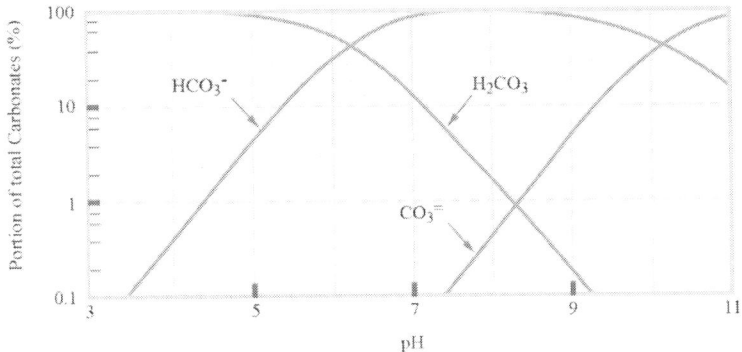

Figure 2.7 Relative proportion of carbonic acid species as a function of pH.

In the idealized case of pure water in equilibrium with a gas containing carbon dioxide, the partial pressure of the gas can be calculated as the mole fraction times the total pressure. This product is referred to as the partial pressure of CO2. The Figure 2.8 below gives the pH of pure water as a function of the partial pressure of CO2 in the gas in equilibrium with the water. Note the significant impact of temperature on the CO_2 solubility.

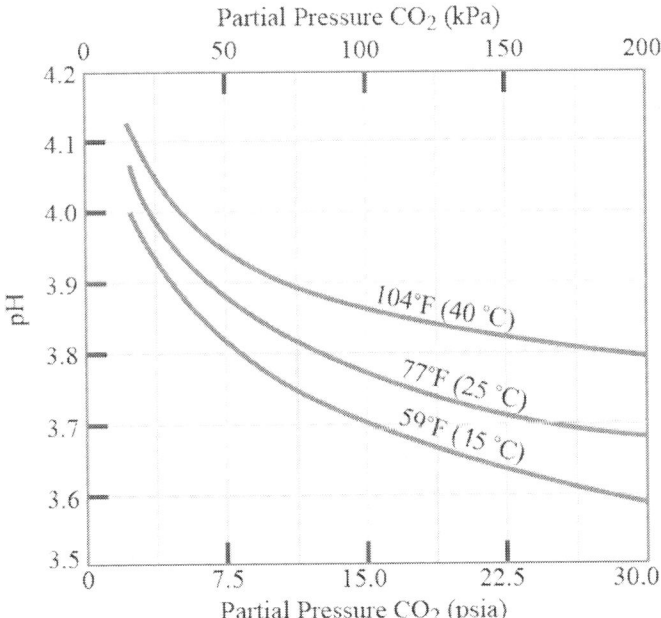

Figure 2.8 The pH of low salinity water as a function of the partial pressure of CO2, for various temperatures.

The above Figure 2.8 provides data for the pH in a low salinity solution containing water, CO_2 and an inert gas. When other constituents are present, carbon dioxide participates in other equilibria which changes the pH relationship. For example, when sodium bicarbonate ($NaHCO_3$) is present, a buffering effect takes place which limits the effect of CO_2 on the pH. Nevertheless, the above relationship is applicable for condensed water as previously discussed.

2.4.2 Hydrogen Sulfide (H₂S)

Hydrogen sulfide is a gas at room temperature and pressure. It smells like rotten eggs. Humans have a high initial odor sensitivity (0.1 ppm in air) which quickly reduces upon exposure. Hydrogen sulfide is readily oxidized to elemental sulfur by oxidants (e.g. air, chlorine or potassium permanganate). Sulfur acts as a colloidal foulant and has a history of not being removed well by conventional multimedia filtration.

Hydrogen sulfide is responsible for stress corrosion cracking of steel. It is toxic. It has an effect on the produced water pH and it has a profound effect on the presence of solids. There are many types of iron sulfide solids that can form in produced water. Most of which have an affinity to acidic compounds and resins and asphaltenes. Together, iron sulfide and asphaltic materials have a synergistic tendency to form a sludge.

The best way to handle produced water that contains H_2S is to use closed, air tight systems and maintain strictly anaerobic conditions in order to prevent the oxidation of H_2S or dissolved sulphides to colloidal sulphur. It is also important that no other oxidant is used in the pre-treatment. The H_2S / sulphide equilibrium depends on the pH of the water. It is advised to keep the pH of the feed water low in order to keep the sulphide ion in its unionized associated hydrogen sulphide form.

Low amounts of sulphides and H_2S can create problems when attempting to remove solids by co-agulation/flocculation. Colloidal sulfur can be created in this process which can slowly plug media filters and any downstream membrane filtration system. Use of coagulation/flocculation is only recommended in the case that H_2S and sulphides are present in water taken from an open intake. The Atlantic coast of Namibia (for example) has seasonal H_2S upwells with open sea water intake levels of H_2S around 1.5 mg/l. In such a case – no other pre-treatment is possible – only highly effective coagulation/flocculation. Figure 2.9. The species balance from the dissociation of H_2S in water as a function of pH is illustrated.

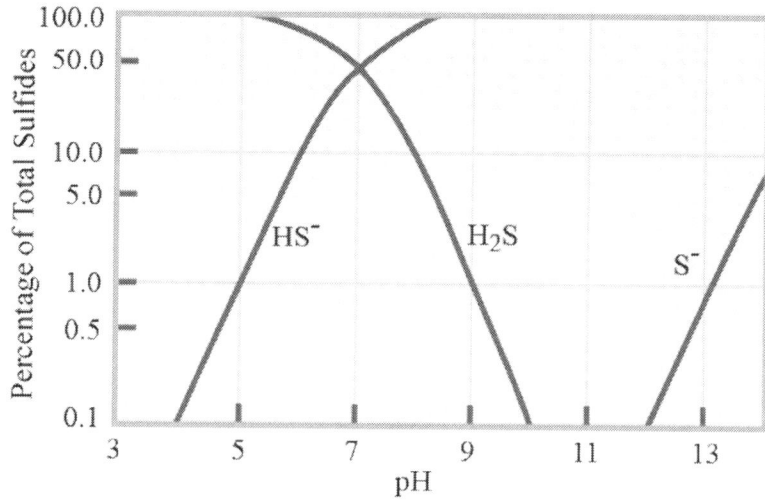

Figure 2.9. Proportion of various hydrogen sulfide species as a function of pH.

Compared to many other gases (such as oxygen, nitrogen, methane), H_2S is very soluble in water due to its polarity and the fact that it dissociates. Like CO_2, when H_2S dissolves, distinct molecules of H_2S are present in the water. These molecules are referred to as aqueous H_2S (aq). Some of these aqueous H_2S (aq) molecules dissociate to form the acid proton and the bisulfide anion (HS^{-1}). The bisulfide anion can further dissociate to form the sulfide divalent anion (S^{-2}).

$$Ka = 8.9 \text{ x } 10^{-8} \text{ mol/L} \qquad pKa = -\log_{10}(Ka) = 7.05 \qquad Ka = [H^+][HS^-] / [H_2S]$$

This pKa value can be determined from the curves in Figure 2.9. It is the pH value at which the HS^{-1} curve and the H_2S curve intersect. This is the pH at which the concentrations of HS^{-1} and H_2S are roughly equal (7.05). At pH values lower than this, most of the dissolved H_2S is in the form of H_2S (aq). Since it is this species that has direct equilibrium with H_2S (gas), this suggests that one way to remove H_2S from solution is to reduce the pH of the water and utilize conventional gas stripping technology to remove the H_2S from the water

$$H_2S(aq) \quad \leftrightarrow \quad H^{+1} + HS^{-1} \qquad\qquad \text{Eqn (2.4)}$$

$$HS^{-1} \quad \leftrightarrow \quad H^{+1} + S^{-2} \qquad\qquad \text{Eqn (2.5)}$$

It is generally accepted in the industry that "sweet" produced water has an H2S concentration less than 5 mg/L. "Sour" produced water is water which has a dissolved H2S concentration greater than 5 mg/L. Although corrosion is not covered in this book, it should be mentioned that the sulfide causes stress corrosion cracking in hardened steel. This is a severe form of corrosion in that the steel actually cracks and fails from the inside of the pipe or vessel.

2.4.3 Oxygen (O_2)

Oxidation/Reduction (Redox): Chemical oxidation occurs when a molecule losses an electron. When this occurs, the oxidation number increases. Chemical reduction is the opposite from oxidation. Reduction occurs when a molecule gains an electron. Whenever oxidation of a molecule occurs, there is a corresponding reduction in another molecule.

Most produced water does not contain dissolved oxygen. At some point in the ancient geologic history of most produced water, prior to burial of the water, oxygen was present. But oxygen is relatively reactive and is consumed by biological processes and through the oxidation of minerals. Thus, the dissolved and suspended components in produced water are typically in a reduced state, in the sense of oxidation/reduction potential.

Once produced water is exposed to oxygen, its nature can change dramatically and for the worse in several regards, including the precipitation of iron compounds and the partial oxidation of certain hydrocarbons as was discussed above. Introduction of oxygen laden water into the gas/oil/water process stream is detrimental because it oxidizes the iron in solution in the water with the possible precipitation of iron oxide solids and creates aggressive, damaging corrosion due to the formation anodic sites on internal metal surfaces. Precipitated iron solids can contribute strongly to the stabilization of emulsions.

A few common sources of oxygenated water entering into a separation system are by way of the introduction of oxygen saturated surface water, the use a seawater sump system, holding tanks for off-spec produced water, an open drain system, or a gas blanketing system that contains oxygen. Such water systems should always be segregated from the production streams. Also, rigorous biological control should be practiced in such systems.

Another common source of oxygenation of produced water is by way of an open API separator system. Such systems may be cost effective as primary water clarifiers and have been used successfully in the industry. However, they are open to the atmosphere and as such must have a means of mechanically collecting the scum that will form on the surface and the sludge that will form on the bottom of the separator. These accumulations will contain solids and sometimes sand, dirt and biological material that is blown into the separator. Most importantly, any of the reject streams from an API separator (scum, sludge and oily reject) should absolutely never be routed back into an upstream section of the oil/water separation train or water treating system that does not already contain oxygen.

2.5 Hydrocarbons in Produced Water

The term 'hydrocarbon' is, strictly speaking, reserved for compounds that are composed exclusively of carbon and hydrogen atoms. Thus, methane, ethane, propane, benzene, toluene, cyclohexane, phenanthrene, and the polycyclic aromatic hydrocarbons (PAH) are all hydrocarbons. Phenol, benzoic acids are not hydrocarbons. They contain oxygen atoms and are classified as organic compounds. All hydrocarbons are organic compounds, but not all organic compounds are necessarily hydrocarbons.

Examples of Hydrocarbons: Within the general class of hydrocarbons, there are groups of compounds such as the alkanes, alkenes, alkynes, and aromatics. Within the alkanes, there are compounds such as methane, ethane, propane, n-butane, iso-butane, n-pentane, iso-pentane, neopentane, etc.. The alkenes include ethylene, propylene, 1-butene, 2-butene, 1-pentene, 2-pentene, etc.. The aromatic compounds include benzene, toluene, ethyl benzene, and the ortho-, meta-, and para-xylene. These are referred to as the BTEX compounds. Other aromatics include naphthalene, phenanthrene, and anthracene as well as many others. Just within the hydrocarbon group of compounds, crude oil contains hundreds if not thousands of individual chemical species.

Hydrocarbon Solubility in Water: Most hydrocarbons have negligible solubility in water. Notable exceptions are aromatic hydrocarbons such as the BTEX compounds. The aromatic compounds have an electronic ring structure that provides moderately strong interaction with, and hence solubility in water. These hydrocarbons do not truly hydrogen bond with water. But they are somewhat polar and thus have an attractive interaction with water beyond the weak nonpolar van der Waals interaction. Benzene solubility in water is about 1700 mg/L at room temperature and pressure. Thus water in equilibrium with a gas condensate containing 1% benzene would be expected to contain about 17 mg/L of benzene in solution. The low molecular weight dissolved hydrocarbons just mentioned (benzene, etc) are generally of particular regulatory concern in most parts of the world. They are however measured and included as part of the Total Oil and Grease.

It should be mentioned that dissolved hydrocarbons can contribute to measured values of dissolved organics. Dissolved organics can be measured by TOC (Total Organic Carbon) and COD (Chemical Oxygen Demand). It is recommended to carry out these tests on filtered and unfiltered samples. The filtered samples will then represent the dissolved portion of the contaminant. Whereas the unfiltered sample provides a measurement of the total contaminant from which the dispersed portion can be calculated. This provides a more detailed understanding of the sample.

Aromatic and polycyclic aromatic hydrocarbon (PAH) have the highest water solubility compared to alkanes, alkenes, and alkynes. Dissolved organics are discussed below in Section 2.6.1 (Dissolved Organic Compounds). Some of the dissolved hydrocarbons can also contribute to measured values of Water Soluble Organics. The Water Soluble Organics are discussed in Section 2.6.3 (Water Soluble Organics). Briefly, WSO are compounds that are extractable by hexane and which do not evaporate at or below approximately 85 C. Because most hydrocarbons are relatively insoluble in water and are instead present as discrete hydrocarbon droplets within the water phase, the treatment of wastewater streams from most E & P operations is focused on the removal of dispersed hydrocarbons, not dissolved.

There are two groups of hydrocarbons that are the focus of monitoring and measurement in the Norwegian sector of the North Sea, and in those other regions that have or are planning to implement the OSPAR convention recommendations. These are the NPD and the PAH compounds.

NPD Compounds: The NPD compounds are:

- naphthalene
- phenanthrene
- dibenzothiophene.

The first two of these compounds are also members of the PAH class of compounds listed below. Typically naphthalene and phenanthrene tend to be the most abundant of the PAH compounds. Thus, they are sometimes referred to separately. The third compound, dibenzothiophene (DBT), is not strictly speaking a hydrocarbon since it contains sulfur. The NPD compounds are often analyzed in the North Sea and monitored due to their relatively high environmental impact, as measured by the Environmental Impact Factor.

PAH Compounds: The sixteen PAH compounds identified by the U.S. EPA are:

- naphthalene
- phenanthrene
- acenaphthylene
- acenaphthene
- fluorene
- anthracene
- fluoranthene
- pyrene
- benz(a)anthracene
- chrysene
- benzo(b)fluoranthene
- benzo(k)fluoranthene
- benzo(a)pyrene
- indeno(1,2,3-c,d)pyrene
- dibenz(a,b)anthracene
- benzo(g,h,i)perylene

The PAH compounds are the most carcinogenic of produced water hydrocarbons or organics. They are one of the classes of compounds responsible for the carcinogenicity of cigarette smoke. The higher molecular weight PAH compounds are of serious regulatory concern in the Norwegian sector of the North Sea because they have a tendency to bio-accumulate. In other words, they can be ingested by small crustaceans and not metabolized. The crustaceans are eaten by fish where again the PAH compounds are not metabolized. The PAH compounds thus make their way up the food chain and eventually wind up in the fish that are eaten by humans. In some parts of the world, such as Norway, there is strong emphasis on eliminating these compounds from overboard discharged produced water [11, 24].

In addition, some of the higher molecular weight PAH compounds are only sparingly soluble in low MW alkane hydrocarbon solvents. In water analysis where a hydrocarbon solvent such as pentane or hexane is used, the PAH compounds do not fully dissolve in the solvent. Thus, some of the PAH compounds may be present in the produced water as dispersed entities which are not fully measured by analytical procedures based on solvent extraction.

Environmental Impact Factor (EIF) for PAH Compounds: Environmental risk assessment studies have shown that the aromatic components in produced water constitute a major contribution to the Environmental Impact Factor (EIF). The EIF is discussed in Section 1.6.4 (Regulations, and Disposal and Reuse Options). Recent studies have shown that alkylated phenols and Polycyclic Aromatic Hydrocarbons (PAH) have finite partitioning into both the oil and water phases, as shown in the table below.

Table 2.3 Partitioning of PAH components between oil and water for two different oil in water concentrations. Partitioning percentages are given in mass percent [25].

	10 mg/L		100 mg/L	
EIF Group	oil %	water %	oil %	water %
naphthalenes	57	43	81	19
2 – 3 ring PAH	67	33	92	8
4 – 6 ring PAH	61	39	94	6
C0 – C3 phenols	0.05	99.95	0.5	99.5
C4 – C5 phenols	4	96	29	71
C6 – C9 phenols	64	36	95	5

2.6 Organics (besides hydrocarbons) in Produced Water

In this Chapter, the contaminants of produced water are presented in terms of three general categories: hydrocarbon, organic, and inorganic. Hydrocarbons were just previously discussed. Strictly speaking, hydrocarbons are a sub-class of organic compounds. Hydrocarbons are those organic compounds that are composed only of C and H.

Organic compounds are discussed in this section. Organic compounds are composed of carbon and various elements from the upper right hand side of the periodic table, plus hydrogen, and includes C, H, N, O, and S. Organic molecules can also contain the halides (F, Cl, Br, etc). When organic molecules are dissolved in water, ionization only occurs for certain polar groups such as RCO_2H (acid), R_3NH, RSO_3H, and a few others. Ionization greatly increases the solubility of a compound. Generally, higher solubility leads to higher TOC (Total Organic Carbon) and possibly higher Total Oil and Grease. All things considered, the presence of these ionizing organic compounds restricts the disposal options and increases the extent of water treatment required for surface discharge.

In the case of organic contaminants, many individual constituents cannot be thought of as definitively dissolved or definitely dispersed. They may form small clusters such as dimers in the case of carboxylic acids. Further, many of the oxygen containing organic compounds, particularly those with acid groups, are surface active. They migrate to the oil / water interface and it is difficult to determine if they are dispersed or dissolved. Aggregates of these compounds referred to as micelles can form at concentrations above the CMC (Critical Micelle Concentration). Identifying which phase these compounds reside can impact the water treatment method recommended to remove them.

In the past, produced water specialists have tended to downplay the importance of the dissolved compounds. Also, most legislative bodies do not emphasize the importance of dissolved hydrocarbons as much as they do dispersed hydrocarbons. But it is now recognized in some parts of the world, such as the Norwegian sector of the North Sea, that the environmental impact of produced water is more a function of the quantity of particular classes of compounds rather than the total oil content. Therefore, in the subsequent material an effort is made to discuss the chemistry of both dissolved and dispersed contaminants.

There are several papers on the composition of produced water [10 - 16]. In the late 1980's and early 1990's an effort was undertaken by the oil and gas industry, regulating bodies, and industry associations (e.g. American Petroleum Institute) to gain a better understanding of the chemistry of produced water and to derive a rough understanding of its toxicity. References [1384, 1586] provide a good summary of this work. An update, published in 2011 is given in [14].

2.6.1 Dissolved Organic Compounds

Dissolved Organics – Functional Definition: In order to properly design water treatment equipment, it is important to know whether produced water contaminants are dissolved or dispersed. Different sets of technologies are required depending on this distinction. Contaminant material is considered to be dissolved if it passes through a 0.45 micron filter. Particulate solids or dispersed oil is retained on the filter. This is an arbitrary criterion. Nevertheless, it is universally adopted in the E&P industry. The 0.45 micron filter, made of Teflon or cellulose acetate is commonly used for many other purposes and is readily available. One of the problems with this criterion is that drops of low viscosity oil (condensate) may pass through the filter whereas drops of heavier oil do not. Nevertheless, the 0.45 micron filter provides a simple and readily available test and has been accepted as the industry convention for distinguishing between dissolved and dispersed organics and hydrocarbons.

TOC (Total Organic Carbon) and DOC (Dissolved Organic Carbon): Once the sample has been filtered using a 0.45 micron filter, then a simple and commonly used test to measure the concentration of organic compounds is the dissolved organic carbon (DOC) test. DOC is measured by converting all of the organic material in produced water to CO_2 and then measuring the CO_2 that is produced. TOC is the "total organic carbon" content measured using the same test as DOC but run on the unfiltered sample. The filtered sample will not contain dispersed oil, at least in principle. Whereas the unfiltered sample will contain both dissolved and dispersed oil. ASTM D 2579-78 discusses these test methods. It references information relating the TOC to other measures of water quality such as biological oxygen content (BOD), and Chemical Oxygen Demand (COD), which are both used to characterize produced water discharged into rivers lakes, streams, and other biologically sensitive areas.

DOC is often used to assess the biological nutrient content of injection water for water flooding. High DOC content water will likely result in significant reservoir souring (conversion of sulfates to H_2S) if sulfate is present. Seawater, collected at least a few hundred meters offshore, typically contains about 0.5 to 5 mg/L of dissolved organic matter, as measured by DOC. Produced water can contain a few mg/L to many hundreds of mg/L of DOC.

Individual Species versus Class of Compounds: Identification of dissolved organic compounds in produced water is usually made from two perspectives. One perspective is to determine the concentration of specific relatively low molecular weight species such as benzene, toluene, the xylene isomers, ethyl benzene, phenol, benzoic acid, and the Short Chain Fatty Acids (SCFAs) such as formic acid, acetic acid, propanoic acid, and butanoic acid. Another name for these compounds is Volatile Organic Compounds (VOCs). This perspective is known as speciation. It involves the identification of individual species. The second perspective is to identify classes of compounds. Various classes can be defined and include saturates, aromatics, resins, water soluble acids, naphthenic acids, and asphaltenes.

CHNOS Test: Further identification of dissolved organic compounds in produced water can be made by functional group analysis or CHNOS (C, H, N, O, S pronounced "cheenose") analysis. Functional group analysis usually focuses on the following groups: carboxylic acids, substituted phenols (phenol, o-cresol, resorcinol, hydroquinone, etc), amines, amino acids, and to lesser extent functional groups such as alcohols, ethers, and esters. In general, functional group analysis focuses priority on those groups that result in surface activity, transition metal binding, and aqueous solubility of the organic compound.

Examples of Dissolved Organics in Produced Water: In Table 2.4, an example is given of dissolved organics from four North Sea platforms [10]. These components were measured directly without using an extraction solvent. As discussed previously, and below, the use of extraction solvents does not give an accurate measure of dissolved components. The acids were measured directly in the water phase using a technique called isotachophoresis. As shown, there is a significant concentration of organic acids in each of the samples.

Table 2.4. Dissolved Organics Measured in Produced Water from Four North Sea Platforms, Utvik [10]. All values are given in mg/L.

Organic Compound Class	North Sea Platforms			
	Troll (mg/L)	Oseberg C (mg/L)	Oseberg F (mg/L)	Brage (mg/L)
Organic acids	798	717	1,135	757
Phenols	0.6	11.0	1.5	6.1
BTEX	2.4	5.8	8.3	9.0
NPD	1.32	1.60	1.27	0.93
PAH (not NPD)	0.11	0.08	0.15	0.07
THC	33	60	44	58

At the bottom of the table, the measured value of THC (Total Hydrocarbon Content) as determined by the old OSPAR (1997) protocol is given. The protocol for this measurement is: Freon extraction and quantitative IR measurement at the wave number of the aliphatic C-H stretch. Note that the sum of the concentration of the dissolved organics is much higher than the measured THC value. This is due to the fact that the Freon extraction step does not capture all of the dissolved organics. Solvent extraction involves phase partitioning of the dissolved organics between the produced water and the solvent. The more water soluble a component is, the less it will partition into the solvent and the lower will be the measured concentration determined by such tests as the old OSPAR protocol, and the EPA-1664 test [23], to name just two examples.

Phenols: There are two groups of phenols. One group consists of lower molecular weight compounds, including phenol itself. These are mostly water soluble and somewhat volatile. The other group is higher in molecular weight, more highly substituted and are relatively non-volatile, and less water soluble.

An example of a waste water containing phenol is given here. The waste water had a phenol concentration of 160 mg/L. This represented total phenols including phenol, various methyl-phenols (also known as cresols), some chloro-phenols, and other higher molecular weight phenols. This water was treated using chlorine dioxide, at a high dosage of 3,000 mg/L, which is discussed in Section 5.8

(Oxidation). The major by-product of the treatment was benzoic acid which is readily biodegraded in a bio-reactor. Phenol can be treated in a bio-reactor but the bacteria required for digestion must be cultivated for the specific waste water and fluctuations in phenol content can cause the bio-reactor to underperform.

Amines: Due to the many problems caused by amines in the refinery, the subject of amines is discussed here. In a refinery amines can:

- Impede oil/water separation

- Can cause foaming in the FFU

- Inhibit nitrification in the Biotreater

- Be difficult to convert in the Biotreater

- Have a considerable toxity

There are essentially three kinds of amines that are found in upstream oil and gas.

1. **Corrosion inhibitors.** These are the so-called film forming amines and the quaternary amines. Both are surface active. These amines will cause foaming in the refinery. Some of the film forming amines oil soluble and surface active. Some of the film forming amines are water soluble and essentially all of the quaternary amines are water soluble. In that case, they will only be present in the crude when the crude oil has some BS&W. These amines will be soluble in the BS&W. In either case, these amines will find their way to the biotreater whether in the produced water or in the Desalter (crude oil dehydration unit). The quaternary amines are usually toxic to the bacteria in the biotreater, unless those bacteria have been fed a steady diet of quat.

2. **Hydrogen sulfide scavengers.** The typical chemical used is called triazine. Sour crude oils will contain this form of amine. It is not surface active. Has a low foaming tendency. It is not particularly toxic and it can either be oil soluble or water soluble depending on the particular chemistry used. When triazine is present in the BS&W, that water will have a relatively high pH. This could cause carbonate salts ($CaCO_3$, $MgCO_3$, $FeCO_3$) to precipitate.

3. **Biocides.** Chloramines are sometimes used as biocides in upstream produced water. Again, this produced water can get into the BS&W of the crude oil.

Ion exchange resins can remove amines from produced water. The produced water should first be pre-treated to remove dispersed oil and oily solids that will foul the resin. Also, oxidation can be used to convert the amine to nitrate which will act as a nitrogen source for the biotreater. Oxidation though is limited when there is dissolved organics in the water since the dissolved organics will consume the oxidant making it too expensive.

2.6.2 Organic and Naphthenic Acids, and Alcohols in Produced Water

Organic acids in the crude oil and the produced water are mostly carboxylic acids. Roughly speaking, there are two categories of carboxylic acids in crude oil and produced water, the water-soluble acids and the oil-soluble acids (naphthenic acids).

Water-Soluble Acids: The short chain fatty acids (described below) include formic (one carbon atom), acetic (C2), propionic (C3), etc., which are highly water soluble. These are low molecular weight acids in which the acid group may or may not be protonated, depending on the acidity of the molecule and the pH of the water. In the ionized (unprotonated carboxylate anion) form the molecule will have zero partitioning into the oil phase. In the protonated form, the higher carbon number acids in the series may partition partially into the oil phase [26]. As discussed, the volatile members of this series of acids do not contribute directly to most regulatory oil-in-water measurements, such as EPA-1664. They are too water soluble to be extracted using a hydrocarbon solvent, and they are too volatile to remain in the sample during the solvent evaporation step. However they do contribute to the stability of oil-in-water emulsions. They are also detected in most IR measurements of oil-in-water. Due to this, they sometimes contribute to a significant discrepancy between IR measurements, and the gravimetric EPA-1664 test. This is discussed in greater detail in Chapter 7 (Sampling and Analysis).

Naphthenic (Oil-Soluble) Acids: At the opposite end of the molecular weight range, there are multifunctional, typically aromatic, molecules that have a carboxylic acid group, or several, for which the acid is only one heteroatom functional group among several. These molecules are typically soluble in the crude oil, where they are in the protonated form, and may also have some solubility in water at typical mildly acidic pH values (e.g, pH 4 to 7). Depending on the molecular structure, they may be surface active.

Characterization: These two classes of organic acids (naphthenic and water soluble) span a wide range of molecular weights and chemical structures in between these two extremes. Water and oil characterization techniques help to determine if these compounds are present and various correlations help to determine if the mixture will stabilize or destabilize an emulsion. Both types of acid (water soluble and oil soluble) tend to have a pKa of about 5 [27]. The pKa is the pH at which a mono-acid will have half of its molecules ionized and half protonated.

As discussed below, TAN (Total Acid Number) is a convenient means of characterizing the oil-soluble acids. This can be carried out on the crude oil sample, or on various fractions of the oil (SARA fractions discussed below). TAN can also be carried out after washing the crude oil with water of various pH values [27]. Starting with a neutral pH, the water wash will extract from the oil the most acidic fractions. As pH increases, the lesser acidity fractions will be extracted. This can be done quantitatively.

For the water soluble acids, there are a few characterization methods that are useful. The first is to run an ion chromatography test on the water directly. IC can quantitatively detect C1 through C4, and in some cases, C5 carboxylic acids. A simple characterization method is to carry out a hexane extraction of the water over a range of pH values. The hexane extractant can then be analyzed by GC/MS for speciation. Or, a simpler test is to simply run a gravimetric analysis of the extractant. The lower pH samples will have higher values of gravimetric results due to the protonation of the acids and the increase in hydrocarbon solubility.

The characterization of carboxylic acids in crude oil and produced water is complex because there are a very large number of molecules that partition into both oil and water. For example, heptanoic acid partitions into both crude oil and water in equal proportions at 55° C, and pH 4 [28], which is a common pH for produced water containing dissolved CO_2 under low process pressure.

Sources of Carboxylic Acids in Crude Oil: In order to understand the sources of carboxylic acids in produced water, it is necessary to understand their sources in the oil. As discussed previously in Section 2.2.4 (Biodegradation), there are three main sources [8] of carboxylic acids in crude oil:

a) original plant and animal organic matter (particularly heavy crude oil from geologically young formations)

b) biodegradation

c) bacteria themselves (cell wall of bacteria).

Thus, crude oil containing a high concentration of carboxylic acid is either young (not sufficiently degraded), or has been biodegraded. The later process can occur at any time in the history of the organic deposit depending on the temperature and the presence of water and nutrients [6].

Bancroft's Rule: As will be discussed in some length in Chapter 3, Bancroft's Rule is an empirical observation which relates emulsion type to surfactant solubility. Generally speaking, surfactants are not highly soluble in either the oil or the water phase. They have a strong affinity for the oil/water interface. But, they do tend to have at least some partial solubility in one or the other phase. Bancroft's Rule states that "*a surfactant that is more soluble in oil than in water will stabilize a water-in-oil emulsion. Conversely, a surfactant that is more soluble in water than in oil will stabilize an oil-in-water emulsion*". Here we are more concerned about oil-in-water emulsions and are therefore more interested in the smaller chain-like acid compounds because they are more soluble in water than in oil.

Although the above statements are generalizations, and oversimplifications, they give us a starting point for the discussion of organic acids, starting with the volatile, highly water soluble, short chain acids and proceeding to the higher molecular weight, oil soluble and more complex acids.

Short to Medium Chain Length Carboxylic Acids: The natural abundance of the alkyl carboxylic acids diminishes rapidly with chain length. The figure below gives an approximate distribution of acids in a number of North Sea produced waters.

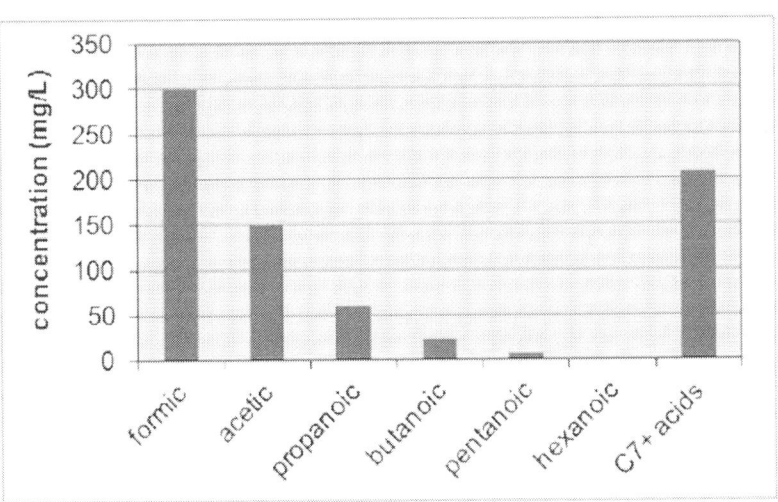

Figure 2.10. Typical values for the concentration of specific acid compounds in produced water. Values for C1, C3-C7+ are from Utvik (24). The value for acetic acid was not reported by Utvik and was estimated independently.

Most of these values in Figure 2.10 are reported by Utvik [10]. However, in the Utvik paper, there is no reported value for acetic acid. An estimate for acetic acid was made based on roughly a dozen samples from the North Sea that were gathered for a separate study conducted by a major operator. The intent of that study was to develop an improved understanding of reservoir souring associated with the use of produced water for water flooding.

Much of the compositional data available for the acid content of produced water comes from data taken in the early 1990's for North Sea offshore platforms [14, 15]. The amount of information related to acids and the importance of such measurements has diminished over time due to the general perspective in the North Sea that carboxylic acids, per se, do not pose a high environmental impact. The perceived importance of acids in the Gulf of Mexico has not diminished, however, as it has been determined that higher MW acids can, in some cases, make a significant impact on the EPA-1664 Total Oil & Grease [23] test results. Carboxylic acids have been measured and reported for a number of Gulf Coast produced waters and some of this data is summarized in Table 2.5 below. The acids measured range from C1 (formic) to C7 (heptanoic). As discussed below the lower molecular weight acids do not contribute to the EPA 1664 test results. As discussed by Stephenson [13], the short chain acids (C1 through C6) have too low of a solubility in the extraction solvent (hexane) to have an impact on the EPA-1664 test. They are also too volatile to survive solvent evaporation. But hexanoic and certainly heptanoic acid and higher molecular weight acids do contribute to the test results. As shown in the table, in some cases these acids are present at relatively high concentration.

Table 2.5. Carboxylic acids anions in formation waters from the Louisiana Gulf coast [29]. Values in mg/L.

No.	Butanoic	Pentanoic	Hexanoic	Heptanoic
1	45	30	11	ND
2	197	215	106	32
3	114	99	107	83
4	73	120	62	99
5	40	64	32	20
6	92	106	42	27
7	311	20	31	19

Naphthenic Acids: Use of the term "naphthenic acids" is widespread in the oil and gas industry, particularly in the downstream sector. In some instances it is synonymous with carboxylic acids containing one or more saturated rings [8]. In other cases it is used to indicate only the higher molecular weight fraction of carboxylic acids. In general, there is not much consensus on what the term means.

One of the more widely referenced [30] early works on naphthenic acids dates back to 1955 and is likely responsible for the widespread use of the term. In those days, precise analytical methods to determine composition were not available. In fact, only two broad categories of organic acids were at the time known to be present in crude oil, and in produced water. One category was the straight and branched paraffinic acids. Today we refer to these as volatile (or short chain) fatty acids. There were only two compounds identified in the other category, cyclopentanoic acid and cyclohexanoic acid. Since these were both saturated cyclic compounds (a.k.a. naphthenes) they were given the name naphthenic acids. Thus the two categories that emerged in the mid-1950's to describe all organic acids in crude oil and produced water were:

- paraffinic acids – straight and branched paraffinic acids

- naphthenic acids – saturated cyclic acids and all other acids

Arla et al. [31] gives the definition of naphthenic acids that will be used in this book. Naphthenic acid is any organic in the crude oil that contains the RCOOH group where R refers to the naphthene group, consisting of the cyclopentane or cyclohexane derivates, or aliphatic chains or aromatics. Sjoblom and co-workers [27] also use a similar definition and comment that "the term naphthenic acids originally implied that the acids contained naphthenic rings, today it comprises cyclic, acyclic, and aromatic acids in crude oils." This definition has been cited in the Kirk-Othmer Encyclopedia of Chemical Technology 1995 Ed, p. 1017-29). For purposes in this book, naphthenic acids are any acids that are preferentially soluble in the oil. Those acids that are preferentially soluble in the produced water, are not naphthenic acids, and will be referred to by the specific chemical class (e.g. n-alkyl mono-carboxylic acid) or by the chemical name.

Currently it is recognized that there can be an enormous number of different acids in some crude oils [8, 32, 33]. Figure 2.11 gives a breakdown of water soluble organics found in California Midway Sunset crude oil. The relative abundance of relatively long chain (surfactant-like) acids is shown. In [33] Seifert and Howells identified 1,500 acids with boiling points ranging from 250 to 350 C in the same crude oil. Their main interest was to identify those acids that are interfacially active.

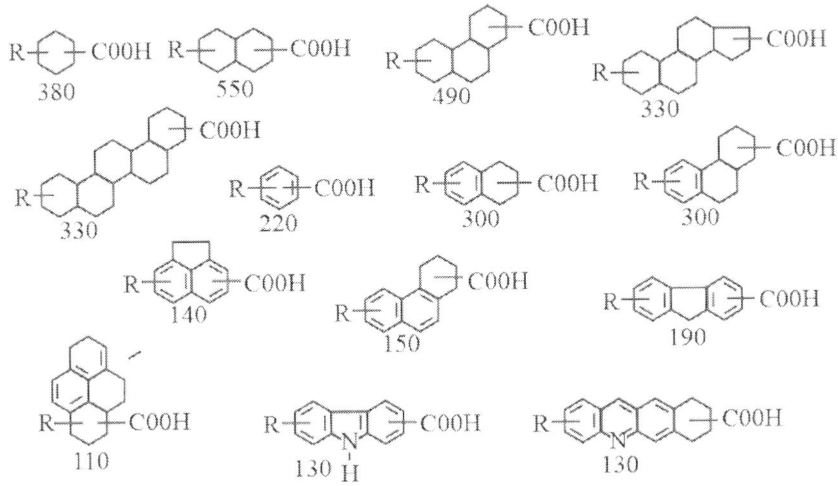

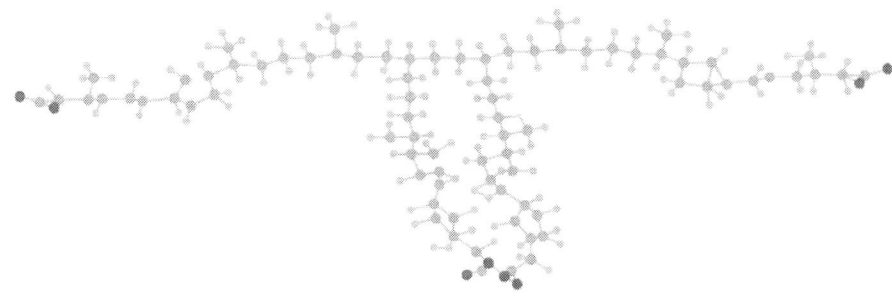

Figure 2.11. Structure elucidation of the monofunctional acids in a particular crude oil, not including volatile fatty acids. Numerical values are mg/L in the whole crude oil for the entire class of acid shown. The total concentration of acids shown here is 3,750 mg/L. The crude oil is California Midway Sunset, a pliocene crude oil [32]

ARN Acids & Calcium Naphthenates: A class of compounds of major importance, due to their detrimental effect, are the tetra-acid naphthenates [34]. It is estimated that naphthenate deposits are present in 10 % or more of North Sea systems, 30 % of South East Asian, and 20 % of West African production systems [34]. These compounds are also known as Arn acids. Arn (Ørn) is the Norwegian word for eagle [35]. When the molecular structure of these compounds was first elucidated, one of the leading research scientists said that they look like an eagle. One molecular structure is given in Figure 2.12.

**Figure 2.12 Three dimensional molecular structure of an Arn acid [36].
Arn (Ørn) is the Norwegian word for eagle [35].**

ARN acids are a family of compounds having four carboxylicacid groups, also containing four to eight unsaturated rings, with molecular weights in the range of 1,227 to 1,235 g/mol [37]. For many years, deposits of so-called calcium soaps had been known in the industry. Much later, the ARN acid structure was elucidated [37]. The importance of the ARN acids was demonstrated with the unintended shutdown of Chevron's Kuito field in the 1990's due to calcium naphthenate scaling. The industry responded with significant activity directed at the analysis and understanding of this important class of compounds. Analytical tests were developed to detect naphthenates [37], and phase equilibrium models were developed to estimate their stability. It has since been recognized that naphthenates are not only responsible for well publicized shutdowns, but they are also responsible for a much more pervasive occurrence of water treating difficulties. The severity of the problems associated with naphthenate soaps depends on the concentration of these compounds, and the pH of the produced water. Suffice for now to say that these compounds can, in some circumstances cause significant water treatment problems.

Table 2.6. Naphthenate and carboxylate compounds that contribute to water treatment problems.

Property	Calcium Naphthenate	Sodium Carboxylates	Calcium Carboxylates
As seen in facility	Sticky solids that harden when cool or exposed to air	viscous sludge that does not harden	Viscous sludge that does not harden
Chemistry	Ca ions dominate but Fe and Mg ions also occur	Na ions dominate	Ca ions dominate
Impact on oil dehydration	Moderate WiO emulsifiers	Strong WiO emulsifiers	Strong WiO emulsifiers
Impact on produced water	Major impact on produced water quality	Strong impact on produced water quality	Strong impact on produced water quality

Alcohols: The main alcohols found in produced water are:

- n-alykl and branched alkyl alcohols such as methanol, ethanol, iso-butanol, etc.

- phenol, and substituted and chain-like derivatives

2.6.3 Water Soluble Organics (WSO)

The term Water Soluble Organic (WSO) refers to an analytical result when using the EPA-1664 analysis of oil-in-water concentration. It will be discussed in considerable detail in Section 7.4 (Measurement of Oil in Produced Water). It is worthwhile to also discuss it briefly here in relation to dispersed and dissolved organic contaminants. The most important point to be understood about Water Soluble Organics is that they are not the same thing as dissolved organics. This is explained below and in Section 7.4.3 (Relation Between Dispersed, Dissolved, TPH, TOG, and WSO) and in Chapter 21 (Applications – Dissolved and Water Soluble Organics).

EPA-1664: Total Oil & Grease (TOG) and Water Soluble Organics (WSO): In this book, as well as in most of the US-based oil and gas industry, TOG and WSO are defined in terms of the EPA 1664 analytical method [23, 38]. That method is summarized here for reference.

Summary of EPA 1664 Method:

- A 1-L sample of oily water is acidified to pH less than 2.

- The sample is then serially extracted three times with 30 mL n-hexane in a separatory funnel. The total quantity of hexane used in the extraction is 90 mL.

- The extract is dried by draining the solvent extract through a bed of sodium sulfate in a filter funnel.

HEM: Hexane Extractable Material: the hexane extraction solvent is distilled from the extract. The distillation temperature is somewhat at the discretion of the laboratory [see Section 11.4 of reference 23], and is generally slightly above the boiling point of hexane which is 68.7 C. The remaining material is referred to as HEM. The HEM is desiccated and weighed.

SGT-HEM: Silica Gel Treated HEM: the HEM generated in the above step is redissolved in n-hexane. An amount of silica gel proportionate to the amount of HEM is added. The intention of the silica gel is to remove polar components. After contact, the solution is filtered to remove the silica gel, the solvent is distilled from the extract, and the SGT-HEM is desiccated and weighed.

SGA-HEM: Silica Gel Adsorbed HEM: the weight of material adsorbed by the silica gel. It is calculated as the difference: HEM – SGT-HEM.

Here TOG is equivalent to the n-hexane extractable material (HEM) of the EPA 1664 method. Likewise, water soluble organics (WSO) is equivalent to the fraction of the HEM that is adsorbed onto silica (SGA-HEM). The TPH components are those HEM components that do not adsorb on the silica gel. Those components are said to be "treated" by the silica gel. To summarize:

TOG = HEM (mg/L n-hexane extractable material)

TPH = SGT-HEM (mg/L silica gel treated, n-hexane extractable material)

WSO = SGA-HEM = TOG – SGT-HEM = (mg/L silica gel adsorbed, n-hexane extractable material)

In all cases for EPA 1664, the n-hexane extraction step is followed by evaporation of the n-hexane. This results in vaporization of not only the n-hexane but also volatile components such as C6 and C7 hydrocarbons (including benzene) and the protonated form of formic and acetic acid. Since the sample is acidified to pH < 2, almost all of the formic, acetic and in fact all organic acids will be protonated and, if extracted into the hexane solvent, will be lost during the solvent evaporation step.

It is important to note that neither TOG, nor WSO nor TPH is equivalent to the dispersed oil content of an oily water sample. Further, WSO is not a direct or even indirect measure of dissolved organics in produced water. The term, Total Petroleum Hydrocarbon (TPH), is not entirely appropriate either. Strictly speaking, a hydrocarbon compound is composed only of hydrogen and carbon. According to IUPAC there are four classifications of hydrocarbons (alkanes, unsaturated alkanes, cycloalkanes, and aromatics). The term TPH, as it is used in the oil and gas industry consists of much more than just hydrocarbons. It contains polar organic molecules as well.

Water Soluble Organic compounds are defined as a solubility / volatility class. Individual chemical compounds can be identified, but the fundamental definition of WSO is given in the previous discussion as the fraction of the dissolved and dispersed hexane extractable material that does not flash off during solvent evaporation, and which sticks to silica gel.

This may seem like a complex and unwieldy definition. In fact it is. However, it is not entirely unique to use this kind of definition for oil and water chemistry. In the oil and gas industry, a well-known similar case is asphaltene, which is a solubility class of compounds. Asphaltenes are those compounds that precipitate upon addition of hexane (or pentane or heptane depending on the method specification) to crude oil, and which dissolve upon subsequent separation and contact with methylene chloride. In other words, the asphaltenes as a chemical aggregate group are only defined in terms of their solubility. Table 2.7 below lists the different classes of typical water soluble organics. They are defined here in terms of their abundance, solubility, and volatility.

Table 2.7 Six classes of compounds typically identified as water soluble organic compounds.

WSO Class	Characteristics
Phenol: C6 to C9	highly water soluble typically do not extract into hexane
Phenol: C10+	partition into dispersed oil phase extract to some extent into hexane
BTEX	high to moderate water solubility, extraction into hexane is pH dependent highly volatile
Acids: C1 to C4	very water soluble even at pH=2, typically do not extract entirely into hexane phase partitioning (water/hydrocarbon) strongly dependent on pH high adsorption tendency onto silica gel volatile (referred to as the volatile fatty acids)
Acids: C5+	somewhat water soluble depending on pH and carbon number extract into hexane at low pH phase partitioning (water/hydrocarbon) strongly dependent on pH high adsorption tendency onto silica gel main contributor to WSO in produced water with high WSO

2.6.4 Resins and Asphaltenes (SARA Crude Oil Fractions)

The most commonly cited definition of asphaltenes is that they are a "solubility class" of crude oil as determined by a SARA analysis. Asphaltenes are defined as the fraction of crude oil that precipitates when n-pentane (or n-hexane) is added in sufficient quantity, and which will dissolve in toluene (or xylene or carbon disulfide). The resin, and other fractions of the crude oil, is determined as part of the SARA analysis by gel chromatography. Since the SARA analysis is widely used to define the various fractions of crude oil, it is discussed first. After that, the properties and geological origin of the SARA fractions are discussed. Asphaltene and resins play an important role in determining the fouling tendency of the crude oil, its tendency to coalesce, and the surface chemistry of oil droplets suspended in water. The surface chemistry impacts the class of flocculating agents and flotation aids that are likely to be effective in removing oil from produced water.

SARA Analysis: A SARA analysis determines the weight percent of four fractions in crude oil: Saturates, Aromatics, Resins, and Asphaltenes. In the SARA analysis, the first step is a topping procedure which vents the light hydrocarbons and establishes the "liquid" content of the whole crude oil [39]. The next step is a titration with an alkane such as n-pentane or n-heptane [40, 41]. At some point in the titration the asphaltenes will drop out of solution. When separated, the asphaltene fraction of crude oil is dark brown and tacky. Asphaltenes are the highest molecular weight fraction of crude oil. They are soluble in polar or aromatic solvents such as pyridine, carbon disulfide, carbon tetrachloride, xylenes, toluene, or benzene. The most commonly cited definition of asphaltenes is that they are a "solubility class" as determined by the SARA analysis. Asphaltenes are defined as the fraction of crude oil that will precipitate when n-pentane (or n-hexane) is added in sufficient quantity, and which will dissolve in toluene (or xylene or carbon disulfide).

The portion of the original crude oil that does not precipitate when alkane is added is referred to as Maltene. The Maltene sample is added to a column packed with silica gel [39]. Hexane is added to the column and the least polar components of the crude oil elute with the added hexane. These components are referred to as "saturates." The next step in the SARA analysis is to run benzene or toluene. The fraction that elutes is referred to as the "aromatics." Finally, the "resins" elute with benzene / methanol (or chloroform / methanol or trichloromethane). The solvents are indicated in the Figure 2.13 below. The solvents are evaporated using a nitrogen drying bath at 40 C. The final amount of each fraction is determined gravimetrically. In some cases, HPLC is used instead of gel chromatography [42].

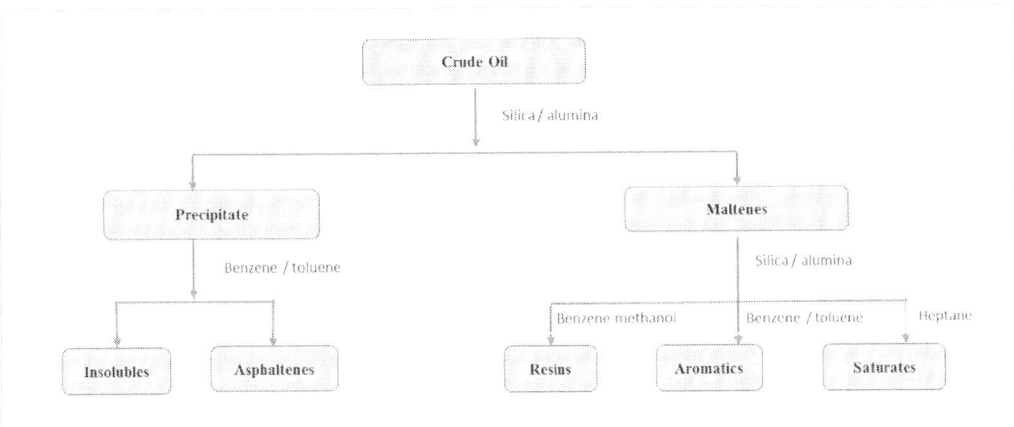

Figure 2.13 Schematic diagram of the different crude oil fractions that can be separated from crude oil by application of different solvents and liquid chromatography [40].

The SARA characterization method has the advantage that it is widely used and therefore there is a significant body of SARA data on crude oils from around the world. But there are several disadvantages. One problem, discussed below, is that the so-called aromatics are not necessarily true aromatics. The "saturates" are probably true saturated hydrocarbons. But the aromatics could contain acids in their protonated form, and other slightly polar compounds that elute off of a silica gel column with toluene as a solvent. There is measured data to support this possibility [31]. Another problem with SARA analysis is that the resins are poorly characterized. As discussed below, resins are an important and diverse group of compounds that can have a substantial impact on stabilizing emulsions. For example, it is believed that the naphthenic (oil soluble) acids contribute primarily to the resin fraction of a SARA analysis. Nevertheless, SARA is used throughout the industry. In the next few sections, the resins and asphaltenes are discussed. This is followed by a more detailed discussion of various compounds within the resins. The section is concluded with some comments as to why the resins and asphaltenes are so important in water treatment.

Asphaltene Chemistry: Asphaltenes are well known in the industry for causing production loss through plugging and forming thick deposits in equipment and piping. From x-ray and other spectrographic methods, it is known that asphaltenes tend to have several fused aromatic rings that give the molecule a plate-like structure. Chemically, they are composed of high molecular weight polycyclic, condensed, aromatic rings with several side chains (see Figure 2.14). Some pendant groups may be ionized in water. The plate-like structure promotes association of several molecules together into a stack (like clay particles). The pendant groups promote hydrogen bonding, acid/base, and electron exchange interaction between asphaltene molecules.

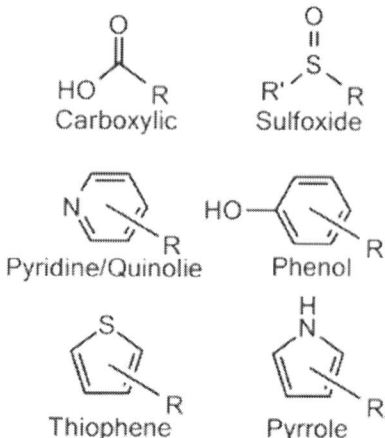

Figure 2.14 Chemical functional groups (side groups) typically found on asphaltene molecules. The R group represents the rest of the asphaltene molecule. In the case of the aromatic groups, the ring can be fused into a larger ring structure or it can be a pendant group.

Geologic Origin of Asphaltenes: The geologic origin of asphaltenes has been studied. Asphaltenes, like all petroleum fractions, is derived from kerogen which is referred to as source material for petroleum. Kerogen is defined as sedimentary organic matter that is not soluble in common organic solvents and which does not flow under geothermal conditions. Over time kerogen degrades due to pressure, elevated temperature and bacterial action. This process is referred to as maturation and it results in simplification of the molecular structure of the material. The more mature the petroleum, the more liquid-like and the lower the molecular weight. Throughout the maturation process liquid petroleum products seep out of the kerogen and migrate through the porous rock until they become trapped in rock formations. The asphaltene fraction of crude oil is the large, high molecular weight fraction of oil that is the most similar to the solid-like kerogen source material. It is the highest molecular weight fraction of the kerogen that still flows. Since it is the least matured fraction it is compositionally similar to the source material (kerogen).

Geologic Types of Kerogen and Asphaltenes: Kerogen, and hence asphaltenes, have been classified. Three or four types of kerogen are identified [99]. Type I is derived from algal and/or bacterial organic matter that has been deposited in lake (lacustrine) environments. Type II kerogen is derived from marine delta regions and coastal muddy marine (sapropelic) environments. Type III kerogen are mainly from onshore flood plains where higher plant debris such as shrubs, bushes, trees, etc. are found. These geologic origins provide some clues as to the composition, molecular structure and molecular weight of the associated asphaltenes. In many Type II kerogen, for example, the organic sulfur content would be relatively high while the organic sulfur content of Type I and Type II kerogen is relatively low. The organic sulfur is often in the form of substituted thiophene groups, as discussed above. Generalizations as to chemistry of these types is not possible since there is variation within each type. However, once a particular asphaltene has been identified, then other fields having the same type of kerogen / asphaltene will likely have similarities in chemistry.

Asphaltene Precipitation: Several studies [43, 44] have shown that asphaltenes exist in crude oil in one of the following states:

- dissolved – molecularly dissolved at high pressure and surrounded by resin molecules

- colloidal dispersion – as grouping of associated asphaltene molecules surrounded by resin molecules

- precipitated as a suspension of solid particles.

It is often the case that asphaltenes are dissolved in the crude oil under reservoir conditions. As fluids move out of the reservoir and into the production system, precipitation of asphaltenes can occur. The variable with the greatest influence is pressure. As the pressure of the fluid goes down, through the production process, the cohesive energy density of the crude oil decreases as well. The chemical structure of the crude oil molecules does not change; rather the affinity of the crude oil for asphaltene decreases. Precipitation can occur early in the process before the oil reaches the well bore, or it can occur much later such as in the separators, or anywhere in between. At a pressure just above the bubble point of the crude oil, asphaltene solubility reaches a minimum. As asphaltene solubility decreases (as the surrounding oil becomes less asphaltene loving), the asphaltenes tend to form aggregates, or tend to migrate to the water / oil interface. Prior to precipitation, there is a close association between asphaltenes and resins in the bulk crude oil. Resins help to stabilize the asphaltenes and keep them dissolved. Since both asphaltene and resin molecules have polar groups (carbonyl, carboxyl, amine, etc) it is possible that a large multi-functional asphaltene molecule is surrounded by resin molecules that interact via hydrogen bonds, acid / base or electron exchange interactions. Due to the compositional complexity and number and types of functional groups involved, the solubility of asphaltenes varies widely from one crude oil to another.

Asphaltene precipitation takes place when the crude oil loses its capability to disperse and stabilize the asphaltene molecules. Asphaltene stability depends on the composition of the crude oil, temperature, and pressure. Under static reservoir conditions, asphaltenes normally are held in a stable suspension by resins [42]. The relatively high pressure of the reservoir maintains a high cohesive energy density in the crude oil which helps keep asphaltene in solution. The decrease in fluid pressure that is generally associated with oil production decreases the solubility of asphaltenes in the crude oil and may cause them to flocculate and precipitate out of suspension and adsorb onto the rock or pipe surfaces. They also have a natural tendency to migrate to the oil / water interface due to the presence of ionizable chemical functional groups, particularly the carboxylic acid group.

Asphaltene Stability: Asphaltene phase stability is a major concern in flow assurance of production systems [45], and water treatment systems as well. A stability diagram can be constructed from field observation of asphaltene deposition, together with the SARA analysis of composition. One such stability diagram is shown in the Figure 2.15 below. As discussed, aromatic and resin material, as defined by the SARA analysis, tend to stabilize asphaltenes in solution and reduce their tendency to precipitate. As the concentration of these components goes down, the value of the x- and y-axis goes up. Some of these crude oils then fall into the region of the diagram where the asphaltenes are unstable, as verified by reported problems in the field. This type of information is useful in designing water treatment systems or diagnosing performance issues with the systems since the precipitation of asphaltene can lead to or exacerbate water treatment problems.

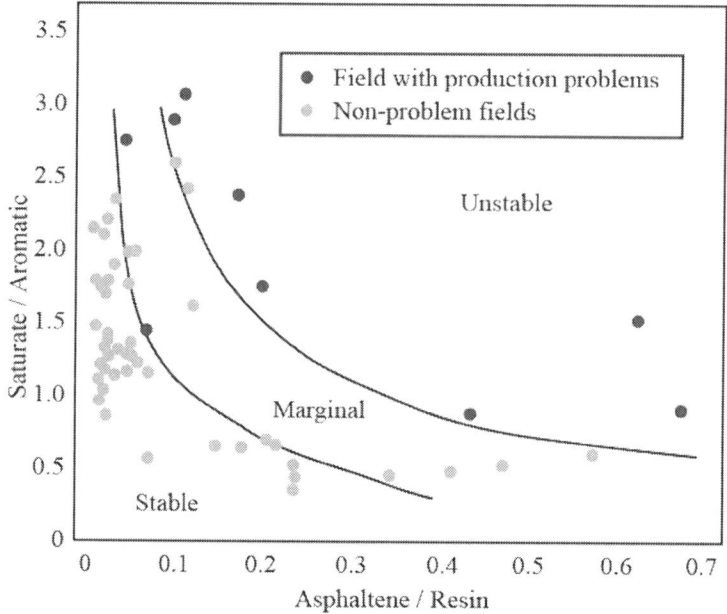

Figure 2.15 Stability diagram based on SARA composition analysis [45].
The axes are concentration ratios where concentration is expressed in wt % from a SARA analysis.

In addition to the stability diagram based on SARA analysis, the Colloidal Instability Index (CII) can provide a quantitative measure of the likelihood of asphaltene precipitation [100]:

$$CII = \frac{Sat + Asp}{Res + Aro}$$

Eqn (2.6)

Where all quantities are expressed in weight fraction. When the CII > 0.9 the asphaltenes are unstable. They have a tendency to precipitate and to migrate to the oil / water interface, as well as to stick to pipe and vessel walls. From a water treatment standpoint, it is also helpful to know if the asphaltenes are marginally stable. Marginally stable asphaltenes also have a tendency to migrate to the oil / water interface. Values of CII greater than 0.7 and less than 0.9 are marginally stable.

Regardless of the mechanism causing the asphaltene to precipitate, the result is a plugging effect that inhibits or reduces oil production. Precipitation of asphaltene particles may also provide nuclei for paraffins to stick to. In many cases of gunky deposits, the "gunk" is a combination of asphaltene and paraffin. When such a combination forms a precipitate, it is often extremely sticky. Due in part to its "stickiness", it will adsorb solid mineral particles, if they are present. Thus, gunky deposits are often associated with inorganic material such as formation solids, salts, and iron oxides. Of course, if such deposits form, then an emulsion of dispersed gunky drops of oil in water will also likely be present in the flowing fluids. Particles formed of the conglomeration of hydrocarbons and inorganic materials will have a range of specific gravities. Some particles will have a high mineral content and be heavier than the brine. Other particles will have more hydrocarbon content and be lighter than the brine. There will often be a fraction of particles that have a specific gravity close to that of the brine. These particles will be neutrally buoyant and will not be separated by primary separation equipment. Only flotation and various media will be effective to separate these particles from the produced water.

Particle Size of Precipitated Asphaltenes: The diameter of precipitated asphaltene particles has been measured for a limited number and type of crude oils. The objective of these measurements is to model the rate of asphaltene particle migration to the pipeline wall and the rate of particle deposition [46-48]. Knowledge of the particle size distribution is required for particle deposition modeling. These results are shown in the Figure 2.16 below. As shown, the particle diameters are in the range of 1 to 20 micron. The data shown in the figure are for asphaltene precipitation in the oil phase. But, depending on the polarity of the molecules, these particles may migrate to the oil/water interface or be distributed into the water phase by exposure to high shear such as in a pump. The particle size distribution will depend on a number of factors such as the concentration of asphaltenes, the speed with which the asphaltenes precipitate, and the fluid flow characteristics when they do precipitate. All of these factors have an effect on the agglomeration rate and final particle size. Although a wide range of particle sizes would be expected, the data in the figure show that at least in some cases, the particle sizes can be quite small. This is a difficult range of particle size to remove from produced water.

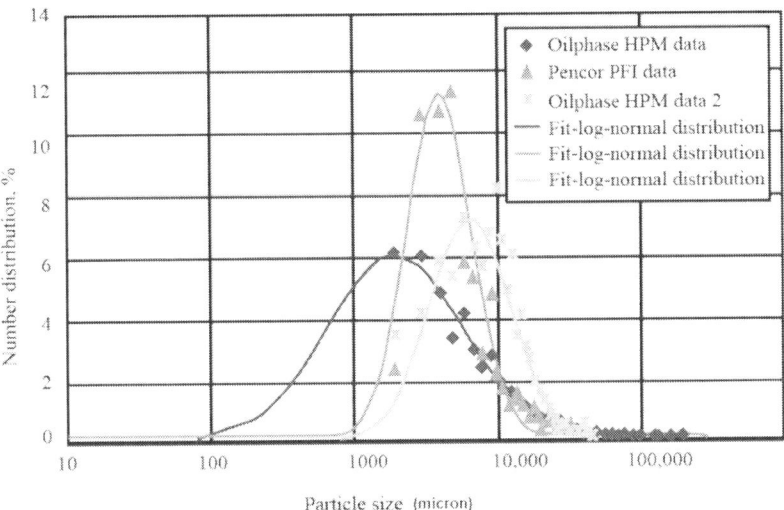

Figure 2.16 Number distribution of asphaltene particles as a function of the particle diameter (microns). As shown, the particle diameters are in the range of 1 to 20 micron. Asphaltene precipitates in the oil phase but, depending on the polarity of the molecules, these particles may migrate to the oil/water interface. This is a difficult range of particle size to remove from produced water. See [46-48].

Role of Asphaltenes in Water Contamination: The mechanism of asphaltenes in exacerbating the challenges of water treatment is somewhat complex but the net result is nevertheless profound. It is covered in detail in Chapter 3, Emulsions. Suffice for now to say that there is an enormous difference between a field that produces oil with a low concentration of stable asphaltenes, versus a field that produces an oil with a high concentration of asphaltenes that are not stable. In the former case, oil droplets will, without other mitigating factors, coalesce and recover from sources of shear, and will therefore be relatively easy to separate from water.

In the later case, oil droplets will be sensitive to shear, generating a high concentration of small drops that will not coalesce. The more polar components of the asphaltenes will migrate to the oil/water interface. When asphaltene molecules accumulate at the oil / water interface they tend to form a rigid elastic network. As discussed by Sheu and Mullins in their treatise on asphaltenes [43], this actually increases both the interfacial tension and the elasticity of the interface. Surface pressure isotherms [49] of asphaltenes spread on a water surface form an incompressible hard film on the surface. The interface becomes elastic like a balloon. Deformation of the interface is resisted, which inhibits droplet coalescence and preserves small droplet size throughout a facility.

In addition, asphaltenes will stick to solids particles making them partially oil wet and perhaps neu-trally buoyant. These oily solids in turn will accumulate on surfaces and cause fouling of equipment. Stable emulsions will form in the separators at the oil/water interface. Generally speaking, water treatment will be more difficult.

Resins: The resin fraction of crude oil is soluble in n-pentane and n-hexane. In the SARA analysis they are separated from the saturate and aromatic crude oil fractions by chromatographic methods. Resins and aromatic compounds both act to stabilize the asphaltenes, preventing to some extent their tendency to precipitate upon crude oil pressure reduction.

Resins are such a broad class of chemical compounds, it is difficult to generalize about their proper-ties. However, some things can be said. Resins tend to be lower molecular weight than asphaltenes and tend to have polar functional groups. The oil soluble naphthenic acids are part of the aromatic and resin fraction, as are the longer chain fatty acids. In one study, a particular Athabasca bitumen was found to contain 20 % asphaltene. The resins in the oil were found to contain 1,200 mg/L of naphthenic acids. This is a significant concentration of acidic species. Most of the acids in Athabasca bitumen are C21 to C24 tri-cyclic terpenoid acids. Naphthenic acids, like the whole resin fraction, are found to soften the oil/water interface where asphaltenes have been absorbed.

Table 2.8 Measured molecular weight for the resin and asphaltene fractions of three Norwegian crude oils [50].

Crude oil	resin MWt (g/mol)	asphaltene MWt (g/mol)
B1	1700	4200
B2	1200	1300
G	900	1400

As shown in the table, there is considerable variation in molecular weight in the two fractions (resin and asphaltene) from one crude oil to another. Note that the molecular weight of the resin fraction of the B1 crude oil is higher than that of the asphaltene fraction of the B2 crude oil. Variation in the characteristics of these fractions from one crude to another is generally observed. It is also noted that the molecular weight of asphaltenes was a controversial issue for almost 20 years due to aggregation of asphaltenes [27].

Some particular resins, though surface active, were not found to significantly stabilize water-in-oil emulsions by themselves. In fact, the presence of resins has been found to reduce the tendency of asphaltenes to stabilize w/o emulsions. Two mechanisms are thought to occur. It is thought that resins can solvate asphaltene molecules making them more soluble in the bulk crude oil. This would reduce their tendency to congregate at the oil/water interface. It is also observed that the resins have higher surface activity than asphaltenes and thereby displace the asphaltenes from the interface. This is discussed in greater detail in Section 4.4.3 (Surface Excess Concentration & Surface Activity).

Naphthenic Acids in the Resin Fraction: Depending on the particular crude oil, and on the details of the SARA fractionation, some of the naphthenic acids can be found in the so-called "aromatic" fraction [31]. This is easier to understand if it is recalled that the aromatic fraction is that fraction that elutes from a silica gel column upon the addition of toluene (or benzene/toluene) as the eluent. Thus, the so-called aromatic fraction of a SARA analysis is not necessarily composed strictly of aromatic compounds. But most studies find that asphaltenes and resins account for most of the polar groups in a crude oil. It is believed that most of the naphthenic acids are found in the resin fraction [42]. Thus, most surface active compounds come from these two fractions, though not all of the compounds found in these fractions are surface active.

Crude Oil Aging: One of the interesting observations of crude oil aging is that water-in-oil emulsion stability tends to increase in the initial stages of aging but then reaches a maximum from which it then starts to decrease until after about a month there is no further change in the emulsion stability [21]. The later decrease in emulsion stability was at first unexpected. Now it is generally understood that both the asphaltene and resins undergo a gradual oxidation that makes them more surface active with time. As the asphaltenes become more surface active they tend to increase emulsion stability. However, as resins become more surface active they tend to displace asphaltenes from the interface which prevents the elastic network of associated asphaltene molecules from forming.

CHNOS of Resins and Asphaltenes: While the composition and molecular weight of asphaltenes and resins varies considerably from one crude oil to another, there are certain trends [p. 474 of 76]. Roughly half of the carbon atoms in most crude oil asphaltene fractions are aromatic carbons (50 % of the carbon atoms). Likewise, for the asphaltene fraction of most crude oils, the H / C ratio is about 1.15 (number of H atoms / number of carbon atoms). A representative elemental composition for resins and asphaltenes is given in the Table below.

**Table 2.9 Elemental composition of resins and asphaltenes
from a half dozen crude oil analyses [51 - 54].**

	resins	asphaltenes
H / C atomic ratio	1.2 to 1.48	1.0 to 1.2
atomic O %	0.8 to 2.0	0.3 to 4.9
atomic S %	0.4 to 5.1	0.5 to 10.3
atomic N %	0.35 to 0.65	0.6 to 2.6

The Figure 2.17 below provides industry-wide data that can be used for comparison with new samples. This is sometimes helpful in determining whether a new sample is unique in its content of saturates, aromatic or resins and asphaltenes.

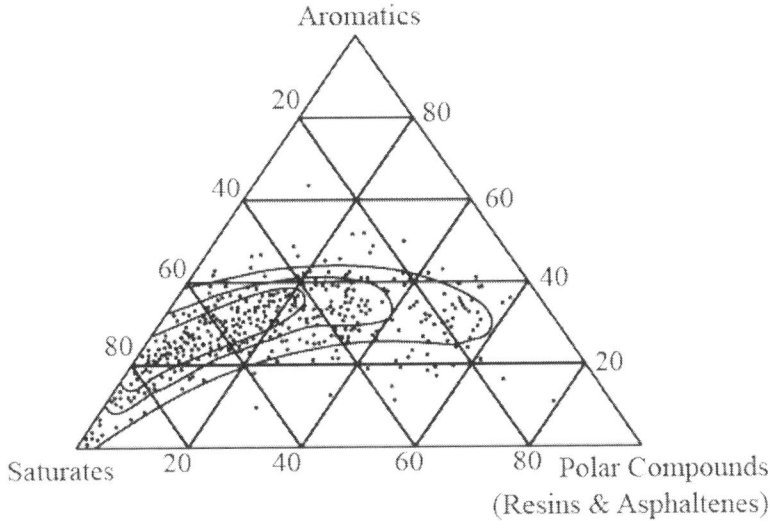

Figure 2.17 Composition map of the saturates, aromatics, and resins/asphaltenes of many crude oils.
These data are useful for determining the uniqueness of a crude oil. Data from [51].

2.6.5 Naturally Occurring Surface Active Constituents

In this Section, a summary is given of the naturally occurring surface active compounds in produced water. As discussed previously, any chain-like molecule that has a polar group on one end and which is attached to a non-polar chain or group will have some degree of surface activity. It is also the case that non-chain molecules with multiple fused aromatic groups and pendant polar groups will have a degree of surface activity. Such molecules will accumulate at the oil/water interface and reduce the oil/water interfacial tension, and in some cases create an elastic membrane-like structure at the interface. The details of this process will be discussed in the chapter on Emulsions.

In fact, crude oil is loaded with a large number of different molecules that have some degree of surface activity. Given the great diversity of crude oil composition from one formation to another, there is likewise a great diversity in the degree of surface activity of one crude oil / produced water mixture to another. This fact, which is often overlooked in water treatment in the oil and gas industry, is a major contributor to the seemingly unique behavior of each producing reservoir.

In Table 2.10, several examples are given of organic acids which are surface active. Note that the smallest members of the series, formic (C1), acetic (C2), and propionic (C3) are not given as examples. They are really too short to have appreciable surface activity. They do reduce the interfacial tension between oil and water that contains them, however, they are not nearly as surface active as the higher molecular weight (longer chain) homologues. This is explained in greater detail in Section 4.4.3 (Surface Excess Concentration & Surface Activity).

Table 2.10 Organic acids which are surface active

Compound Class	Example	Interface Characteristics
short-chain acids	acetic acid	Essentially no surface activity due to the short Oleophilic chain
longer-chain acids	Hexanoic acid and larger. Part of the Resins in a SARA analysis.	Highly surface active. Moves to interface rapidly. Will displace other less surface active compounds at the interface.
naphthenic acids	Long-chain, multiple acid groups, cyclic structures. Part of the Resins in a SARA analysis.	Most are oil soluble but some, smaller and more polar compounds are water soluble. More surface active than asphaltenes. In sufficient quantity can push asphaltenes away from the oil/water interface and reduce the elasticity (soften the interface) in so doing.
resins	Hexane soluble material, but distinct from alkanes and aromatics. Includes longer-chain acids and naphthenic acids.	Most resins are oil soluble but some, more polar compounds are water soluble. More surface active than asphaltenes. In sufficient quantity can push asphaltenes away from the oil/water interface and reduce the elasticity (soften the interface) in so doing.
asphaltenes	Hexane insoluble / CS2 soluble material	Depending on the particular molecular structure, will be soluble in oil or not soluble in either water or oil. Moves to interface but will not displace other more surface active molecules such as hexanoic acid. But, in sufficient concentration, will form a strong elastic membrane at the interface that prevents oil droplet or water droplet coalescence. Solid asphaltene particles will also stabilize an oil/water interface.
phenols	benzophenol	Somewhat surface active. Usually not implicated as a barrier to coalescence.

As discussed by Sjobolm and co-workers [50, 21], of the four fractions determined by SARA analysis (saturates, aromatics, resins, asphaltenes), only the resins and asphaltenes are surface active. SARA is a whole-crude analysis which means that all of the compounds in the crude oil will be measured in one of the four fractions. In discussing the chemistry of crude oil it is necessary to provide a finer distinction that just the four SARA groups. Thus, short-chain acids, longer-chain acids, and the higher molecular weight acids, which are all part of the resins, as referred to separately.

The presence of polar and ionizable groups such as carboxylic acid, amine and sulfoxide will give a certain degree of surface activity to any molecule in which they are present. By performing liquid-liquid extractions of the crude oil using a sequence of various solvents, it is possible to separate a number of different fractions of the crude oil. SARA analysis extractions will separate the crude oil into four fractions. Other fractions within each of the SARA groups can be separated as well. The fractions can then be studied for molecular weight, interfacial characteristics, and so on. When this is done, it becomes evident that there is a wide variety of resin and asphaltene types and that they differ markedly from one crude oil to another. This is important to keep in mind since the type of resin and asphaltene can have a significant effect on the performance of the water treatment system. It is also important to keep in mind that the procedure for a SARA analysis includes some flexibility. For example, depending on the particular laboratory that does the analysis, pentane, hexane, or heptane can be used to precipitate asphaltenes, and different solvents are used in the silica gel chromatography.

Fractionation studies of this kind have determined that the resins tend to be more polar and surface active than the asphaltenes. They tend to accumulate more quickly at the interface and form a film. Resins form a film that is more compressible than that formed by asphaltenes [49] which form a rigid, incompressible and strong interfacial film. The asphaltenes create a stable interfacial film that reduces drop-drop coalescence. In some cases, the combination of asphaltene and resins contribute to the formation of the most stable emulsions [21].

It is worthwhile to point out that resins and asphaltenes are significantly more soluble in the oil than they are in the water. Thus, they contribute to the stability of water-in-oil emulsions to a greater extent than to oil-in-water emulsions. The reasons for this are discussed at some length in Section 4.4.5 (Bancroft Rule).

2.7 Inorganic Constituents in Water

In this section, typical inorganic constituents in produced water and hydraulic fracturing flow back water are reviewed.

2.7.1 Dissolved Inorganic Constituents (mineral and metal ions)

In this section, the important anions and cations found in produced water are discussed in relation to their origin, impact on water quality, toxicity, and options for their removal. Some of these constituents are specialized in the sense that they are only found in a small number of facilities or regions around the globe and their handling and removal require somewhat specialized approaches and technologies. The heavy metals, for example, do not have a separate section elsewhere in the book. Therefore, this section discusses more than just their chemistry and includes their occurrence and separation as well.

Metals and Heavy Metals: Stephenson [13] compiled data for the occurrence of heavy metals from overboard discharge of produced water in the Gulf of Mexico, in the early 1990's. Given the timeframe of his work, all of the samples would have been from shallow water platforms. His results are given in the Table below. Note that the concentrations are expressed as micrograms per liter (mg/L), which is 1/1000 of mg/L.

Table 2.11 Heavy metals measured in produced water in shallow water Gulf of Mexico [13]

Metal	Average (µg/L)	Standard Deviation (µg/L)	Maximum (µg/L)	Minimum (µg/L)
Cd	27	12	98	0
Cr	186	68	390	0
Cu	104	180	1,455	0
Pb	315	670	5,700	2
Ni	192	307	1,674	0
Ag	63	17	152	12
Zn	170	253	1,600	17

These compounds can occur in produced water in a range of forms (ionic, elemental, organometallic). Most technologies are only able to remove certain forms and not necessarily all forms of these metals. Thus, careful sampling and analysis must be carried out in order to determine the form of the metal. Once the form is known, water treatment technology can be selected based on the mechanism of removal, taking into account the form of the contaminant. Fuast and Ali [55] provide an excellent summary of technologies to removal heavy metals from water. Some of those techniques are given here.

In Section 5.5.4 (Chemistry of Carbamates and Chelating Agents), the chemical treatment of produced water to remove heavy metals is discussed. Dithiocarbamates (DTC) are particularly effective. This class of chemical compounds functions as both chelating agent and as flocculating agent. As a chelating agent, the DTC forms a complex with the dissolved or dispersed metal. As DTC molecules proceed to bind the metal and form complexes, the concentration of these complexes increases until a floc is formed. The floc can then be removed by settling or flotation.

In Section 14.1.4 (Electrocoagulation – Representative Performance Data) it is mentioned that electrocoagulation is effective at removing heavy metals. In fact, in the industrial water industry, this is the main use for EC. Heavy metals removal using EC is still dependent on the chemistry of the water and the pH. These compounds can occur in produced water in a range of complexes and compounds. It appears that EC can remove the colloidal and suspended forms as well as a significant portion of the elemental, ionic and dissolved forms. This is somewhat expected given that flocculation chemicals have similar capability, though at high dosage. The group II alkali earth ions ($Mg+2$, $Ca+2$, etc) can be removed to a limited extent. Other dissolved components that contribute to TDS cannot be removed to any great extent. This agrees with the mechanisms that are known to operate in an EC unit.

In Section 13.8 (Ion Exchange) removal of arsenic, heavy metals, nitrates, radium, uranium and other elements from the produced water is discussed. Generally speaking, ion exchange requires pre-treatment to remove suspended material that would otherwise foul the ion exchange resin bed.

Anions and Cations: Cations are ions with a positive (+) charge. Anions are ions with a negative (-) charge. The sharing of electrons by anions and cations creates uncharged species with electroneutrality. For example, the calcium ion is a divalent cation and will combine with two monovalent chloride anions to form the electrically neutral salt known as calcium chloride. A balanced water analysis will have the same number of positive (cation) equivalents as negative (anion) equivalents. An "equivalent" is the molar concentration of a particular element or charged entity divided by the number of charges which the element or entity as in solution. Thus a solution with 100 mg/liter of calcium ion, Ca^{++}, (At.Wt. = 40), for example, contains 100/40 x 2 = 5 milliequivalents (meq) of Ca^{++}. If the same solution contained 390 mg/liter of HCO_3^- (Mol. Wt. = 61), then it would contain 390/61 x 1 = 6.4 meq/liter of HCO_3^-.

Once the dispersed oil and suspended solids, and oil-wet solid (Small solids with oil-wetted surfaces. Oil-wet solid may be organic or inorganic in character) has been removed from a sample by filtration, the dissolved components can be analyzed. As described previously, in addition to the dissolved inorganic entities (charged or neutral) there are dissolved hydrocarbons and dissolved organics that also play a significant role in water chemistry and the performance of produced water treating equipment. Those components were discussed above. In this section, we discuss the inorganic anions and cations which include the dissolved minerals, metallic cations, transition metal cations, and anions such as the halides, carbonates, sulfate, sulfides, etc.. With respect to produced water treating, most of the cations and anions are of importance because of their tendency to form mineral precipitates which can then become partially oil wet and act to stabilize oil in water emulsions.

The salinity of water is typically expressed as the concentration of Total Dissolved Solids (TDS), indicating the quantity of dissolved inorganic salts present in the water.

The salinity of produced water is an important consideration for a number of reasons. For example:

- For surface disposal, the salinity of the discharge water should not be significantly different from the salinity of the receiving environment. This should encompass both the discharge of saline water into fresh waters and the discharge of fresh water into saline waters (e.g. onshore).

- Salinity affects the density of the water which will in turn influence the design of gravity based separation equipment and the dispersion characteristics of disposed waste water.

- Changes in the composition or process conditions of saline waters (e.g. temperature changes) may lead to the precipitation of inorganic salts and scale formation.

- The salinity of the water will add to the corrosion potential of the water stream, influencing the selection of materials for process equipment.

Aluminum (Al): Aluminum has relatively low water solubility and is not typically found at appreciable concentrations in produced water. However, it is sometimes used to coagulate oil droplets and solids particles. For this purpose, various forms of aluminum can be used including the simple salts $Al_2(SO_4)_3$ (aluminum sulfate a.k.a. alum) and $AlCl_3$, as well as the stabilized polyelectrolyte polyaluminum chloride. These compounds are effective at reducing the charge-charge repulsive barrier that exists on the surface of small suspended particles such as clay, formation sand, and precipitated minerals. Much more discussion is given in Section 4.3.2 (Electrical Double Layer and Zeta Potential).

Ammonium (NH_4^{+1}) and Ammonia (NH3): Non-ionized ammonia becomes partially ionized in water to form the ammonium ion and hydroxide ion. The extent of ionization depends on pH, temperature, and the ionic strength of the produced water. At higher pH the ammonia species is prevalent. At lower pH, the ammonium ion is prevalent. Ammonia and ammonium exists in an equilibrium at varying relative concentrations in the general pH range of 7.2 to 11.5. Simple ammonium salts are highly soluble and do not typically cause a scaling problem. Short chain and long chain ammonium compounds (where a hydrocarbon entity replaces one of the hydrogens attached to the nitrogen atom in the ammonia molecule) are often used as corrosion inhibitors.

Ammonium is typically not found in reservoir produced water and aquifer well water. This is due to bacterial action, which would have converted ammonia (or ammonium) to the transitory nitrite (NO_2) ion which is then oxidized into the more prevalent nitrate (NO_3) ion.

Ammonium is found in surface water sources at low levels (up to 1 ppm as the ion), generally the result of biological activity and the breakdown of organic nitrogen compounds. Surface water sources can be contaminated with ammonium from septic systems, animal feed lot runoff, or agricultural runoff from fields fertilized with ammonia or urea. Thus, if such waters are used for shale hydraulic fracturing then ammonia may be present in the flow back water. Ammonium is prevalent in municipal waste facilities with levels up to 20 ppm. Another source of ammonium is the result of adding ammonia to chlorine to form biocidal chloramines.

Arsenic (As): Arsenic exists in produced water in two oxidation states As (III) and As (V) with the reduced ion being prevalent in water which has not been exposed to air or another oxidant. As(III) is present as arsenite: AsO_3^{-3} and As (V) is present as arsenate: AsO_4^{-3}. Arsenite is the product of dissociation of arsenous acid: H3AsO3. Arsenate is the product of dissociation of arsenic acid H3AsO4.. Generally arsenate salts are much less soluble than arsenite salts. So arsenic is typically oxidized to the +5 state before being removed by precipitation, for example by Fe^{+3}. Ferric arsenate is about 10,000X LESS soluble than $BaSO_4$.

Elemental arsenic and its various compounds and salts are classified as toxic and as group 1 carcinogens. Various compounds containing Arsenic are widely found throughout the earth's atmosphere and crust. The Current limit (2005) for arsenic in potable water is 10 ug/l for most municipalities and for some agricultural uses.

The ion exchange resins used for removal of arsenic are strong basic type. The basic components of the resin are aromatic ring and functional group of quaternary amines. Two types of quaternary amine resins are used. Type I with ammonium - methyl groups and type II where one of the methyl group is replaced with ethanol group. Strong base ion exchange resin removes arsenic from water according to the following equations:

$$R\text{-}Cl + H_2AsO_4\text{-} = R\text{-}H_2AsO_4 + Cl^- \qquad \text{Eqn (2.7)}$$

$$2R\text{-}Cl + HAsO_4^{-2} = R\text{-}HAsO4 + 2Cl^- \qquad \text{Eqn (2.8)}$$

Regeneration is carried out with excess of chloride ion using either NaCl or HCl

$$R\text{-}H_2AsO_4 + Cl^- = R\text{-}Cl + H_2AsO_4^- \qquad \text{Eqn (2.9)}$$

Barium (Ba): Barium is a divalent cation in water. It can be problematic in produced water, particularly when sulfate is present. The solubility of barium sulfate ($BaSO_4$) is very low, on the order of a few mg/L. It is important that barium be measured with instruments capable of 0.01 ppm (10 ppb) minimum detection levels.

BaSO4 tends to precipitate as a scale which is difficult to remove from pipes, valves, instruments and vessels. The solubility of BaSO4 is low but increases with salinity (TDS) and increases with temperature at relatively low temperature and decreases with changes in temperature at higher temperatures, depending on TDS.

Bicarbonate (HCO_3^{-1})/ Carbonate (CO_3^{-2}): Bicarbonate and carbonate are part of the CO_2 / carbonic acid / bicarbonate / carbonate group of compounds. This group of compounds is important for two main reasons. The first is that they have a strong effect on the pH of the produced water. The second reason is that magnesium carbonate, iron carbonate and calcium carbonate can cause scaling and mineral precipitation problems in produced water. The solubility of calcium and magnesium carbonate is relatively low and decreases with increasing temperature and pH. Bicarbonate anion is commonly found in produced water. When bicarbonate is present, the water is said to be "buffered" because it acts to neutralize any added hydrogen ions, thus stabilizing the pH of the water. The concepts of hardness and alkalinity are tied to the concentration of bicarbonate, and other species. All of this is discussed in some detail in Section 2.12 (pH, Alkalinity and Hardness). The bicarbonate ion dissociates into carbonate (CO_3^{-2}) which will react with calcium, magnesium, iron, barium and strontium to form scale mineral precipitates. Bicarbonate is often the major constituent of alkalinity. Carbonate is one component of alkalinity and its concentration is in a balance with bicarbonate between the pH range of 8.2 and 9.6. At a pH of 9.6 and higher, there is very little carbon dioxide or bicarbonate in the water so all of the alkalinity is in the carbonate form. High pH can occur, for examp, in systems where H2S is chemically scavenged.

Boron (B): From the standpoint of produced water reuse, boron can be a problematic compound. Boron is an essential ingredient in animal diet. It naturally occurs in most seawater up to about 5 ppm and at lower levels in brackish waters where inland seas once existed. Yet above a narrow range of concentration it can be toxic. Suggested limits of boron are 0.5 mg/L for human consumption and 1.0 mg/L for irrigation. Thus, for produced water reuse intended for agricultural application, boron must be measured and, if necessary, controlled.

Boron (in the form of Borate ion) is often used to cross-link the polysaccharides used in hydraulic fracturing fluids. Thus, for hydraulic fracturing flow back water intended for reuse, boron must also be measured and controlled.

Boron does not typically form scaling or fouling compounds. In general, the relative concentrations of borate and boric acid are dependent on pH, temperature and salinity. At relatively high pH, boron is found in the form of the borate monovalent anion **$B(OH)_4^-$**. At relatively low pH, boron is found in the form of the non-ionized boric acid **$B(OH)_3$**. This chemistry is exploited in some removal processes. For example, the non-ionized boric acid is too small to be efficiently removed by RO membranes. Also, not having a charge makes it more permeable through a reverse osmosis membrane. The borate anion, on the other hand, is just large enough to be rejected by an RO membrane, and having a charge makes it less permeable through typical reverse osmosis membranes. Thus, by raising the pH of produced water, RO membranes can be used to eliminate it.

Bromide (Br^{-1}) : Bromide salts are sometimes used in drilling and completions. Bromide salts are more expensive than chloride salts, but bromide salts are heavier and are used for well control when drilling and completing in higher pressure formations. Bromide salts, like chloride salts are very soluble.

Chloramines: Non-oxidizing chloramines have come into frequent use for the treatment of seawater in systems where sulfate rejection or reverse osmosis membranes are used. It many applications it is used continuously. Most sulfate rejection and RO membranes have reasonably high tolerance toward chloramines.

Chloramines in the form of monchloramines (NH_2Cl) are produced by adding ammonia (NH_3) to chlorinated water to produce the following reaction:

$$NH_3 + HOCl = NH_2Cl + H_2O$$
Eqn (2.10)

If the dosage ratio is not stoichiometric, there can be either residual free chlorine or free ammonia. The residual free chlorine would need to be neutralized by reaction with sodium bisulfite or with carbon filtration in order to protect the downstream membrane filter. But this can also result in dechloramination with a resulting increase in ammonia gas or ammonium ion levels. Caution is required because the increased presence of sodium bisulfite, ammonia or ammonium can facilitate biofilm growth if all the chloramines were removed. Ammonia is known to be corrosive to any downstream non-stainless steel metal fixtures. The passage of chloramines into a membrane filter permeate is relatively high, and has been observed at up to 80% of the feed level. The passage of ammonia through an RO membrane 100% since it is a gas molecule dissolved in water and is nonionic. The ammonium cation is a monovalent ion and is rejected by RO membranes.

Chlorine (Cl_2): The benefits of chlorine are that it is an effective biocide, is inexpensive, and controls the volume of any biofilm mass. Chlorine comes in various forms. The most common are the industrial or household bleaches in the form of 3% to 5% Sodium Hypochlorite (NaOCl). Due to its hazardous nature, Gaseous chlorine is typically reserved for use at large municipal and industrial water systems. Chlorine is used for the oxidation and subsequent removal of iron and manganese.

Chlorine Dioxide (ClO_2): Chlorine dioxide is effective as a biocide and partial oxidant in produced water. One advantage of ClO_2 is that it is a weaker oxidant than HOCl, HOBr and O_3 and thus less corrosive. Also, ClO_2 gas penetrates biofilm better and degrades the biofilm material better than the hypochlorite compounds. Because of the chemistry and reactivity differences between Cl2 and ClO_2, it has been reported that approximately one quarter of the dose of ClO_2 is required to maintain an effective disinfectant concentration compared to Cl2. Additionally, since it is gaseous, it will not be rejected by an RO membrane and will thus pass into the permeate at the same concentration as the feed. The typical sulfanil RO membrane used to desalinate seawater has a higher tolerance for ClO_2 than for chlorine and hypochlorite.

Chromium: There are two forms of chromium in produced water and in surface water, Cr III (trivalent, insoluble) and Cr VI (hexavalent, soluble). The soluble form of chromium is the subject of the 1990s film Erin Brockovich, which was based on a true story. It is toxic and carcinogenic. In drinking water, the limit of Cr VI is 10ppb and the limit for Total Cr is 50 ppb (mg/m^3 or micro-grams/liter). The chromium content of stainless steel is what differentiates it from other types of steel. It imparts corrosion resistance to steel. Chromium, unlike metals such as iron and nickel, does not suffer from hydrogen embrittlement. Nevertheless, if the produced water is corrosive, then detectable chromium ion concentrations (a few mg/L) will be found in the produced water from the steel. Chromium metal is passivated by oxygen, forming a thin protective oxide surface layer. This layer is very dense, and prevents the diffusion of oxygen into the bulk metal. Iron and plain carbon steels do not form this passive layer and thus are more subject to corrosion. The passivation can be enhanced by short contact with oxidizing acids like nitric acid. Nitric oxide treatment is often applied in closed circuit hot water and steam systems on offshore platforms. Passivated chromium resists corrosion down to a moderately acid pH (roughly 3.5). Reducing agents such as bisulfite have the opposite effect. Thus, overtreatment using oxygen scavengers should be avoided as they may increase the tendency of stainless steels to corrode.

Copper (Cu^{+2}): Copper is a divalent cation similar to cadmium, lead, and calcium. Well waters or surface waters may contain very low levels of Copper, generally less than 0.001 mg/L. Potable water systems for offshore are often constructed of copper piping. When this is the case, particular attention must be paid to bacterial contamination since bacterial corrosion will dissolve significant and detrimental amounts of copper into the potable water system.

Magnesium (Mg^{+2}): As a divalent cation, magnesium can account for a significant fraction of the hardness in a brackish water. It has a concentration five times higher than calcium in sea water. The solubility of magnesium salts in the neutral pH range is high and typically does not normally cause a scaling problem. However, at high pH, Magnesium solubility drops sharply.

Mg^{++} is usually present in produced water at much lower concentration than calcium ions. Magnesium carbonate will co-precipitate with calcium carbonate and therefore increase the mass of carbonate scale. Magnesium forms soluble ion pairs with sulfate ion which decreases the concentration of free sulfate ion in solution and thus decreases the scale tendency for other sulfate minerals such as CaSO4.

Manganese (Mn): Mn+2 (manganous ion) is the form of manganese present in water that is anaerobic. If this water is maintained in this reduced form, the manganese is soluble. In many systems, the presence of Mn^{++} in produced water along with Fe^{++} is an indicator that the soluble iron is from corrosion.

MnO_2 (manganese dioxide): Manganese, like iron, can be found in organic complexes in surface waters. In the oxidized state, it is insoluble and usually in the form of black manganese dioxide (MnO_2) precipitate. Drinking water regulations limit manganese to 0.05 ppm due to its ability to cause black stains. Dispersants used to control iron fouling can be used to help control manganese fouling as well. The oxidation reaction to convert Mn^{++} to the insoluble form is as follows:

$$Mn^{2+} + \tfrac{1}{2} O_2 + H_2O \rightarrow MnO_2(solid) + 2H^+ \qquad \text{Eqn (2.11)}$$

Mercury: In 1973, a catastrophic failure of aluminum heat exchangers occurred at the Skikda LNG plant in Algeria. It was determined that mercury corrosion of the aluminum heat exchangers caused the failure. Since 1973, mercury has caused numerous aluminum exchanger failures in gas processing facilities. The mechanism is now relatively well understood. When mercury comes into contact with an aluminum metal surface, the aluminum diffuses into the mercury droplet where it is rapidly converted to Al_2O_3, which has limited mechanical strength. By this mechanism, metallic mercury actually bores a hole into the aluminum. Thus, the presence in and removal of mercury from natural gas is an important subject in producing natural gas [56]. The presence of mercury in natural gas is also a good indicator of the presence of mercury in produced water.

Mercury values for condensate are given in the following table [57]. Typical mercury values for crude oil on the world markets are given by the following:

- 38% of all crude on the world markets contain < 1 ppb

- 90% of all crude on the world markets contain < 10 ppb

- The median mercury content is ~ 1.5 ppb

- The mean mercury content is ~ 7.3 ppb

Price penalties for mercury content are not typically applied unless the mercury content is above 50 ppbw.

Table 3. Regional Average Level of Mercury in Condensate [57]

Location	Concentration (ppbw)
Gulf of Thailand	400 – 1,200
Indonesia	10 – 500
Malaysia	10 – 100
South America	50 – 100
North Africa	20 – 50
Africa	20 – 1,000
Western US	1 – 5

Mercury in Northern Europe is found in the Groingen gas field which produces mostly from the Rotliegend formation in the North East of the Netherlands [58, 55]. In South East Asia merrucry is found in East Natuna and the Arun field in Indonesia, and the Gulf of Thailand [59]. Unocal operations in the Gulf of Thailand have dealt with mercury for decades. They report that the high temperature of the reservoirs has decomposed naturally occurring mercury to the elemental form. Most of the mercury found in the water phase is in this form [59].

Mercury occurs in nature in the following forms: mineral ($HgCl2$, HgS, HgO, Hg_2Cl_2), elemental (Hg^0), +1 ionic (mercurous), +2 ionic (mercuric), and as organomercury compounds. The sampling and analysis of mercury in produced fluid phases is important in order to select the proper treatment technology, and is reasonably well understood [55]. As elemental mercury is very volatile element, dangerously toxic levels are readily attained in air. Organic mercury compounds are also extremely toxic [55]. Small amounts of mercury spillage can be cleaned up by addition of sulfur powder. The resulting mixture should be disposed of as a toxic waste. Elemental mercury reacts with iron oxide and iron sulfide corrosion scale to form a mercury-rich layer on such corroded surfaces. Such reactions are not known to impact corrosion rates and thus should be inconsequential to pipeline integrity. In aqueous solution, mercuric salts will form the mercuric cation. Organic mercury forms also exist and consist of two main groups of compounds: the organometalic compounds designated by the formula R-Hg-X, and the organo mercury compounds designated by R-Hg-R. For both species, R is an organic group, of which methyl (-CH3) is prominent, and X is an mono-valent inorganic anion, such as chloride or other halide. The organometalic species R-Hg-X includes monomethylmercury compounds. The most prominent dialkylmercury compound is dimethylmercury H3CHgCH3 but diethyl-, ethylmethyl-, and phenyl- species are also known to exist.

Fish and shellfish have a natural tendency to concentrate mercury in their bodies. Species of fish that are higher in the food chain, such as swordfish, mackerel, and tuna accumulate higher concentrations of mercury than non-predatory fish. This is because mercury is not metabolized and is instead stored in the muscle tissues of fish. When a predatory fish eats another fish, it consumes and ends up storing the entire mercury contents of the fish. Thus species that are high on the food chain concentrate mercury levels that can be ten times higher, or more, than the species they consume. This process is called biomagnification. The first occurrence of widespread mercury poisoning by biomagnification, was also one of the worst environmental disasters in history. It occurred in Minamata, Japan, when a fertilizer and petrochemical company polluted the Minamata Bay for several years before being linked to mercury poisoning in the region. It is estimated that over 3,000 people suffered various deformities, severe mercury poisoning symptoms or death from what became known as Minamata disease.

Much of the knowledge about elemental and ionic mercury removal from produced water comes from the industrial waste water industry [55]. In a very well organized and written report for the US EPA, Patterson reviews various treatment options [60]. Precipitation, coagulation/precipitation, coagulation/flocculation and modified activated carbon adsorption are the most widely applied technologies in the industrial water treatment industry. Besides the modified activated carbon, these technologies can be classified as chemical treatment followed by separation. They can be applied with no, or only moderate, pre-treatment to remove dispersed oil.

Modified activated carbon, if used, is generally applied as the last step in a process. It is a consumptive media. While it is very effective, it can be costly if used without pre-treatment to remove dispersed contaminants that will consume the capacity of the carbon. Ion exchange can also be used in a similar fashion where it is applied for lower concentrations and for water that is not heavily contaminated with sticky organics or other fouling materials. Electrocoagulation is also used for specialized cases. While effective at deep removal of suspended solids and dissolved multivalent ions, the capital equipment cost is high.

Most heavy metal ions form insoluble sulfides when some form of sulfide is added to the water such as sodium sulfide, sodium hydrosulfide, magnesium sulfide. Sulfide addition is usually applied together with pH adjustment when necessary, flocculation followed by settling, filtration or flotation for solids separation. Such sulfide processes can achieve separation efficiencies up to 99.9 % with feed concentrations as high as 10 mg/L. Minimum sulfide concentrations appear to be in the neutral pH range. Sulfide treatment typically has difficulty achieving mercury concentrations in the range of less than 100 mg/L. For lower concentrations, modified activated carbon can be used. For deep mercury removal, granulated activated carbon (GAC) can be soaked in a carbon disulfide solution (CS2) and dried. The modified GAC can be used by itself with good removal of mercury down to 0.2 mg/L, depending on the application and conditions. However this direct application, without initial mercury removal, can consume large amounts of modified GAC. Therefore the sulfide treatment process (with precipitation and solids removal) is often followed by modified GAC when high (a few mg/L) concentrations of mercury must be removed down to sub-microgram levels. The effectiveness of modified GAC can be further improved by addition of chelating agents such as EDTA (ethylene diamine tetracetic acid), or tannic acid.

CETCO, an oilfield services and technology company, developed a form of activated clay impregnated with sulfur for the removal of elemental and ionic mercury. Since CETCO already possessed activated clay technology (Crudesorb®) for the removal of oil from produced water, the new media would provide both oil and mercury removal in one step. The media are referred to as MercSorb MR1, MR2, and MR4. MR1 is designed to remove organo-mercury and colloidal mercury from produced water. Crudesorb MR2 is designed to remove all four valence states of mercury. Crudesorb MR4 is designed to remove all four states of mercury with emphasis on metallic and ionic mercury.

Nickel (Ni 2+): A divalent cation. Nickel is considered to be a heavy metal. It can be found in ground and surface waters ranging in concentration from 1 ppb to 71 ppb (0.001 ppm to 0.071 ppm).

Nitrate (NO3 -): A monovalent anion. Nitrate salts are highly soluble. Nitrate, along with ammonia gas and ammonium, is a nitrogen-based ion whose presence is tied with nature's nitrogen cycle.

Phosphate: Phosphate occurs in produced water from natural (reservoir) sources and from manmade sources such as scale inhibitors. Orthophosphate (PO_4^{-3}) is a trivalent anion. It is the typical, inorganic form of phosphorus found in wastewaters at concentrations between 3 ppm and 10 ppm. Phosphate can combine with calcium to form a calcium phosphate $Ca_3(PO_4)_2$, scale. Three potential phosphate scales include: 1) tricalcium phosphate, 2) Hydroxylapatite, and 3) Strengite. The most probable source of phosphate scaling is tricalcium phosphate which has a high order of reaction as shown below:

$$3 \, [Ca] + 2 \, [PO_4] \;\rightarrow\; Ca_3(PO_4)^2 \qquad\qquad \text{Eqn (2.12)}$$

$$\text{Saturation level} = \frac{[Ca]^3[PO_4]^2}{Kspc} \qquad\qquad \text{Eqn (2.13)}$$

Tricalcium phosphate saturation level calculations are fifth order. Small changes in free phosphate concentration, or even smaller changes in calcium concentration, significantly affect the calculated saturation level. Most inhibitor treatments can prevent phosphate scale formation even if not specifically directed towards tricalcium phosphate. A reduction in pH can decrease tricalcium phosphate saturation level to an acceptable level. When tricalcium phosphate saturation level rises a reduction in pH and/or the use of tricalcium phosphate specific antiscalants is recommended. Unlike most other compounds, but similar to most carbonate compounds, the solubility of calcium phosphate becomes lower as temperature increases.

Sodium Absorption Ratio (SAR): SAR is a somewhat specialized subject but one that is extremely important in produced water reuse in agricultural applications [61]. SAR is a function of the ratio of sodium to the sum of calcium and magnesium cations as shown in the following:

$$SAR = \frac{[Na+]}{\sqrt{\dfrac{[Ca2+] + [Mg2+]}{2}}}$$

Eqn (2.14)

where the concentration are in meq/l. SAR values of greater than 12 are considered very high. The specific SAR value at which soil damage begins depends on the nature of the soil itself. The optimum conductivity and SAR must be determined on a site-by-site basis. Acceptable SAR values depend on the end use for the produced water; criteria for SAR values are also controlled to a high degree by each state. In many cases, SAR numbers of less than six are beneficial for the use in treatment systems. From the equation it can be seen that reducing the SAR from high values to acceptable levels can be accompanied through processes that either decrease sodium or increase magnesium or calcium.

Sodium Bisulfite (NaHSO3) SBS: A reducing agent used as a preservative or for declorination or deoxygenation. For purposes of de-chlorination, the ratio of sodium bisulfite (SBS) to total chlorine is 1.8 ppm SBS to 1 ppm Cl2. For deoxygenation, 6.5 mg of $NaHSO_3$ reacts with 1.0 mg of dissolved oxygen. The continuous use of SBS for de-chlorination or deoxygenation may increase the rate of biofouling by providing an additional food source for sulfate reducing bacteria. The sulphite in sodium bisulfite exerts an oxygen demand on the water which is being dosed.

Strontium (Sr): A divalent cation. The solubility of strontium sulfate is low and can contribute to the deposition of sulfate scales. Coprecipitation of $SrSO_4$ with $BaSO_4$ is common.

Sulfate (SO4 2-): A divalent anion. The solubility of calcium, barium and strontium sulfate is low. The formation of sulfate scales with one or more of these cations is common. The solubility of these sparingly soluble salts is lower with decreasing temperature. The recommended upper limit for sulfate in potable water is 250 ppm based on taste issues. $SO_4^=$ is a food source for Sulfate Reducing Bacteria (SRBs). SRBs reduce sulfate to sulfide ($S^=$) which results in the formation of H_2S.

2.7.2 Suspended Inorganic Solids:

Solids are an important topic in produced water treatment. A brief introduction is given here. Chapter 18 (Applications – Solids) provides practical guidance and options for dealing with solids. Suspended solids in produced water span the range from the very small, < 1 micron, to relatively large, several hundred microns. Because of this size range and the corresponding variation in the surface properties of these solids, their impact on water treatment, emulsions, oil/water interfaces is complex and requires careful consideration before a water treatment technology is selected for solids separation and handling.

Solids in hydrocarbon and produced water streams can originate from the formation, by precipitation of inorganic scale forming minerals, or precipitation of organic materials. Typical solids in produced water systems include:

- formation fines (calcite, silica, alumina-silicate minerals);

- mineral scales such as carbonate (calcium carbonate, magnesium carbonate, iron carbonate), sulfate (calcium sulfate, barium sulfate, strontium sulfate), or sulfide (iron sulfide);

- corrosion products (hematite and magnetite);

- salt (halite);

- mineral/organic combinations (sodium and calcium naphthenates) [34];

- organic solids from the oil (asphaltenes, waxes);

- organic solids from the production chemicals (water clarifier, deoiling compounds, polymer floc);

As far as oil in water emulsions are concerned, the most important properties of produced water solids are the size distribution, the overall quantity of solids, and the surface wetting characteristics. These topics are discussed in the following sections.

Oil Wet Solids: As discussed in greater detail in Section 4.4.8 (Surface Wetting), the properties of partially oil-wet solids are different from those of oil droplets or solids that are completely oil- or water-wet. The density of a partially oil-wet particle depends on the mass ratio of oil to solids in each particle, which can vary significantly. The combination of a light component (oil), plus a heavy component (solid) will result in a specific gravity decrease for the solid particle. When this occurs, separators and hydrocyclones will be less effective because there is less density difference to drive the relative movement of the particle from the water. Small neutrally buoyant particles can stay suspended for months in a desk top sample.

From a water treating standpoint, the presence of oily solids can be devastating. Once a solid stabilized emulsion forms, it is particularly difficult to separate the components from each other and to separate the oil-wet solid from water, due to the shift in density. The more successful technologies for removing such oil-wet solid include flotation, chemical treatment, and deep bed filtration. Chemical treatment is discussed in Section 5.3 (The Mechanisms of Coagulation, Flocculation and Coalescence), and Section 5.7 (Dispersants).

The fluids in some geographical areas are particularly suitable to forming emulsions which are stabilized by small, partially oil-wetted solids. The Organic component of oil-wet solid tends to have a relatively high asphaltene content, from a few to several percent. Resins are also somewhat high in the oil-wet solid, varying in the range of a few percent to over 15 %. Due to their relatively high aromaticity and associated resin content the asphaltenes tend to be marginally stable. While this means that while asphaltene precipitation may not be a problem in the reservoir, near well bore or tubing locations, asphaltenes can and do nevertheless precipitate and contribute to the stability of both water in oil and oil in water emulsions. This, together with the relatively short residence times found in offshore and other remote facilities can lead to a relatively challenging situation regarding oil and water separation.

Chemical treatment of oil-wet solid has been effective in many cases but chemicals for this purpose must be applied with care. The injection of an acid, for example, is intended to displace the oil from solid particle surfaces and return the solid particles to a water-wet state which then facilitates gravity separation and eventual discharge. However, acidic produced water will almost certainly cause corrosion problems. Also, the presence of bicarbonate in the water increases the amount of acid required to reduce the pH of the produced water – which results in both logistic and cost issues for this strategy. The injection of so-called wetting agents can have a positive effect to separate the oil from the solid surface. But this too must be done in a limited manner otherwise emulsions will be stabilized due to the surfactant nature of the wetting agent.

Flocculating agents are also used with success to treat solids. When a floc forms, the larger diameter of the floc increases the terminal settling velocity of the particle (Stokes Law) and this counter balances the reduction in terminal settling velocity due to a reduction in the solid-liquid density differential.

Flotation can be quite effective in separating neutrally buoyant, solids-stabilized emulsions from produced water, through the attachment of bubbles which then create the density difference needed for separation. The gas bubbles stick to the solids themselves and to the oil-wet surfaces of the solids, carrying them to the surface of the water where they are floated over a spillover weir. This is described in greater detail in reference [62] which discusses problems encountered on a deep water platform offshore Brazil.

2.7.3 NORM (Naturally Occurring Radioactive Material)

The earliest reports of NORM in the oilfield are from the 1930's when elevated levels of radium were reported in Russian oilfields. In the 1980's elevated levels of NORM-containing scale were reported on production platforms in the North Sea. In 1986, high levels of NORM were reported in scales from some oilfields in Mississippi and the Gulf of Mexico. In recent years, one of the challenges in the Marcellus shale has been high concentrations of NORM in the produced water.

In the oil and gas industry, the term, Naturally Occurring Radioactive Material (NORM) is strongly associated with Radium isotopes. But this is not strictly correct. The term NORM is also used extensively outside of the oil and gas industry and refers to all naturally occurring radioactive materials where human activities have increased the potential for exposure compared with the unaltered situation. This does not necessarily mean that the concentration has been enhanced. It does mean that through mining, oil and gas production, or other activities, the radioactive material has been transported, conveyed, or otherwise changed in a way that increases the potential for human exposure. Examples of activity that generates NORM include mining and burning coal, making and using fertilizers, and of course oil and gas production. Thus, NORM potentially includes all radioactive elements found in the environment such as the long-lived radioactive elements such as uranium, thorium, potassium and any of their decay products, as well as radium (Group II element) and radon (inert gas element). All of these elements are present in the Earth's crust, at greater or lesser concentrations depending on location.

The term NORM is not intended to include radioactive material has been deliberately concentrated for subsequent use. This is a different class of material altogether. If the concentration of the radionuclides has been increased (enhanced through purification or other means), then the term Technologically-Enhanced NORM (TENORM) is used. Examples of this activity would be in the nuclear power industry or the manufacture of atomic and hydrogen bombs. It is easy to understand why a separate classification is necessary for this material.

In the oil and gas industry, the radium isotopes are of greatest importance. The most common radium isotopes, Ra-224 and Ra-226, decompose with half-lives of 3.6 days and 1,600 years respectively. These nuclides do not emit penetrating radiation. Instead, they emit relatively weak alpha particles to form isotopes of the inert gas element radon, Rn-222 and Rn-220, which are themselves radioactive also emitting alpha particles. Both radium and radon isotopes do not cause a hazard as long as they remain on the inside surfaces of the equipment. Radon is itself a gas even at the low temperature processing conditions of gas plants. Radon boils at 62°C. Thus, it concentrates in the ethane and propane section of a fractionator where the decomposition products can form scales and accumulate over time. When equipment is dismantled, after prolonged exposure to Radon containing gas streams, or Radium containing produced water streams, personnel exposure to these radioactive materials becomes a potential hazard. This is typical of NORM experience in the oilfield [63] with for example, the handling of hydraulic fracture flow back water where NORM containing scales are found in piping and holding tanks.

Radium is a chemical element with atomic number 88. Only about four Radium isotopes are found naturally in any significant concentration. Table 2.12 gives the four naturally occurring radium isotopes. It is a member of the alkaline earth metals (group II) which include magnesium, calcium, strontium, and barium. Like the other alkaline earth metals, radium ionizes in water to a +2 oxidation state (Ra^{+2}). It is not easily complexed. It typically forms ionic salts such as $RaCl_2$ and $RaCO_3$, and RaSO4.

The high temperatures and pressures of oil and gas reservoirs can result in radium (as well as Thorium-232 and Uranium-238) dissolving in the reservoir fluids. During oil and gas processing, as the temperature is lowered, these elements can deposit on the inside of piping and processing equipment. Radium typically deposits as carbonate, but can also deposit as chloride and bromide. It is this situation that causes the most problems in the oilfield since the piping containing this material must be disposed of according to the regulations of the region.

Radium has been studied due to its hazard to human health upon long term or high dosage exposure. Of the naturally occurring isotopes, Ra-226 is the most radiotoxic for several reasons:

- it has a long half-life

- it has a high relative abundance due to the abundance of its parent U-238

- it emits a high flux of alpha particles

- it has similar chemical properties to calcium which means that precipitation of calcium carbonate or calcium sulfate is sometimes associated with precipitation of Ra-226

- ingestion of radium may lead to its assimilation in the body in calcium related structures such as bones and teeth

Table 2.12 Details and dose conversion factors [Sv/Bq] of the four naturally isotopes of radium

	228_{Ra}	226_{Ra}	224_{Ra}	223_{Ra}
Half life (t1/2)(ICRP, 1983)	5.75 years	1620 years	3.66 days	11.44 days
ICRP dose conversion factor (Adult)*	6.9×10^{-7}	2.8×10^{-7}	6.5×10^{-8}	1.0×10^{-7}
Parent of decay chain	232_{Th}	238_{U}	232_{Th}	235_{U}
ICRP Dose conversion factor for parent of decay chain	2.3×10^{-7}	4.5×10^{-8}	2.3×10-7	4.7×10^{-8}
* Dose conversion factors (DCF) convert actual activity of a radionuclide someone has ingested into an effective committed radiation dose (in Sv) that will be received from the given exposure over their lifetime. The DCF's given are for adults, taken from ICRP Publication 72.				

While radium is the typical NORM component, other radioactive compounds can be present. To understand the difference between these radio nucleotides, it is helpful to discuss the different decay particles. Radioactive decay of Uranium-238 takes place by emission of alpha, beta and gamma particles. The alpha particles are not energetic enough to penetrate the human skin and so do very little harm. The beta particles are just able to penetrate human skin but will typically be absorbed by the steel of piping and separator vessels. It is only upon ingestion or inhalation of material that emits alpha or beta radiation that these radio nucleotide particles cause significant harm. But conditions which would lead to such inhalation would have other serious hazards as well. The gamma radiation will penetrate steel and human skin and is therefore hazardous.

The term radioactivity indicates the number of radioactive atoms that disintegrate in a time period and is measured in units of curies. One curie is defined as that amount of any radioactive material that decays at a rate of 37 billion disintegrations per second. This is equal to the amount of radiation emitted by 1 gram of Ra-226. Thus, 5 pCi/L is equivalent to 5 pico-grams of Ra-226 in one liter of waste water.

The contaminant levels allowed in effluent discharge are most often expressed in terms of radiation level or radioactivity. For example, the USEPA Maximum Contaminant Level (MCL) of Ra-226 in drinking water is 5 pCi/L. This is also the limit adopted by many regional EPA offices as a discharge limit. In that case, 5 pCi would be allowed per 1 liter of effluent.

Radioactivity limits for municipal landfills are set by states, and range from 5 to 50 pCi/g. In Pennsylvania, the maximum allowable radiation level (from Ra-226) for disposal of solids as nonhazardous waste is 25 pCi/gr (per gram of waste). This is equivalent to 25 pico grams of Ra-226 per gram of solid waste, a very small amount. In fact, some concentrated hydraulic flow back fluids in the Marcellus shale contain over 10,000 pCi/L of radioactivity [64].

The EPA administers the Underground Injection Control (UIC) program. The details of this program are given in Section 1.6.4 (Regulations, Disposal and Reuse Options). As far as NORM is concerned, Class II wells are allowed to be utilized to dispose of hydrocarbons, well completion fluids, brine, and fluids contaminated with NORM. As discussed by Basu [65], for offshore production, overboard discharge of NORM containing fluids is covered under NPDES (National Pollutant Discharge Elimination System) permitting. On shore, regulations vary from state to state.

Table 2.13 below gives the measured composition for flow back water from several wells in the Marcellus shale, as well as one well from the Barnet shale [66]. Upon careful inspection of the data, it is noted that the concentration of Ra-226 is not given. Instead, a radioactivity value is given in the row labeled Ra-226. It is not clear if all of the reported radioactivity is from Ra-226. Typically Ra-226 is only one of several radio nucleotides in a sample. In any case, the reported values are significantly above landfill limits.

Table 2.13 Produced Water Compositions from Pennsylvania Marcellus Shale Gas Wells.

	Well 1	Well 2	Well 3	Well 4	Well 5	Well 6	Well 7	Design Case
County	Bradford	Bradford	Bradford	Butler	Tioga	Washington		
pH[b]	7.3	6.3	5.4	5.8	5.9	6.2	6.6	7.0
TDS	98,294.0	155,705.0	199,242.0	68,439.0	149,188.0	122,562.0	124,421.0	132,460.0
Na^+	26,500.0	38,200.0	51,800.0	19,200.0	39,000.0	32,300.0	33,900.0	35,000.0
Mg^{++}	460.0	840.0	1,290.0	570.0	1,000.0	800.0	1,170.0	800.0
Ca^{++}	5,560.0	10,280.0	13,120.0	5,360.0	13,000.0	8,700.0	10,880.0	9,500.0

County	Well 1	Well 2	Well 3	Well 4	Well 5	Well 6	Well 7	Design Case
	Brad-ford	Brad-ford	Brad-ford	Butler	Tioga	Wash-ington		
Sr^{++}	2,030.0	3,670.0	4,580.0	1,290.0	2,600.0	2,340.0	1,750.0	2,500.0
Ba^{++}	6,580.0	13,200.0	11,600.0	32.0	3,500.0	5,800.0	147.0	6,200.0
Fe^{++}	26.0	74.0	123.0	55.0	32.0	75.0	47.0	50.0
Mn^{++}	1.5	2.5	3.4	1.7	2.7	4.3	1.2	3.0
Cl^-	57,120.0	89,429.0	116,713.0	41,845.0	90,014.0	72,525.0	76,493.0	78,407.0
SO_4^-	< 10	< 10	< 10	57.0	< 5	< 50	< 100	0.0
SiO_2	16.7	11.0	13.0	29.0	39.0	18.0	33.0	0.0
Hard-ness as $Ca++$	9,167.0	17,196.0	20,727.0	6,899.0	16,860.0	12,782.0	13,653.0	13,772.0
$^{226}Ra^c$	5,400.0	7,600.0	4,200.0	4,600.0	5,600.0	820.0	2,300.0	5,000.0
TSS	202.0	282.0	500.0	62.0	520.0	210.0	898.0	0.0
Turbid-ity[d]	78.0	399.0	1,160.0	17.4	192.0	45.0	164.0	0.0
TOC	< 10	11.8	11.8	72.0	151.0	160.0	88.0	0.0
PW Type[f]	III	III	III	I	III	III	I	III

(all quantities mg/L except where noted)
a - Produced water from Barnett Shale (TX
b - Dimensionless
c -pCi/Liter
d - Turbidity units: NTU
e - Adjusted to force ion balance (prior analyses found other anions were < 1% of the chloride on a molar basis)
f - Produced Water Type

See Section 13.8 (Ion Exchange) for a discussion of how to remove radium from produced water.

2.7.4 Iron Compounds

In produced water iron is found in one of two forms, the water-soluble ferrous iron has a + 2 valence state. The water-insoluble ferric iron has a + 3 valence state.

Fe2+ ferrous (water soluble): In non-aerated well waters ferrous iron behaves much like calcium or magnesium hardness in that it can be removed by softeners or precipitation. Precipitation can be prevented by the use of a chelating chemical. If the produced water is maintained in the reduced form, the iron is quite soluble and should not be a problem except if CO_2 is present so that $FeCO_3$ precipitation is a possibility. The operators should review their water chemistry for potential iron scale that may form, such as ferrous carbonate (siderite) and ferrous sulfate. Scaling potential will depend on the concentration of iron, bicarbonate, sulfate, ionic strength, temperature, and pH. Soluble iron can be removed by iron filters, softeners or lime softening.

The introduction of air into produced water with soluble ferrous iron can occur by placing the water in tanks without gas blanketing, through leaky pump seals, mixing with air-saturated surface water, or other means. This will result in the oxidation of the soluble ferrous iron to insoluble ferric iron and precipitation of the ferric iron as ferric hydroxide or, in some cases, iron sulfide.

Fe^{+3} ferric (water insoluble): Insoluble ferric iron oxides or ferric hydroxides, being colloidal in nature, will precipitate and foul equipment. Sources of insoluble iron are aerated well waters, surface sources, and iron scale from unlined pipe and tanks. Insoluble iron can be removed by iron filters, lime softening, softeners (with limits), ultrafiltration (with limits) and multimedia filtration with polyelectrolyte feed (with limits). Precautions are required with the use of potassium permanganate in manganese greensand iron filters in that potassium permanganate is an oxidant that could damage a downstream RO or NF polyamide membrane. Precautions are also required with a cationic poly-electrolyte (e.g., Polyaluminum chloride) in that they also can irreversibly foul a negatively charged polyamide membrane. The reaction to the insoluble form is as follows:

$$Fe^{2+} + \tfrac{1}{4}\,O_2 + \tfrac{5}{2}\,H_2O \;\rightarrow\; Fe(OH)_3(solid) + 2H^+ \qquad\qquad \text{Eqn (2.15)}$$

This oxidation of iron from the soluble to the insoluble followed by sedimentation and filtration can be used to remove the iron. The iron can be oxidized through aeration as shown in the formula above, or by oxidizing agents such as $KMnO_4$, Cl_2, O_3, or ClO_2.

As discussed below, iron can also form complexes with organic material which may also promote precipitation and the formation of floccs which may be sticky. It should be noted that in specific circumstances, ferric iron flocc is formed intentionally in order to clean water in an induced gas flotation vessel. In some cases, the presence of iron can create a bio-fouling problem by being the energy source for iron-reducing bacteria. Iron-reducing bacteria, or any bacteria for that matter, can cause the formation of a slimy biofilm that can cause fouling problems and can create suspended solids that are less dense than the iron particles themselves, and so making it harder to remove iron particulate by settling.

Some water treatment systems operate with ferric chloride used as a coagulant. The ferric ions com-plex with the colloidal or organic material in the feed and grow into floc which can be removed by fit-for-purpose settling tanks or induced gas flotation. Problems with the introduction of ferric chloride occur when excess iron is not removed by the appropriately designed treatment equipment.

Iron Sulfide: So-called iron sulfide is more appropriately considered as a class of compounds rather than a single compound, some of which are amorphous and some of which are crystalline. In nature and in produced water it must be kept in mind that the most thermodynamically stable compound (e.g., crystalline pyrite, FeS_2) is not always the compound that will be formed first. For some iron sulfide compounds the rate of transition from one compound or form to the more stable form can take weeks if not months, depending on the conditions.

In many instances, the analysis of produced water does not differentiate between one form of iron sulfide and another. If iron sulfide is reported, it is usually reported as 'iron sulfide' or as FeSx. In most cases, this is sufficient since the origin of iron sulfide is predominantly from biological processes, reservoir souring, pipeline contamination, etc.. However, there are some cases where iron sulfide originates from other sources and in those cases, it is helpful to determine which form of iron sulfide is present since this can give further insight as to origin and how to remediate it. Therefore it is helpful to review the chemistry of the different forms of iron sulfide.

It is also helpful to review the surface chemistry of iron sulfide since adsorption of organic com-pounds such as acids and bases leads to the formation of agglomerates that contribute to organic scale buildup and fouling of process equipment. Iron sulfide is one of the most problematic solids in produced water treating. Iron sulfide, like iron oxide is very insoluble. As is typical of the E&P

industry, problematic systems are often given descriptive if not colorful names. In the case of solids formed from iron, the designated name is "Schmoo." In gas fields, Schmoo causes severe operating problems in compressors, pipelines, stabilizer columns, gathering systems, and storage facilities. According to the Gas Machinery Research Council, Schmoo is the least understood and most prominent contamination problem in pipelines and gas compression equipment.

In the material below both the chemistry of formation and the surface properties of iron sulfide is reviewed. At least seven different solids consisting of only iron and sulfur are known to occur naturally at moderate temperatures of the earth and marine environment. The most common of these are listed in Figure 2.18 and in Table 2.14 below.

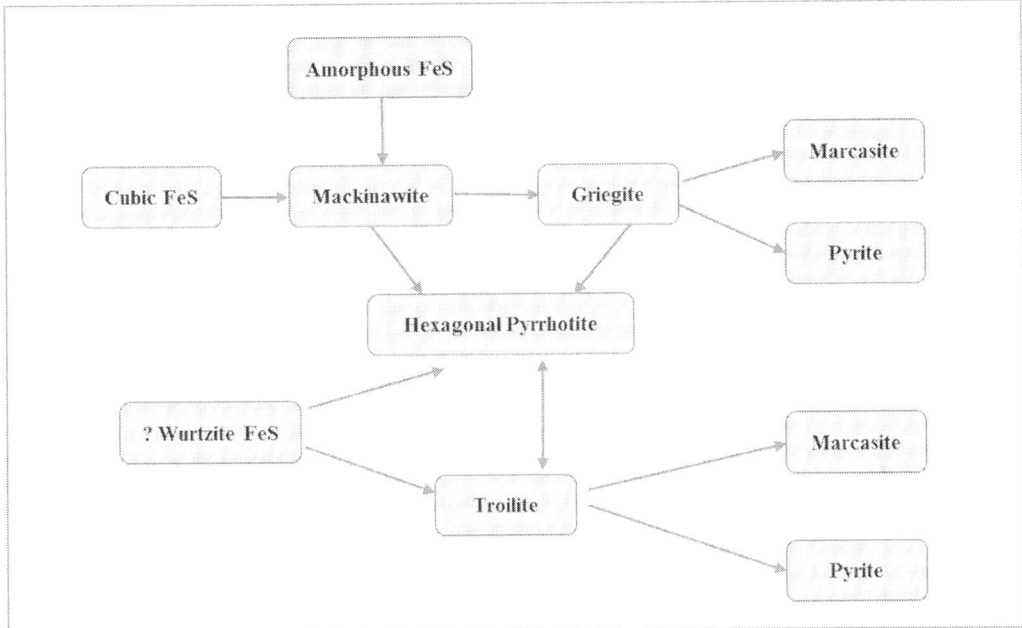

Figure 2.18 Diagram showing the phase relations between different iron sulfides at moderate temperature in the FeS system. Solid arrows indicate transformations that have been experimentally verified. Reproduced from Devey [67].

A variety of other iron sulfide phases have been suggested in addition to those listed above, most commonly by analogy with isomorphic Fe-O materials [68].

Pyrite (cubic FeS_2) has a high thermodynamic stability. It is the common end product in the geological transformation of many of the other less stable iron sulfides [69]. However in many cases the transformation from one form of iron sulfide to pyrite either does not occur rapidly or requires specific conditions to advance at all. Below 100 C, mackinawite is the direct precursor of pyrite. However, an oxidant is required for the transformation to occur [70]. Pyrite can also be formed in the absence of oxygen from greigite in aqueous solutions around pH 6 and temperatures above 60°C [71]. In general, pyrite is the most abundant sulphide mineral found in marine sediments. However, pyrite does not nucleate significantly from solutions at temperatures typical of sedimentary environments [72].

Table 2.14 Solids phases of iron sulfides [68]

Material	Composition	Structure	Properties	Natural abundance
Mackinawite	FeS_m	tetragonal $P4/nmm$	metastable material that is the major constituent of the FeS precipitated from aqueous solutions	Widespread mineral in low-temperature aqueous environments
Cubic FeS	FeS_c	cubic $F43m$	Highly unstable phase formed before FeS_m	Not found naturally
Troilite	FeS_t	hexagonal $P62c$	Stoichiometic end member of the $Fe_{1-x}S$ group	Mainly found in meteorites
Pyrrhotite	$Fe_{1-x}S$	Monoclinic, for example, $A2/a$; hexagonal $P6/mmc$	Nonstoichiometric stable group where x > 0.2; monoclinic form is approx. Fe_7S_8; hexagonal form is approx. $Fe_{10}S_{11}$	Most abundant iron sulfides in the Earth and solar system; rare in marine systems
Smythite	Fe_9S_{11s}	hexagonal $R3m$	Metastable phase related to the $Fe_{1-x}S$ group	Rare mineral mainly found in hydrothermal systems usually associated with carbonates
Greigite	Fe_3S_{4g}	cubic $Fd3m$	metastable $Fe^{II}Fe^{III}$ sulfide; the thiospinel of iron	Fairly widespread mineral particularly associated with fresh water systems
Pyrite	FeS_{2p}	cubic $Pa3$	stable iron (II) disulfide known as 'fool's gold'	The most abundant mineral on the Earth's surface
Marcasite	FeS_{2m}	Orthorhombic $Pnnm$	metastable iron (II) disulfide	Locally common mineral in hydrothermal systems and in sedimentary rocks

Although it is somewhat rare, it is possible that iron sulfide in produced water has its origin in the reservoir rock itself. In this case, particles of iron sulfide, including pyrite, would be produced just as particles of sand, carbonate, and clay are sometimes produced as a result of the hydrodynamic forces in the near wellbore. Mackinawite and Greigite are the main solid Fe-S phases that may be encountered from a geological origin in produced water. Consultation with a geologist who is familiar with the mineralogy of the producing reservoir is recommended.

A more common origin for iron sulfide is from reaction of hydrogen sulfide with iron in a pipeline. The initial precipitate formed from iron and sulfide ion in an anaerobic solution is commonly referred to as nanocrystalline mackinawite (FeS) or amorphous iron sulfide (FeS_{am}). A precipitate forms very rapidly (within 1s or even 10 ms) after mixing Fe^{2+} and S^{2-} solutions [68].

As reviewed by Rickard and Luther [68], Mackinawite is very reactive with oxygen, which makes it a difficult compound to study [73]. When synthetically produced under lab conditions, nanocrystalline mackinawite is only produced under strictly anaerobic conditions. If any oxygen contamination occurs, amorphous iron sulfide (FeS_{am}) is produced [74]. Over the course of about two months, in aqueous aging studies at room temperature, amorphous iron sulfide transformed into a mixture of Mackinawite and Greigite [74]. The complete transformation to well-crystallized Mackinawite requires up to two years in aqueous solutions at 25°C [75]. Based on these findings, it is likely that the Fe-S solids in produced water consists mostly of amorphous iron sulfide, as mentioned. A comparison of the XRD detected mineralogy of a solids sample with an EDX/XRF analysis of the same solid may be helpful in determining the presence of amorphous iron sulfides. But care is required in interpreting the data since the oxidation of reactive iron sulfides as described above can complicate the interpretation of the data.

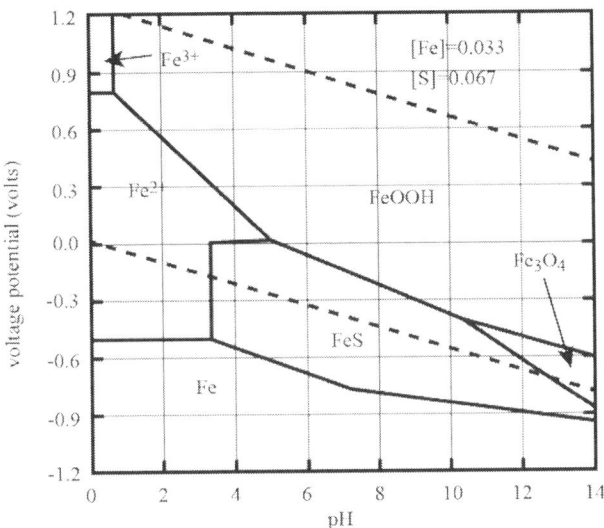

Figure 2.19 Pourbaix diagram of the iron, sulfur, and oxygen system.

The stability of various iron sulfide compounds depends not only on temperature and pH, but also on the concentration of oxygen (or the application of an electrical potential as in Electrocoagulation). Thus, the phase diagram should include oxidation potential as is the case in a Pourbaix diagram. Pourbaix Diagrams are explained in some detail in Section 14.2.1 (Pourbaix Diagram). The Pourbaix diagram for iron-sulfide-water is shown in Figure 2.19 [76]. As shown, in a non-oxidizing environment (-300 mV < Eh) FeS is stable over a wide pH range. Below a pH value of about 3.5, and above an oxidation potential of -500 mV, the more stable form is ferrous iron (Fe^{+2}) and sulfide (S^{-2}). One way to dissolve iron sulfide is to lower the pH. This is true in the field as well as in the lab. A common lab test for the presence of iron sulfide is to first filter out any solid particulates, then add a few drops of acid to the solid particles. If iron sulfide is present, it will dissolve in the acid and form HS^{-1} and H_2S. Some of the H_2S will vaporize giving off an odor of rotten eggs. Another test to determine if iron sulfides are present is to gently heat them in a glass beaker which is covered with a watch glass. Iron sulfides will decompose and emit elemental sulfur upon being heated and the elemental sulfur vapor will collect on the underside of the watch glass as a whitish powder.

So-called pyrophoric iron sulfide must be handled with care since it can initiate the spontaneous combustion of hydrocarbons, or other combustible material.

One comment that must be included is to point out that a typical Pourbaix diagram, like the one shown, only describes a relatively pure system where there are no other chemical or ionic species, except for the ones shown in the diagram. This is unrealistic when dealing with produced water which has a number of contaminants (such as salts) at relatively high concentration. Nevertheless, a Pourbaix diagram does give a rough idea of which species will be stable.

There is a relative lack of data characterizing the surface and morphological characteristics of amorphous iron sulfide [77, 78]. Since sulfide formation in natural environments is usually associated with bacterial sulfate reduction, Herbert et al. [79] studied its precipitation in the presence of sulfate-reducing bacteria (SRB). They described the precipitate as disordered mackinawite. The particles showed a complex structure, with "rosette" shaped aggregates of 1 to 2 μm in diameter. The surface composition seemed to be close to greigite (i.e. $Fe^{(3+)}_2Fe^{(2+)}S_4^{(2-)}$). However, no actual greigite was identified by XRD.

Aggregation of amorphous FeS in solution was also noted by Cornwell and Morse [77], who found that higher ionic strength solutions enhance the aggregation of these phases. This is consistent with the electrical double layer model which is discussed in Section 4.3.2 (Electrical Double Layer and Zeta Potential).

In addition to iron sulfide, several other metal-sulfide compounds can be formed. For background information, the most typically found metal-sulfide compounds are:

Galena:	PbS
Sphalerite:	ZnS with substitutions of Fe
Chalcopyrite:	$CuFeS_2$
Covellite:	CuS
Arsenopyrite:	FeAsS

Schmoo: There have been many reports on the composition of Schmoo, and its precise definition is the source of ongoing debate. The composition varies considerably depending on water composition and the use of various production chemicals such as corrosion inhibitors. For our purposes, we classify Schmoo as an oily solid oil-wet solid formed fundamentally from iron solids such as iron sulfide and to lesser extent iron oxides.

Iron sulfide is a major constituent in Schmoo and it is the compound most responsible for its formation. Schmoo is a malodorous, tar-like black substance which may cause problems like coating of produced water piping [80]. Schmoo is formed in liquid systems where there is crude oil with some asphaltene content. In dry systems, mainly composed of gas, iron sulfide may precipitate but it is unlikely that Schmoo will form. In system with high GOR, iron sulfide may be present as "oil-coated dust." This represents an intermediate composition between that of dry iron sulfide particles, and tar-like Schmoo globs. There is more discussion of this subject in Section 4.4.8 (Surface Wetting) which focuses on wetting of solids and formation of large masses of contaminant particles.

Unique Properties of Iron Solids: It is generally true that when a solution changes conditions rapidly, such that a relatively insoluble species reaches supersaturation rapidly, precipitation of the species will form many small particles rather than fewer large particles. This is particularly true of the iron compounds. Solids formed from iron tend to have size distributions that average less than a micron in diameter. It is typical in cases such as this that over time, the smallest particles redissolve and other particles, acting as seed crystals, will grow larger. The evolution of particle size distributions is just one of several phenomena that contribute to the changing character of produced water samples over the first several hours or days of their acquisition.

When iron solids are a suspected problem, it is particularly important to practice good sampling technique. If the produced fluids have relatively high CO_2 content (> 1 mole %), then loss of CO_2 upon sampling must be avoided. Loss of CO_2 will increase the pH and cause an initially clear sample to turn black due to precipitation of iron sulfide at the higher pH. In that case, iron sulfide may be stable at the lower pH of the in-situ fluid. Further, if oxygen is allowed to enter the sample then precipitates of iron oxides will form, again where they may not have been present in the process fluids under process conditions. Release of CO2 or intrusion of oxygen will also shift the dissolved CO2/bicarbonate equilibria which will affect subsequent analysis of carbonate stability.

In addition, solids formed from iron are easily oil wet, thus making them partially hydrophobic and partially hydrophilic. This property causes them to act like a surfactant and accumulate preferentially at the oil/water interface. Further, various surface active compounds (acids, asphaltenes, corrosion inhibitors) also tend to bind to the surface of iron solids thus making them attract oil to an even greater degree. These properties of iron solids have been exploited in the cleanup of produced water systems through the use of dithiocarbamate (DTC) chemicals which are described in Section 5.5.4 (Chemistry of Carbamates and Chelating Agents). However, from what has been said thus far regarding iron solids, one can imagine that the use of dithiocarbamates must be undertaken with care. Given their small size, large numbers, and wetting properties, solids formed from iron are extremely effective at stabilizing oil in water emulsions. DTC stabilized emulsions tend to be viscous and sticky. Special consideration is required for the design and configuration of water treatment equipment when this class of chemicals is expected to be utilized in a water treatment system.

2.7.5 Silica and Silicon Compounds

Compounds in which the element silicon occurs can be present in produced water in several different forms. For most produced water handling and treatment operations, the presence of silicon compounds is of minor consequence. However, in steam flood applications, the presence of silicon compounds can have significant consequences. In this section, the chemistry of silicon compounds is discussed.

Different types of silicon compounds: Generally speaking, there are four broad categories of silicon (Si) containing compounds of interest. They are:

- crystalline silica (SiO_2), which has several forms (polymorphs) of which α-quartz is the most common;

- the silicates, which are ionic salt compounds such as sodium alumino silicate, and sodium iron silicate;

- reactive silica (SiO_2), which is the dissolved form of silicon dioxide;

- amorphous silica (SiO_2). There are many forms of amorphous silica of which the most familiar is glass.

Silica: Silica (silicon dioxide – SiO_2) is a complex chemical species. It has many forms which differ from each other in the way the silica molecules are arranged, i.e. the way they are packed together. The most insoluble form of silica is quartz, which has a crystalline, and therefore dense, structure. Quartz itself has a wide variety of forms but for our purposes, we are only concerned with generic quartz (α-quartz).

The colloidal form of silica is also present in many forms. It is present in a polymerized form with multiple units of $SiO2$. Or it is present in complexes with organic compounds or inorganic compounds such as those containing aluminum or calcium oxides.

The form which is predominant depends not only on the thermodynamic conditions but also on the previous conditions of the water. In other words, silica often occurs in a metastable or non-equilibrium state. It occurs as an anion, as non-ionic molecules, and as suspended colloidal particles.

A silica analysis of water will often just report the total elemental silica (Si) in a water sample. This is somewhat misleading since such an analysis will give no indication of which form the Si is in. The "Total Silica" content of a water sample is composed of "Reactive Silica" and "Unreactive Silica".

Reactive Silica: Reactive silica (e.g. silicates SiO_4) is dissolved silica that is slightly ionized and has not been polymerized into a long chain. Reactive silica, though it has anionic characteristics, is not counted as an anion in terms of balancing a water analysis but it is counted as a part of total dissolved solids (TDS). Reactive silica solubility increases with increasing temperature, increases at a pH less than 7.0 or more than 7.8, and decreases in the presence of iron which acts as a catalyst in the polymerization of silica. Dissolved, or reactive, silica has several different forms. The following reactions are representative of the chemical equilibria which occur:

$$SiO_2 + H_2O \rightleftharpoons H_2SiO_3 \qquad \text{Eqn (2.16)}$$

$$H_2SiO_3 + H_2O \rightleftharpoons H_4SiO_4 \qquad \text{Eqn (2.17)}$$

All of these species are dissolved. The silica on the right hand side of both equations is referred to as meta silicic acids. These acids are very weak and deprotonate according to the following equilibria at relatively high pH [81]:

$$H_2SiO_3 \rightleftharpoons H^{+1} + HSiO_3^{-1} \qquad \text{Eqn (2.18)}$$

$$HSiO_3^{-1} \rightleftharpoons H^{+1} + SiO_3^{-2} \qquad \text{Eqn (2.19)}$$

$$H_4SiO_4 \rightleftharpoons H^{+1} + H_3SiO_4^{-1} \qquad \text{Eqn (2.20)}$$

These reactions are shifted far to the left over a wide pH range. For example, at pH of 8.5, only 10 mole % of the dissolved silicon present is in the ionized form. The rest is protonated. The pKa of the final reaction is between 9 and 10 and depends on the concentration of these metasilicic acids. In general, pK1 is equal to the pH only when the concentration of the conjugate acid and the concentration of the conjugate base are equal. This will take place at the inflection point for the titration curve for a weak acid (Henderson-Hasselbalch equation).

The above reaction plays a role in alkalinity measurements since it is pH dependent. For example, at pH 9.7, for every 100 milliequivalents (meq) of silicate, 58 meq of alkalinity are contributed to the total alkalinity. Thus, for high pH water supplies that contain appreciable amounts of silica, this equation must be included in the alkalinity calculations in order to create a properly balanced water analysis.

Unreactive or Colloidal Silica: Unreactive silica is polymerized or amorphous colloidal silica. Colloidal silica, with sizes as small as 0.008 micron can be measured empirically by the SDI (Silt Density Index) test. Particulate silica compounds (e.g. clays, silts and sand) are usually 1 micron or larger. Polymerized silica, which is composed of silicon dioxide as the building block, exists in nature (e.g. quartzes and agates). Silica, in the polymerized form, also results from exceeding the reactive silica saturation level.

Silica, a very weak anion, is not used to calculate the ionic balance of cations and anions (though it is used in the calculation of TDS). It is generally thought that silica (SiO_2) is the final and most stable form of silico-oxygen acid polymerization. First, monosilicic acid (H_4SiO_4) polymerizes by dehydration to form Si-O-Si anhydride bonds:

$$n\ Si(OH)_4 \rightarrow (OH)_3Si\text{-}O\text{-}Si(OH)\ \text{dimer}$$

Then, oligomers (a finite number of subunits linked) are formed. Finally, colloidal polymers and particulate silica (crystalline) will be formed. Amorphous silica represented by the formula SiO_2 is more accurately represented by the formula of $(SiO2)n$ where n is large and generally indeterminate number.

In crystalline silica, the tetrahedral structure is continued over a large range, forming a well-ordered lattice (crystal). In amorphous silica, this long range order is absent because at random intervals some atoms are missing a neighbor to which it would be able to bind.

Effect of iron and aluminum: Silica scaling depends strongly on the presence of aluminum or iron. It has been reported that when $Al3+$ and $Fe3+$ coexist in the pretreated feed water, silica is precipitated below its saturation concentration. To prevent this kind of precipitation, both $Al3+$ and $Fe3+$ must be less than 0.05 mg/L in the feed water. Since $Al3+$ and $Fe3+$ salts are used for coagulation in municipal and other industrial water processing, frequent and accurate measurements of these ions are needed even though the feed water itself does not contain high levels of aluminum and iron ions. If colloidal silica and silicates are present in produced water destined for steam generation, then a flocculation/filtration process and/or a fine grade prefilter (1 μm or less) may be required to remove them.

Silica Solubility: The solubility of both amorphous silica and quartz increases as temperature increases up to a high temperature (350 C), as shown in Figure 2.20 [82].

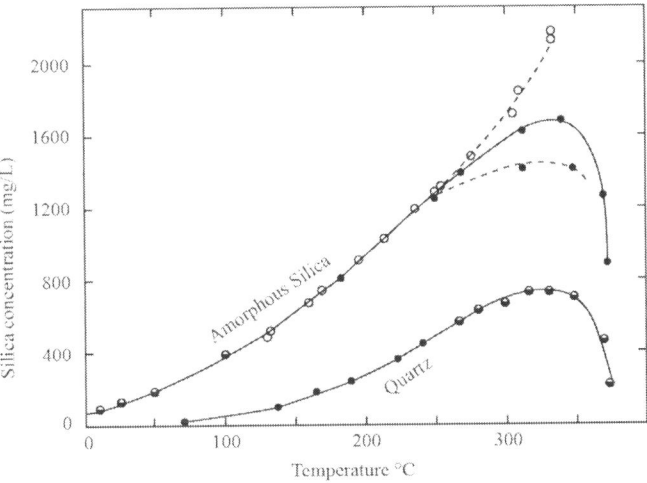

Figure 2.20 The solubility of amorphous and crystalline silica as a function of temperature [82].

Steam Flood: The solubility of silica also depends on the pH of the solution as shown in Figure 2.21 [83]. It is worthwhile to mention at this point how the properties of these compounds affect steam generation in a steam flood. For example, the concentration of silica in the produced water from a steam flood operation can be estimated from a knowledge of the bottom hole temperature and the above solubility curves for quartz. In other words, quartz solubility generally determines the silica concentration downhole, in equilibrium with the reservoir minerals [84]. As the oil/water mixture ascends the well bore, the solubility of amorphous silica determines whether silica precipitates or not. Quartz formation is not rapid enough to contribute to scale formation. Since amorphous silica is more soluble than quartz, it is rare to experience silica precipitation in the well bore. In the steam generator, the amorphous silica solubility curve is the controlling variable and is aided by the higher water pH that generally results from softening, and the fact that the solubility increases with higher temperature. It should be noted that silica scales in OTSGs (Once Through Steam Generators) are often relatively soft and easily removed by pigging the tubes. However, if the water also contains Mg^{++} or Fe^{++}, then a harder scale can form which is more difficult to remove from steam tube walls.

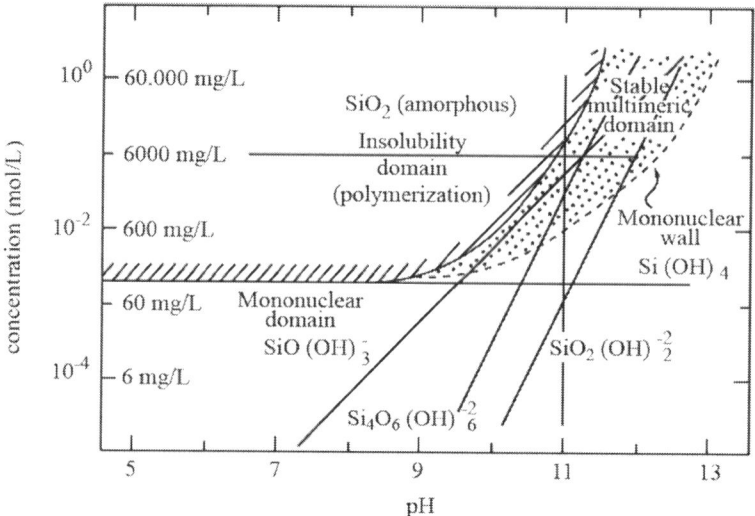

Figure 2.21 Silica solubility as a function of pH. Note the many different forms of silica [83].

The most common scales in these operations are the silicates, such as sodium iron silicates or magnesium and calcium silicates. These compounds form rapidly once their solubility limit has been exceeded.

Operating Limits for Silica: Generally speaking, the industrial cooling water industry uses 75 mg/L silica as the upper limit to prevent scaling. For high pressure steam generators that will feed turbines, the upper limit of silica recommended varies with the operating pressure/temperature of the boiler. For a 300 PSIG boiler, the total silica in the water should not exceed 125 mg/liter. For a 900 PSIG boiler, the total silica in the water should not exceed 20 mg/liter. For OTSG operation, a generally accepted total silica limit of 100 mg/liter is preferred, but many operators utilize water with 200 mg/liter or more total silica and accept the cost of steam generator down time for regular cleaning.

Table 2.15 shows representative steam generator feed water composition for several Canadian and California (US) operators [85].

Parameter (mg/L unless otherwise specified)	Cold Lake, Alberta	Ft. Kent, Alberta	Marguerite Lake, Alberta	Peace River, Alberta	Kern River, California
Ca	48	290	100	1800	-
Mg	5	40	30	20	-
Na	1743	4500	2500	70	215
Total Hardness ($CaCO_3$)	141	890	374	4582	132
CO_3	15	N.D	N.D	N.D	15
HCO_3	304	410	275	1700	292
SO_4	111	525	175	50	53
Cl	3094	7253	3850	1900	199
S	40	0.01	-	-	N.D
NO_3	-	0.34	-	-	N.D

Parameter (mg/L unless otherwise specified)	Cold Lake, Alberta	Ft. Kent, Alberta	Marguerite Lake, Alberta	Peace River, Alberta	Kern River, California
B	10	-	-	-	1
Ba	-	-	-	-	-
SiO$_2$	250	236	-	200	125
Fe	0.8	0.1	-	-	0.1
TDS	6015	13300	7300	5700	844
pH (units)	8	8.1	7	8.5	8.3
Conductivity (umhos/cm)	-	17000	-	-	1190
Suspended Solids	68	447	-	-	
Oil (insol. Or suspended)	400-2000	220	-	-	200
N.D - Not detected					

2.8 Biological Constituents

In many produced water treatment systems bacteria can be found in high numbers e.g. > 1 x 10^3 colonies / mL). Problems associated with bacteria:

- foul smelling

- microbial induced corrosion (MIC)

- Acid producing bacteria can contribute to general or pitting corrosion in a pipe or vessel

  o sulfate reducing bacteria will generate hydrogen sulfide

As mentioned previously, Produced water has low oxygen content both in the reservoir and throughout most of the production and water treatment system. Thus, in the upstream oil and gas industry one of the major distinctions that is made regarding bacteria is whether they are aerobic or anaerobic.

In fact, from a microbiological standpoint there are more precise definitions:

- Aerobic, which require oxygen to grow

- Obligate anaerobes, which are harmed by the presence of oxygen

- Aerotolerant organisms, which cannot use oxygen for growth, but tolerate its presence

- Facultative anaerobes, which can grow without oxygen but use oxygen if it is present

A short list of some of the commonly encountered bacteria and their response to oxygen:

Bacteria	Response to oxygen
Sulfate reducing bacteria (SRB)	Obligate anaerobe
Acid forming bacteria (AFB)	Facultative anaerobe
General heterotrophic bacteria (GHB)	Aerobic, facultative anaerobic
Slime forming bacteria (SFB)	Aerobic

Although SRB are listed as obligate anaerobes, they can survive in an aqueous environment that contains dissolved oxygen. They have the ability to go into a dormant state that protects them from oxygen attack. Also, it should always be kept in mind that environments containing dissolved oxygen may have deposits or an accumulation of solids that become starved of oxygen within the deposit that allow SRB and other obligate anaerobes to become active.

From the standpoint of sampling and analysis, bacteria are either stuck to the pipe and vessel walls, to the surfaces of mineral solids, or they are suspended in the flowing fluid:

- planktonic (floating)

- sessile (sticking to pipe or vessel wall or other solids surfaces)

This is a strictly an operational definition of bacteria in the sense that most bacteria encountered will have colonies that stick to the pipe wall and some colonies that float in the flowing water stream. Many bacteria types and species will stick to the pipe and vessel walls but will also sluff off into the flowing water stream. Thus, even bacteria that are predominantly sessile can be detected by taking a sample of the flowing water. Of course, if a sessile type of bacteria is detected in the flowing water then there are many more colonies stuck to the vessel or pipe wall.

2.8.1 Bacteria Contamination

Bacterial contamination becomes a greater problem in the presence of oxygen since aerobic bacteria are between 10 to 1,000 times more active than anaerobic bacteria. Aerobic bacteria grow significantly faster, and multiply significantly more frequently than anaerobic bacteria. Aerobic Bacterial activity can contribute to corrosion (especially under deposit corrosion, and to the generation of sticky biopolymers which contribute to the stability of oil-in-water emulsions and which can contribute to pad formation in separator vessels.

2.8.2 Typical Produced Water Bacteria and Their Characteristics

This section is rather brief for such an important subject. A more in-depth treatment is given in Chapter 6 (Petroleum Microbiology and Bacteria Control).

1. **Sulfate Reducing Bacteria**

   - Are anaerobic

   - contribute to FeS (black water, small microns, floaters, etc.) formation

   - Can exist in Oil pads (increasing potential to affect produced water without notice)

   - Affect overboard water O&G (oil and grease)

   - Can negatively affects water injection quality

   - Contribute to Biofouling

- Contribute to MIC (Microbiological Induced Corrosion)
- Generate H_2S (poisonous gas) – foul smelling

2. **Acid Producing Bacteria**

- Can be Anaerobic or Aerobic
- Contribute to Acid corrosion - MIC
- Produce nutrients for SRB

3. **Iron Related Bacteria**

- May be Anaerobic or Aerobic
- Are a Broad class with two major groups:
  - Iron-oxidizing
  - Iron-reducing
- Contribute to Odor problems
- Contribute to the formation of Red water

4. **Heterotrophic Bacteria**

- Aerobic
- Facultative (able to live in aerobic and anaerobic conditions)
- Primarily used for screening

5. **Slime Forming Bacteria**

- Aerobic
- Glue that protects other bacteria underneath in oxidative conditions
- Not commonly tested in offshore
- Used for Screening bacteria

6. **Archaea**

- Unicellular microorganisms that are genetically distinct from bacteria and eukaryotes, which often inhabit extreme environmental conditions.
- Methanogens (microorganisms that produce methane)
- Archaeoglobus are sulfate reducing Archaea.
- Must use Molecular Microbiological Methods (MMM) to detect

Bacterial Test Methods

Sampling and analysis of produced water in order to determine the concentration and types of bacteria is covered in Section 6.5 (Bacteria Sampling and Analysis).

- Serial Dilution Method

  o NACE Standard: TMO194-2014 Standard Test Method "Field Monitoring of Bacterial Growth in Oil and Gas Systems"

  o Official standard method to obtain a MPN (Most Probable Number) of bacteria in a media that encourage growth

  o Test results in 7 – 28 days

- ATP Analysis

2.9 Treating Chemicals - Chemistry

A wide variety of treatment chemicals may be present in produced water streams. Typical examples include corrosion inhibitor, scale inhibitor, demulsifiers, hydrate inhibitors, biocides, flotation aids etc. Other incidental chemicals may also be present such as detergents or solvents used for wash down or cleaning purposes. The purpose of this section is to present the chemistry of these compounds. Discussion about how these chemicals are selected, and their performance in the system is presented in Chapter 5.

The effect of treating chemicals can be significant in the selection and design of water treatment equipment. Many treatment chemicals are surface active and may stabilize small oil drops in the water phase. The corrosion inhibitors used in gas production operations and the demulsifiers used to assist oil dehydration can sometimes enhance the stability of emulsions.

Some treatment chemicals may be incompatible with other chemicals. Examples are demulsifier chemicals interfering with subsequent deoiling chemicals, or deoiling chemicals reacting with dilute polymers present in the water as a result of enhanced recovery schemes.

2.9.1 Corrosion Inhibitors

Most corrosion inhibitors are composed of amphoteric molecules. They are designed with a polar head group that sticks to the metal surface and a nonpolar tail which makes the metal surface somewhat hydrophobic. This molecular architecture is effective as a corrosion inhibitor. However, it also means that typical corrosion inhibitors are also effective at stabilizing emulsions and preventing oil drop coalescence. In fact, one of the most effective class of corrosion inhibitors is referred to as the Film Forming Amines. As a general rule, any compound that forms a film will also stabilize an emulsion. Nevertheless, it is generally possible to select a corrosion inhibitor that is both effective in inhibitoting corrosion and which does not greatly enhance the formation of emulsions. This selection process is discussed in Section 5.13 (Corrosion Inhibitors – Use and Impact on Water Treatment).

2.9.2 Hydrate Inhibitors

The thermodynamic hydrate inhibitors such as methanol, monoethylene glycol (MEG), and triethylene glycol (TEG) cause several water treating problems. The focus of this section is on methanol for which more information and data are available than the glycols. The problems described here for methanol are also encountered for the glycols to a greater or lesser extent depending on the particular problem. An attempt is made to point out the relative effect of the glycols compared to methanol, where that information is available. Treatment options for produced water containing methanol are presented. Also discussed is the affect of methanol on analytical methods for oil-in-water concentration measurement.

Thermodynamic Hydrate Inhibitors: Methanol and the glycols are used extensively to inhibit the formation of gas hydrates. The presence of these compounds reduces the activity of water and hence decreases the temperature at which gas hydrates form, are stable, and melt. Reducing the activity of water is a bulk thermodynamic effect. As such, the concentrations at which these so-called thermodynamic inhibitors are effective is in the range of a few percent to a few tens of percent. From a water treatment perspective, these are very high concentrations. Thus, the effect of thermodynamic hydrate inhibitors on water quality can be significant.

Mutual Solvents: Methanol is a thermodynamic hydrate inhibitor. It is also a mutual solvent. This means that it increases the solubility of organics (such as aromatic compounds) in water. As discussed previously, acids and polar organics are abundant in crude oil and produced water. Depending on the other properties of the oil and water, these compounds partition in the two phases. Upon addition of methanol, the partitioning of these compounds is shifted to a higher concentration in the water. This is what is meant by saying that methanol is a mutual solvent.

Problems Caused by Methanol: Methanol lowers the density of the water phase, and it lowers the interfacial tension between oil and water. This is primarily due to the mutual solvent properties of methanol and not due to any significant accumulation of methanol at the oil/water interface. As a result, small oil drops are more easily formed due to shear through valves and pumps. The small oil drops, and reduced density difference between oil and water make it more difficult to separate oil droplets in bulk separators and hydrocyclones.

Smaller oil drops do have a tendency for coalescence, albeit usually at a slower rate than is the case for larger oil droplets with higher droplet-droplet collision energies. The lower interfacial tension, without interfacial accumulation or stiffening of the interface, does allow for coalescence of the oil drops. This coalescence can be enhanced further by the use of coalescing media which has been demonstrated to be effective in removing small drops of oil in the presence of methanol. In a study conducted by NATCO [86], a coalescing filter was effective at increasing the droplet sizes of condensate dispersed in the methanol/brine, producing crystal clear fluid with residual condensate dispersed as moderately sized droplets.

Methanol also decreases the solubility of mineral solids in water. This can be a consequence of the lowered thermodynamic activity, due to the presence of methanol in the water. It can also be a consequence of the formation of other hydrated forms of mineral solids in the presence of methanol. The solubility of essentially all mineral species is decreased. This often leads to problems with the unexpected or intermittent accumulation of oily solids in the water treatment system.

Another problem that can occur is reduction of the effectiveness of water treatment chemicals. This is due to reduced ionic strength of the water-containing methanol. Most water treatment chemicals are polyelectrolytes. As such they are selected for a given water chemistry and salinity. When methanol is used, it is typically used for startup and shutdown conditions. This means intermittent use and there is no practical way that water treatment chemicals can be tailored for this type of application. Thus, when the produced water contains a high concentration of methanol, the polymers will typically not be as active and may "salt out" or precipitate.

Methanol – Oil Measurement Problems: The presence of methanol will also affect the results of oil-in-water measurements [86, 87]. Methanol itself is detected in a typical bench top IR. The concentration detected will depend on the calibration of the IR machine which depends on the crude oil characteristics of the platform. IR detection methods require a calibration which relates the molar concentration of C-H bonds (mol C-H bonds/L) to the gravimetric concentration of organic compounds (mg organics/L). As a rough estimate, IR methods detect about 1/3 of the methanol gravimetric concentration. In other words, if methanol is 1,000 mg/L, a typical IR machine will report about 300 mg/L. Gravimetric measurement of oil-in-water, where an extraction solvent and an evaporation step is used, typically does not detect methanol [87]. This is because methanol does not extract efficiently into the hydrocarbon solvent, and because the boiling point of methanol (64.7 C) is lower than the evaporation temperature (typically 70 C) for the extraction solvent.

Methanol – Oxygen Problems: Another problem associated with the use of the thermodynamic hydrate inhibitors is that they can act as carriers for oxygen. Because they are used in percentage concentrations, large volumes of these inhibitors are required. Eliminating oxygen during storage of these thermodynamic inhibitors can sometimes be challenging. Gas blanketing of the methanol storage tank is typically employed but maintaining a gas blanketing system seems to be challenging for many offshore operators. When oxygen is dissolved in the methanol, and the methanol is subsequently injected into the production system, oxygen is then carried into the production system as well. Besides corrosion of the production system, which is a severe consequence by itself, oxygen also has the detrimental effects discussed previously in Section 2.4.3 (Oxygen).

As shown in the table below, the solubility of oxygen in methanol is quite high. Roughly 100 mg/L of oxygen will dissolve in methanol when it is in contact with atmospheric air, at 25 C. Some rough order of magnitude calculations can give an idea of the potential problem. If a nitrogen blanket is poorly applied such that it contains 96 % nitrogen and 4 % percent oxygen, then the methanol will contain 16 mg/L of oxygen (412 mg/L x 0.04 = 16.48 mg/L). If the methanol is dosed into the produced water at a concentration of 10 % then the water will contain 1.9 mg/L of oxygen (1,900 ppb) which is a very high concentration in terms of corrosivity. It is also high in terms of oxidation potential for precipitation of iron and other metals. Furthermore, if there is even a small amount of H_2S in the water, H_2S can be oxidized to elemental sulfur which can precipitate and degrade water quality as previously discussed.

Table 2.16 Solubility of oxygen in methanol (mg O2 / L methanol) [88, 89].

Source of information	Partial pressure O2 (kPa)	T (K)	Solubility O2 in MeOH (mg O2/L MeOH)
IUPAC	101.3	248.15	458
IUPAC	101.3	273.15	432
IUPAC	101.3	298.15	412
IUPAC	101.3	323.15	404

Toxicity: Methanol is readily biodegradable with relatively low toxicity. Published toxicity data gives a water flea (Daphia Magna) LC50 0f 13,250 mg/L (48 hr test) and brine shrimp (Artemia Salina) LC50 0f 43,600 (48 hr test).

In the deep water GOM, bioassay tests are performed at approximately 1.5% effluent concentration (67/1 dilution) per EPA NPDES specifications. For a MeOH in water concentration of 50% by volume, a 67/1 dilution yields a MeOH concentration of about 6,000 mg/l, well below the measured LC50 (the concentration lethal to 50% of organisms that would not have died naturally).

OSPAR has MeOH on their list of substances used and discharged offshore which are considered to "Pose Little or No Risk" (PLONAR) to the Environment. Placement on this list allows MeOH to be discharged overboard in the North Sea without regulation.

The Offshore Continental Norwegian Shelf (OCNS) uses the OSPAR Harmonized Mandatory Control Scheme (HMCS) which ranks chemical products according to Hazard Quotient (HQ), calculated using the Chemical Hazard and Risk Management (CHARM) system. The CHARM rating for MeOH is a minimum HQ of >0 and maximum HQ of <1, which corresponds to a Gold color banding, the lowest environmental hazard rating. OCNS also classifies MeOH as a Class E chemical, again putting MeOH in the lowest potential environmental hazard group. Offshore discharges of a Class E chemical from a point source (such as a FPSO) are limited to 1000 metric tons annually. For MeOH, this corresponds to approximately 8,000 bbl/year. There are currently no MeOH discharge regulations set by the BOEMRE or the United States Environmental Protection Agency (EPA) for the GOM.

There are two general groups of low dose hydrate inhibitors, the anti-agglomerates and the kinetic inhibitors. Most of the anti-agglomerates are surface active. They are designed such that one end of the molecule sticks to the hydrate (clathrate) surface and the other end is oleophilic. This reduces the growth rate of the hydrate crystal and it reduces the tendency of hydrate particles to stick together. The net result is that the hydrate forms a suspension instead of a blockage.

2.9.3 Triazine H_2S Scavengers

One of the options for removing H_2S from produced hydrocarbon gas is to inject an amine-based chemical scavenger. Commonly used H_2S scavengers contain a triazine-type chemical as the active ingredient. Examples are tri(hydroxyethyl) triazine and trimethyltriazine (see Figure 2.22). The tri(hydroxyethyl)triazine is the more water soluble of the two, and is generally favored when it is likely that the scavenger will come in contact with produced water. The trimethyltriazine is the more oil soluble and is often used when the scavenger will contact predominantly oil-based produced fluids. This compound does have finite partitioning into both oil and brine. Both of these triazine examples contain a similar amine structure and therefore undergo similar hydrolysis and scavenging reactions. The product of these reactions include primary amines. The scavenging reactions for tri(hydroxyethyl) triazine are shown in Figure 2.23. As shown, electrophilic attack of the nitrogen groups by hydrogen sulfide results in sulfur substitution into the cyclic structure. Also, a primary amine (ethanolamine in the example) is produced in each step of the sequential reaction.

There have been reports of the spent, sulfur containing triazines decomposing in higher temperature produced waters with the consequent release of elemental sulfur. The elemental sulfur, as discussed previously, has a tendency to plug filters and can accumulate in injection well sand face pores, reducing their permeability and making it difficult to restimulate the wells.

Figure 1: Structures of Triazine based scavengers

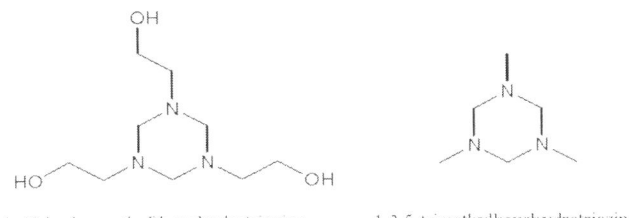

1,3,5-Tris (2-hydroxyethyl)hexahydrotriazine   1,3,5-trimethylhexahydrotriazine.

Figure 2.22 Structures of triazine based scavengers.

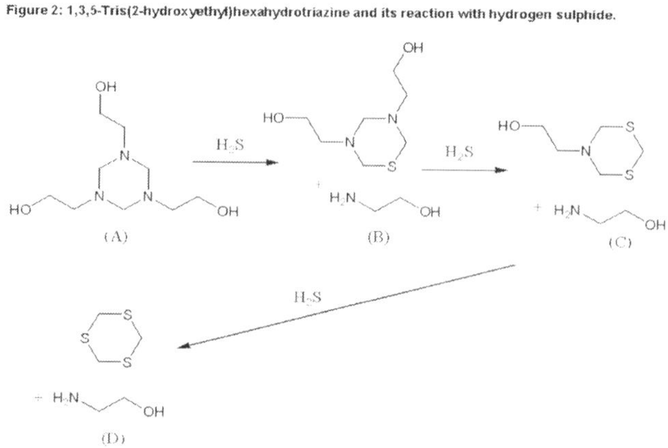

Figure 2: 1,3,5-Tris(2-hydroxyethyl)hexahydrotriazine and its reaction with hydrogen sulphide.

Figure 2.23 Structures of a triazine based scavenger and its reaction intermediates and products.

Carbonate Scaling: While triazine chemical scavengers can be effective at removing H_2S, and have health, safety and environmental advantages compared to other scavenger chemicals such as aldehydes or hydrazine, they act to raise the pH of produced water with which they are mixed such that scaling from carbonates becomes a potential problem [90]. The formation of calcium carbonate, magnesium carbonate and iron carbonate scales can occur over a wide range of temperatures, pressures, and carbonate concentrations whenever the pH of produced water is increased.

In many process configurations there is no alternative but to allow the mixing of spent H_2S scavenger solutions with produced water. This problem can occur at the location where the spent triazine is injected, as well as downstream of that location. Both the unreacted scavenger and its reaction products contain alkaline amines and will therefore raise the water's pH. In many systems the primary triazine reaction products are transported in the gas process system and are dumped into the oil/water or water process system at a later stage. If either the scavenger or its reaction products are mixed with produced water containing carbonates, the increase in the pH will increase the probability of carbonate scale formation.

Scale Inhibitors: In cases where the unreacted or reacted scavenger comes in contact with produced water, extensive laboratory work is typically required to select an appropriate scale inhibitor. It is usually necessary in such studies to understand in detail the chemistry of the produced water and its pH and alkalinity characteristics.

Spent Triazine: Another problem associated with the use of triazine inhibitors is the generation of a particular undesirable reaction byproduct. Complete reaction of the scavenger results in a cyclic sulfur compound (trithiane) that is insoluble in either hydrocarbon or brine. If the compound is allowed to form, it will precipitate on pipe walls and vessel internals.

Scavenger is therefore normally applied in excess of the stoichiometric ratio, and the contact time between scavenger and H_2S containing gas is shortened in order to prevent over-reaction. A typical minimum dose rate is 1 mol of triazine to 2 moles of sulfide. Both of these actions help prevent the formation of the insoluble reaction product, while still achieving the desired scavenging target. A consequence of this practice is the presence of unreacted scavenger in the product which has an impact on the water's pH and carbonate precipitation. Thus, the product of typical scavenger applications in the oil and gas industry is a complex mixture of unreacted triazine, partially reacted triazine, and primary amines. As discussed previously, the chemistry of these compounds in the produced water must be studied and understood in order to prevent undesirable consequences.

2.10 Drilling and Completion Fluids (including hydraulic fracturing flowback fluids)

2.10.1 Hydraulic Fracturing Fluids

The composition of hydraulic fracturing flow back fluids is of great interest to the shale industry because of the difficulty of disposing of flow back water in some regions and because of the scarcity of water and the need to recycle in other regions. An obvious place to start the discussion is with the composition of the fracturing fluids that are pumped into the well. From a high level standpoint, there are only a few types of fluid [1]: slick water, linear gels, cross-linked gels, and combinations of these. Figure 2.24 below is a generalized illustration of what type of fluids are used with the various types of shales found in the major US basins. Within each of these main varieties of fracture fluids there are a number of additives that have a significant effect on the properties and composition of the fluids. Additives include oxidative and organic biocides corrosion inhibitors, iron chelating agents, wetting agents, enzymes, and so on. Table 2.17 below contains a brief description of the various classes of chemicals included in hydraulic fracture fluids.

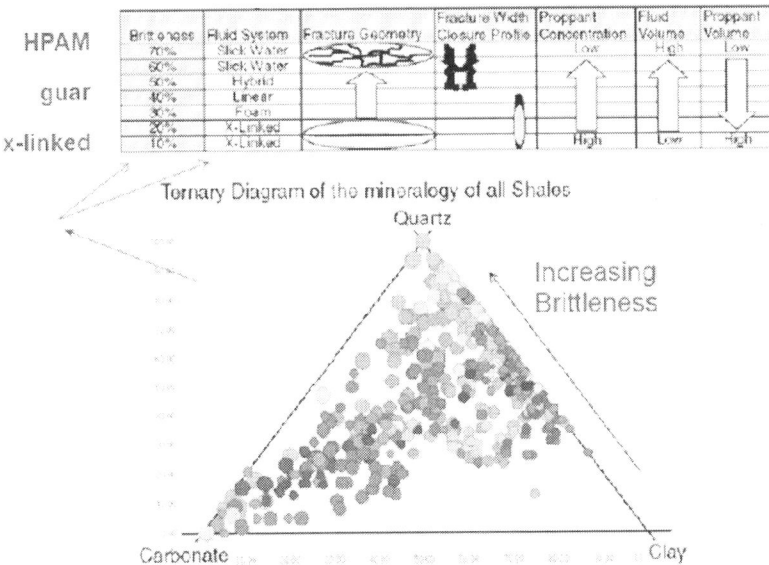

Figure 2.24 The type of shale determines the type of hydraulic fracturing fluids required. The higher the clay content, the greater the need for proppant with guar or cross-linked guar as polymer [91].

Table 2.17 Fracturing fluid additives in hydraulic fracturing fluids.

Additive Type	Main Component(s)	Purpose	Common use of main
Diluted acid (15%)	Hydrochloric acid or muriatic acid	Help dissolve minerals and inititate cracks in the rock	Swimming pool chemical cleaner
Biocide	Glutaraldehyde	Eliminates backeria in the water that produce corrosive by-products	Disinfectant to sterilize medical and dental equipment
Breaker	Ammonium per-sulfate	Allows a delayed breakdown of the gel polymer chains	Used in hair coloring, as a disinfectant, and in manufacturing of household plastics
Corrosion inhibitor	N, n-dimethyl formamide	Prevents corrosion of the pipe	Used in pharmaceuticals, acrylic fibers and plastics
Crosslinker	Borate salts	Maintains fluid viscosity as tempature increases	Laundry detergents, hand soaps and cosmetics
Friction reducer	Polyacrylamide	'Slicks' the water to minimize friction	Water treatment and soil conditioner. Makeup remover, laxatives and candy
Gel	Guar gum or hydroxyethyl cellulose	Thickens the water in order to suspend the sand	Cosmetics, toothpaste, sauces, baked goods and ice cream
Iron Control	Citric acid	Prevents precipitation of metal oxides	Food additive, flavoring in food and beverages
KCl	Potassium chloride	Creates a brine carrier fluid	Low-sodium table salt substitute
Oxygen scavenger	Ammonium bi-sulfite	Removes oxygen from the water to protect the pipe from corrosion	Cosmetics, food and beverage processing, water treatment
pH adjusting agent	Sodium or Potassium carbonate	Maintains the effectiveness of other components, such as crosslinkers	Washing soda, detergents, soap, water softner, glass and ceramics
Proppant	Silica, quartz sand	Allows the fractures to remain open so the gas can escape	Drinking water filtration, play sand, concrete, brink mortar
Scale inhibitor	Ethylene glycol	Prevents scale deposits in the pipe	Used in household cleansers, de-icer, paints and caulk
Surfactant	Isopropanol	Used to increase the viscosity of the fracture fluid	Glas cleaner, antiperspirant and hair color

Note: The specific compounds used in a given fracturing operation will vary depending on company preference, source water quality and site-specific characteristics of the target formation. The compouns shown above are representative of the major compounds used in hydraulic fracturing of gas shales.

2.10.2 Hydraulic Fracturing Flow Back Fluids:

Wolford [92] studied the analytical methods used to measure TOC and COD in hydraulic fracturing flow back fluids. He determined that high concentrations of chloride and alkaline earth cations (Ca, Ba, and Sr) interfere with the measurement of COD, particularly using the colorimetric technique. He developed a modified titration endpoint method which gave reasonable results when compared against known standards. Values of COD reported in the literature have been as high as 19,000 mg/L, with many samples in the range of 5,000 to 8,000 mg/L. On several samples, side-by-side comparisons of the colorimetric results with those of the modified titration results showed that the colorimetric method gave COD values roughly 4 times higher than did the modified titration method. None of the other analytical measurements, such as TSS, were found to correlate with the colorimetric results, whereas the modified method was found to correlate to other methods.

Table 2.18 Measured COD values or flow back water from the Marcellus [92].

Sample	COD (mg/L)	TOC (mg/L)	COD / TOC
Flow back A	509	78	6.5
Flow back B	1,111	303	3.7
Flow back C	1,069	551	1.9

COD measured by modified titration method.
TOC measured as non-purgeable organic carbon

As discussed in Section 7.5.3 (The Ratio COD / TOC) the theoretical ratio of COD / TOC for sugars is 2.7. For hydrocarbons, such as methane, the ratio is in the range of 5.3. Flow back A has a ratio above this theoretical maximum. There is most likely some degree of analytical error. Based on this limited data, it is difficult to determine the ratio of hydraulic fracturing polymer versus hydrocarbon. However, the COD and TOC values are within the expected range.

2.11 Toxicity

The subject of this chapter is the chemistry of produced water and hydraulic flow back water. The environmental impact of the disposal of this water is closely related to its chemistry. As discussed, the chemistry is a combination of a multitude of naturally occurring minerals, hydrocarbons, organics, and a number of man-made production chemicals.

Stephenson [13] has compiled several studies carried out in the late 1980's that were aimed at determining which classes of compounds found in produced water have the greatest environmental impact on surface water. He frames the discussion in terms of three types of oil: paraffinic, asphaltenic, and gas condensate. Water produced with paraffinic oil often has a high concentration of simple fatty acids. Water produced with asphaltenic crude oil tends to have a high concentration of naphthenic (oil soluble) acids. Water produced with gas condensate tends to have high concentrations of BTEX. The BTEX compounds tend to be toxic (benzene is also carcinogenic) but they are also volatile and tend to not assimilate into the surface water food chain. Soluble compounds tend to persist in surface water for a long time. The simple fatty acids, in the ionized form are not toxic, many of which are already found in the natural environment. The acid form of these compounds is toxic but the acid form would not be stable at environmental pH values. Most napthenic acids are biodegradable. They are very complex as a group of compounds and much less can be said about them as a class of compounds. Polynuclear aromatic hydrocarbons are essentially non-soluble and have been found to have a high Environmental Impact Factor.

But there are several other factors that are involved in determining the environmental impact of produced water such as:

- temperature of the produced water and receiving water

- the extent to which volatile compounds vaporize due to temperature, pH and salinity of the waters

- the extent of biological activity that would assimilate the contaminants and allow them to enter the food chain

- the ability of the microbes to break-down the contaminants rather than pass them up the food chain

- the effect of the microbes on oxygen availability and other effects (metabolic byproducts)

- the amount of precipitation

- the extent of mixing and dilution at the injection site

Thus, the toxicity of the compounds themselves is only a part of the story. In the U.S. and several other countries, the impact of overboard discharge is determined on the basis of whole-fluid testing. As discussed in Section 1.6.4 (Regulations, and Disposal and Reuse Options), in Norway there is a much more-chemical species oriented approach where the Environmental Impact Factor is calculated based on chemistry and chemical properties. Specific classes of compounds (e.g. Polynuclear Aromatic Hydrocarbons – PAH; and phenols) are measured and restricted by regulations.

U.S. Gulf of Mexico Overboard Discharge: In 1991 the U.S. EPA published a notice which eventually became a requirement for approval for an NPDES permit. The notice proposed adding aquatic chronic toxicity limits for produced water overboard discharge. The required analytical method is based on the No Observable Effect Concentration (NOEC) [93]. The method involves a seven day test using two species: mysid shrimp, and sheepshead minnow. A mixing zone is calculated based on water depth and flow rate of discharged water. Serial dilutions are used to determine the NOEC.

2.12 pH, Alkalinity, and Hardness

This section discusses the concepts of pH, alkalinity, and hardness. All of these quantities are related to each other. While a detailed knowledge of these properties is not always needed, it is worthwhile understanding the concepts as an aid to troubleshooting the performance of a water treatment process.

Water samples from oil and gas fields contain many weak acids and bases, such as:

- Carbonic acid ($H_2CO_3 = CO_2 + H_2O$), derived from dissolved CO_2

- Water soluble organic acids, including the C1-C6 organic acids, also known as "volatile fatty acids" (VFA)

- Higher molecular weight acids such as C7 – C20+ which contribute to the Water Soluble Organics measurement

- Hydrogen sulphide (H_2S), when the field is sour or when Sulphate Reducing Bacteria (SRB) are present

- Boric acid ($B(OH)_3$)

- Silicic acid (H_4SiO_4)

- Ammonia (NH_3)

The concentrations of these various acids and bases vary substantially from one reservoir to another across the globe and depend on the origin of the water. The contribution of these acids and bases to the pH, alkalinity, and harness are discussed here.

The pH of a water sample is a common measurement. However, there can be a significant difference between the pH of a water sample measured in the lab, and the pH of produced water in a process stream. The presence of volatile gases (CO_2, H_2S) and the oxidation effect of air can change the pH of a sample quickly by one or more pH units. Because of this, many specialists feel that although knowledge of pH in the process is usually very important, the measurement of pH in a degassed lab sample is not very useful. Instead of measuring the pH in the lab, the alkalinity of the water and the concentration of acid gases in the process gas should be measured. From these values, an accurate pH can be calculated using equilibrium thermodynamics via the use of commercially available software. This method of calculating the process pH by first measuring alkalinity of the water and gas composition is discussed here.

2.12.1 pH

Accurate measurement of pH is actually very difficult for produced water. Unless a calibrated, in-line probe is used, a sample of the water must be taken. Whenever a sample is taken from the process, the pH immediately changes due to pressure and temperature reduction and flashing of the CO_2 and H_2S from the water into the air. If the sample is taken to a field lab, located at the facility, for pH measurement, this flashing can be significant and temperature will also likely decrease. It is common for the pH to increase by a unit or more compared to the pH of the process fluid, depending on the water chemistry and the process and sampling and measurement conditions. If the sample is transported to an offsite lab, the pH can vary in an unpredictable way due to further changes related to precipitation of components.

Sample Turns Green, Orange or Black: Flashing of CO_2, and the related increase in pH, is often the cause of water samples turning green, orange or black after being acquired. As previously discussed, as the CO_2 flashes, the pH goes up, and iron sulfide which was soluble at the lower pH then precipitates and turns the sample black. If the sample turns green, yellow or orange, this is commonly an indication that iron oxides (and hydroxides) have formed and have precipitated from the combined effect of oxygen intrusion into the sample and the pH increase. The green color is typical of ferrous hydroxide which forms under certain pH and oxygen levels. Most common is the orange color associated with ferric hydroxide. All of the iron precipitates form very small particles and hence they tend to stay suspended for several days if not weeks. Ferric hydroxide may actually float for several days or weeks. It has many waters of hydration and it tends to form a floc that traps oil and vented gas. In some (typically onshore) water treatment systems these flocculating properties are used to remove suspended solids.

Accurate pH Determination: If carbonate scaling is a potential problem, or if amine H_2S scavengers are to be applied, then an accurate pH value for the produced water is quite important. For this purpose the CO_2/alkalinity method for calculating a water's pH was developed. This is required not only to obtain an accurate determination of pH, but also as input to thermodynamic scaling programs which allow the prediction of buffering capacity and shifts in pH under various processing conditions. The use of thermodynamic modeling of produced water chemistry allows one to evaluate any number of potential locations where carbonate scaling could be a problem. For the CO_2/alkalinity method, the alkalinity of the water, the CO_2 content of the gas, temperature, pressure and the composition of alkalinity contributing species are measured [94]. A thermodynamic software program is then used to back-calculate the pH of the water in the process system at operating conditions.

This method makes use of the fact that alkalinity does not change as CO_2 is flashed from the sample. Alkalinity is discussed in Section 2.12.2 below.

pH is not a measure of the actual acidity or alkalinity in a water sample. ASTM D 1293-78 discusses the measurement of pH while ASTM D 1067 discusses the measurement of actual acidity or alkalinity. The pH of the waste water is subject to regulation in some countries which typically require the pH of discharged waste waters to be between 5.5 and 9.0

Alkaline waste waters with a pH in the range of 8-10 may react with components of the hydrocarbon stream to form surfactant-like molecules. Examples of these molecules are hexanoic acid, the so-called naphthenic acids, benzoic acid, etc.. At alkaline pH the organic acid group is unprotonated and the organic acid is in its ionic form As such these molecules are far more surface active than if they were protonated and electrically uncharged. In their ionic form, organic surfactants can contribute to the stability of emulsions.

2.12.2 Alkalinity

The term "alkalinity" is often used, and misused in discussions about produced water characteristics. Alkalinity is by definition an expression of the quantitative ability of an aqueous solution to neutralize acid (i.e., hydronium ions). Roughly speaking, alkalinity is the sum of the concentrations of bicarbonate anion (HCO_3^{-1}), plus carbonate anion (CO_3^{-2}), plus hydroxides, plus organic acid anions. As discussed below, only the salt form of these species (e.g. NaHCO3, CaCO3) contribute to alkalinity. The species derived from CO_2 equilibria (e.g. H_2CO_3) do not contribute to alkalinity.

Alkalinity has an impact on mineral precipitation and scaling, the pH, corrosion processes, and is relevant for alkaline surfactant flooding. Alkalinity is a direct measure of the buffering capacity of the solution.

- The following example events change alkalinity:

  - dissolution/precipitation of CaCO3 (changes CO_3^{-2} concentration)
  - dissolution/precipitation of NaHCO3 (changes $HCO3^{-1}$ concentration)
  - dissolution/precipitation of FeS (changes HS^{-1} concentration)
  - dissolution/precipitation of $Mg(OH)_2$ (changes OH^{-1})
  - corrosion of Fe(s) to Fe^{+2} (consumes two H_3O^{+1} ions) or Fe^{+3} (consumes three H_3O^{+1} ions)

- The following example events do not change alkalinity:

  - absorption or evolution of CO_2 (CO_2 is not an anion of a weak acid)
  - dissolution/precipitation of NaCl or other simple salts (Cl^{-1} is not related to a weak acid)
  - dissolution/precipitation of $BaSO_4$ (SO_4^{-2} is not related to a weak acid)

The main source of alkalinity in oilfield brine is NaHCO3 (sodium bicarbonate). When alkalinity is present at a concentration of 100 mg/L or more, there is significant buffering capacity. One of the distinguishing features of condensed water is that it has low alkalinity, since it has very little sodium bicarbonate content. The pH of condensed water is almost entirely a function of the CO2 equilibria and, as a result, the pH can be relatively low (e.g. 4 to 5).

Alkalinity is a Conserved Quantity: As fluids are produced (rise up the well bore, pass through the choke, and flow into and through the facility) the CO2 and H2S content (if present) of the water changes. Generally speaking, the amount of CO2 and H2S in the water decreases. Since alkalinity does not change as the CO_2 or H_2S content of the water changes, alkalinity will not change as a result of flashing of these or other acids with process changes. Providing that there are no scale deposition processes, alkalinity can be measured by analyzing the produced water at several locations in a facility with the same result. If the measured value of alkalinity appears to change across a process, then either the sampling / measurement method is flawed, or there are important chemical transformations occurring that were previously unknown (such as scaling).

Alkalinity – Chemical Definition: Alkalinity will be defined from two perspectives: chemical and mathematical. Both are useful in developing a full understanding of alkalinity. The chemical definition of alkalinity highlights its acid-neutralizing capacity. It is the sum of all the titratable bases and its measured value may vary significantly with the end-point pH used. Because the alkalinity of many surface waters is primarily a function of carbonate, bicarbonate, and hydroxide content, it is often taken as an indication of the concentration of these constituents. The actual measured value of alkalinity includes contributions from organic acids, borates, phosphates, silicates, or other bases if these are present:

$$\begin{aligned}
[\text{Alkalinity}] = \; & [HCO_3^-] + 2\,[CO_3^{2-}] \\
& + [HCOO^-] + [CH_3COO^-] + [C_2H_5COO^-] \\
& + [B(OH)_4^-] \\
& + [H_3SiO_4^-] + 2[H_2SiO_4^{2-}] + 3\,[H_2SiO_4^{2-}] + 4[SiO_4^{4-}] \\
& - [NH_4^+] \\
& + 2[PO_4^{3-}] + [HPO_4^{2-}] \\
& - [H^+] + [OH^-]
\end{aligned}$$

$$\text{Eqn (2.21)}$$

Other acids or bases can be added if present. The symbol [xx] refers to the concentration of substance xx in units of moles/Liter.

CO_2 Equilibria: With respect to the CO2-based equilibria (shown below), a careful distinction needs to be made regarding pH buffering and alkalinity. As stated already, the CO2-based compounds do not contribute to alkalinity. However, they do contribute to the pH buffering capacity of an aqueous solution. Using the definition of alkalinity given immediately above allows one to understand why the CO2 derived species do not contribute to alkalinity. Consider the addition of CO2 to water as described by the following equilibria:

$$CO_2 + H_2O \rightleftharpoons H_2CO_3$$

$$H_2CO_3 \rightleftharpoons H^+ + HCO_3^{-1} \qquad\qquad \text{Eqn (2.22)}$$

$$HCO_3^{-1} \rightleftharpoons H^+ + CO_3^{-2}$$

A major fraction of CO2 will combine with water to form carbonic acid (H2CO3), as shown in the first equilibrium. A fraction will remain as molecular CO2 in the water. The carbonic acid will partially dissociate to form H+ and HCO3-1, as shown in the second equilibria. The latter two species cancel each other out in the summation of concentrations given for alkalinity. The bicarbonate ion does not add to the acid neutralizing capability of the water because it is already balanced, one-to-one, with a hydrogen ion formed when the neutral species dissociates. A similar situation occurs with the third equilibria.

Another way to understand this situation is to consider the calculation of alkalinity for a solution of CO_2 in water, with no other components. For this system, the alkalinity is given by:

$$[Alkalinity] = \left\lfloor HCO_3^{-1} \right\rfloor + 2\left\lfloor CO_3^{-2} \right\rfloor - \left\lfloor H^{+1} \right\rfloor \qquad\qquad \text{Eqn (2.23)}$$

For the same system, the concentration of hydronium ions is given by:

$$[H^{+1}] = \left\lfloor HCO_3^{-1} \right\rfloor + 2\left\lfloor CO_3^{-2} \right\rfloor \qquad\qquad \text{Eqn (2.24)}$$

This last equation requires a little bit of thought to verify. Substituting the right hand side of this equation into the equation for alkalinity shows that the alkalinity for this system is zero. Thus there is no net change in alkalinity due to the addition of removal of CO_2 in an aqueous solution.

Buffer Capacity of CO2 Solutions: However, CO_2 solutions do have buffering capacity. If protons are added from another source (such as mineral acid- e.g. HCl), then the CO_2 equilibria shifts to the left, protons are consumed, CO_2 is formed some of which will evolve from the solution as a gas, and there is less of a change in pH then there would have been without the CO_2 equilibria. This is not to say that the pH of CO_2 solutions is high. On the contrary it can be quite low, e.g. 4 to 5. But the addition of other acids has a diminished effect on pH. Thus, CO_2 equilibria contribute to buffering capacity but not to alkalinity.

Alkalinity Measured by Titration: Though methods vary in their detail, alkalinity is typically measured as the amount of strong acid (e.g. HCl) required to titrate a water sample to the bicarbonate inflection point, pH = 4.5. The method of Tomson is recommended [94, 39]. Acid is added to the water, under a nitrogen purge, and the pH decrease is monitored. As more acid is added, the anions listed above are neutralized, and the CO_2 present in the sample is driven into the vapor phase. The molar amount of acid required to reach a certain low pH value is the alkalinity. The titration end point is important in both understanding the concept of alkalinity and obtaining the correct result from a titration, particularly where high partial pressures of CO_2 are involved.

Alkalinity can be reported as M-Alkalinity or P-Alkalinity. M-Alkalinity measures the total alkalinity in water based on an acid titration. 'M' alkalinity is the sum of concentrations of all the bases that will react with acid as the pH of the sample is reduced to the methyl orange endpoint of between 4.2 to 4.4 (Note: the actual pH for the methyl orange endpoint is affected by the ionic composition of the water and can vary from 3.7 to 4.7). For measurement consistency, the value of alkalinity as determined to an actual end point of pH = 4.5 (preferred). It is the total alkalinity which is the sum of hydroxide, carbonate, organic acids, and bicarbonate concentrations.

'P' alkalinity is the sum of concentrations of all the bases that will react with acid as the pH of the sample is reduced to the phenolphthalein end point of between 8.2 and 8.4 using a Phenolphthalein Pink indicator endpoint 8.3. It includes all the hydroxyl and half the carbonate content. If the water sample is initially at a pH below this end point, then the sample has zero P-alkalinity.

Units of Alkalinity (mmol/l or mEq/l): Alkalinity is usually reported in units of mEq/L or mmol/L. mEq/L which stands for milli-equivalent per liter which is the same as mmol/L for single-charge ions. Sometimes, alkalinity is reported as its calcium carbonate ($CaCO_3$) equivalent. When alkalinity is reported as a $CaCO_3$ equivalent, it 'pretends' that the only alkaline species in the water is the carbonate ion (CO_3^{2-}). CO_3^{2-} 'counts twice' in the alkalinity definition because it has a +2 charge (see above). The Molecular Weight of $CaCO_3$ is 100 g/mol or 100 mg/mmole. But since the calcium ion and the carbonate ion have a charge of +2 and -2 respectively, the Equivalent Weight of $CaCO_3$ is 100/2 = 50. Thus the following equation can be used to convert Total Alkalinity expressed as meq/L to mg/L $CaCO_3$ where meq = milli-equivalents:

Alkalinity [mg $CaCO_3$/L] = (100/2) x Total Alkalinity {expressed as meq/L}

It should be noted that this is different from the hardness reported as a $CaCO_3$ equivalent. The hardness reports the concentration of Ca-ions and Mg-ions as an equivalent amount of $CaCO_3$.

Alkalinity - Mathematical Definition: In anticipation of later sections, it is useful to provide an alternative mathematical definition of alkalinity. The explanation given here is based on several sources [95-97, 94], plus discussions with experts at Scaled Solutions Laboratories [98]. This discussion provides a rigorous explanation as to why alkalinity is a conserved quantity.

It is a fundamental principle of chemistry that solutions are electrically neutral in the absence of an applied electric field. This principle is expressed by the following equation which states simply that the total *number* of positive charges associated with cations in the water will be equal to the total *number* of negative charges associated with anions in the water:

$$\sum_i z_i m_i = 0$$

Eqn (2.25)

in which m_i is the molar concentration of ion species i, and z_i is the charge of ion species i. Thus z_i in the equation can be a positive or a negative number (e.g., +2 or -2). This equation is known as the charge balance equation. It can be written for the general case with cations on the left hand side and anions on the right hand side:

$$m_{H^+} + m_{Na^+} + m_{K^+} + 2m_{Ca^{+2}} + ... = m_{Cl^{-1}} + 2m_{SO_4^{-2}} + m_{HCO_3^{-1}} + 2m_{CO_3^{-2}} + m_{OH^-} + ...$$

Eqn (2.26)

This equation includes all ionic species in solution. There is a certain subset of ions that are "conservative." This will provide the defining concept of alkalinity. The conservative ions are those ions whose concentrations are not affected by changes in pH, pressure, temperature, or shifts in chemical equilibria (provided there is no precipitation – as discussed below). Examples are: Na^+, K^+, Ca^{+2}, Mg^{+2}, Cl^{-1}, SO_4^{-2}, and NO_3^{-1}. These conservative ions typically dissociate completely in solution, but this is not a requirement. Ions in solution may be found as either unassociated ions, the ionic constituents of salts (e.g. NaCl, $CaCl_2$), or the counter ions of acids or bases (e.g. CH_3CO_2Na, $NaHCO_3$, NaOH).. If the above equation is rearranged by collecting all the conservative species on the left hand side, then a useful quantity is obtained.

$$\sum_i^{\{cc\}} z_i m_i + \sum_i^{\{ca\}} z_i m_i = -m_{H^+} + m_{HCO_3^{-1}} + 2m_{CO_3^{-2}} + m_{OH^-} + m_{HS^1} + m_{NH_4^{+1}} + ...$$

Eqn (2.27)

Where the first summation on the left hand side is taken over the set of conservative cations (designated as "cc"), and the second summation is taken over the set of conservative anions (designated as "ca").

Thus the quantity on the left hand side of equation 5 is itself a conservative quantity and is referred to as the alkalinity of the brine. Note that the right hand side is identical to the previous "chemical definition" of alkalinity. As shown, alkalinity can be calculated from knowledge of the ionic concentrations in the brine. Species such as simple salts (NaCl, CaCl$_2$) do not contribute to the value of alkalinity since the concentration of cations (positive sum) is balanced by the concentration of anions (negative sum) and it is not possible for these salts to contribute to the acid neutralizing power of the brine.

For a particular brine, the value on the left hand side is equal to a number, usually expressed in units of meq/L, where meq is milli-equivalents as defined above. The value of alkalinity is independent of temperature, and pressure. The residual alkalinity of a solution is dependent upon pH. The bicarbonate content of a particular brine, for example, varies with pH because some of the bicarbonate is neutralized as acid is added to the water. Yes, this changes the chemical composition of the water but still, the pH dependence exists. The pH does depend on the constituents and their concentrations, but does not depend on any factors that influence the dissociation of weak acids and bases. The quantity on the left hand side is independent of the complex chemical and phase equilibria which govern weak acid and base dissociation. Therefore, the quantity on the right hand side is also independent of the complex chemical and phase equilibria. This independence makes alkalinity a fundamental property of the solution, and an important concept in the study of brine scaling.

Another important aspect of alkalinity that is now somewhat more elucidated is the relationship to pH. According to the equation above, the left hand side is a constant that only changes when compounds are added or taken out (e.g. by precipitation) from solution. Thus, the alkalinity does not change with temperature, for example. However the individual terms on the right hand side do change with temperature due to the effect of temperature on the various acid/base equilibria. As the individual concentrations change, on the right hand side, the concentration of protons (and therefore pH) changes in response. The concentration of protons changes such that the sum of all other changes (a.k.a., alkalinity) is constant. Again, alkalinity is a fundamental property of a solution.

As just explained, alkalinity is a useful quantity due to the fact that it is a fundamental characteristic of the solution, can be measured, can be calculated from knowledge of other quantities, and does not change as process conditions change (e.g. temperature or pressure). This definition of alkalinity highlights an important point – alkalinity has both mathematical appeal (as the sum of conservative ions), and a physically intuitive character (as the amount of acid required to titrate a solution). Both of these aspects of alkalinity are important considerations in the analysis of the composition of a solution.

2.12.3 Hardness

Generally speaking, hardness is the sum of the concentration of the calcium and magnesium ions, though any other polyvalent cation can also contribute to hardness. Hardness in water can result in scale formation in process equipment. The release of dissolved CO_2, during depressurization may also alter the hardness characteristics of water due to, for example, the precipitation of Ca^{++}.

Hardness is reported as "mg/liter of CaCO$_3$". Hardness is calculated based on the measured concentrations of Ca^{++} and Mg^{++} in solution. The derivation of this calculation is as follows:

ppm CaCO3 = (ppm of ion) x [equivalent weight of CaCO3 / equivalent weight of ion]

therefore…

   Ca as ppm CaCO3 = (ppm of Ca) x [50.0 / 20.0]

   Mg as ppm CaCO3 = (ppm of Mg) x [50.0 / 12.2]

   Hardness = Ca as ppm CaCO3 + Mg as ppm CaCO3 = 2.5 x [Ca^{++}] + 4.1 x [Mg^{++}]

If the origin of hardness is known, total hardness can be broken down into carbonate hardness and non-carbonate hardness. The hardness that comes from calcium and magnesium carbonate is referred to as carbonate hardness. The hardness that comes from calcium and magnesium chloride, sulfate, and silicate is referred to as non-carbonate hardness. It must be pointed out that this distinction cannot be made from just a water analysis alone unless all of the anions are matched with corresponding cations to give the mineralogical content of the water. There are software programs that will do this with the push of a button.

2.13 Precipitation and Scaling Tendency

Mineral precipitation and scaling are often discussed as if they are the same thing. They are related but are distinctly different. Scaling tendency and scale formation are subjects within flow assurance where the important variable is the thickness of an adherent scale layer on the inside of a pipe, separator, or other piece of process equipment. Scale formation is the culmination of precipitation with adherence to a vessel, pipe, valve, or any other permanent & stationary surface,. It is only when a scale forms that flow assurance becomes a risk.

While inhibition, mitigation and prevention of scale formation are important in water treatment, precipitation is also important. In other words, in water treatment it is important to understand and, if necessary, to prevent nucleation and growth of mineral precipitates. Most scale inhibitors act by preventing the deposition or growth of scale mineral precipitates on a stationary surface. They do not prevent the minerals from precipitating, although they at times do effect a delay in the onset of such precipitation.

Thus, a flow assurance specialist may use similar tools to that of an upstream water treatment specialist. However, the water specialist is more concerned with mineral precipitation than with eventual scale formation. The most facile way to develop an understanding of or a prediction for mineral precipitation and scale tendency is to use a commercial water chemistry thermodynamic modeling software program.

When using a commercial chemistry modeling program, it is very important to "benchmark" your calculations against measured data. This ensures that the calculations are being done correctly and that the results are being interpreted correctly.

Interpretation of the output from commercial chemistry modeling programs can still be a challenge that can only be overcome by benchmarking the calculation against known results. The Figure 2.25 and 2.26 below show how the solubility of $BaSO_4$ and the solubility of various mineral phases of $CaSO_4$ change with temperature. These figures illustrate the importance of conducting water chemistry modeling at the correct (process) temperature and not just at ambient temperature.

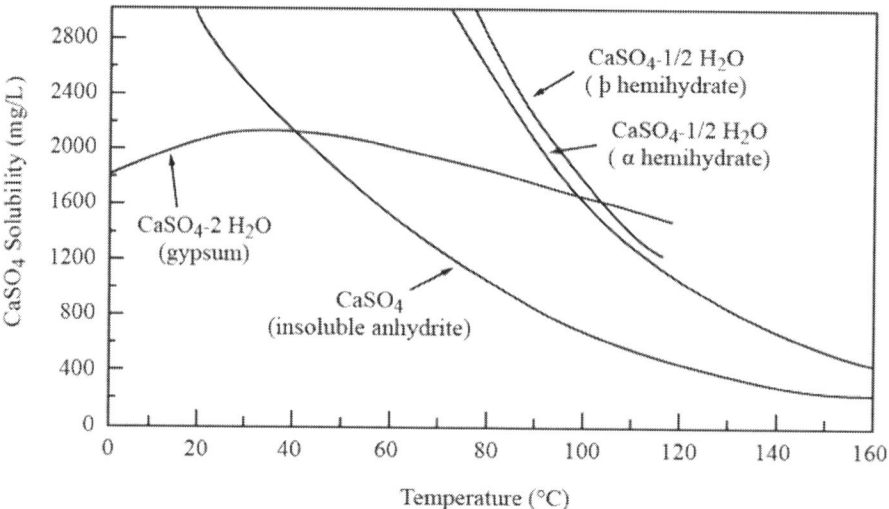

Figure 2.25 Calcium sulfate solubility as a function of temperature and pressure

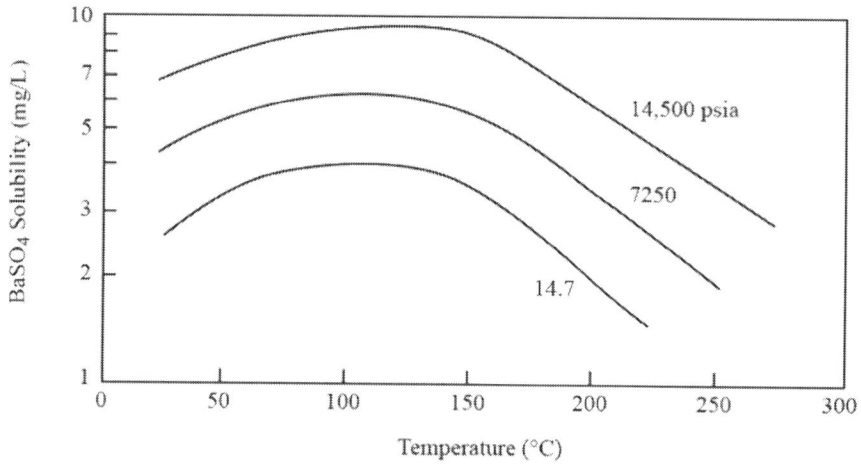

Figure 2.26 Barium sulfate solubility as a function of temperature and pressure.

The figures show data for pure water with no other source of chlorides or sulfates. Note the extremely low solubility of BaSO4 over a wide range of temperature and pressure. Figure 2.27 below shows how the solubility of $CaCO_3$ changes with the partial pressure of CO_2 in the gas with which the water is saturated. Note the significant impact of temperature on the CO_2 solubility. As temperature increases the solubility of calcium carbonate decreases, for a given partial pressure of carbon dioxide The impact of carbon dioxide partial pressure is also significant. As the partial pressure of carbon dioxide increases, the pH decreases which shifts the carbonic acid equilibria and favours the solubility of calcium carbonate.

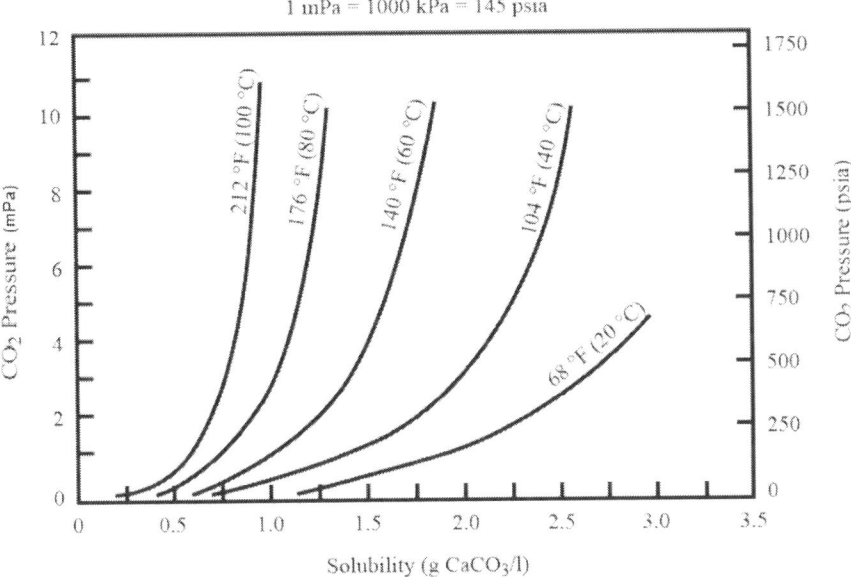

Figure 2.27 Solubility of calcium carbonate as a function of temperature and CO2 partial pressure.

2.13.1 Langelier Scaling Index:

The Langelier Stability Index or Scaling Index (LSI) is a calculated number that gives a rough guide to the thermodynamic potential for calcium carbonate to precipitate from a brine with a relatively low ionic strength (e.g., TDS < 5000 mg/liter). A positive value (LSI > 0) indicates a potential for calcium carbonate to precipitate. If a solution is just saturated in $CaCO_3$, the calculated value will be LSI = 0. A negative LSI value indicates that the solution is under-saturated in $CaCO_3$ and thus mineral precipitation will not take place.

LSI is calculated by the following equation:

$$LSI = pH - pHs$$

Eqn (2.28)

Where:

LSI: Langelier Stability Index
pH: measured pH
pHs: the pH at which the water would be saturated with calcium carbonate

$$pHs = pCa + pAlkalinity\ M + C_1$$

Eqn (2.29)

where:

pCa = - log [moles/liter of Ca^{++}]

pAlkalinity M = -log[equivalents/liter of total alkalinity as measured by titration to pH 4.5]

C_1 is a constant which depends on TDS and temperature but whose value is ≈ 1.0

Since the solubility of carbonate mineral is higher at lower pH, it is easy to see why LSI should be negative in order to prevent precipitation.

In practice, LSI can vary from 0 to 0.5 without an appreciable risk of precipitation. Besides pH, temperature has a strong effect on LSI. Carbonate minerals have higher solubility at low temperature. Thus, as temperature goes down the LSI goes down as well.

When calculating the value for pHs for the LSI equation, it is important to include the activity coefficient for the calcium and the bicarbonate ions in the calculation since it is only free, unassociated Ca^{++} that contributes to the $CaCO_3$ scale tendency. The Solubility Product equation for calcium carbonate is written as follows:

$$K_{sp} = \gamma_{Ca++} [Ca^{++}] \, \gamma_{CO3=} [CO_3^{=}] \qquad \text{Eqn (2.30)}$$

Where γ is the activity coefficient for each ion. For example, if $\gamma = 0.7$, then 70% of that ion would be present in solution as free, unassociated ion. The calculation of the activity coefficient is based on the ionic strength of the brine.

As the total chloride content of a brine increases, the amount of calcium that is associated with chloride also increases with the formation primarily of $CaCl^+$. Calcium which is tied up in this complex ion is not available for contributing to $CaCO_3$ solid precipitation. This explains why the carbonate mineral scale potential for divalent ions decreases as the TDS of a brine increases.

2.13.2 The Role of Solids in Stabilizing Emulsions

Given this discussion of scaling tendency, it is appropriate to make a few comments on the role of solids in stabilizing emulsions. Pickering emulsions are stabilized by very fine solids. In addition to these, small partially oil wet solids have a tendency to accumulate at the oil/water interface and stabilize oil-in-water emulsions. Many of the solids found in produced water systems, particularly the mineral precipitates and corrosion products are initially water-wet. In some cases, the solid particles remain water wet and are discharged in the overboard water stream. Some regions of the world have a limit on the amount of solids that can be discharged. In other regions there is no limit, however the turbidity caused by discharge of solids may have an environmental impact as well as being visually undesirable in areas populated by the general public.

The solids may settle in the bottom of separators where they occupy volume and thus reduce the residence time for process fluids. Such solids can contribute to under deposit corrosion or become habitat for bacteria. Solids can also contribute to pipe-clogging precipitates on the walls of vessels and piping when they form scale which is a major threat to flow assurance.

Oily Solids – Surface Activity: Often though, a fraction of the particles will become at least partially oil wet by attachment of sticky crude oil components such as asphaltenes, waxes, naphthenates and naphthenic acids, or by attachment of various production chemicals including corrosion inhibitors and water clarifiers. A solid particle that is initially water wet and which becomes partially oil wet will then be surface active. Thus, any component that is itself moderately surface active has the potential for wetting the solid surface and enhancing its emulsion stabilizing properties. In produced hydrocarbon fluids, potential wetting agents include injected chemical demulsifier chemicals, organic acids, naphthenic acids, asphaltenes, resins, waxes, and other polar compounds.

Small Solids – Small Oil Drops & Tight Emulsions: In general, the smaller the solid particles, the tighter the emulsion (smaller emulsion drops) that can form and be stabilized. As a general rule, particles that are predominantly water-wet will stabilize an oil-in-water emulsion. Likewise, particles that are predominantly oil-wet will stabilize a water-in-oil emulsion.

Solids Stabilized Emulsions:

- Small solids (fines, scale particles, asphaltene particles) can stabilize an emulsion

- The smaller the solid particles, the smaller the drops of oil in water that can be stabilized

- The greater the amount of solids, the more oil in water emulsion that can be stabilized

- Dispersed solids contribute more to the stability of oil in water emulsions (rather than agglomerated solids)

- Solids that are interfacially active (accumulate at the interface) contribute more to the stability of an emulsion

- The wettability of the solids will determine the type of emulsion:

  - if solids are oil-wet, then they will stabilize a Water in Oil (W/O) emulsion

  - if solids are water-wet, then they will stabilize an O/W emulsion

  - most oil field solids start out as water wet (carbonate, barium sulfate, sand, clay, iron sulfides, hematite, etc)

  - asphaltenes can adsorb onto solids and make them oil-wet

  - other organics can change the wettability of solids, both natural (wax, resins, naphthenates, acids, soap) and synthetic (glycols, MeOH, acids, lube oils)

  - FeSx and schmoo are interfacially active at low concentrations

2.14 Concluding Remarks

The chemistry of produced and hydraulic fracture flow back water is a complex subject. There is a direct connection between the chemical composition of the produced fluids, the properties of the fluids, and the water treatment equipment and processes that are required to treat the water. A great amount of detail is required in order to support subsequent chapters where the discussion concerns the separation of contaminants and the purification of produced water. After the discussion of fluid mechanics (stokes Law and other subjects), and emulsions, the practical application of these subjects is provided in Chapter 4: Characterization.

As mentioned, in order to understand the composition of produced water, all four phases of the produced fluids (gas, oil, water and solids) must be characterized. The impact of CO_2 and H_2S on the pH through the system, and the consequent effect that pH has on mineral precipitation, surface activity of acid compounds in the crude oil and water, and the contribution that Water Soluble Organics make to the regulatory measured properties were discussed. The concentration and stability of asphaltenes and their properties were discussed at some length due to their importance in determining coalescence of oil droplets, the formation of oil-wet solids, and the overall ease of oil/water separation. Also discussed was the precipitation tendency of mineral ions which can also contribute to the formation of oily solids and the difficulty of removing such solids.

Since there is no one set of technologies that can be used for all types of produced water, the chemical nature of produced water must be taken into account. This chapter is one of the fundamental or background chapters that set the stage for a discussion of what is actually measured in the various regulatory and monitoring methods used to determine an oil-in-water concentration. Also this chapter sets the stage for the discussion of emulsions. The naturally occurring and man-made surfactants in produced water play a large role in stabilizing the emulsions that must be resolved (or broken) in order to clean the water. When a problematic produced water is encountered, it is worthwhile to determine if there is a naturally occurring or man-made (production chemical) responsible for stabilizing the emulsion. Identification of which chemical constituents can lead to stable emulsions is discussed briefly in this chapter, and more extensively in Chapter 3.

References to Chapter 2

1. B.T. Hlidek, M.J. Economides, T. Martin, **_Modern Fracturing—Enhancing Natural Gas Production._** Energy Tribune Publishing (2008).

2. A.B. Carpenter, "Origin and Chemical Evolution of Brines in Sedimentary Basins," SPE-7504 (1978).

3. Volkova (need to find this reference)

4. P. Webb, O. Kuhn, "Enhanced scale management through the application of inorganic geochemistry and statistics," SPE-87458 paper presented at the 6th International Symposium on Oilfield Scale, Aberdeen (May 2004).

5. T. Frankiewicz, C. Kittipong, L. Lien, "The use of ultra-filtration and reverse osmosis to remove hydrocarbons and heavy metals from oilfield produced water," paper presented at the Produced Water Society, Houston (2001).

6. I.K. Zherebtsova, N.N. Volkova, "Experimental study of the behaviour of trace elements in the process of natural solar evaporation of Black Sea water and Sasyk-Sivash brine," Geochem. Int., v. 3, p. 656 – 670 (1966).

7. W. Meredith, S.-J. Kelland, D.M. Jones, "Influence of biodegradation on crude oil acidity and carboxylic acid composition," Org. Geochem., v. 31, p. 1059 (2000).

8. B. Skaare, J. Kihle, T. Torsvik, "Biodegradation of crude oil as potential source of organic acids in produced water," Chapter 4 from Produced Water, Environmental Risks and Advances in Mitigation Technologies, Ed.: K. Lee, J. Neff, Springer Science (2011).

9. J.G. Speight, High Acid Crudes, Gulf Professional Publishing (GFF), an Elsevier Company (2014).

10. J. Walsh, T. Frankiewicz, "Treating produced water on deepwater platforms: Developing effective practices based upon lessons learned," SPE 134505 (2010).

11. T.I.R. Utvik, "Chemical Characterization of Produced Water From Four Offshore Oil Production Platforms in the North Sea," Chemosphere, Vol. 39, No. 15, pp. 2593-2606 (1999).

12. T.I.R. Utvik, J.R. Hasle, "Recent knowledge about produced water composition and the contribution from different chemicals to risk of harmful environmental effects," SPE – 73999 and reprinted in JPT December (2002).

13. R.P.W.M. Jacobs, R.O.H. Grant, J. Kwant, J.M. Marquenie, "The composition of produced water from Shell operated oil and gas production in the North Sea," from Produced Water – Technological / Environmental Issues and Solutions, Ed.: J.P. Ray, F.R. Engelhart, Environmental Science Research, v. 46, Plenum Press, p. 13 (1992).

14. M.T. Stephenson, "Components of produced water: a compilation of industry studies," SPE 23313, paper printed in the Journal of Petroleum Technology, Society of Petroleum Engineers, p. 548 May (1992).

15. K. Lee, J. Neff, <u>Produced Water, Environmental Risks and Advances in Mitigation Technologies,</u> Springer Science (2011).

16. J.P. Ray, F.R. Engelhart <u>Produced Water – Technological / Environmental Issues and Solutions,</u> , Environmental Science Research, v. 46, Plenum Press, p. 13 (1992).

17. R.M. Allen, K. Robinson, "Environmental aspects of produced water disposal," SPE – 25549, paper presented at the SPE Middle East Oil Technical Conference and Exhibition, Bahrain (1993).

18. R.M. Pashley, M.E. Karaman, <u>Applied Colloid and Surface Chemistry</u>, Wiley, England (2004).

19. M. Yang, personal communication (May 2014).

20. M. Yang, "Oil in Produced Water Analysis and Monitoring in the North Sea," SPE 102991 (2006).

21. H.P. Ronningsen, J. Sjoblom, L. Mingyuan, "Water-in-crude oil emulsions from the Norwegian continental shelf. 11. Ageing of crude oils and its influence on the emulsion stability," Coll. Surf. A, v. 97, p. 119 (1995).

22. M.-H. Ese, J. Sjoblom, H. Fordedal, O. Urdahl, H.P. Ronningsen, "Aging of interfacially active components and its effect on emulsion stability as studied by means of high voltage dielectric spectroscopy measurements," Coll. Surf. A, v. 123, p. 225 (1997).

23. M.-S. Lin, J.M. Chaffin, M. Liu, C.J. Glover, R.R. Davison, J.A. Bullin, "The effect of asphalt composition on the formation of asphaltenes and their contribution to asphalt viscosity." Fuel Sci. Tech. Int., v. 14, p. 139 (1996).

24. United States Environmental Protection Agency, Method 1664, Revision A: n-Hexane extractable material and silica gel treated n-hexane extractable material by extraction and gravimetry, EPA-821-R-98-002 (1999).

25. International Association of Oil & Gas Producers, "Aromatics in Produced Water: Occurrence, Fate & Effects, and Treatment," Report No. 1.20/324 January (2002).

26. A. Descousse, K. Monig, K. Voldum, "Evaluation study of various produced-water treatment technologies to remove dissolved aromatic compounds," SPE – 90103, paper presented at the SPE Annual Conference and Exhibition, Houston (2004).

27. M.A. Reinsel, J.J. Borkowski, J.T. Sears, "Partition Coefficients for Acetic, Propionic, and Butyric Acids in a Crude Oil / Water System" J. Chem. Eng. Data v. 39, p. 513 – 516 (1994).

28. P.V. Hemmingsen, S.Kim, H.E. Pettersen, R.P. Rodgers, J. Sjoblom, A.G. Marshall, "Structural characterization and interfacial behavior of acidic compounds extracted from a North Sea Oil," Energy & Fuels, v. 20, p. 1980 (2006).

29. M.A. Reinsel, J.J. Borkowski, J.T. Sears, "Partition Coefficients for Acetic, Propionic, and Butyric Acids in a Crude Oil / Water System" J. Chem. Eng. Data v. 39, p. 513 – 516 (1994).

30. D.B. MacGowan, "Carboxylic acid anions in formation waters, San Joaquin basin and Louisiana Gulf coast, USA – implications for clastic diagenesis," Appl. Geochem., p. 687 (1990).

31. J. Lochte, E.R. Littmann, The Petroleum Acids and Bases, (1955).

32. D. Arla, A. Sinquin, T. Palermo, C. Hurtevent, A. Graciaa, C. Dicharry, "Influence of pH and water content on the type and stability of acidic crude oil emulsions," Energy & Fuels, v. 21, p. 1337 (2007).

33. W.K. Seifert, "Carboxylic acids in petroleum and sediments," Chapter 1 of Progress in the Chemistry of Organic Natural Products, Springer (1975)

34. W.K. Seifert, W.G. Howells, "Interfacially active acids in a California crude. Isolation of carboxylic acids and phenols," Analytical Chem., v. 41, p. 554 (1969).

35. "Naphthenate Deposits, Emulsions Highlighted in Technology Workshop," Journal of Petroleum Technology (July 2008).

36. A.G. Shepherd, A mechanistic analysis of naphthenate and carboxylate soap-forming systems in oilfield exploration and production, thesis submitted to Heriot-Watt University (2008).

37. H. Mediaas, K.V. Grande, J.E. Vindstad, "Efficient management of calcium naphthenate deposition at oil fields," Statoil presentation to TEKNA (26 Sept 2007).

38. T.D. Baugh, K.V. Grande, H. Mediaas, J.E. Vindstad, N.O. Wolf, "The discovery of high-molecular weight naphthenic acids (ARN Acid) responsible for calcium naphthenate deposits," SPE 93011, paper presented at the SPE International Symposium on Oilfield Scale, Aberdeen (2005).

39. P.R. Hart, "Removal of Water Soluble Organics from Produced Brine without Formation of Scale," SPE 80250 (2003).

40. E. Tegelaar, "New Guidelines for the Modified IP 143 / 57 Method," Baseline Resolution, Inc. (1999).

41. N. v. d. T. Opedal, NMR as a tool to follow destabilization of water-in-oil emulsions, PhD thesis submitted to the Norwegian University of Science and Technology, Trondheim, Norway (2011).

42. P.K. Kilpatrick, P.M. Spiecker, "Asphaltene emulsions," Chapter 30 in Encyclopedic Handbook of Emulsion Technology, Ed.: J. Sjoblom, Marcel Dekker, New York (2001).

43. J. Sjoblom, N. Aske, I.H. Auflem, O. Brandal, T.E. Havre, O. Saether, A. Westvik, E.E. Johnsen, H. Kallevik, "Our current understanding of water-in-crude oil emulsions. Recent characterization techniques and high pressure performance," Adv. Colloid Int. Sci., v. 100, p. 399 (2003).

44. E.Y. Sheu, O.C. Mullins, Asphaltenes – Fundamentals and Applications, Plenum Press, NY (1995).

45. S. Acevedo, M.A. Ranaudo, G. Escobar, L.B. Gutierrez, X. Gutierrez, "A unified view of the colloidal nature of asphaltenes," from E.Y. Sheu, O.C. Mullins, Asphaltenes – Fundamentals and Applications, Plenum Press, NY (1995).

46. A. Stankiewicz, Proc. 3rd International Conf. on Petroleum Phase Behavior and Fouling, AIChE, New Orleans, Paper 47C, pp. 410-416 (2002).

47. K. Akbarzadeh, J. Ratulowski, T. Davies, "The importance of deposition measurements in the simulation and design of subsea pipelines," SPE-115131, paper presented at the SPE Annual Technical Conference and Exhibition, Denver, CO (2008).

48. K. Akbarzadeh, J. Ratulowski, T. Lindvig, T. Davies, Z. Huo, G. Broze, R. Howe, K. Lagers, "The importance of asphaltene deposition measurements in the design and operation of subsea pipe-

lines," SPE-124956, paper presented at the SPE Annual Technical Conference and Exhibition, New Orleans, LA (2009).

49. M. Grutters, K. Ramanathan, D. Naafs, E. Clarke, Z. Huo, D. Abdallah, S. Zwolle, A. Stankiewicz, "Integrated discipline approach to conquer asphaltene challenges in on-shore Abu Dhabi fields," SPE-138275, paper presented at the Abu Dhabi International Petroleum Exhibition & Conference, Abu Dhabi, UAE (2010).

50. M.-H. Eso, X. Yang, J. Sjoblom, "Film forming properties of asphaltenes and resins. A comparative Langmuir – Blodgett study of crude oils from North Sea, European continent and Venezuela," Coll. Polymer Sci., v. 276, p. 800 (1998).

51. H. Fordedal, Y. Schildberg, J. Sjoblom, J.-L. Volle, "Crude oil emulsions in high electric fields as studied by dielectric spectroscopy. Influence of interaction between commercial and indigenous surfactants," Coll. Surf. A, v. 106, p. 33 (1996).

52. J.D. McLean, P.K. Kilpatrick, "Comparison of precipitation and extrography in the fractionation of crude oil residua," Energy and Fuels, v. 11, p. 570 (1997).

53. J.G. Speight, "Petroleum Asphaltenes. Part 1 – Asphaltenes, Resins and the Structure of Petroleum," from Oil & Gas Science and Technology, v. 59, p. 467 (2004).

54. R. Cimino, S. Correra, A.D. Bianco, "Solubility and phase behavior in hydrocarbon media," Chapter 3 from E.Y. Sheu, O.C. Mullins, Asphaltenes – Fundamentals and Applications, Plenum Press, NY (1995).

55. P.M. Speiker, K.L. Gawrys, P.K. Kilpatrick, "Aggregation and solubility of asphaltenes and their subfractions," J. Colloid Int. Sci., v. 267, p. 178 (2003).

56. S.D. Faust, O.M. Aly, Chemistry of Water Treatment, 2nd Ed., Ann Arbor Press, MI (1998).

57. K. Suresh Kumar Reddy, A. Alshoaibi, C. Srinivasakannan, "High efficient metal sulfide based porous carbon matrix for mercury removal," SPE – 177701, paper presented at the Abu Dhabi IPEC, November (2015).

58. M. Wilhelm, A. McArthur, "Removal and Treatment of Mercury Contamination at Gas Processing Facilities," SPE 29721, paper presented at the SPE / EPA Conference, Houston (1995).

59. M. Zettlitzer, H.F. Scholer, R. Eiden, R. Falter, "Determination of elemental, inorganic and organic mercury in North German gas condensate and formation brines," SPE – 37260, February (1997).

60. N. Pongsiri, "Initiatives on mercury," SPE – 54523 (1999).

61. J.W. Patterson, "Capsule Report – Aqueous Mercury Removal," The US EPA, Office of Research & Development, Report No. EPA/625/R-97/004, Was., DC (1997).

62. T. Hayes, paper presented at the 11th Annual International Petroleum Environmental Conference, Albuquerque, NM, October (2004).

63. C. Yang, M. Galbrun, T. Frankiewicz, "Identification and resolution of water treatment performance issues on the 135 D platform," SPE – 90409, paper presented at the Society of Petroleum Engineers Annual technical Conference and Exhibition, Houston (2004).

64. D.L. Bernard, "NORM's in oil and gas production facilities," Presentation to the Produced Water Society, Houston (1993).

65. V.J. Brown, "Radionuclides in Fracking Wastewater, Managing a Toxic Blend," Env. Health Perspectives, v. 121, p A51 (2014).

66. S. Basu, "A review of the chemical characteristics of Frac / Flowback / Produced water," paper presented at the Workshop on Water Management in the Marcellus Shale, Atlantic City, NJ (2011).

67. J.M Silva, H. Matis, W.L. Kostedt, V. Watkins, "RPSEA Final Report 08122-36.Final. Produced water pretreatment for water recovery and salt production,"

68. Devey, "Computer Modelling Studies of Mackinawite, Greigite and Cubic FeS," Thesis, University of London, department of chemistry (2009)

69. Rickard, D; Luther,D.W., (2007) 'Chemistry of Iron Sulfides', Chem. Rev. 2007, 107, 514-562

70. S. Hunger, L.G. Benning, "Greigite: A true intermediate on the polysulfide pathway to pyrite," Geochemical Transactions 8, Article No.: 1 (2007).

71. L.G. Benning, R.T. Wilkin, H.L. Barnes, "Reaction pathways in the Fe-S system below 100°C," Chemical Geology 167 (1-2), pp. 25-51 (2000).

72. I.B. Butler, D. Rickard D, "Framboidal pyrite formation via the oxidation of iron (II) monosulfide by hydrogen sulphide', Geochimica et Cosmochimica Acta, v. 64, p. 2665–2672 (2000)

73. M.A.A. Schoonen, H.L. Barnes, "Reactions forming pyrite and marcasite from solution: I. Nucleation of FeS2 below 100°C," Geochimica et Cosmochimica Acta 55 (6) , pp. 1495-1504 (1991).

74. J.W. Morse, T. Arakaki, "Adsorption and coprecipitation of divalent metals with mackinawite (FeS)," Geochimica and Cosmochimica Acta, v. 57, p 3635-3640 (1993).

75. D. Csákberényi-Malasics, J.D. Rodriguez-Blanco, V. Kovacs Kis, A. Recnik, L.G. Benning, M. Posfai, "Structural properties and transformations of precipitated FeS," Chemical Geology, p. 249-258 (2012).

76. Rickard D.T., (1995) 'Kinetics of FeS precipitation, Part I. Competing reaction mechanisms', Geochim. Cosmochim. Acta, v. 59 (1995).

77. D. Wei, K. Osseo-Asare, "Particulate pyrite formation by the Fe+3 / HS-1 reaction in aqueous solutions: effects of solution composition," Coll. Surf. A, v. 118, p. 51 – 61 (1996).

78. Cornwell J.C., Morse J.S., "The characterization of iron sulfide minerals in anoxic marine sediments," Marine Chemistry, v. 22 (1987).

79. M. Wolthers, L. Charlet, P.R. van Der Linde, D. Rickard, C.H. van Der Weijden, 'Surface chemistry of disordered mackinawite (FeS) ', Geochimica et Cosmochimica Acta, v. 69, p. 3469 (2005).

80. R.B. Herbert, S.G. Benner, A.R. Pratt, D.W. Blowes, "Surface chemistry and morphology of poorly crystalline iron sulfides precipitated in media containing sulfate-reducing bacteria," Chemical Geology, v. 144, p. 87–97 (1998).

81. W.M. Bohon, D.J. Blumer, A.F. Chan, K.T. Ly, "Novel Chemical Dispersant for Removal of Organic/Inorganic "Schmoo" Scale in Produced Water Injection Systems," Corrosion 98, Paper no.73 (1998).

82. D. Peairs, "Silica over-saturation, precipitation, prevention and remediation in hot water systems," Cal Water Technical Note.

83. R.O. Fornier, J.J. Rowe, "The solubility of amorphous silica in water at high temperatures and high pressures," American Mineralogist, v. 62, p. 1052 (1977).

84. US patent 7438129, assigned to W.F. Heins, "Water treatment method for heavy oil production using calcium sulfate seed slurry evaporation," GE Ionics (2008).

85. R.W. Bowman, L.C. Gramms, R.R. Craycraft, "High-silica waters in steam flood operations," SPE Production & Facilities Journal, v. 15, p. 123 (2000).

86. J. Gaertner, "Produced water – types of oil treating," CANMET publication (2010).

87. T.C. Frankiewicz, M. Zaouk, "Removing hydrocarbons from methanol-water mixtures," Presentation to the Produced Water Society, Houston (2002).

88. J.C. Robinson, J. Keathley, P. Wilks, "The effects of methanol on the measurement of oil and grease in produced water with portable meters," Presentation to the Produced Water Society, Houston, TX (2003).

89. Ref: IUPAC Solubility Data Series, v. 7, Oxygen and Ozone, Pergamon Press (1981)

90. C.B. Kretschmer, J. Nowakowska, R. Wiebe, "Solubility of oxygen and nitrogen in organic solvents from – 25 C to 50 C," Ind. Eng. Chem., v. 38, p. 506 (1946).

91. N. Goodwin, J.M. Walsh, R. Wright, S. Dyer, G.M. Graham, "Modeling of the effect of triazine based sulfide scavengers on the in-situ pH and scaling tendency," SPE 141583 (2011).

92. R. Rickman, M. Mullen, E. Petre, B. Grieser, D. Kundert, "Practical use of shale petrophysics for stimulation design optimization: All shale plays are not clones of the Barnett Shale," SPE – 115258, paper presented at the SPE Annual Technical Conference and Exhibition, Denver, CO (2008).

93. R. Wolford, "Characterization of organics in the Marcellus Shale flow back and produced waters," MS Thesis, Penn. State Univ., Dept Civil & Env. Eng. (2011).

94. C.M. Moffitt, M.R. Rhea, P.B. Dorn, J.F. Hall, J.M. Bruney, "Short-term chronic toxicity of produced water and its variability as a function of sample time and discharge rate," p. 235 – 244 from Produced Water. Technological/Environmental Issues and Solutions, Ed.: J.P Ray, F.R. Engelhardt, Plenum Press, NY (1992).

95. M.B. Tomson, A.T. Kan, G. Fu, X. Wu, "Simultaneous analysis of total alkalinity and organic acid in oilfield brine," SPE 93266 (2005), paper presented at the SPE International Symposium on Oilfield Chemistry, Houston (2005).

96. C.D. Stiz, D.K. Barbin, B.J. Hampton, "Scale control in a hydrogen sulfide treatment program," SPE 80235.

97. J.I. Drever, "The Geochemistry of Natural Waters," Second Edition, Prentice Hall, New Jersey (1988).

98. W. Stumm, J.J. Morgan, Aquatic Chemistry," Second Edition, Wiley-Interscience, New York (1981).

99. N. Goodwin, S. Dyer, "Study 0659 Interim Report 2, v. 1.4. Introduction to scale prediction modeling – methodology and initial field calculations," (25-Jan 2010).

100. J.S.S. Damste, T.I. Eglinton, J.W. de Leeuw, P.A. Schenck, "Organic sulfur in macromolecular sedimentary organic matter: I. Structure and origin of sulphur-containing moieties in kerogen, asphaltenes and coal as revealed by flash pyrolysis," Geochim. Et Cosmo. Acta, v. 53, p. 873 (1989).

101. T. Al Mubarak, M. Al Khaldi, M. Al Mubarak, M. Rafie, H. Al-Ibrahium, N. Al Bokhari, "Investigation of acid-induced emulsion and asphaltene precipitation in low permeability carbonate reservoirs," SPE – 178034, paper presented at the SPE Saudi Arabia Section ATCE, Al-Khobar, Saudi Arabia (2015).

CHAPTER THREE

Fluid Mechanics

Chapter Three Table of Contents

3.0 Introduction

Fluid mechanics plays a major role in the design and operation of water treatment systems. It impacts the settling of solid particles and the rise of oil droplets and bubbles through water. It is important to know the rate at which these processes occur in order to design equipment that performs effectively in a given application. Fluid mechanics plays a role in the turbulent shearing of oil drops. The shearing in pipelines and flow lines is much different from the shearing in pumps, which is different from the shearing in valves. Pipelines and flow lines typically involve mild turbulence which actually promotes drop-drop coalescence. Valves and pumps typically shear oil droplets to very small diameters. Finally, the fluid flow patterns through tanks and vessels and the resulting fluid residence times are controlled by fluid mechanics.

The chapter begins with a discussion of Stokes Law including the calculation of rise velocity of a single isolated droplet surrounded by a continuous phase of water. This is obviously an idealized situation. In the case of a single isolated drop, the presence of other suspended matter is ignored. Nevertheless, this is a convenient place to start since it allows the simple introduction of a number of essential concepts in fluid mechanics. These concepts include Stokes Law, Stokes Factor, Reynolds Number, and Weber Number.

Following the discussion of a single drop, the presentation is generalized to consider a large number of droplets with a distribution of droplet diameters. The concept of droplet diameter distributions is described. This is an important topic since the performance of gravity-based separation equipment depends, to a great extent, on the droplet size distribution.

The subject of droplet shearing is discussed in some detail. The discussion starts with an in-depth presentation of turbulence and the formulas that allow calculation of turbulence energy dissipation rate. Using these formulas, the maximum droplet diameter that can exist in a flow field with a specified turbulence intensity is calculated. The application of these concepts to the design of low shear valves and pumps is presented in Section 3.5 (Shearing of Oil Droplets).

The topics of collision rate and coalescence rate are discussed in Chapter 4 (Emulsions). That material draws heavily on the formulas for turbulence energy dissipation rate which are established here in this chapter. The final subject of the chapter is multi-phase flow. This is a complex subject but it has relevance to shearing in flowlines and in the sizing and design of water treatment process equipment.

The material in this chapter is intended to provide useful formulas for calculating various fluid mechanics effects. Also, it is intended to provide some understanding of turbulence which is important for the design and sizing of water treatment equipment now and will become even more important as engineers develop and implement low shear valves and pumps in the future.

3.1 Continuum Mechanics

In Chapter 2 (Chemistry) there is a focus on the molecular composition and characteristics of produced water. In this chapter and the chapters to follow, there is considerable discussion of separation of oil and solids from water on the basis of the density difference between the dispersed phase (oil droplets, solid particles) and the continuous phase (water). In primary separators this density difference manifests itself as a buoyant force which makes the oil droplets rise and makes most solids fall. In order to understand settling, creaming, shearing and coalescence, it is more practical to consider a larger scale of dimensions and forces beyond the molecular scale that was the focus of water chemistry discussed in Chapter 2.

Molecular Scale: Molecular dynamics is a computational method that is used to simulate the motion of and interactions between molecules and their environment. The computations require knowledge of atomic mass, potential energy between the atoms of the molecules, potential energy between molecules, kinetic energy, and the number of molecules in the container. These computations are said to "simulate" the motion of molecules and can be graphically visualized. A video of a molecular dynamics simulation shows the molecules moving through space, tumbling, crashing into each other and into the walls of their container. On the molecular scale the fluid appears chaotic.

Continuum Properties: Observing the same fluid from the standpoint of a human, it is easy to convince ourselves that the fluid is uniform and that it can be described accurately by a few properties such as density, pressure, temperature. This is the continuum scale. On this scale, the "average manifestation" of the molecules and not the molecular motion itself are of greatest interest. For many purposes, average properties are adequate to describe the fluid characteristics.

In continuum mechanics, properties such as viscosity, interfacial tension, shear stress and turbulence intensity are defined in order to account for the aggregate action of molecules at surfaces and in bulk phases. These properties are "useful fictions," so to speak. While being practical, they must nevertheless be defined in order to avoid confusion and inaccuracy.

Colloidal Scale: Colloid and interface science is generally concerned with dispersions or suspensions having diameters less than one micron and larger than several hundred nanometers. At the small end of this range the particles display chaotic movement due to the bombardment of the surrounding molecules (Brownian motion). This is a type of motion that is described neither by molecular mechanics (because the dispersed particles are too big), nor by continuum mechanics (because it does not account for individual molecular motions).

Combined Molecular and Continuum Approach: When dealing with small droplets of oil or sand dispersed in water, as is the case with produced water treatment, neither the molecular scale, nor the macroscopic scale alone are entirely adequate. The molecular scale remains too small for practical calculations. The macroscopic scale, by itself, does not provide important insight that can be gained by considering molecular interactions.

As will be shown, for some of the calculations carried out in this text, both the molecular and the continuum perspectives are used. For example, surface tension is a continuum property. The formal definition of surface tension does not require consideration of molecular interactions. However, the origin and meaning of surface tension can only be understood by considering the attractive interactions between molecules. There are other examples. In order to understand and remember the sign of a pressure change across a curved interface (Young-Laplace equation) it is helpful to consider these molecular interactions. Understanding the role of interfacial tension in the calculations of droplet shearing or coalescence is facilitated by considering the effect of molecular interactions.

An objective of this text is to achieve a practical approach where the details of molecular interactions, or formal definitions of continuum properties does not get in the way of providing insight and solving problems in equipment design and troubleshooting.

Dimensionless Numbers: Dimensionless numbers help to establish the range of size, time scale, forces, and momentum that are important in the discussion of fluid mechanics principals. This discussion establishes what is meant by a "small" droplet, "high" shear, or an "energetic" collision between droplets. All of the terms are relative to the dimensions of, and forces between small droplets and particles dispersed in water. The units and scales are quite different from the macro-scale units which petroleum engineers deal with on a daily basis. A 6 foot diameter separator might be average for a given application, whereas a 100 micron diameter droplet might be average as well. These two contrasting size scales are discussed in a way that allows an engineer to have a sense of their relative importance and to understand how these vastly different size scales interact and play their respective roles in the sizing, design, and evaluation of water treatment equipment.

This is the role of dimensionless numbers. The Reynolds number, for example, is a familiar dimensionless number. It indicates the relative magnitude of convection forces versus viscous forces in describing momentum transfer in a fluid. A high Reynolds number indicates a turbulent flow situation. Another dimensionless number that is even more familiar is π. It provides the ratio of various geometrical quantities. For example, $\pi / 4$ is the ratio of the area of a circle to that of a square. Likewise, $\pi / 2$ is the ratio of the perimeter length of a circle to that of a square. These ratios can be calculated for large or for small objects and the ratios stay the same. They give significant meaning to the number π. In this chapter, two important dimensionless numbers are introduced, the Reynolds number and the Weber number. They are used extensively in produced water treatment.

3.2 Droplet Rise Velocity, Stokes Law, and Related Quantities

In this section, the properties of a single droplet in a continuous fluid are discussed. These properties include the rate of rise (creaming rate) of a single droplet in a continuous phase, given by Stokes Law. Other properties discussed include the Stokes Number, Stokes Factor, Reynolds Number, Capillary Number, and Weber Number. Most of the equations discussed in this section involve viscosity. Since the units of viscosity are confusing to most engineers, this section begins with a brief review.

Units of Viscosity: Almost all engineers know that water has a viscosity of about 1 centi-Poise. While this is easy to remember, most engineers do not recall what a Poise is or how it relates to other units of measure. In SI units the viscosity of water near room temperature is 1 milli-Pascal second (0.001 Pa s). This allows conversion directly into SI units since most (international) engineers remember that one Pa is equal to one N/m² (Newton per meter squared).

$1\ cP = 1\ mPa{\cdot}s = 0.001\ Pa{\cdot}s = 0.001\ kg/(m\ sec)$.

In the above notation, mPa·s refers to a milli-Pa·s (1×10-3 Pa·s). Values of water viscosity can be interpolated from the following table.

Table 3.1 Viscosity of water as a function of Temperature.

Temperature (°C)	Viscosity (Pa·s)	Viscosity (cP)
20	1.003×10−3	1.003
30	0.7978×10−3	0.7978
40	0.6531×10−3	0.6531
50	0.5471×10−3	0.5471
60	0.4658×10−3	0.4658
70	0.4044×10−3	0.4044
80	0.3550×10−3	0.3550
90	0.3150×10−3	0.3150

The viscosity of other fluids is given in the table below.

Table 3.2 Viscosity of common fluids.

Fluid	Viscosity (Pa s)	Viscosity (cP)
Honey	2–10	2,000–10,000
Molasses	5–10	5,000–10,000
Chocolate syrup	10–25	10,000–25,000
Ketchup	50–100	50,000–100,000
Peanut butter	1,000	1,000,000

3.2.1 Oil Density vs Temperature

Since the difference between the density of oil in a droplet which is dispersed in produced water and the density of the water (or brine) is an important factor in Stokes Law, a method is presented here for calculating the density of an oil phase at process conditions. In most cases, the pressure of the system will not have a significant impact on oil density. However, some reduction of the lab-measured dead oil viscosity under "live" or gas-saturated conditions may be noticed.

To correct the density of an oil from the temperature at which it is measured to a density at 60oF, the procedure specified in ASTM 1250 is used. The procedure refers to a set of density tables. If the density vs temperature tables are inconvenient or not readily available, then the following equation can be used with the expansion factor "K" derived using Figure 3.1.

$$P\ (ToC) = \rho\ (20oC)[1 - K(ToC - 20)] \qquad \text{Eqn (3.1)}$$

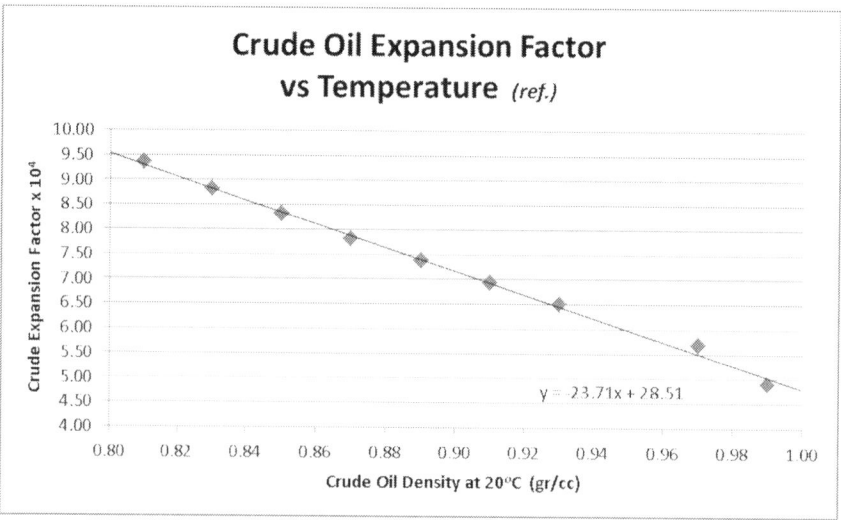

Figure 3.1 The Crude Oil Expansion Factor, "K" is plotted vs the measured density of the crude at 20°C.

3.2.2 Reynolds Number

The Reynolds number is a dimensionless number. It is defined as the ratio of the convective force to the viscous force in a fluid under a given set of flow conditions. As such, it can be used to determine if a particular instance of fluid flow is turbulent or not.

The Reynolds number can be derived by many approaches. It was originally derived a couple hundred years ago by men who built small scale replicas as a means to demonstrate the magnitude of fluid forces. The studies were aimed at understanding the eddies created by the wake of a boat, the steady or turbulent flow through a canal, flow over a dam, storms in the atmosphere, and air flow across a sail of a boat, and so on. By matching the Reynolds number of the replica to that of the full scale object, they were able to obtain measurements in the lab that corresponded to those of the full scale object. It is indeed an important tool in understanding fluid motion. It can also be derived by converting the Navier-Stokes equation into dimensionless form. When all of the variables are expressed in their non-dimensional form, there is a single group of parameters left over. This group of parameters is the Reynolds number. It is dimensionless.

As will be discussed in other sections, some of the formulas for settling, shearing of droplets, and coalescence can only be applied accurately over a specified range of the Reynolds Number. For example, the rise velocity of a bubble can be calculated by Stokes Law for small values of the Particle Reynolds Number ($N_{RE} < 1$), a quantity which is dimensionally different from the Fluid Reynolds Number – as will be explained below. For larger values of the Particle Reynolds Number a different formula must be used. For fluid flow in straight pipe, the transition from laminar flow to turbulent flow occurs at around a Fluid Reynolds number of 2,200. Below that value the pressure drop can be accurately calculated using one formula. Above that value, turbulence must be taken into account and a different formula must be used.

<u>Reynolds Number for a Particle, Droplet or Bubble in Water:</u> For a spherical object such as a droplet of liquid or a spherical solid particle, the Particle Reynolds number is defined as:

$$N_{RE} = \frac{du\rho_w}{\mu_w}$$

Eqn (3.2)

where:

u = fluid / particle relative velocity (m/sec)
r_w = density of water phase (kg/m3)
d = diameter of droplet (m)
m_w = viscosity of water phase (Pa sec)

For our purposes, the continuous fluid will always be water. The equation applies to solid particles, oil droplets, or gas bubbles dispersed in water. The velocity in this equation (u) is the <u>relative velocity</u> between the particle (solids particle, oil droplet or gas bubble) and the bulk fluid (water). When SI units are used, no conversion factors are required. An example of the calculation for a 100 micron oil droplet with a relative velocity of 0.00167 m/sec (10 cm/minute) in water is given by the following:

$$N_{RE} = \frac{(100 \times 10^{-6}\ \text{m})(0.00167\ \text{m/sec})(1000\ \text{kg/m}^3)}{0.001\ \text{kg/m sec}} = 0.167$$

This is a relatively small value. Since it is less than one, it signifies that Stokes Law is applicable, as will be discussed shortly. In SI units, it is straightforward to confirm that the Reynolds number is dimensionless. As discussed in the section below on Stokes Law, the unit of viscosity (Pa sec) is kg /

(m sec). Thus, the units in the denominator are kg / (m sec). By cancelation, the units in the numerator are also kg / (m sec). Therefore the number is dimensionless.

When cgs units are used, a conversion factor is required and the equation is written as:

$$N_{RE} = 1.67x10^{-4} \frac{du\rho_w}{\mu_w}$$

Eqn (3.3)

where:

u	= oil droplet rise velocity (cm/min)
r_w	= density of water phase (g/cm3)
d	= diameter of oil droplet (microns)
μ	= viscosity of water phase (cP)

Using the same values as in the equation above gives:

$$N_{RE} = 1.67x10^{-4} \frac{(100 \text{ micron})(10 \text{ cm/min})(1 \text{ g/mL})}{1 \text{ cP}} = 0.167$$

It is straightforward to confirm that this equation results in the same value for the Reynolds number as the previous equation. Since this is a small value and is considerably less than unity, Stokes Law is applicable. For larger values of the droplet Reynolds number, there will likely be some turbulence present. A modified version of Stokes Law is required for that case. This is discussed in greater detail below. Next we discuss the Reynolds number for pipe flow.

Reynolds Number for Pipe Flow: For a cylindrical pipe, with circular cross-section, the Reynolds number is defined as:

$$N_{RE} = \frac{Du\rho_w}{\mu_w}$$

Eqn (3.4)

where:

u	= fluid velocity (m/sec)
r_w	= density of water phase (kg/m3)
D	= diameter of pipe (m)
m_w	= viscosity of water phase (Pa sec)

When SI units are used, no conversion factors are required. An example of the calculation is given by the following, where the fluid is traveling at a speed of 3 m/sec, the pipe is 6 inch (0.15 m), the fluid viscosity is 1 cP, and the fluid density is 1 gr/cm³.

$$N_{RE} = \frac{(0.15 \text{ m})(3.0 \text{ m/sec})(1000 \text{ kg/m}^3)}{0.001 \text{ kg/msec}} = 300,000$$

This value for the Reynolds number in a pipeline indicates that the flow is turbulent. The transition from laminar to turbulent flow occurs at around 2,200. Such high values of the Reynolds number for pipe flow are relatively common in transporting produced water through pipes in a facility. From a piping standpoint, a value of 300,000 is relatively high. It indicates that there will be significant pressure loss. When such high values are calculated, it is necessary to calculate the turbulence energy dissipation which will indicate whether the turbulence, on the small scale of droplets, is mild or severe. In the former case (mild turbulence) coalescence of oil droplets will occur. In the later case (severe turbulence) droplets will break apart. The calculation of turbulence energy dissipation is discussed below. In SI units, as used in the equation, it is straightforward to confirm that the Reynolds number is dimensionless. As discussed in the section below on Stokes Law, the unit of viscosity (Pa sec) is kg / (m sec).

3.2.3 Stokes Law Settling (and Creaming) Rate

Solid particles are generally, though not always, heavier than water and therefore sink. Oil droplets are generally lighter than water and therefore rise (cream). For a spherical contaminant, falling or rising, there will initially be a brief period of acceleration after which the particle reaches what is referred to as terminal velocity. At the terminal or steady state velocity, the viscous drag force is equal in magnitude to the gravitational body force. These two forces act in opposite directions. The velocity is steady due to the balance of these two forces. The viscous drag force on a spherical droplet, particle or bubble having diameter "d" is given by:

$$F_{drag} = 3\pi d u \mu$$
Eqn (3.5)

The buoyancy force of the droplet or particle in water is given by:

$$F_{gravity} = \pi d^3 g(\rho_w - \rho_o)/6$$
Eqn (3.6)

where:

u	= oil droplet velocity (m/sec)	
r_w	= density of water phase (kg/m³)	
r_o	= density of contaminant or dispersed phase (kg/m³)	
d	= diameter of oil droplet (m)	
μ	= viscosity of water phase (Pa sec = N sec / m²)	
g	= the gravitational constant (9.81 m/sec²)	

The balance of these two opposing forces determines the terminal velocity of the particle or droplet. By equating these two equations, Stokes Law is obtained.

Stokes Law Equation: The following equation is typically referred to as "Stokes Law." The equation is valid as written when all parameters are expressed in SI Units. The equation gives accurate results for laminar flow around a spherical object. This is discussed later.

$$u = \frac{g(\rho_w - \rho_d)d^2}{18\mu}$$
Eqn (3.7)

Where:

u	= oil droplet (terminal) rise velocity (m/sec)	
r_w	= density of water phase (kg/m³)	
r_o	= density of contaminant or dispersed phase (kg/m³)	
d	= diameter of contaminant (m)	
μ	= viscosity of water phase (Pa sec = N sec / m²)	
g	= the gravitational constant (9.81 m/sec²)	

When SI units are used, no unit conversion factors are required. An example of the calculation is given by the following for a 100 micron oil droplet (Density = 800 kg/m³) in water (Density = 1100 kg/m³) :

$$u = \frac{(9.81 \text{ m/s}^2)(1100 \text{ kg/m}^3 - 800 \text{ kg/m}^3)(100\text{x}10^{-6}\text{m})^2}{18(1\text{x}10^{-3} \text{ kg/ms})} = 1.64\text{x}10^{-3} \text{ m/s}$$

Eqn (3.8)

u = 9.81 cm / minute

When cgs units are used, the following conversion factor is required and the equation is written as:

$$u = 3.27\text{x}10^{-3} \frac{(\rho_w - \rho_o)d^2}{\mu}$$

Eqn (3.9)

where:

u = oil droplet rise velocity (cm/min)
r_w = density of water phase (g/cm³)
r_o = density of oil phase (g/cm³)
d = diameter of oil droplet (microns)
μ = viscosity of water phase (cP)

Substituting the previous values gives the expected result:

$$u = 3.27\text{x}10^{-3} \frac{(1.1 \text{ g/cm}^3 - 0.8 \text{ g/cm}^3)(100 \text{ micron})^2}{1 \text{ cP}} = 9.81 \text{ cm/minute}$$

Eqn (3.10)

Stokes Law only applies in the case where the particle Reynolds Number is roughly unity or smaller. As the particle diameter declines, so does the Particle Reynolds number. However, at some point, as the diameter of a particle declines, the Particle Reynolds number is no longer a good predictor for the validity of Stokes Law. For very small particles, e.g., less than 1 to 5 microns, Brownian motion effects become significant and can inhibit the rise or fall of a particle or droplet. Also, if the fluid velocity becomes sufficiently high, then other hydrodynamic principles relating to the suspension and dispersion of particulate in a flowing fluid become important and will effectively inhibit the rise of an oil droplet or the fall of a solid particle in a vessel or tank. Finally, in bulk liquid, convection currents can become a limiting factor which enhances or inhibits the ability of a particle or oil droplet from achieving its theoretical net upward or downward terminal velocity in a vessel or tank.

Stokes Law for a Bubble: If Stokes Law is applied to a gas bubble in water, then the density difference between water and the gas bubble becomes essentially the density of the water itself. If mixed units are used, then the following numerical result is obtained:

$$u = 5.45\text{x}10^{-5}d^2$$

Eqn (3.11)

where:

u = gas bubble rise velocity (cm/sec)
d = diameter of gas bubble (microns)

For example, a 100 micron bubble will rise 0.5 cm / sec.

Stokes Law Applications: As will be demonstrated presently, it is possible to use Stokes Law directly in the field using simple bottle samples and a timer to estimate the size distribution for an ensemble of oil droplets dispersed in a water sample. No calculation is necessary.

As shown, a droplet of oil, having a diameter of 100 micron, having a density of 0.8 g/mL, in water having a density of 1.1 g/mL, and water having a viscosity of 1 cP (0.001 Pa sec), will rise at a rate of 9.81 cm / minute. Based on this analysis, a rough approximate result is that a 100 micron droplet will rise 10 cm in one minute.

This rule of thumb is easy to remember: 100 microns / 10 cm / 1 minute. This rule of thumb leaves out a lot of detail, which will be added below. But it provides a simple useful example for a quick evaluation of the water clarification capability of a vessel or tank which can be useful for design engineers, equipment specialists, and surveillance engineers. Anyone who looks at a vessel drawing or at a vessel itself should be able to quickly determine, roughly, how much time is required for a 100 micron droplet to reach the oil / water interface. It is worthwhile to develop a physical intuition about the rise velocity of oil drops. This rule of thumb is a good starting point.

A 100 micron droplet is visible, if the oil is dark and if the water is relatively clear. The distance, 10 cm (4 inches) is easily visualized or marked on a sample jar. One minute is a convenient time interval for observing a sample. If a sample is taken, with a depth of roughly 10 cm (4 inch) and a layer of oil forms at the top within a minute, it is roughly true that the droplets that rose to the top were 100 micron or larger.

As a practical field exercise, take a water sample and observe how dark it appears – or take a photo. Observe how the sample clarifies over a 1 minute time period and how the sample color has changed. Then note the sample color again or take another photo. From these observations, one can, with a little practice, estimate the fraction of droplets that are 100 microns in diameter and larger. By taking repeat samples and samples at several process locations, one can judge how the oil droplet size changes as a result of process equipment. It may in some cases also be possible to estimate the oil concentration in the whole sample after some settling has occurred. With an understanding of Stokes Law and a few astute observations, one can learn a lot about the water quality in a system without the use of sophisticated instrumentation that requires training, special handling, permitting, and approval.

3.2.4 Stokes Law for Higher Reynolds Numbers

In some applications the oil droplet or solid particle moves rather quickly through the fluid. In that case, it is necessary to account for the presence of turbulence around the particle.

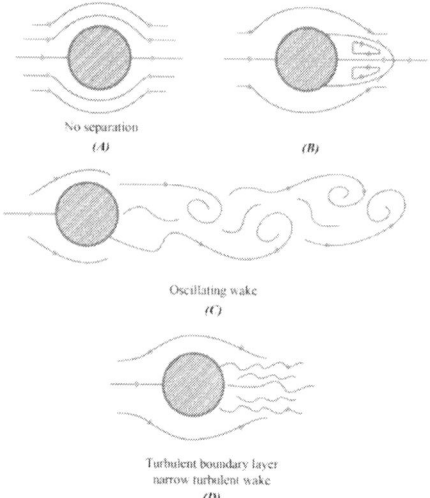

Figure 3.2 Flow patterns around a non-deformable sphere. (A) shows no separation of streamlines around the sphere. This is laminar flow where Stokes Law applies. All other flow patterns involve some deviation from Stokes Law.

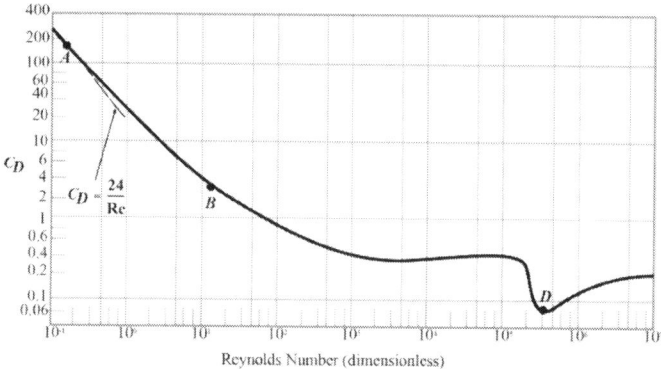

Figure 3.3 Drag coefficient for flow around a sphere.

For the case where the particle does not have a spherical shape or the Reynolds number is greater than unity, a different form of Stokes Law is used:

$$u = \left[\frac{3}{4} \frac{gd(\rho_w - \rho_d)}{\rho_w C_D} \right]^{1/2}$$

Eqn (3.12)

Where:

u	= oil droplet rise velocity (m/sec)
r_w	= density of water phase (kg/m³)
r_d	= density of oil droplet or discrete particle (kg/m³)
d	= diameter of oil droplet or discrete particle (m)
μ	= viscosity of water phase (Pa sec = N sec / m²)
g	= the gravitational constant (9.81 m/sec²)
C_D	= the Drag Coefficient (see below)

If the Reynolds number is less than unity, then $C_D = 24/N_{RE}$ and Stokes Law is recovered. If the Reynolds Number is greater than unity or the shape is not spherical, then Figure 3.3 can be used to obtain a value for C_D.

The final pertinent equation for this section is the Richardson-Zaki equation for hindered settling.

$$\text{Vhindered} = \text{Vterm}(1 - C)3.5 \quad (2.4 < \text{exponent} < 4.6) \qquad \text{Eqn (3.13)}$$

Where:

C = Vol. Fraction of discreet phase

This equation is not generally considered in the evaluation or sizing of a 3-phase separator or skim tank but is nevertheless conceptually important. The equation explains why droplets or particulate are slow to settle through an interface which is composed of a high density of dispersed liquid droplets. It also explains why the oil-water interface in a vessel or tank is slow to contract. The velocity of droplets during Hindered Settling are listed here to illustrates how rapidly the Terminal Velocity for settling declines as the volume fraction of the dispersed phase particulate increases in an interface emulsion. Equation (6) also helps to explain why solids which are settling out of an oil phase in a separator tend to become concentrated in an interface emulsion.

C = 0.2 è Vhind = 0.46 Vterm
C = 0.5 è Vhind = 0.09 Vterm.
C = 0.7 è Vhind = 0.01 Vterm

The effects of changes to the viscosity of an interface emulsion are also important to consider when evaluating the impact which a high concentration of dispersed particulate or oil droplets has on the calculation of a terminal settling velocity. These considerations, including the concepts of non-Newtonian viscosity and phase inversion points for an emulsion are discussed in Chapter 3 (emulsions) of this book. One effect a demulsifier or chemical which effectively collapses oil/water interface emulsions can have is to alter the phase inversion point for where the emulsion transitions from oil continuous to water continuous. Using the above values as an example, an interface emulsion which inverts at a 70% oil droplet concentration has a Vhindered which is 9X slower than Vhindered for an interface emulsion which inverts at a 50% oil droplet concentration.

3.2.5 Stokes Factor

The Stokes Factor is an important parameter in assessing the driving force for separation due to density and viscosity of the fluids. The Stokes Factor is a function of the fluid densities and the water viscosity which is primarily a function of temperature. It is not dependent on the oil droplet size. The following equation defines the Stakes Factor:

$$S_F = \frac{(\rho_w - \rho_d)}{\mu_W} \qquad \text{Eqn (3.14)}$$

Where:

S_F = Stokes Factor (sec/m^2)
r_w = density of water phase (kg/m^3)
r_o = density of oil phase (kg/m^3)
μ_W = viscosity of water phase (Pa sec)

When SI units are used, no unit conversion factors are required. An example calculation is given by the following:

$$S_F = \frac{(1100 \ kg / m^3 - 800 \ kg / m^3)}{(1x10^{-3} \ kg /(ms))} \qquad \text{Eqn (3.15)}$$

$$S_F = 300,000 \ sec/ \ m^2$$

The Stokes Factor has the units of sec/m2. These are non-intuitive units. Bear in mind however, that in order to calculate the creaming (or settling) velocity (m/sec), the Stokes Factor must be multiplied by the gravitational constant (m/sec^2), multiplied by the droplet diameter (m), and divided by the numerical constant (18). When this is done, the Stokes Law settling velocity is obtained. Thus, the Stokes Factor is only the fluid physical property part of the driving force for droplet movement. The Stokes Factor only has relevance when comparing one fluid to another, or when comparing the effect of water temperature and hence viscosity. The Stokes Factor is used from this point onward to account for fluid density and water viscosity. The effect of fluid properties, including temperature, on separation performance can be captured through calculation of the Stokes Factor.

Water Residence Time vs Stokes Factor: In designing a separator for oil/water separation, the geometrical dimensions of the vessel are determined from the flux rate required to resolve the dispersion band and the residence time required to perform initial oil dehydration and water deoiling. Together these parameters determine the volume of the liquid section, the height of the oil/water interface and the height of the oil/gas interface, and the cross sectional area of the liquid section of the vessel.

3.3 Particle and Droplet Size Distributions

Droplet size distribution is an important topic since the performance of most separation equipment depends on the distribution of droplet sizes going into the equipment. To have a good understanding of this important topic it is necessary to provide some precise explanations, and to give concrete numerical examples with the details of the calculations shown explicitly. First, a set of measured data is presented. The data has been manipulated a bit to smooth out the inevitable scatter and to provide a somewhat idealized set of numbers to work with. Nevertheless, the data shown are realistic. Various averages are calculated. Also, a distribution of droplet size is constructed from the measured data. Then, a set of mathematical distribution functions are presented. These distribution functions are mathematical representations of droplet size data. These math formulas have parameters that can be adjusted such that they mimic or represent the measured distribution of droplet sizes. They are used extensively in predicting the performance of water treatment equipment in a process system.

Application of Droplet Size Distributions: Instrumentation to measure droplet size has been available for laboratory and field use for several decades. In the 1980's early work was carried out at the Orkney Water Treatment Center using full scale hydrocyclone and flotation equipment. Droplet size distributions going into the equipment and coming out of the equipment was measured and reported. Over the next couple decades many laboratory and field studies have significantly advanced our underatnding of how the droplet size distribution changes throughtout the production process.

Distribution functions for oil droplet diameter, and solid particle diameter are used extensively in designing and troubleshooting equipment. This will be demonstrated in subsequent chapters. For now, the mathematical properties of droplet size measurements and distribution functions are presented.

Number Distribution versus Volume Distribution Droplet Size: One of the important concepts that is presented in this section is the concept of volume distribution which is distinctly different from number distribution. The number distribution has little importance to the operator who must remove oil. The operator needs to know the separation efficiency as a function of the volume distribution. The later tells the operator what size of droplet is associated with a certain volume (or concentration) of oil in water. Misunderstanding of these two quantities happens often and can lead to equipment installations that perform as specified according to the vendor, but do not perform as required by the operating company.

Instruments that Measure Droplet Diameters: There are various instruments used to measure droplet diameter. These instruments are important for troubleshooting where they may be used during a site visit to conduct a survey across the water treating process system. They can also be installed in permanent locations in order to provide feedback to the operators on a continuous basis. Operators then typically make adjustments to process operating parameters. In this sense, they are part of a manually operated control strategy. In any case, installation of such continuous monitoring equipment is strongly advocated. Actual equipment is discussed in Chapter 7 (Sampling and Analysis). For now, the results of droplet diameter measurements are discussed without delving into the details of the instrumentation and equipment used to measure it.

Numerical Example: There are several mathematical concepts that must be covered. The discussion is facilitated by working with a hypothetical but realistic data set as given in the table below. For illustrative purposes, it is advantageous to use a simple set of data that is easy to manipulate. Most of the calculations that will be discussed can be carried out by inspection.

The first step in analyzing the data is to describe what is typically reported. First, all of the particle size analyzers report the number of particles in a range of diameter values for a certain sample volume. This is much easier to understand by referencing the table below.

Column A gives a set of droplet diameters. These droplet diameters define what are referred to as bins. For example, a particle with a droplet diameter of 12 microns would be counted in the 15 micron bin. The 15 micron bin includes all droplets with a diameter between 10 micron and 20 micron. Particles with a diameter of precisely 20 micron are counted in the next larger bin, the 25 micron bin. Calculations using values in the 15 micron bin represent an average of the low (10 micron) and high (less than 20 micron) values of this bin. Bin size (in this case 10 micron) is something that can be chosen.

Column B is a simple calculation of the volume of a droplet having a diameter equal to the bin size diameter. The volume of the droplets will be used in the calculation of the total oil contained in this sample.

Column C is the number of droplets per mL of sample that have a diameter that falls into the bins in Column A. For example, there are 300 droplets that have a diameter that is greater than or equal to 10 micron and less than 20 micron. The average of these two limits is 15 micron, so the instrument reports that there are 300 droplets that have an average diameter of 15 microns.

Column D is the volume of oil associated with each bin of droplet diameters. It is calculated using columns A, B, and C. Column B (volume per drop) times Column C (number of droplets per mL) gives the volume of oil associated with droplets of bin size given by Column A. This result is given in Column D. Some unit conversions are required to calculate Column D. The product of column B times column C gives the volume of oil in units of cubic microns/mL. Dividing this by 1×10^6 gives the volume in units of microliter of oil per liter of oily water. This is equivalent to ppmv. Column D provides a very useful result. It is literally the parts per million of oil that is associated with droplets of various bin diameters. It tells the operator how small of oil droplets need to be removed in order to achieve the desired oil concentration.

Column E is the cumulative oil concentration. Column E, the final column of numbers, is associated with the very important question of what diameter of droplets must be removed to achieve the target oil in water concentration. For example, if the target oil in water concentration is 47 ppmv, then all droplets larger than 50 micron must be removed (50 micron is the upper limit of diameter for the 45 micron bin). If all droplets of 50 micron and larger are removed, then the total oil concentration will be 47 ppmv. Likewise, if all droplets of 40 micron and larger are removed, then the remaining oil concentration will be 18 ppmv.

Table 3.3 Hypothetical Droplet Size Data

Column A	Column B	Column C	Column D	Column E
drop diameter (micron)	volume of each drop (cubic micron)	number of drops per mL of water sample	vol of oil per mL of sample (ppmv)	cumulative vol oil (ppmv)
0	0.0E+00	0	0.0	0
5	6.5E+01	100	0.0	0
15	1.8E+03	300	0.5	1
25	8.2E+03	500	4.1	5
35	2.2E+04	600	13.5	18
45	4.8E+04	600	28.6	47
55	8.7E+04	500	43.6	90
65	1.4E+05	300	43.1	133
75	2.2E+05	100	22.1	156
Total		**3000**	**156**	

Number Average: From an inspection of the data in Table 3.3, it can be seen that the average droplet diameter is 40 microns (Column C). This is also the median for the number distribution of droplet sizes. In other words, there are as many droplets with diameters greater than 40 microns as there are droplets with diameters less than 40 microns. The number average is usually written as D_{N50} or as D_{50}.

Volume Average: However, the median volumetric droplet size is closer to 52 microns (column D). Half of the volume of the oil is in droplets of 52 micron diameter and larger. It is common to have the median values for number size distributions and volume size distributions be substantially different. The volume average is usually written as D_{V50}. The volume average should not be written as D_{50} since that would allow confusion with the number average.

The figure below provides a graphical display of the bin data contained in the table. Column C is plotted as the y-axis. Column A is plotted as the x-axis. Table 3.3 is a tabulation and Figure 3.4 is a graphical representation of what is referred to as the "number distribution." As suggested by the name, the number distribution is a distribution of the number of droplets in a given volume having a specified diameter.

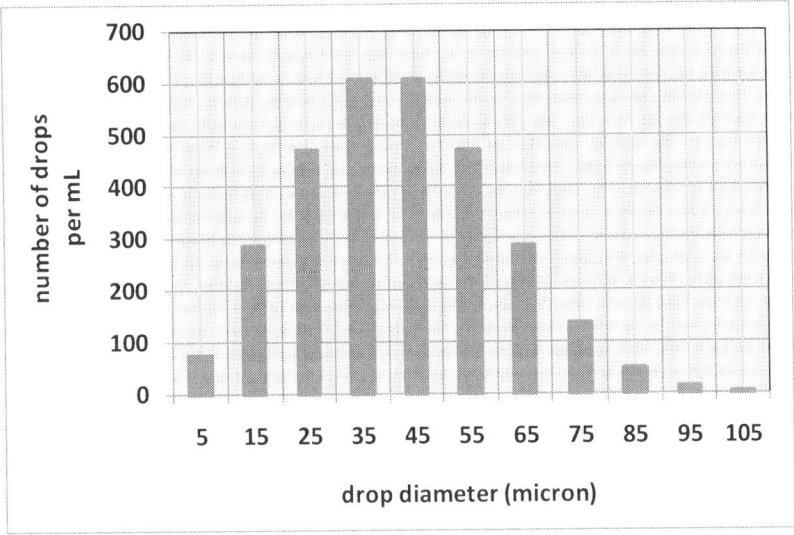

Figure 3.4 Graphical display of the data shown in Table 3.3. As shown, for example, there are 471 droplets per mL of water that have a diameter between 20 and 30 micron. These droplets are represented in the figure as a single bar centered at 25 micron. The total number of droplets per mL of sample can be calculated by simply adding the values of each bar.

An alternative way to report the results is given in the next figure. In that figure, the percent of number of droplets is reported. A smooth curve is drawn through the points.

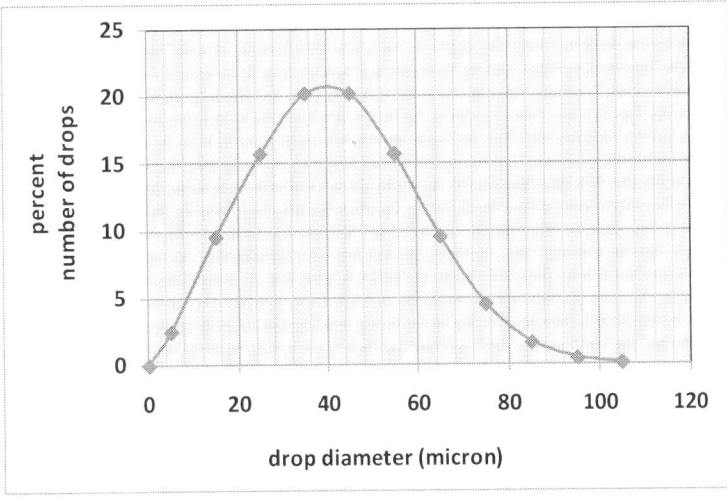

Figure 3.5 Graphical display of the data shown in the table. A smooth curve has been drawn in terms of the percentage of drops. It must be kept in mind that in going from a bar chart (Figure 3.4) to a smooth curve of percentage the total oil concentration cannot be calculated from a percentage curve.

Characteristic Parameter (e.g. D_{10}, D_{V50}, mean, etc): In addition to the information given in the figure, typical results will include the number of droplets in a specified volume, the volumetric concentration (ppmv), and various measures of the distribution such as D_{10}, D_{50}, D_{90}, D_{V10}, D_{V50}, D_{V90}, the mean and the median diameter. All of these quantities are discussed in detail below.

Volume Distribution: An important measure of droplet size distribution is the contribution that each droplet makes to the volume of oil. The figure above shows a hypothetical droplet size distribution expressed in terms of the ppmv (μL oil/L water or mL oil/m³ water). This volume distribution is based on the number distribution of the previous figure. In this hypothetical example, the instrument measured the number of droplets in the range 0 to 9.99 microns and reported the volume of oil at a size of 5 microns. Similarly, it measured the number of droplets in the range above 10 micron to 19.99 micron, and reported the volume of oil at 15 micron diameter and so forth.. The formula used to calculate the volume of oil is:

$$v_d = n_d \left(\frac{\pi}{6} \right) d^3$$

Eqn (3.16)

where:

n_d = number concentration of droplets of diameter d (number of drops/mL)
d = diameter of droplet (micron)
v_d = volume concentration of droplets of diameter d (cubic microns oil / mL water)

Note that n_d is the number concentration of droplets which was previously discussed. Of course, the value of v_d must be multiplied by 1x10⁶ in order to express the result in ppmv (μL oil/L water or mL oil/m³ water).

In effect, the instrument measures the volume concentration of droplets for various diameters. In this example, the total number of droplets is 3,000 droplets / mL. The total volume of droplets measured in a 1 L sample is 156 micro L, which is equivalent to 156 ppmv.

Number Average (Mean) Diameter: The number average or mean can be calculated by adding all of the values of droplet diameters and dividing by the number of droplets. This is the basis of the common formula given below:

$$D_{50} = <d> = \frac{1}{N} \sum_{i=1}^{N} d_i$$

Eqn (3.17)

where:

N = total number of droplets in the sample
d_i = diameter of the ith droplet (micron)

There is another way to calculate the average which is easier to use once a distribution has been generated. There are several different types of distribution that can be generated. The simplest distribution and the one that contains the most information is a simple bar chart, as was shown in Figure 3.4 above. To calculate the average from such data the formula below is used:

$$D_{50} = <d> = \frac{1}{N} \sum_{i=1}^{M} m_i d_i$$

Eqn (3.18)

where:

M = total number of diameter bins
N = total number of droplets
d_i = bin diameter; diameter of the ith bin (micron)
m_i = number of droplets in the ith bin

The median diameter is defined as that diameter for which half of the droplets have a larger diameter and half of the droplets have a smaller diameter.

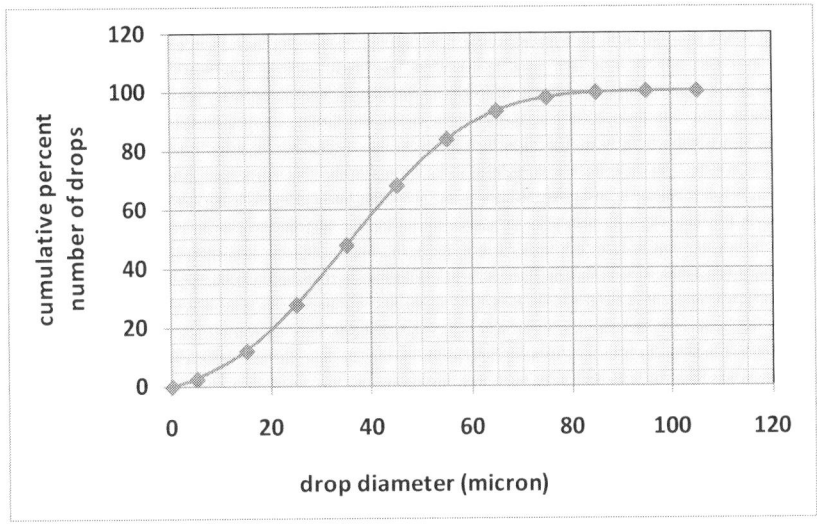

Figure 3.6 Graphical display of the data shown in the table.
The cumulative percent number of droplets is plotted versus the droplet diameter.

Volume Average (Mean) Diameter: The volume average or mean can be calculated by adding all of the volumes associated with each diameter and dividing by the number of droplets. This is the basis of the common formula given below:

$$D_{V50} = \frac{1}{N} \sum_{i=1}^{N} v_i = \frac{1}{N} \frac{\pi}{6} \sum_{i=1}^{N} d_i^3 \qquad \text{Eqn (3.19)}$$

where:

N = total number of droplets in the sample
d_i = diameter of the ith droplet (micron)
v_i = volume of droplets of diameter d_i

In the next figure, the cumulative distribution of oil volume per mL of produced water is plotted as a function of droplet diameter. This is an important diagram since it indicates the oil droplet diameter that must be separated in order to achieve the target effluent concentration of oil. For example, if the target is 100 ppmv, then the droplet diameter that must be removed is 55 micron. Likewise, if the target is 35 ppmv, then the oil droplet diameter that must be removed is 40 micron. In other words, if all droplets greater than or equal to 40 micron are removed, then the remaining droplets will have an oil concentration of 35 ppmv. The D_{V50} value for this distribution can be easily read from this figure by locating the 50 % point on the y-axis (75 ppmv), and reading across to the curve and then down to the x-axis to find 50 micron.

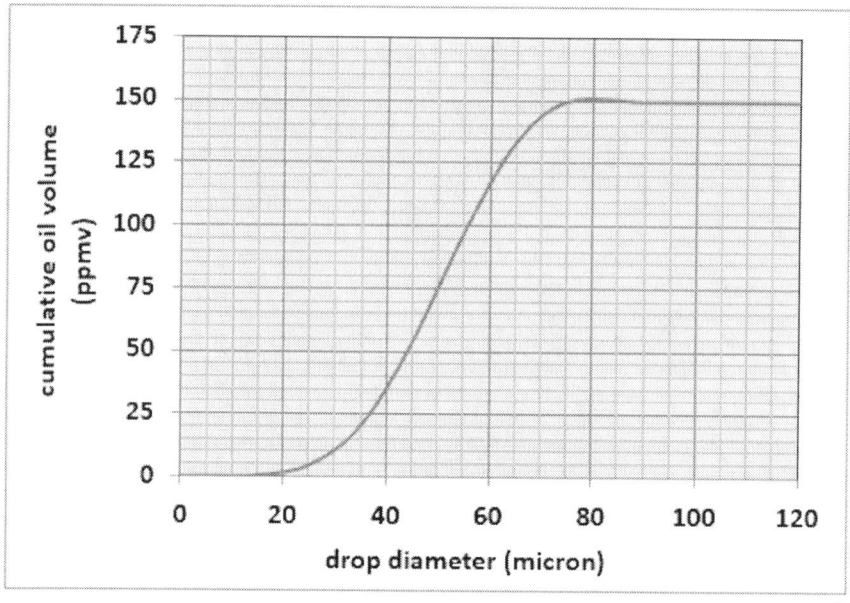

Figure 3.7 Graphical display of the data shown in the table.
The cumulative percent number of droplets is plotted versus the droplet diameter.

The final diagram is a comparison of droplet number distribution compared to droplet oil volume distribution.

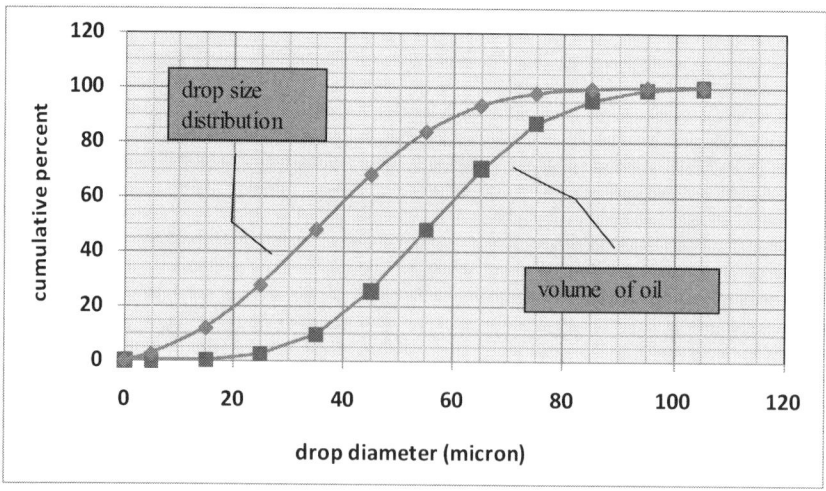

Figure 3.8 Graphical display of the data shown in the table.
The cumulative percent number of droplets is plotted versus the droplet diameter.

Sauter Mean Diameter: In some studies of droplet size distribution, the Sauter Mean diameter is used. It is defined below. Compared to the volume average, it reduces the contribution of the large drops to the mean.

$$d_{32} = \frac{1}{N} \sum_{i=1}^{N} \frac{d_i^3}{d_i^2}$$

Eqn (3.20)

3.3.1 Log-Normal Distribution

The Log-Normal distribution is often used for oil droplet distributions in coalescence and shearing studies. The Log-Normal distribution is given by the following formula.

$$f(x) = \frac{1}{\beta\sqrt{2\pi}}(1/x)\exp[-(\ln x - \alpha)^2/2\beta^2]$$

Eqn (3.21)

x	independent variable (e.g. droplet diameter)
a	adjustable parameter
b	adjustable parameter
$f(x)$	probability distribution of x values

The parameters α and β are adjustable. They are adjusted in order for the calculated distribution to match the measured distribution. The mean of x values is given by:

$$\mu = \exp[\alpha + \beta^2/2]$$

The standard deviation of x values is given by:

$$\sigma = \mu\exp[\beta^2 - 1]$$

Example of Log-Normal Distribution: An example of the Log-Normal Distribution is given in the figure below. As shown, the Log-Normal distribution has an extended tail for high values of x (e.g. droplet diameter). Unlike the throwing of dice, which conforms to a bell-shaped or Normal Distribution, oil droplets are subjected to non-linear forces in nature. Convective energy transport, viscous shear, mixing intensity, and so on are nonlinear and in some cases highly nonlinear processes.

This non-linear or logarithmic description is most clearly visualized by plotting the logarithm of the distributed variable (x). As shown in the figure below, when this is done, the Log-Normal distribution has a decidedly normal shape (as in similar to a Normal Distribution). This is what gives the log-normal distribution its utility in the modeling of oil droplet diameters.

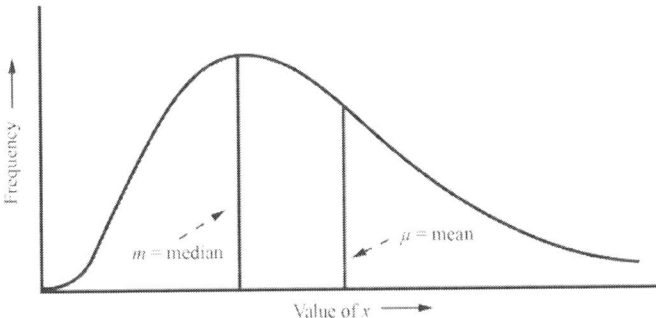

Figure 3.9 Log-normal distributions

Cumulative Distribution: The Log-Normal distribution cannot be integrated in closed form. The Log-Normal distribution must be integrated numerically.

3.3.2 Rosin-Rammler Distribution

Another distribution function that is often used in modeling oil / water dispersions is the Rosin-Rammler distribution. The Rosin-Rammler is a versatile mathematical function due to the fact that the exponent of the independent variable is an adjustable parameter. The shape of the distribution function is therefore flexible. It can be adjusted to have a long nose for the distribution of large drops, or have a long tail for distribution of small drops. It was originally developed to model the crushing of coal particles and hence, tends to model accurately the distribution of small droplets generated by shearing processes. A comparison of Rosin-Rammler to the Log-Normal distribution is given in [1] for the case of oil droplets shearing in a centrifugal pump. Rosin-Rammler more accurately represented the oil droplet data in the nose and tail, whereas the Log-Normal did better near the mean. A comparison of Rosin-Rammler, Log-Normal and the Weibull distribution is given in [2]. The literature relating to drop-droplet coalescence utilizes either the Log-Normal distribution [3] or the Rosin-Rammer distribution [4].

The Rosin-Rammler distribution is given by the following formula.

$$f(x) = \frac{b}{a}[(x/a)^{b-1}]\exp[-(x/a)^b] \qquad \text{Eqn (3.22)}$$

x	independent variable (e.g. droplet diameter)
a	adjustable parameter
b	adjustable parameter
$f(x)$	probability distribution of x values

The parameters a and b are adjustable parameters. Unlike the Normal Distribution, the Rosin-Rammler distribution is not written using the mean and standard deviation. Those quantities are calculated independently from the following formulas.

The mean of x values is given by:

$$\mu = a\,\Gamma(1+1/b) \qquad \text{Eqn (3.23)}$$

As shown, the gamma function (Γ) is required to calculate the mean. The standard deviation is a complicated formula and is not given here.

Example of Rosin-Rammler Distribution: The Rosin-Rammler is a versatile mathematical function due to the fact that the exponent of the independent variable is an adjustable parameter. The shape of the distribution function is therefore flexible. It can be adjusted to have a long tail for large droplet diameter, or have a long tail for small droplet diameters. Karabelas [4] used the Rosin-Rammler distribution to correlate the droplet size distribution for water droplets suspended in oil, in pipe flow where shear and coalescence occur simultaneously.

Cumulative Rosin-Rammler Distribution: Another feature of the Rosin-Rammler equation is that it can be analytically integrated. The integral is given by the formula:

$$F(x) = 1 - \exp[-(x/a)^b] \qquad \text{Eqn (3.24)}$$

x	independent variable (e.g. droplet diameter)
m	mean of x values
s	standard deviation
$F(x)$	cumulative function x values

The cumulative distribution can be calculated either by numerical integration or by the formula above. In fact, it is useful as a safeguard against errors to calculate it both ways. The simplicity of the cumulative distribution formula and the versatility of the distribution function make the Rosin-Rammler a popular equation in modeling droplet diameter distributions.

3.3.3 Separation Efficiency as a function of Particle (or Droplet) Size:

In the design of water treatment equipment, the separation efficiency is often given as a function of the particle or droplet size. This stands to reason. Larger particles settle-out faster. They are easier to separate in general. Separation efficiency is expressed as a fraction which varies between 0 and 1, or as a percentage which varies between 0 and 100%. A value of unity represents complete separation, thus a value greater than unity is not physically possible. If the calculated $S(d)$ is greater than unity, then $S(d)$ is set equal to unity. An example of a Separation Efficiency Curve is given in Figure 3.10 (see the green line). According to equations 4 and 5, the separation efficiency curve for a gravity separator depends on the oil and water density, the water viscosity, the water residence time, and the volume and height of the water leg.

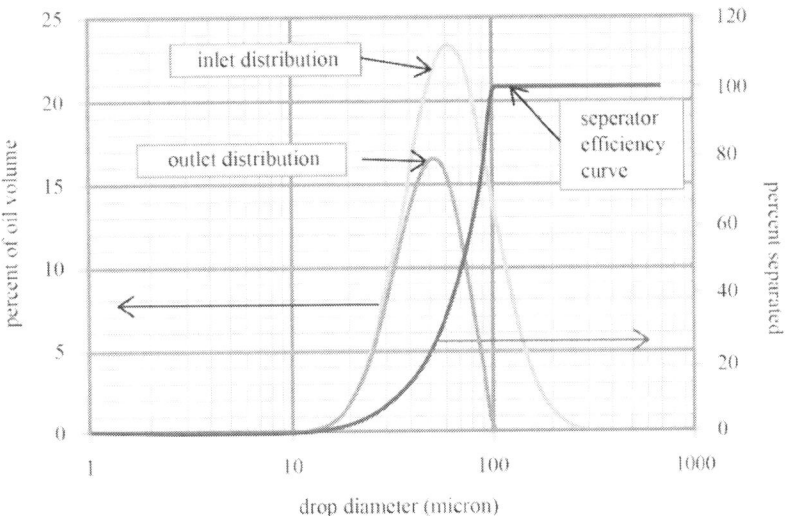

Figure 3.10 Example of a Separation Efficiency Curve and its effect on the droplet diameter distribution. In this example, the separator efficiency curve is for a gravity separator, and is calculated using Stokes Law. Note the sharp cut-off at 100 micron. This defines the Dmax, the smallest droplet for which there is 100% separation. In this example, the Dmax is 100 micron.

The separation efficiency is typically a measured quantity. However, as will be shown, there are accurate models that allow calculation of separation efficiency for certain types of equipment. Once the separation efficiency curve is known, and the inlet oil droplet size distribution is known, then the effluent droplet size can be calculated as follows.

$$F(d)_E = F(d)_F (1 - S(d))$$

Eqn (3.25)

$F(d)_E$ oil droplet diameter distribution of the effluent
$F(d)_F$ oil droplet diameter distribution of the feed

In Figure 3.10, the Separation Efficiency Curve is based on Stokes Law settling for oily water in a primary separator. The Separation Efficiency Curve for Stokes Law settling has a characteristic maximum droplet diameter. For this particular example, this value is 100 microns. In general this value depends on the residence time, fluid viscosity, and density difference between the oil and water, i.e. all of the Stokes Law variables. Note that, as shown in the figure, the maximum droplet diameter is 100 micron (green curve) and the outlet droplet diameter distribution (blue curve) has no oil droplets of 100 micron or larger.

3.3.4 Equipment Sizing based on Fluid Density Difference:

The fluids in the liquid settling zone of the vessel can be thought of as having three distinct zones. The upper-most zone is composed of an oil continuous liquid containing dispersed water drops. The water droplets settle according to Stokes Law. The bottom-most zone is composed of a water continuous liquid containing dispersed oil drops. The oil droplets cream (rise to the top) according to Stokes Law. Between these two layers, there is typically a complex interface emulsion where the transition between oil-continuous emulsions and water-continuous emulsions takes place.

Stokes Law is applied to model the separation of oil droplets from water in the water leg of gravity based separators.

In a separator, the physical situation is the following. The oil/water interface is the location where the lower water continuous emulsion layer inverts to the upper oil continuous emulsion layer. The design intent in a separator is to provide a region of low turbulence, with parallel streamlines and uniform velocity where oil droplets can rise according to Stokes Law. If oil droplets rise fast enough, they will arrive at the oil / water interface and be carried over the oil spill-over weir into the oil bucket and be discharged with the oil. Thus, the probability of separation is a function of the height, viscosity, and density gradient of the oil/water interface.

For a three-phase gravity separator the separation efficiency can be calculated according to:

$$S(d) = u(d)t_T / h$$

Eqn (3.26)

$S(d)$	separation efficiency as a function of the oil droplet diameter d
$u(d)$	oil droplet rise velocity (m/sec), a function of d.
t_T	theoretical residence time (sec), calculated as $t_T = V / Q$
h	height of the oil/water interface (m)
V	volume of water leg (m³)
Q	volumetric flow rate of water (m³/sec)

In the equations above, the separation efficiency is expressed as a fraction which varies between 0 and 1. A value of unity represents complete separation, thus a value greater than unity is not physically possible. If the calculated $S(d)$ is greater than unity, then $S(d)$ is set equal to unity.

In a separator, the physical situation is the following. The bottom of the dispersion band defines the oil / water interface. The level of the oil / water interface is set by the operators and is detected and controlled by the vessel level control system. Under the dispersion band, water forms a continuous phase. Oil droplets are dispersed within the water. The design intent in a separator is to provide a region of low turbulence, with parallel streamlines and uniform velocity where oil droplets can rise according to Stokes Law. If oil droplets rise fast enough, they will arrive at the oil / water interface and be carried over the spill-over weir into the oil bucket and be discharged with the oil. Thus, the probability of separation is a function of the rise velocity, the residence time, and the height of the oil/water interface.

At this point, it is important to point out that there are a number of approximations implied in the approach when applied to modeling gravity separators. These equations are only valid for an ideal set of conditions which almost never occur in practice. To account for real-world effects, the residence time is multiplied by a Hydraulic Efficiency Factor which has been assigned a value of 0.7. This factor accounts for the following set of non-ideal conditions. First, the actual residence time is a distribution function and not a single value. Part of the fluid follows a streamline from the inlet to the discharge. This fraction of fluid will have a short residence time. Another part of the fluid gets caught in swirls and eddies, with longer residence time. Internal devices such as perforated plates are intended to reduce such effects but they typically do not eliminate them. Most separator vessels have formation sand in the bottom which reduces the volume available for residence time. Also, drop-droplet coalescence is not treated explicitly. It has been observed that a Hydraulic Efficiency Factor of 0.7 is reasonable.

Other separation equipment can be modeled using the equations above, such as hydrocyclones, and flotation units. As discussed below, a modified version of Stokes Law is applied to model hydrocyclones and centrifuges. In those cases, the Separation Efficiency Curve derived for the particular equipment would be used. More detail is given in subsequent chapters. Using the above approach, it is possible to calculate the oil droplet diameter distribution at various points within the water treating system. Combining these tools allows a complete model for the water treating system.

3.4 Turbulence

High intensity turbulence, like that in a valve or a high speed centrifugal pump, causes oil droplets to break apart. This gives rise to smaller droplets which are more difficult to separate from produced water. In such situations it may be necessary to reduce the sensitivity of the fluids to shear by adding a chemical treatment.

Mild turbulence can have the opposite effect. In pipelines, and flow lines, mild turbulence promotes drop-droplet coalescence and actually leads to an increase in the average droplet size. It is important to have at least a rough estimate of which turbulent regime (mild / coalescing versus intense / shearing) will be induced by various selected pieces of equipment.

This section provides the basic understanding of turbulence that will be used in other sections to develop models for droplet and particle collision rates, for droplet coalescence, flotation, and shearing in valves and pumps. All of these subjects are related through the turbulence energy dissipation rate (ε), which is the basic measure of turbulence intensity and which can be used to determine if the turbulence will cause shearing or coalescence.

It turns out that a direct approach to the subject of oil droplet shearing and coalescence can be provided. The material below starts from relatively little background of turbulence and ends with practical calculations and correlations. The non-expert will be able to appreciate the complexity of the subject, obtain a functional understanding of those aspects that are relevant, ignore the very large part of the subject that is not directly relevant, and make calculations that are reasonably accurate.

For this purpose, turbulence is presented as an empirical subject. Mathematical models and correlations are presented, of course, but only to the extent that they provide useful organization of the empirical information and allow calculations of physical properties of interest. Thus, very little theoretical development is presented. In particular, it is not possible to give background on Fourier Transforms, correlation functions, or vector calculus. These are useful subjects in their own right. Their omission here is only in the interest of providing a practical and direct approach to understanding the effects of turbulence on oil drops.

For our purposes, we consider only the removal of oil from water in a water treatment system. Removal of water from oil, as in oil dehydration, will be discussed elsewhere. This discussion has four objectives:

1. to provide simple formulas that can be used to recognize when shearing is likely to be a significant problem;

2. to provide formulas that help determine if mild turbulence is likely to mitigate upstream shearing;

3. to give practical advice on how to reduce shear in pumps and valves, how to mitigate the effect of shear;

4. to introduce ε (turbulent energy dissipation rate) as a measure of turbulent shearing, and to provide formulas that allow the calculation of maximum droplet size as a function of e.

This discussion is intended to remind engineers that the decisions made for upstream design can have dramatic consequences on downstream separation equipment performance. Also, the effect of smaller oil droplets on water treatment equipment is not discussed in detail. That effect varies depending on the type of equipment, be it gravity settling, a hydrocyclone, or a flotation unit. Suffice to say that the smaller the droplet diameter, the poorer the separation efficiency and the poorer the treated effluent water quality.

Turbulence in Oil & Gas Facilities: Turbulence and the shearing of oil droplets is one of the features of oilfield water treatment that distinguishes it from other industrial water treatment applications. Upstream oil and gas commonly involves high pressure fluids coming into a facility, and low pressure treated produced water leaving. The pressure cascade which is used to separate gas from oil, and to condition the oil to its target vapor pressure inevitably results in intense shearing of fluids through valves. In addition, recovery of oil separated from produced water usually involves pumps for recycle of oily water. Many other industrial water treatment applications, such as pulp and paper or food and beverage involve contaminants such as oil, grease, sticky tar-like materials, suspended polymer, oily or polymer coated solids particles. But oilfield water treatment is unique in that it involves all of those difficult contaminants with the added complication of shearing.

Sources of Shear: There are many sources of shear. Fluids are initially subjected to shear in the reservoir, near the well bore, where pressure drop, and the narrowing radius of the flow creates an acceleration of fluid. Lifting techniques, if they are used, such as electrical submersible pumps create turbulence to varying degrees. The well head choke is another, sometimes intense, source of shear. The list goes on and includes the control valves and pumps used in the facility itself.

In general, the wells and facilities (chokes, valves, pumps, etc.) cause shearing. Pressure reduction through valves, from the well head through the facility causes shearing. Pumping of fluids within the facility also causes shearing. To some extent, shear is an inevitable consequence of processing conditions. Fluid pressure and flow rate must be controlled through a facility in order to achieve the objectives of the facility including:

1. gas conditioning (dew point, water content)

2. oil conditioning (vapor pressure, water content)

3. oil and gas pumping, if required, for export.

4. water treatment for disposal or reuse (remove oil and solids).

Degassing of crude in order to prepare it for transport and storage is generally carried out in stages (successively lower pressures) in order to minimize the energy required to compress the gas that is generated. Almost all upstream facilities make use of this so-called pressure cascade. Hydrate management in some cases imposes constraints on pressure control of the fluids. For example, during cold startup, or even during continuous operation, it may be necessary to choke the fluids at a subsea well head so that the fluids flow from the well head to the boarding valve on the facility at a sufficiently low pressure that avoids operation within the hydrate formation region. While pressure control of the hydrocarbons is essential, there are design and operating details which can reduce the shearing that is applied to the fluids. Also, coalescence can be enhanced through relatively simple process changes.

Small-Scale Turbulence is Simplified: Before applying these ideas to water treatment, some background information about turbulence is worthwhile. Within the broad subject of turbulence, only a small portion of written material deals with turbulent shearing of oil droplets in water. It is a specialized application within turbulence, which is itself a very complex and specialized subject. However, one of the features that helps to simplify this complex subject is the fact that for small suspended droplets, only the small eddies are important. Small oil droplets, on the order of 1 to 50 micron diameter, which are the greatest challenge in oilfield water treatment, are mostly influenced by collision with eddies of roughly the same size. It turns out that the kinetic energy of these small eddies is relatively independent of how the turbulence is generated. Whether the turbulence is generated in a pipe, pump, or valve, the kinetic energy of the small eddies only depends on a single parameter regardless of all other features of the flow. That parameter is the turbulence energy dissipation rate (ε). It is basically a measure of the intensity of the turbulence. Once the turbulence energy dissipation rate has been calculated, it can be used to predict a number of properties of the small eddies. The eddy collision frequency, energy of those collisions, the drop-droplet collision frequency, all can be calculated using simple formulas that depend on the turbulence energy dissipation rate. On a small scale, the turbulence generated in any of these devices looks the same, for a given value of turbulence energy dissipation rate. Excellent models exist for the calculation of the breakage of droplets as a function of this parameter. Thus, it turns out that small-scale turbulence is actually rather straightforward.

Turbulent Fluctuations: For our purposes, the fundamental observation of turbulence is the following. Above a critical value of the Reynolds Number, the fluid velocity is no longer steady. Instead, the velocity fluctuates both with respect to time, and with respect to direction. Thus, the starting point for this presentation is a description of velocity fluctuations.

Fluctuating velocity is the fundamental property of turbulent fluid flow. In turbulent flow there is a macro-flow characterized by an average velocity. The average velocity has both a magnitude and direction. In addition to the average velocity, turbulent flow is characterized by seemingly random velocities superimposed on the average velocity. Their direction and magnitude are governed by mass balance and momentum transfer. The statistical properties of these fluctuations can be predicted to some extent. The following equation defines the fluctuating velocity:

$$\underline{u} = <\underline{u}> + \underline{u}'$$ \hfill Eqn (3.27)

Where

$\underline{u}$	= fluid velocity, a vector quantity (m/sec)
$<\underline{u}>$	= time average or ensemble average fluid velocity, a vector (m/sec)
$\underline{u}'$	= velocity fluctuation, a vector (m/sec)

The above is a vector equation, as indicated by the underscore on all of the variables. As such, it represents fluid speed and direction. All of the variables in the equation are functions of the location within the fluid, and of time. Depending on the flow geometry, the quantities may vary significantly from one point to another, and they vary in direction and time.

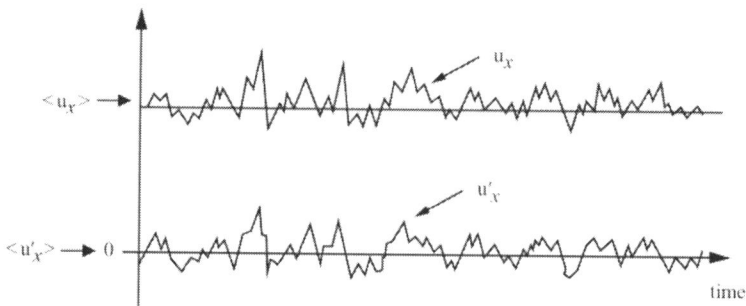

Figure 3.11 Fluid speed in the x-direction as a function of time. The fluid speed in the x-direction is shown as the upper data set. The time average speed is shown as the dash line through the upper data set. The fluctuating component of the fluid speed is shown as the lower data set. Note that the time average of the fluctuating speed is zero.

Equation 3.32 is represented in Figure 3.11, for the x-direction of flow. For the data in the figure, there is no average flow in the y or z directions. For steady flow, the average fluid velocity does not change with time. If the direction of the average fluid velocity and the coordinate system are aligned, then there is only one main component of $< \underline{u} >$. However, even if this is the case, the fluctuations, represented by $\underline{u}'$ can and typically do occur in all directions. Since this is the case, then $\underline{u}$ also has components in all directions. It should be physically apparent that the quantity $< \underline{u}' >$ is zero. It is mathematically obvious from inspection of equation 3.32. Physically, it makes sense that it should be zero on the basis that the fluctuations are random in all directions and over time. By first taking the square of the fluctuating velocity and then taking the average, a nonzero quantity is generated. Physically this quantity is related to the stress exerted on the droplet by the fluctuating or turbulent component of the velocity.

The magnitude and direction of fluctuations in a turbulent flow are sometimes described as being random. Actually the term "random" is not an accurate description. If the fluctuations were truly random, the fluctuations would be far less complicated than they actually are. The behavior of random variables was elucidated a few hundred years ago and is well characterized by simple mathematical relationships.

Fluctuations of velocity in turbulent flow depend on many factors including the flow geometry, the location of a particular point within the flow, the direction of observation, the size scale of observations, the time interval of observations, and the intensity of the turbulence. From a continuum mechanics standpoint, such fluctuations are not deterministic. They are random. Thus, it is not possible to predict at any point in time the direction and magnitude of the next fluctuation. However, fluid fluctuations do have certain average properties that can be predicted. Fluctuations are stochastic which means that statistical measures, such as averages, standard deviations, and correlations can be analyzed, modeled, and predicted. One of the key objectives of the study of turbulence is to predict the magnitude, direction and frequency of fluctuations for a given flow situation.

Shearing of production fluids creates tight oil/water emulsions including small droplets of oil in water and small droplets of water in oil. The greater the shearing, the smaller the droplets, and the more difficult the subsequent process of separating the oil from the water. Small droplets rise very slowly and are often not adequately separated in a given residence time. This can overwhelm downstream equipment unless additional steps are taken such as increasing chemical dosage, adding or increasing heat, or removal of the emulsion for separate treatment. This will increase operating and capital expenses.

Energy Spectrum: The Energy Spectrum is a property of a particular flow geometry (cylindrical pipe, airplane wing, bright abutment in a river, etc), flow rate and fluid properties (viscosity, density, etc). It is calculated as the Fourier Transform of the correlation of longitudinal fluctuations. It is precisely defined in terms of mathematics. Roughly speaking, it gives the energy per unit mass of fluid associated with eddies of a particular size. It is discussed here because of the physical insight that it provides. It is also discussed because it helps to introduce the parameter ε, the turbulence energy dissipation rate

Energy Spectrum - Large versus Small Size Eddies: The figure below is a plot of the normalized One-Dimensional Energy Spectrum versus the normalized wave number. Both the x-axis and the y-axis values are normalized. The x-axis is wave number (k) times the Kolmogorov length scale (η). Wave number is proportional to 1 / wavelength. Thus, the quantity kη is dimensionless. A large wave number corresponds to a small wavelength. Thus, the right hand side of the x-axis corresponds to small eddies and the left hand side corresponds to large eddies.

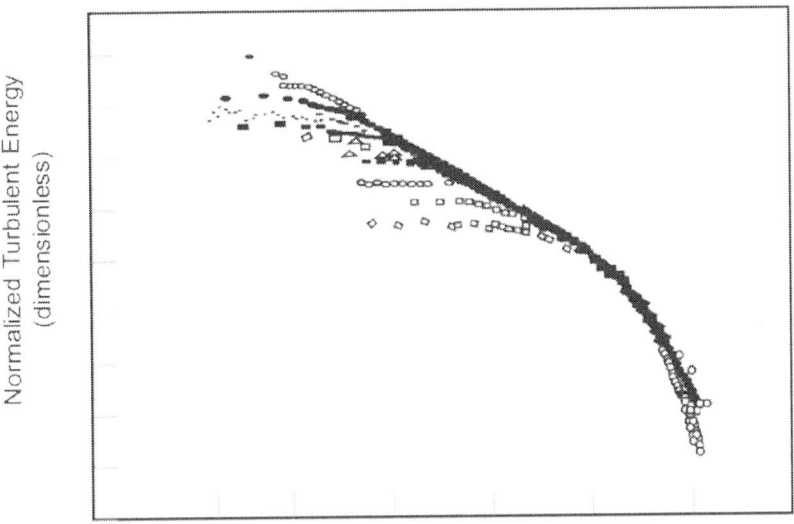

Figure 3.12 Energy spectrum of isotropic turbulence from page 227 of [39].

The energy spectrum for several different flow geometries, fluids, flow rates, etc. is shown in the figure. Both axes are logarithmic. The studies were aimed at understanding the commonalities in the turbulence created by the wake of a boat, flow through a canal, flow over a dam, storms in the atmosphere, and air flow across a sail of a boat, and so on. On the upper left hand side of the figure the energy spectrum for these different flow geometries are somewhat different from each other. The normalized turbulent energy spans three orders of magnitude. For a given wave number on the left hand side, the energy spectrum of the different studies can be differentiated from one another. As mentioned, small wave numbers (left hand side) correspond to large eddies. But on the lower right hand side, all of these results collapse into a single curve. This is an important feature of the turbulent energy spectrum. The consequence is that the size of, and energy contained in small eddies (large wave numbers) is the same regardless of the macroscopic flow geometry, fluid properties, flow rate, etc.. At macroscopic (large) scales, the boundaries (geometry) of the flow are important and the energy contained in turbulent fluctuations differs from one geometry to another, and from one flow rate to another. However, for the smaller scales (right hand side), the flow geometry has little impact on the size and energy of the eddies. In other words, the size and energy of the large eddies depend on the flow geometry but the size and energy of the small eddies do not depend on the flow geometry. This has direct relevance to turbulence in produced water as will be discussed momentarily.

These concepts can be completely and unambiguously developed using Fourier Transforms. While Fourier Transforms are straightforward to calculate, a proper introduction would require more development than appropriate for a book on produced water. Suffice to say that a Fourier Transform is analogous to a series expansion using sine and cosine functions. It is a means to represent spectral (time varying) information, or fluctuating values of a function. High frequency values of some function, for example, are represented by sine and cosine terms that have short wavelength. Lower frequency values are represented by longer wavelength sine and cosine terms in the expansion.

The y-axis of the figure is, roughly speaking, the kinetic energy of fluctuations associate with eddies of a particular size. The size is given by the x-axis, as just discussed. As can be seen from this equation, the quantity E_{11} has the dimensions of m^3/sec^2. The quantity ε is the turbulence energy dissipation rate. It has units of energy per second per unit mass. The quantity v is the kinematic viscosity. It is a property of the fluid. It has units of m^2/sec. Thus, the y-axis is also non-dimensional.

As shown in the figure, normalized values are used which cause the measurements to collapse into a single line (on the right hand side – small eddies), with a number of branches (on the left hand side – large eddies). There is notable branching-off in values for the smaller x-axis values. But starting at about $k\eta = 0.1$ all data collapse into a single universal function. As shown by this figure, the normalization parameters (ε, η, v) provide a very simple way of describing the relationship between turbulent energy and eddy size, for small scale eddies. That relationship is:

$$E_{11} / \left(\varepsilon v^5\right)^{1/4} = c_1 \left(k\eta\right)^{-5/3}$$

Eqn (3.28)

where:

E_{11}	= turbulence kinetic energy per unit mass (J/kg)
e	= turbulence energy dissipation rate (J/kg sec or m^2/sec^3 or W/kg)
n	= kinematic viscosity ($v = \mu/\rho$) (m^2/sec)
k	= wave number (inverse of length 1/m)
h	= size of an average eddy (m)
c_1	= empirical constant

The turbulence energy E_{11} (y-axis) is the kinetic energy contained in eddies of size k (x-axis). The empirical formula given in Equation (3.33) reproduces the universal function (single line) that is observed in the figure for the range of small eddy sizes (large wave numbers k). It suggests that turbulence energy depends on a very simple set of parameters (ε, η, v) for the smaller size scales.

Similarity for Small Size Eddies (Large Wave Numbers): This curve provides empirical evidence of the previous statements made about the large versus small eddies in homogeneous isotropic turbulence. It was stated earlier that the small eddies have size and energy that is relatively independent of the macroscopic geometry of the flow. This can be seen in the figure. Small scale eddies correspond to the large values of $k\eta$. This is where the largest degree of universality occurs. The energy spectrum for many different types of flow collapses to a single curve. At smaller values of $k\eta$, there is significant discrepancy from once set of set to another. These data sets differ from one another in the macroscopic geometry. In some cases, the data are from very large scale wakes, and channels. In other cases the data are from small scale grid turbulence and turbulent jet experiments. The differences only manifest in the large scale (small values of the x-axis).

Figure 3.12 is essentially an empirical observation. Through the use of kinematic viscosity, and the introduction of two parameters, ε and η, a wide range of data can be collapsed into a single curve. These two parameters are fundamental to turbulence and will be discussed in greater length as part of the introduction of formulas to estimate shear intensity of valves and pumps, collision rate between particles and drops, and the forces involved in collisions which might lead to coalescence.

An order of magnitude estimate of the rate of energy transfer can thus be given by the kinetic energy of eddies of size l, divided by the lifetime (k/τ) of eddies of size l:

$$k_l / \tau_l = c_1 < (u')^2 >^{3/2} / l \qquad \text{Eqn (3.29)}$$

The parameter c_1 is a number with a value on the order of unity. This relationship was put forward by Kolmogorov. The physical units are m²/sec³, J/sec kg (energy per unit time per unit mass) or W/kg (watts per unit mass). These are the same units as the turbulence energy dissipation rate (ε), i.e. energy per unit time per unit mass. This equation can be rearranged to give:

$$< (u')^2 > = c_2 (k_l l / \tau_l)^{2/3} \qquad \text{Eqn (3.30)}$$

The next step in developing a useful equation for eddy velocity and lifetime is to relate the above equation to the three parameters used in scaling the Energy Spectrum, ν (kinematic viscosity), ε (turbulence energy dissipation rate), and η (size of an average eddy). This is done by examining the turbulence energy dissipation rate of the smallest eddies where the Reynolds Number is expected to be equal to unity. This is carried out in the next section.

3.4.1 Kinetic Energy in Isotropic Turbulence

In this section, all of the previous sections on turbulence are pulled together to provide an equation that relates the turbulent kinetic energy ($< (u')^2 >$) to the turbulence energy dissipation rate (ε). The resulting equation is a powerful tool since it allows the estimation of the maximum droplet size in turbulence, as well as the collision rate, collision energy, and lifetime of collisions.

While the energy transfer rate and the turbulence energy dissipation rate are not equal, they are at least related to each other. Some of the energy dissipation occurs at the smallest scale of eddies. This size and velocity scale is consistent with a Reynolds number equal to unity. This is expected from knowledge of the physical meaning of a Reynolds number. Recall that when the convective energy is equal to the viscous dissipation energy, the Reynolds number is unity. This is expected to happen at the smallest eddy size. On this basis, the following relations have been put forward:

Size of the smallest eddies:

$$\eta = \left(\nu^3 / \varepsilon\right)^{1/4} \qquad \text{Eqn (3.31)}$$

Velocity of the smallest eddies:

$$V = \left(\nu\varepsilon\right)^{1/4} \qquad \text{Eqn (3.32)}$$

Lifetime of the smallest eddies:

$$\tau = \left(\nu / \varepsilon\right)^{1/2} \qquad \text{Eqn (3.33)}$$

Turbulence energy dissipation rate of the smallest eddies:

$$\varepsilon = v^2 / \tau$$

Eqn (3.34)

Where:

e	= turbulence energy dissipation rate (m²/sec³ = W/kg)	
n	= kinematic viscosity = μ/ρ (m²/sec)	
m	= viscosity of the fluid (Pascal seconds = kg/m sec)	
r	= density of the fluid (kg/m³)	
h	= size of small scale eddies (m)	
t	= lifetime of small scale eddies (sec)	
v	= velocity of small scale eddies (m/sec)	

The Reynolds number of these smallest size eddies can be calculated as follows with the velocity of the smallest eddies given by the equation above.

$$N_{RE} = \frac{(v^3 / \varepsilon)^{1/4} (v\varepsilon)^{1/4}}{v} = \frac{v}{v} = 1.0$$

Eqn (3.35)

These relationships for the small eddies are consistent with the requirement that the Reynolds Number is unity for the small eddies.

It is instructive to calculate some of these quantities. The coalescence efficiency of dispersed droplets in a turbulent flow field depends on the size of turbulent eddies. For a droplet size larger than the Kolmogorov size, η, and smaller than the length scales of the energy-containing eddies or turbulent length scale, [5], which is on the order of 10 mm in diameter. The effect of mean macro-velocity gradient is negligible in comparison to the turbulent fluctuations [6]. As an illustrative calculation, the Kolmogorov, η length scale is calculated for the values:

e	= 10,000 m²/sec³
v	= 1 x 10⁻⁶ m²/sec³

$$\eta = (v^3 / \varepsilon)^{1/4} = \left[\left(1 \times 10^{-6} \, m^2 / sec\right)^3 / 10,000 \, m^2 / sec^3 \right]^{1/4} = 3.2 \text{ micron}$$

Eqn (3.36)

Thus, at a fairly high value of the turbulence energy dissipation rate, the diameter of an average small-scale eddy is about 3 microns. This is an order of magnitude smaller than most oil droplets encountered in the upstream portion of the process. As the fluids flow through the system, most of the larger oil droplets are removed leaving only small droplets.

Returning to the previously derived equation for the kinetic energy of the velocity fluctuations, the Kolmogorov relationships are substituted.

$$< (u')^2 >= c_2 (k_l l / \tau_l)^{2/3} = c_2 (v^2 l / \tau_l)^{2/3} = c_3 (\varepsilon \, l)^{2/3}$$

Eqn (3.37)

This last relationship says that the turbulent kinetic energy per unit mass of the fluid ($<u'>^2$) is proportional to the turbulence energy dissipation rate (ε) times the size of the eddy (l), raised to the 2/3 power. As discussed previously, the size of eddies in this intermediate size scale (*l*) is an independent variable. The turbulence energy dissipation rate (ε) is locally constant in isotropic and homogeneous flow. In general, the turbulence energy dissipation rate does not depend on the size scale. We can choose an eddy size and use the formula above to calculate the kinetic energy per unit mass associated with that eddy size and with the flow configuration that gives rise to the energy dissipation rate.

Kuboi and Otake and their coworkers [7] measured velocity fluctuations in isotropic turbulent flow in order to test the validity of the above equations. Their results are shown in the Figure below.

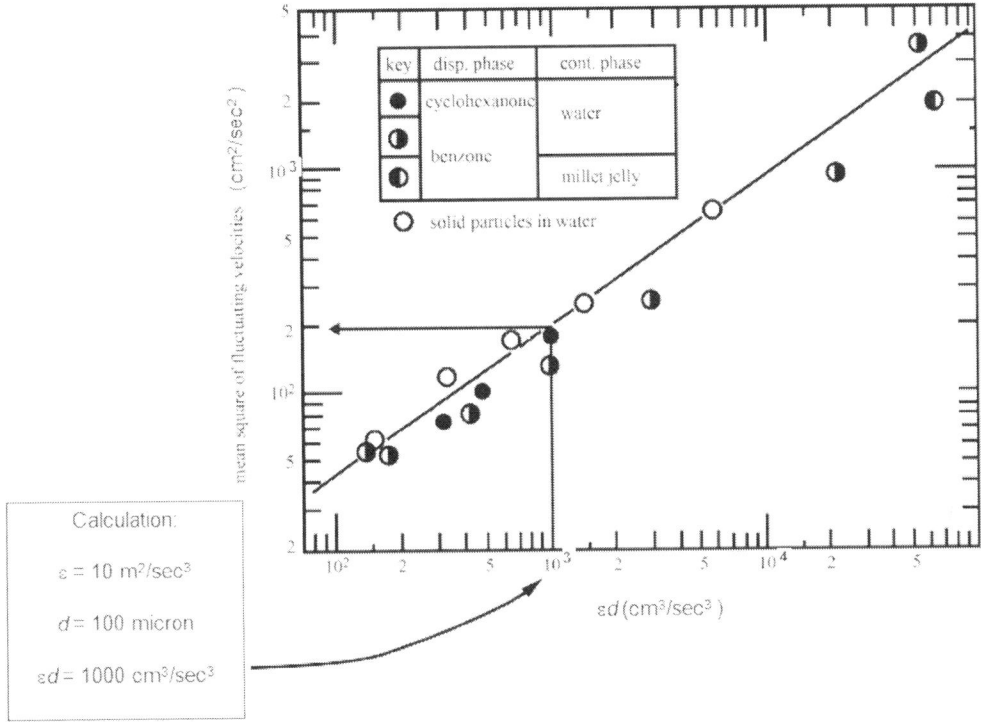

Figure 3.13 Mean-square turbulence velocity as a function of the product of turbulence energy dissipation rate times the diameter of a suspended particle [7].

The Figure includes an illustrative calculation. As a test of the equation, any value of the x-axis can be selected. The selected values are given as:

e $\quad$ = 10 m²/sec³
d $\quad$ = 100 micron
ε d $\quad$ = 1000 cm³/sec³

According to the above equations, the mean square of the fluctuating velocity (shown as the y-axis in the Figure) is given as:

$$< (u')^2 > = 2(\varepsilon l)^{2/3} = 2(1000 \, cm^3/sec^3)^{2/3} = 200 \ cm^2/sec^2 \qquad \text{Eqn (3.38)}$$

The calculated value of 200 cm2/sec2 agrees well with the measured data shown by the arrow pointing to the y-axis in Figure 3.13 above. In these measurements the energy dissipation rate is small (10 m²/sec³). This corresponds to pipe flow for which oil droplet coalescence is the dominant process rather than droplet breakup.

Eddy Properties in Mild Turbulence: According to the formulas above, it is possible to calculate the average velocity of an eddy generated in mild isotropic turbulence such as that found in moderate velocity pipe flow. As will be shown in Section 4.5 (Coalescence of Oil Droplets), mild turbulence gives rise to droplet-droplet coalescence.

For the following calculations, an energy dissipation of $\varepsilon = 1$ m²/sec³ is chosen. For a 100 micron eddy, the formula gives an eddy velocity of 66 x 10⁻³ m/sec (roughly 7 cm/sec). Next, the lifetime of an eddy is calculated.

$$\tau = \sqrt{\upsilon / \varepsilon}$$

τ : turbulence timescale (eddy lifetime) (seconds)

υ : kinematic viscosity (cSt = 1 x 10⁻⁶ m²/sec)

Eqn (3.39)

ε : energy dissipation rate (m²/sec³)

$$\tau = \sqrt{\upsilon / \varepsilon} = \sqrt{(1 \times 10^{-6} \, m^2 / \sec)/(1 m^2 / \sec^3)}$$
$$= 1 \times 10^{-3} \text{ seconds}$$

As shown in the calculations above, it is possible to calculate the average lifetime of an eddy generated in isotropic turbulence, in water, with an energy dissipation of $\varepsilon = 1$ m²/sec³. According to this calculation, a typical eddy has a lifetime of about 1 millisecond.

Combining the last two calculations, velocity and lifetime, it is possible to make a rough calculation of the distance traveled by a typical eddy. The calculated value is 66 x 10⁻⁶ m (66 microns). Altogether, the following gives a summary of the characteristics of these eddies:

Selected values:

energy dissipation rate	= 1 W/kg
eddy diameter	= 100 micron

Calculated values:

velocity	= 66,000 micron / second
lifetime	= 1 millisecond
distance traveled	= 66 micron (during the lifetime of the eddy)

These values suggest a rough physical picture of an eddy in mild turbulence. It is very short lived. It travels relatively fast which means that for a small object it has significant kinetic energy. But since it is short lived, it does not travel very far, only 2/3 of its diameter. The short life and the short mean free path are of course a consequence of the crowding together of many eddies in the fluid. They crash together, change direction, transfer momentum, sometimes they combine with larger eddies or break apart into smaller ones.

This discussion of turbulence started with a definition of fluctuating velocity and ended with the equation above. This equation provides a measure of the average fluctuating velocity, which is the fundamental quantity in turbulence. It also provides a measure of kinetic energy which can be related to collision frequency, collision lifetime, and collision energy.

3.4.2 Turbulence Energy Dissipation Rate (ε)

Energy Dissipation: The typical equation that is used to estimate the energy dissipation (h) in turbulent pipe flow is given by:

$$h = 2f \frac{L}{D} u^2 \qquad\qquad \text{Eqn (3.40)}$$

where:

L	= length of pipe segment between points 1 and 2 (m)
u	= superficial velocity (u=Q/A, where Q is the volumetric flow rate, and A is the cross-sectional area of the pipe (m)
D	= pipe diameter (m)
f	= Fanning Friction Factor (dimensionless)

This is a well-known empirical relationship. The friction factor depends on the Reynolds number of the flow. It can be calculated for straight pipe, elbows and T-junctions. It is discussed in more detail below. The units of energy dissipation (h) are m^2/sec^2, which is equivalent to energy per unit mass (J/kg), as expected. In the next section, the turbulent energy dissipation rate is discussed which has the units of Watt/kg, which is equivalent to J/kg sec, or m^2/sec^3.

Most readers will realize that the energy dissipation term (h, aka friction loss) in Bernoulli's equation encompasses all energy losses (straight pipe, elbows, valves, etc) and for that matter energy inputs due to work done from say pumps. Equation (2) allows the calculation of h for only a couple of these cases. One case of interest here is that of a dispersion of oil droplets in water at moderate oil concentrations (< 1 %). Another case, of less interest is that of a single phase fluid. The same equation can, to a very good approximation, be applied in both cases. Essentially the same energy dissipation will occur, for a given set of turbulent flow conditions, whether the oil is present or not, for moderate oil concentrations. From the standpoint of the fluid, very little energy goes into oil droplet breakage. From the standpoint of the oil droplets, the energy is intense. Almost all of the energy goes into viscous dissipation – thermal energy. Thus an equation written for the case of a single phase fluid (2) can be applied to the case of interest here (oil droplets dispersed in water).

Energy Dissipation versus Energy Dissipation Rate: As discussed above, the turbulent energy dissipation rate is the rate of energy loss due to fluid friction. If the location of point (1) and (2) are chosen such that the flow is relatively similar in terms of flow rate, diameter of pipe, and turbulence intensity, then the situation is relatively easy to analyze. The rate of energy loss is given by h / t, where t is the time required for the fluid to traverse from point (1) to point (2). A simple way to calculate this time is given by t = L / u, where L is the length of pipe between point (1) and point (2); and u is the superficial velocity (u = Q / A). Substituting these quantities into the equation for the rate of energy dissipation gives:

$$\varepsilon = \frac{h}{t} = \frac{hu}{L} = \left(\frac{2f}{D}\right)u^3 \qquad\qquad \text{Eqn (3.41)}$$

where:

ε = turbulent energy dissipation rate (m²/sec³ W/kg – or Watt per kilogram)

h/t = rate of energy dissipation due to fluid friction (W / kg)

This equation will be used below to calculate values of energy dissipation rate. The units are energy per unit mass per unit time: W/kg (J/sec kg). Physically, these units are power per unit mass. The units highlight an important aspect of this equation. The quantity of interest is not a measure of energy. The quantity of interest is energy dissipation _rate_. It is the rate at which fluid turbulence dissipates energy that is important in oil droplet shearing. The greater the rate of energy dissipation, the smaller the oil droplets become as a result of shear.

It is also of note that turbulent energy dissipation will occur whether the fluid is single phase (e.g. pure water), or a dispersion of oil droplets in water, or multiphase (gas, oil, water). The calculation of ε will vary in these situations. The particularly simple case of a dispersion of oil droplets in water was chosen for the illustrative example given above.

In the case of oil droplets dispersed in water, not all of the turbulent energy goes into breaking up the droplets. Almost all of the turbulent energy is dissipated in the form of heat. The mechanism for this dissipation is the fluid friction experienced by the eddies. This fluid friction occurs over all sizes of eddies but the greatest dissipation rate occurs at the small scale eddies. These small scale eddies also happen to smash into the oil droplets and break them into smaller droplets. This is the process modeled here. But for moderate concentrations of oil in water this process has negligible effect on the rate of energy dissipation.

Illustrative Calculations: The equation above can be used to calculate values for the energy dissipation rate for pipe flow. The results for such a calculation are illustrated in the Figure below where values of the energy dissipation rate have been calculated for a straight pipe. A range of flow rates have been used. Two pipe diameters were selected, 4 and 6 inch. The Fanning Friction Factor (f) was estimated as 0.01. To simplify the comparison, a single value of the Fanning Friction Factor was used. For more rigorous calculations, the Fanning Friction Factor should be calculated as a function of the Reynolds Number. For our purposes here, this level of detail is not necessary. Metric units are used. A fluid flow rate of 3 m/sec corresponds to 13,200 BWPD for a 4 inch (0.1 m) diameter pipe. Substituting these values into the equation gives: 2 x 0.01 x 3³ / 0.1 = 5.4 for the turbulence energy dissipation rate. This is the value in the figure (red line). The same calculation can be repeated for the 6" line with the same flow rate and the value for ε is about 0.3 or about 6% of the value for the 4 inch pipe.

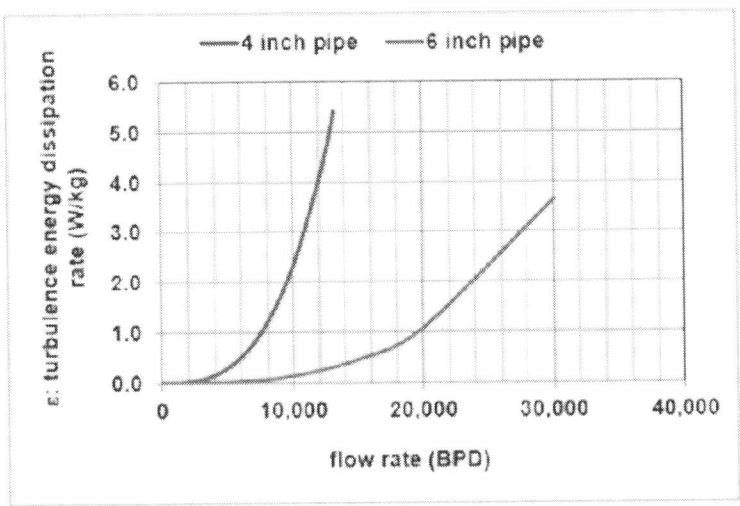

Figure 3.14 Turbulent energy dissipation rate (ε) versus flow rate for two different diameter flow lines. This is an example of mild shear discussed in the text.

Order of Magnitude: As shown by these calculations, the magnitude of energy dissipation rate is between 1 and 10 W/kg, for straight pipe flow. As will be demonstrated shortly, these values are low compared to those for flow through a valve or pump. In fact, it will be shown that these values do not cause droplet breakup. Instead, this level of mild turbulence intensity actually enhances coalescence. This is the basis of the SP-Pack device invented by Ken Arnold [8].

Units of Energy per Unit Mass per Unit Time: As was shown above, the units of turbulence energy dissipation rate are m^2/sec^3 or equivalently, W/kg (J/sec kg). Physically, these units are energy per time per unit mass. This can be verified by evaluating the units in Equation (1.1.x) and keeping in mind that Joules (J) are equal to Newton – meters (Nm), both of which are a measure of energy, and a Newton is equal to $kg\text{-}m/sec^2$. The units highlight an important aspect of this equation. The quantity of interest is not "energy dissipation". The quantity of interest is "energy dissipation <u>rate</u>". It is the rate at which fluid turbulence dissipates energy that is important in oil droplet shearing. The greater the rate of energy dissipation, the smaller the oil droplets become as a result of shear.

While the length (L) from point (1) to point (2) canceled out of the above equation, in the discussion below related to control valves, it plays an important role. It is physically reasonable that length does not appear in the above equation since ε is not total energy dissipation. Instead, ε is the rate of energy dissipation. It measures the intensity of turbulence, rather than the net or total energy dissipation so the distance between the two points is not relevant. This issue will be discussed further below. For now, it is worthwhile to calculate some values in order to get a feel for the order of magnitude of this quantity.

Reynolds Number: It is worthwhile to point out the difference between the energy dissipation rate (ε) and the Reynolds number (N_{RE}). For many engineers, Reynolds number is the quantity that comes to mind in relation to fluid turbulence. Reynolds number provides a good criterion for the transition from laminar to turbulent flow. That transition is governed by the ratio of convective force to viscous resistance. But it does not provide a good measure of turbulence energy dissipation rate. As shown in the figure above, the turbulence energy dissipation rate increases by a factor of about 15 in going from 6 inch to 4 inch pipe at a fixed volumetric flow rate. The Reynolds number only increases by a factor of 3/2 for the same change in pipe diameter, at a fixed volumetric flow rate. As will be discussed below, the droplet size distribution is more directly related to the turbulence energy dissipation rate (ε) than to Reynolds number.

3.4.3 Weber Number

Droplets of oil in water are delicate objects. The only physical force that holds them together is the interfacial tension. A droplet with a high interfacial tension is held together more strongly than a droplet with low interfacial tension. A droplet with low interfacial tension deforms and breaks apart more easily in a shear field. The Weber Number gives a relative measure of the shear forces acting on a droplet in turbulent flow compared to the restraining force. This section gives practical information on the calculation and use of the Weber number, including proper units and sources of values for the physical properties that are required.

The Weber Number is a ratio of two forces. The numerator is the shear force acting across the surface of the drop. The denominator is the only restraining force, the interfacial tension. The greater the Weber Number, the greater the deformation and the greater the tendency of the fluid forces to overcome the interfacial tension and cause the droplet to distort or break apart.

$$N_W = \frac{\tau_e}{\tau_i} = \frac{\text{external turbulent stress}}{\text{internal restorative stress}} \qquad \text{Eqn (3.42)}$$

There are two kinds of forces that are transmitted by the surrounding fluid to a droplet. As will be evident later, there are other forces, such as electrostatic, that are not transmitted by the fluid. For now, we consider only the forces that are transmitted by the fluid. These forces are transmitted over the surface of the droplet so it is natural to discuss them in terms of force per unit area or stress. There are two kinds of stress. The first is shear stress. It is the result of fluid shear. The magnitude of this stress depends on the fluid viscosity and the fluid shear rate.

The second fluid force is a pushing or pulling stress. It is the result of direct head-on collision of the droplet with a flowing fluid. The magnitude of this stress depends on the kinetic energy (velocity and density) of the fluid. Each of these stresses can be compared to the interfacial restraining stress in the form of a dimensionless number.

The Weber Number is calculated as:

$$N_W = \frac{\rho d u^2}{\sigma} \qquad \text{Eqn (3.43)}$$

The symbols are defined as follows. The SI units are given by:

d = diameter of oil droplet (m)
u = mean velocity in the fluid (m/sec)
ρ = density of the continuous phase (kg/m³)
σ = interfacial tension (N/m)

It is straightforward to confirm that the Weber Number is dimensionless. When SI units are used, no conversion factors are required. It is interesting to note that the Weber Number does not depend on viscosity. This is a consequence of the fact that the Weber Number is used only in reference to turbulent flow where viscosity is not dominant. In cases where viscosity is significant, i.e. low Reynolds Number flow, it is more appropriate to use the Capillary Number which is not discussed in this book. SI units of interfacial tension are mN/m (milli-Newtons/meter). Most chemical engineers are familiar with units for interfacial tension of dyne / cm. It just so happens that dyne/cm are equivalent to mN/m.

The Weber Number is typically calculated for turbulent flow situations. As previously discussed, in isotropic homogeneous turbulence, there is a direct relationship between turbulence intensity and average turbulent velocity fluctuations. In turbulent flow it can be shown that the velocity fluctuations are related to the energy dissipation rate per unit mass. This establishes the following relation:

$$< (u')^2 >= (\varepsilon d)^{2/3} \qquad \text{Eqn (3.44)}$$

where:

d = diameter of droplet or bubble (m)
u = mean velocity in the fluid (m/sec)
ε = turbulence intensity (m²/sec³)

The quantity $< (u')^2 >$ has already been extensively discussed in Section 3.4.1 (Kinetic Energy in Isotropic Turbulence). It is calculated by measuring and then tabulating velocity fluctuations (u') as a function of time. The square of the velocity fluctuations are calculated, then the average of the square is calculated. The brackets denote the average. Basically it is a statistical measure of velocity fluctuations. Upon substitution the following relation is derived for the Weber Number.

$$N_W = \frac{\rho d^{5/3} \varepsilon^{2/3}}{2\sigma} \qquad \text{Eqn (3.45)}$$

The following example calculation provides some guidance in the use of this equation.

$$N_W = \frac{(1,000\ \text{kg/m}^3)(100\ \text{x}\ 10^{-6}\ \text{m})^{5/3}(10,000\ \text{m}^2/\text{sec}^3)^{2/3}}{2\ \text{x}\ 0.03\ \text{N/m}} = 1.7 \qquad \text{Eqn (3.46)}$$

As discussed, the Weber Number is dimensionless. The value of the Weber Number ranges from less than one to between ten and 30 or so. When the calculated value is less than unity, it signifies that droplet breakup due to shear is not the dominant process and that coalescence is dominant instead. The conditions evaluated here are for oil droplets of 100 micron, an interfacial tension typical of hydrocarbon / water mixtures that do not contain a high loading of natural of added surfactants.

3.5 Shearing of Oil Droplets

In this section, one of the most important subjects in produced water treating is discussed – shearing of oil droplets. Shearing of the production fluids creates small droplets of oil in water and small droplets of water in oil. For the purpose of water treating, we will only discuss the shearing oil droplets in water. The greater the shearing, the smaller the droplets, and the more difficult the subsequent process of separating the oil from the water. Also, the greater the shearing, the greater the concentration of oil droplets in water.

The Dynamic Equilibrium between Breakage and Coalescence: Even when the overall shear intensity is high, there is a distribution of shear rates in the fluid. In a localized high shear zone, droplets will break apart. In a localized zone of moderate or low shear, droplets may actually coalesce. The two processes of droplet breakup and coalescence may actually occur in close proximity to one another, depending on the source of shear and flow path of the fluid. In most flow situations, an emulsion that undergoes shear experiences both breaking and coalescing processes simultaneously, albeit to different degrees depending on the shear intensity and flow configuration.

As an example, in the flow through a valve, the fluid turbulence is at a maximum as the fluid flows through the narrow region between the valve and the seat. As the fluid moves away from this region, the turbulence intensity begins to decrease. The very quantity that causes droplet shearing, the energy dissipation rate, is also responsible for dissipation of the turbulence. As this dissipation occurs, the energy dissipation rate will gradually reduce to the point that it becomes favorable for droplet coalescence. Droplet coalescence tends to occur under mild turbulence conditions and tends to have a much higher rate for concentrated emulsions of oil in water than for dilute since that leads to high droplet/droplet collision rates. Also, the residence time of fluid in the turbulent region has a significant effect on the extent of shearing and coalescence. It is difficult to provide quantitative models for these effects. Nevertheless, some order of magnitude models are available that help guide design and troubleshooting efforts.

3.5.1 Maximum Droplet Size in Turbulent Flow

Whether or not the shear rate is sufficient to break the droplets depends on the interaction between two stresses (force per unit area). One of these stresses is external to the droplet and is due to turbulence as just described. The external stress deforms the drop. The other stress occurs at the oil/water interface. The interfacial stress due to interfacial tension acts to restore the spherical shape of the droplet and minimize the interfacial area. If the shear rate across the droplet due to turbulence is sufficiently high to overcome the interfacial restorative stress, the droplet will break apart. Thus, for a given shear, there will be a maximum droplet size which is related to the relative strength of turbulent shearing forces (characterized by ε), and surface forces characterized by interfacial tension (σ). The ratio of stress caused by turbulence to restorative stress due to surface tension is known as the Weber number.

In reality, two different regimes of turbulence should be distinguished [5, 12], the turbulent inertial, and the turbulent viscous regimes. In the turbulent inertial regime, the oil droplets are larger than the smallest eddies. In the turbulent viscous regime, the oil droplets are smaller than the smallest eddies. The maximum droplet diameter that can survive in the turbulent flow is determined by a somewhat different set of forces depending on which regime is present. In both regimes the forces that hold the droplet together, and which resist deformation, are the same. In both regimes, interfacial tension (capillary pressure) resists deformation. In the turbulent inertial regime, fluctuations on the hydrodynamic pressure cause droplet deformation and breakup. In the turbulent viscous regime, fluctuations in shear stress causes deformation and breakup. In the discussion below, it is assumed that the fluid is in the turbulent viscous regime (oil droplets are smaller than the smallest eddies) and therefore fluctuations in turbulent shear stress are solely responsible for droplet breakage. This assumption takes into account that most water treatment challenges occur as a result of relatively small oil droplets.

What is meant by "maximum droplet size" needs to be defined since the droplet size distribution is a statistical quantity and there may be a finite probability for very large droplets. Typically, the maximum droplet size is defined as the 95 % cut off of the droplet size distribution. This is the droplet diameter at which 95 % of the droplets are smaller, and 5 % are larger. This 95 % cut off corresponds to an empirically determined maximum Weber number (We_{crit}) which is further discussed below.

In this section we develop the general relation between shear intensity and droplet size distribution [12, 13]. The derivation is based on the Weber Number which was introduced above. The Weber number is a characteristic dimensionless number that gives a measure of the relative magnitude of the turbulent stress which distorts the shape of the droplet divided by the stress which restores the droplet to a spherical shape. The turbulent stress is generated by the fluid turbulence which is external to the droplet. The restorative stress is generated by the interfacial tension which is internal to the droplet. In the extreme case, the turbulent stress can break the droplet into smaller droplets. This is the case that will be discussed here.

Maximum Droplet Diameter: Substituting the previous expressions and rearranging gives an expression for the maximum droplet size in a turbulent flow:

$$d_{max} = 0.6 \left(\frac{\sigma}{\rho_w} \right)^{3/5} / \varepsilon^{2/5} \qquad\qquad \text{Eqn (3.47)}$$

where:

d_{max} = maximum droplet diameter (m)
s   = interfacial tension between the oil and water phases (N/m)

From inspection of this equation, the following conclusions can be drawn. The greater the interfacial tension, the larger the maximum droplet size. Indeed, this is often observed in the field. When the interfacial tension decreases, by addition of production chemicals such as methanol or corrosion inhibitor, the droplet size due to shearing also decreases. This is physically reasonable since interfacial tension is the force holding the droplet together, so to speak. When interfacial tension decreases the droplets are more susceptible to shearing.

Droplet Size versus Shear Intensity: The relationship described by the equation above is illustrated graphically in the figure below for three hydrocarbon/water interfacial tensions. The mixing energies range from 0.0010 m^2/s^3 which is the mixing energy associated with the filling of settling tanks up to 100,000 m^2/s^3 which is the mixing energy that may be imparted by a control valve. It can be seen that the stable droplet size over this range falls from 10,000 microns for low mixing intensity to the order of 1 micron for high mixing intensity.

A chemically stabilized emulsion would tend to generate even smaller droplet sizes due to lower surface tension and inhibition of coalescence. Chemically stabilized emulsions can be formed, for example, by high dosages of certain chemicals, e.g., corrosion inhibitor, demulsifiers, wetting agents, etc., and the passage of the oil/water mixture across shearing elements such as control valves, or by recycle pumps.

Maximum Droplet Diameter versus Shear Intensity: The process of droplet breakup due to shear was quantified in mathematical form by Hinze [12] in 1955, and subsequently improved by Davies [14]. As a point of interest, Hinze was a pioneer in the field of turbulence. He started his career at one of the major oil companies where he developed his ideas about droplet shearing. The graph in the figure below gives the result of his model and shows the range of turbulent energy dissipation rate associated with straight pipe, high shear centrifugal pumps, and valves. Karabelas [4] applied these concepts to the process of coalescence of droplets in mild shear. That topic will be discussed in Chapter 3.

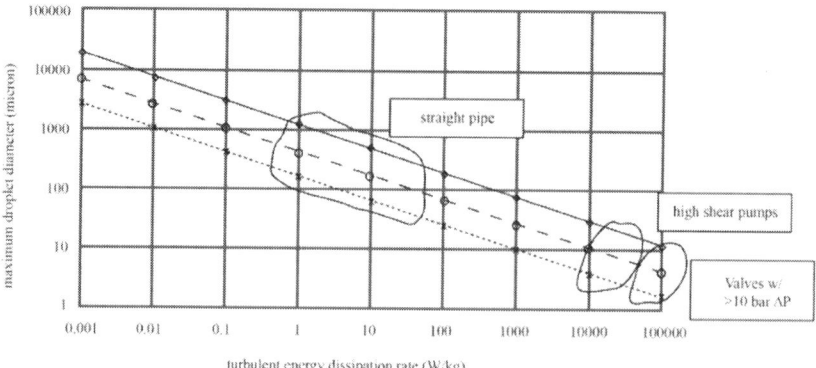

Figure 3.15 The maximum oil droplet diameter (microns) as a function of the turbulent energy dissipation rate (Watts / kg, which is equivalent to m2/sec3). Three values of interfacial tension are considered (1, 5, 30 mN/m), where the larger IFT is shown as the solid line.

In the above equation, the units for interfacial tension are N/m. In figure 3.15, the units are mN/m (milli Newton per meter), which is equivalent to the more familiar units of dyne/cm. In the original Hinze paper [12], the model was compared to experimental data. Good agreement was found for a light (low viscosity) oil in water. It is important to point out that the Hinze model does not take into account the viscosity of the oil phase, nor the concentration of the oil phase (oil cut). Subsequent work on these models do account for these interactions [14]. Also, the Hinze model, and other models for the maximum droplet size, do not take into account drop-droplet coalescence which generally occurs immediately downstream of any intense region of turbulence.

Droplet size as a function of interfacial tension: As shown in Figure 3.15, the droplet size resulting from shear is a strong function of interfacial tension. Low interfacial tension results in small droplets which are generated by shearing of the oil/water mixture. In typical oil/water systems, low interfacial tension can be the result of the following factors:

- naturally occurring polar molecules in the produced fluids that migrate to the interface;

- excessive use of production chemicals such as corrosion inhibitor and/or low dose hydrate inhibitor.

Both of these classes of surface active chemicals have a similar effect. They both reduce the interfacial tension thus making oil droplets more sensitive to shear. Also, the presence of these molecules at the oil/water interface creases an energy barrier to coalescence. Thus, once the small droplets are formed, they remain stable and generally lead to water quality problems. Most applications of production chemicals do not lead to drastic water treatment problems. However, in some cases where asset integrity or flow assurance would otherwise be in jeopardy, high doses of production chemicals is the saving grace for the facility.

Droplet size as a function of viscosity: Hinze discusses the effect of droplet viscosity and continuous phase viscosity on the droplet size resulting from shear [12]. He reviewed the experimental data of Taylor [15]. Although only a few data points were available at the time, the data together with some theoretical considerations led to the following conclusions:

- The energy input required to break a droplet reaches a minimum when the ratio of droplet to continuous phase viscosities are just slightly greater than unity: $\mu_d / \mu_c \cong 1$;

- The size of the droplets formed are smaller when the ratio of viscosities is less than unity: $\mu_d / \mu_c \leq 1$;

In terms of practical application, the viscosity, for moderate API oils, is typically greater than that of water. For relatively heavy oil, the viscosity can be many times greater than that of water. According to Hinze's findings, higher oil viscosity requires greater energy to break the drops, and the resulting droplets size will be larger than that for lower viscosity oil.

Bernoulli Equation: As might be expected, the Bernoulli Energy equation can be used to derive practical equations for the turbulence energy dissipation rate. The Bernoulli equation is written here with subscript (1) indicating the upstream location, and subscript (2) indicating the downstream location. Basically what this equation says is that, along a fluid streamline, energy is conserved as the fluid flows from location (1) to location (2). Any conversion of mechanical energy into heat is accounted for by the energy dissipation term (h), which is explained more completely below and in [9-11].

$$\frac{P_1}{\rho} + \frac{u_1^2}{2} + gz_1 = \frac{P_2}{\rho} + \frac{u_2^2}{2} + gz_2 + h \qquad \text{Eqn (3.48)}$$

where:

- r   = density of the fluid (kg/m^3)
- P_1 = pressure at downstream location (N/m^2)
- u_1 = superficial velocity of the fluid at the downstream location (m/sec)
- z   = elevation of fluid relative to some reference point (m)
- h   = energy dissipation due to fluid friction from point (1) to point (2) (J/kg)
- g   = acceleration due to gravity (9.81 m/sec^2)

Units of Energy per Unit Mass: Each term of the equation has units of Nm/kg which is equal to J/kg (energy per unit mass of fluid). From the standpoint of shearing of oil droplets, the interesting term is h, the energy dissipation due to fluid friction. For flow through a pipe, elbow, or through a pump, energy dissipation is mostly due to turbulence. In reality the energy is not lost, it is converted from mechanical energy into thermal energy. The magnitude of this energy conversion, can be significant from the standpoint of shearing of oil droplets, or even in terms of pressure droplet in some cases. However, from the standpoint of temperature rise, it is rather small, due to the high heat capacity of most liquids.

Calculations for Piping Systems: The equation is a useful engineering equation since it can be used much like the equation for friction loss in piping systems. The Fanning friction factor has been empirically determined for a number of piping geometries, and features as shown in the Table below.

Table 3.4 Head losses caused by changes in geometry and fittings to pipelines

Loss		K	approx. /1/D
Strainer (entrance to a pipeline to a pump)		2.50	113
Bend - with r/D=1/2	22.5o bend	0.20	9
	45° bend	0.40	18
	90° bend	1.00	45
Bend - with r/D=1	22.5° bend	0.10	5
	45° bend	0.20	9
	90° bend	0.40	18
Gate valve	Fully open	0.12	6
	quarter closed	1.00	45
	half closed	6.00	270
	three-quarter closed	24.00	1080

The turbulence energy dissipation rate will in general vary from one location to another in a fluid. Only in the case of homogeneous and isotropic turbulent fluid flow will the turbulence energy dissipation rate (by definition) be constant from one location in a fluid to another location.

Laboratory Study of Shearing: It is worthwhile to mention that both mild and intense shearing can be replicated in the laboratory. This subject will not be discussed in detail except to note some of the references on the subject. Intense shearing is usually generated by using a choke in a flow line. The advantage of using a choke rather than a pump or valve is that the geometry can be easily replicated from one lab to another. Depending on the flow loop design, the pipe diameter and the choke geometry, high shear rates can be achieved in order to study droplet breakup. Van der Zande [1, 16] has used this approach and provides references to most of the relevant literature. At the opposite end of shear rate spectrum, mild shear can be generated in a beaker for small scale studies or in an agitated vessel for larger scale. An early and widely cited study was carried out by Coulaloglou and Tavlarides [17]. They carried out small scale studies in the range of 0.1 to 2.0 Watt/kg of power input. Alopaeus et al. [18] describe the construction and use of a stirred tank and the average energy dissipation rate generated. As discussed by them, and others [19], the average energy dissipation rate is proportional to the diameter of the vessel to the fifth power, and the impellor speed (rpm) to the third power divided by the volume of the liquid in the tank. Kresta and co-workers [3] have analyzed the distribution of energy dissipation rates as a function of location within such tanks.

3.5.2 Shearing Through a Valve

Pressure Drop through a Valve: Returning to Bernoulli's equation, a useful formula can be derived that can be used to estimate the energy dissipation rate for flow through a valve. Assuming that the height of the fluid going into the valve equals that coming out of the valve, $z_1 = z_2$ and assuming that the superficial velocity is the same going in and coming out, $u_1 = u_2$, this leaves only the term involving the pressure drop $(P_1 - P_2)$, and the energy loss (h). The energy loss per unit time is then given by:

$$\varepsilon = \frac{\Delta P}{\rho\, t}$$
Eqn (3.49)

where:

t = time that the fluid experiences intense turbulence (sec)

From inspection of equation (3.56), the following conclusions can be drawn. The greater the pressure drop per unit time, the smaller the maximum droplet size that will be generated. Thus a pressure drop experienced over a short time (e.g. through a control valve) will generate a smaller droplet size than the same pressure droplet experienced over a longer time (e.g. pipeline). Also, a key point is that the pressure profile goes through a minimum which is followed by pressure recovery. If the depth of this minimum can be reduced by clever valve design, then shearing of oil droplets can be reduced as well (see below).

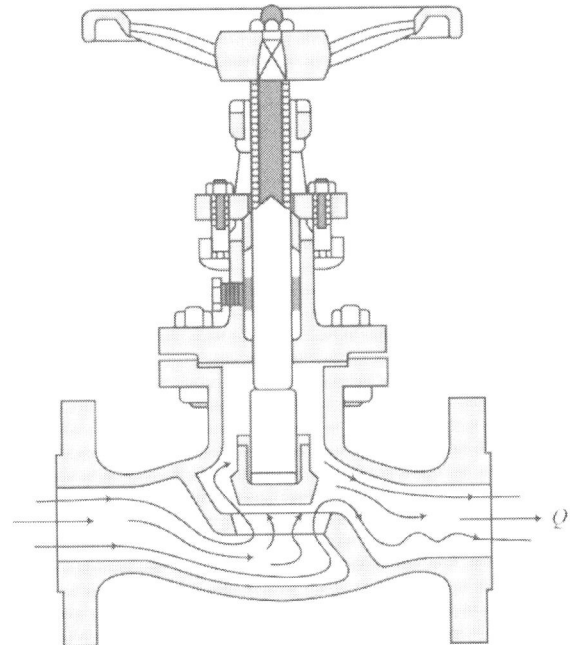

Figure 3.16 Fluid streamlines shown schematically in a globe type control valve.

The quantity 1/t can be replaced with u/L, where L/u is essentially the time it takes a fluid to travel a distance L. The relation between pressure drop, flow rate and pipe diameter for fully developed turbulent flow is then given by:

$$\varepsilon = \frac{u\Delta P}{\rho\, L}$$
Eqn (3.50)

Where:

    e   = turbulence energy dissipation rate (m^2/sec^3; W/kg)

    r   = density of the fluid (kg/m^3)

    DP = pressure droplet (N/m^2)

    L   = distance the fluid travels in the turbulent zone (m)

    u   = superficial velocity (m/sec)

The length of the turbulent zone has been measured empirically for various situations. Based on empirical measurements, van der Zande [ref 1, and p. 15 of ref 10] and Morrison et al. [20] conclude that the length of the turbulent zone (L) is roughly 2.5 times the diameter of the pipe downstream of the valve ($L = 2.5\ D$).

Permanent Pressure Droplet through a Valve: As discussed above, the shear intensity is related to the turbulent velocity fluctuations. An alternative relation, in terms of permanent pressure droplet is possible. Since pressure droplet is an easily measured quantity this is far more useful. In an orifice, choke valve, control valve, or any restriction, the fluid is forced through a relatively small flow volume and therefore the velocity increases. This results in a rapid pressure decrease. In this zone, velocity gradients are present which can result in turbulence with corresponding energy dissipation and a temporary loss of pressure associated with the velocity of the fluid. Once the fluid passes out of the restricted zone, the velocity decreases and the temporary pressure loss related to velocity is recovered. However, the presence of the turbulent zone and its corresponding turbulence energy dissipation, causes a permanent pressure droplet compared to the pressure upstream of the restriction. The pressure drop, which is easily measured, can be related to the turbulence energy dissipation rate and hence to a quantitative measure of shear forces which result in net breakup of oil drops. This is illustrated in the Figure below.

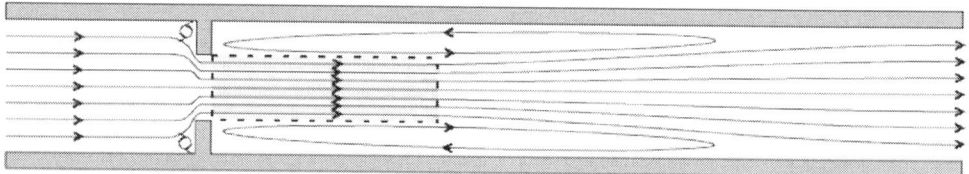

Figure 3.17 Top: fluid streamlines shown for flow through an orifice. In the study of shearing through a control valve, an orifice is often used instead of an actual valve because the fluid dynamics within an orifice are so well characterized.

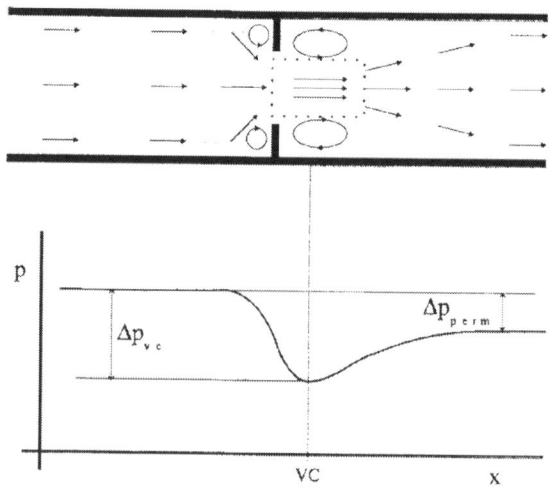

Figure 3.17 Bottom: a schematic diagram of the pressure profile though an orifice.
Of particular interest is the length of the turbulent zone and the permanent pressure drop.

Vena Contracta: In a conventional globe valve, the fluid flows through a narrow gap between the plug and seat. The point at which the gap is the narrowest is referred to as the vena-contracta. The vena-contracta is indicated as VC in the figure above. This is also the point of minimum pressure along the flow path. In the calculations discussed, the vena-contracta pressure does not seem to enter the formulas, at least not directly. The formulas seem to only depend on the permanent pressure drop. In fact however, the pressure profile does have an effect on the turbulence and therefore on the shearing of droplets. The longer and deeper the pressure profile, the greater the shearing effect.

Oversized Valves: The permanent pressure droplet is not the only relevant variable. There is a distinct difference between a small diameter valve that operates say half open, versus a larger diameter valve that must operate nearly closed in order to provide the same permanent pressure droplet for a given flow rate. The smaller valve that operates with a wider gap opening imparts less shear to the fluid. This is often seen in practice. As discussed by Murti and Al Nuaimi [21], so-called "oversized" valves can be replaced with smaller more appropriately sized valves in order to reduce droplet shearing. When this is done, the pressure profile is modified. The pressure droplet at the vena-contracta is reduced. But since the formulas discussed only depend on the permanent pressure drop, this effect is not predicted by these formulas. This is a limitation of the models. A more sophisticated approach that does take this into account is possible but is considerably more complex.

Length of the Turbulent Zone: The size and shape of the turbulent zone has been studied by Laser Doppler Anemometry (LDA) and by computational fluid dynamics (CFD). It has the highest turbulence intensity near the edge of the circular orifice, and the lowest intensity in the middle (along the radial axis). Emerging from the orifice, lines of constant turbulence intensity have a bullet shape. All of these details are important to understanding the shearing of oil droplets. The longer the turbulent zone, the greater the residence time of fluid within the zone. The wider the zone as it traverses along the axis, the more fluid will experience a longer residence time. While all of this is interesting from a fluid mechanical standpoint, for our purposes, a simple formula for the residence time is needed. The most common formula used is that the length of the turbulent zone is equal to 2.5 times the pipe diameter. This is the value that will be used here.

Illustrative Calculation: It is helpful to carry out at least one calculation in order to understand the various quantities in equation 2.5.12

$$\varepsilon = \frac{u \Delta P}{\rho L} = \frac{(3 \,\text{m/sec})(1,000,000 \,\text{N/m}^2)}{(1,000 \,\text{kg/m}^3)(0.25 \,\text{m})} = 11,800 \,\text{W/kg} \qquad \text{Eqn (3.51)}$$

Where:

r	= 1,000 (kg/m³)
DP	= 10 bar (1,000,000 Pa)
D	= 4 inch (0.1 m)
L	= 2.5 x D = 0.25 m
u	= 3 m/sec

The calculated value of energy dissipation for a 10 bar pressure droplet in a valve is roughly 12,000 W/kg. This is a substantial energy dissipation rate. It is significantly higher than the values calculated for pipe flow (Figure 3.15). As will be discussed below, this high value of energy dissipation rate causes oil droplet shearing to droplet diameters of the order of 10 microns or less.

Oman Export Oil Pipeline: There are many empirical examples of the effects of valve shearing on oil droplet size. One of the more interesting examples is the oil export pipeline in Muscat, Oman. All of the oil produced in Oman is produced in the interior of the country and most of it is pumped from the interior to offshore tankers. The storage tanks and pumps for this tanker loading operation are on the coast. Just a couple of kilometers from the coast is a mountain range. The oil from the interior of the country must traverse this mountain range in order to reach the coast. On the interior side of the mountain range there is a pumping station. On the coastal side, the pipeline comes down the mountain slope to a series of storage tanks. The hydrostatic head of the oil decreases significantly as it comes down the side of the mountain.

When the pipeline was installed, a control valve was part of the original installation. This control valve was installed at the bottom of the mountain to control flow rate, reduce the pressure, and arrest the forward momentum. This proved to be problematic. The oil still contained some residual water. In passing through the control valve such tremendous shear was applied that the water formed an emulsion of very small droplets suspended in oil and made it impossible for final dehydration to occur before loading onto the tankers. There was a significant monetary penalty suffered as a function of the water content of the crude oil since that water would eventually drop out as the tanker made its way from Oman to the delivery point. The settled water corroded the crude oil storage tanks in the vessel.

In order to more gently reduce the pressure and prevent an emulsion from forming, a serpentine pipeline was installed to compliment the control valve. The serpentine pipeline has segments of roughly 100 meter long and many 180 degree bends. The point of this pipeline configuration is to allow the dissipation of forward momentum as in the original control valve, but in a longer time span, thus reducing the turbulence intensity and preventing the excessive shearing of oil drops. Of course, this is not always possible. In fact, it is only possible onshore and is only economical in somewhat unique circumstances such as the oil export pipeline in Oman. Today, low shear valves are now available which will be discussed below.

Low Shear Valves: There are fewer options for reducing the shear in valves than there are for pumps. Pressure control is paramount in well control, sand control, hydrate control, oil conditioning, gas gathering and compression, etc. However, there are some options. First, the application of a properly chosen chemical can reduce the effect of valve shearing. The mechanisms is not well understood but is believed to be related to promoting droplet coalescence. In every instance of shearing, there is a gradual reduction of shear and the fluid emerges from the valve or pump. As shear decreases, the fluid enters a zone where the turbulent energy dissipation rate is favorable for droplet-droplet collision and coalescence. Chemicals that migrate to the oil/water interface, force other surface active molecules away from the interface, can actually reduce the effect of shear by promoting coalescence. These chemicals can also be used on their own in moderate shear zones. Location of shearing can also be a factor. If the location of a valve can be moved further upstream from the separation equipment, then there will be additional time for drop-droplet coalescence to occur before separation.

Finally, one last promising area is the development of lower shear valves. There are several candidates being developed and tested. They almost all rely on extending the time over which the fluid experiences the pressure drop. This reduces the turbulent dissipation rate. Since the pressure droplet is fixed by the process conditions, the only variable left is the time.

One promising valve is the Typhoon valve [22-25]. It was designed using hydrocyclone principles where a swirl motion is induced in the fluid as it passes through the pressure reducing zone of the valve. The fluid mechanics of the swirl motion cause the pressure droplet to be much more gently applied and hence less intense.

A field test was conducted on the Statoil Osberg C platform [23, 24]. The process configuration for these tests is as follows. The test valves (Typhoon versus Standard) were installed just upstream of a test separator. The valve was adjusted to various settings (1 – 5 in the figure). Water cut of fluids going into the Test Separator was 50 %. Water retention time in the Test Separator was between 10 and 15 minutes. The pressure droplet across each test valve was 70 bar. The water quality (y-axis: oil-in-water concentration) was measured for the oily water discharge from the Test Separator. Droplet size was not measured. Although measurement of droplet size would have been scientifically interesting, the measurement of oil-in-water concentration after settling in the Test Separator, under controlled conditions, is more reliable offshore. The improvement seen in the oil-in-water concentration using the Typhoon valve indicates that the valve provides less shearing and hence larger oil droplets which rise more quickly in the separator.

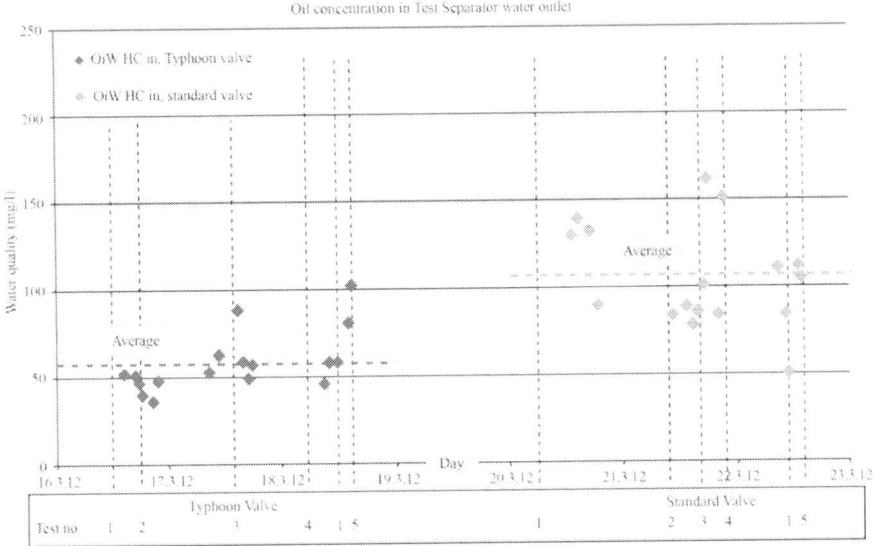

Figure 3.18 Oil-in-water concentration in the water discharge of the Test Separator at the Statoil Oseberg C platform [22]. The water cut of fluids going into the Test Separator was 50 %. Water retention time in the Test Separator was between 10 and 15 minutes. The pressure droplet across each test valve was 70 bar.

In [22], mineral oil (Eureka 10) dispersed in water was run through a cyclonic-based valve. The concentration of oil was 500 ppmv. Particle tracking and laser Doppler techniques were used to determine the velocity at different radial and axial locations. It appears that both tangential and axial velocity is relatively low near the radial center (core) of the device. As demonstrated for hydrocyclone fluid mechanics, the inner or core region rotates like a rigid body, while the outer portion of fluid (closer to the wall) behaves like a free vortex. For an explanation of these concepts see Section 10.2.3 (Tangential Velocity and Vortex Analysis). Given the centrifugal forces induced by the swirling motion, the oil droplets would have migrated to the core and would not have experienced as much turbulence intensity as the fluid closer to the wall, and particularly near the wall.

Table 3.5 Droplet Diameter

Location / Valve	Median (micron)	Mean (micron)
Inlet	11.5	42.6
Outlet choke valve	6.3	39.1
Outlet cyclonic valve	13.6	46.2

Results for droplet diameter are given in the Table 3.5. The mean diameter is based on the number average, and not the volumetric average. Also the median is the value of droplet diameter for which half the number (and not volume) of droplets are larger, and half the number are smaller. As shown, the cyclonic valve significantly reduces the effect of shearing. In fact, the cyclonic valve has a higher mean droplet diameter than the inlet fluid. This, and other data reported [26], indicates that the cyclonic valve actually has a coalescing effect on the droplets. There is a dramatic improvement in the median value of the droplet diameter upon use of the cyclonic valve, instead of the choke valve. Pressure droplet is not reported and it is not entirely clear whether the tests were carried out at the same pressure droplet through the different valves. Also, no properties are reported for the mineral oil or mineral oil / water mixture.

Typhonix has developed a low shear control valve based on cyclonic principles. The idea for the design originated in careful study of the cyclonic fluid flow that occurs in hydrocyclones. The heart of the design is the use of a cage that has tangential holes in its side, as shown in the figure. The tangential holes induce a tangential motion in the fluid which has similar pressure droplet to that observed in a hydrocyclone.

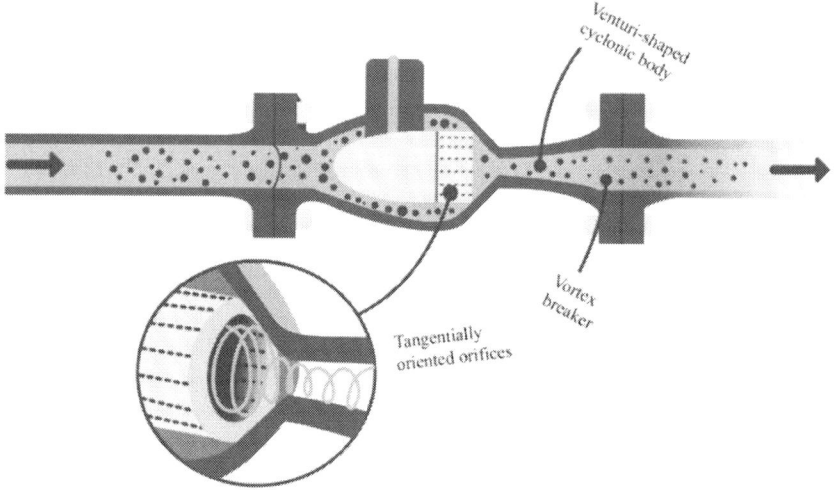

**Figure 3.19 Valve body showing the cage (left) and vortex breaker (right).
The section between these two devices is referred to as the venturi section [23].**

Conventional choke valve replaced by a Typhoon valve [22]:

Poor produced water quality from this well is assumed to be a consequence of heavy choking (150 bar ΔP). Based on on-field droplet size measurements, from this particular well, and on laboratory experiments performed with the Typhoon Valve, it is estimated that *hydrocyclone efficiency* increases from **79 to 94%** by replacing the existing choke valve with a Typhoon Valve. In this estimate it is assumed that the produced water from the well is treated directly by hydrocyclones.

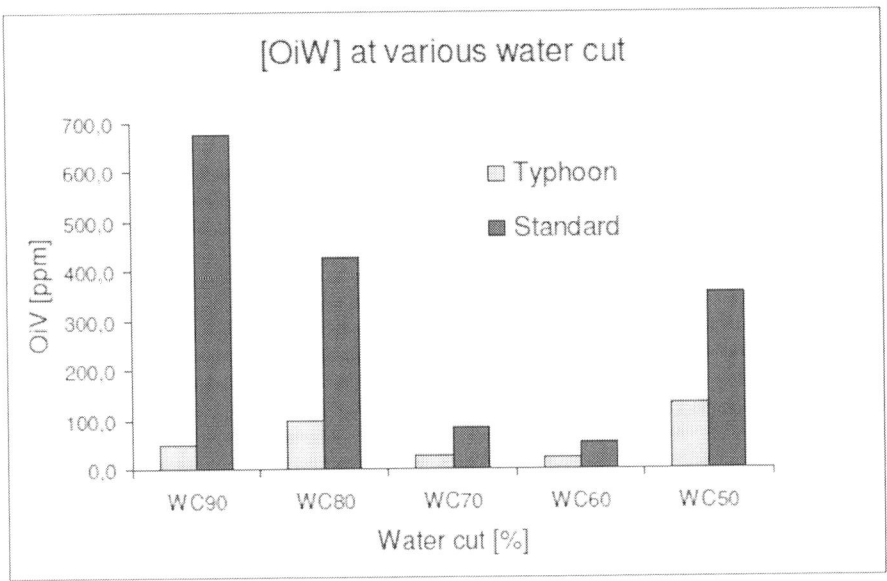

Figure 3.20 Oil-in-water concentration at valve discharge for a 10 bar pressure droplet across each valve. Three phase fluid was fed to each valve (standard gate valve versus Typhoon cyclonic valve). Gulfaks crude oil and Gulfaks hydrocarbon gas were used. The water cut is shown. The gas / liquid ratio (GLR) was 0.5. Testing was carried out at the Statoil Multiphase Flow Loop test facility in Porsgruun [23].

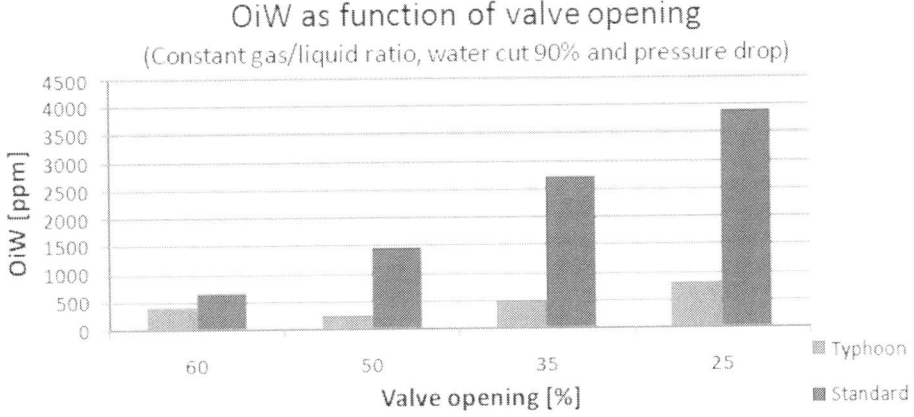

Figure 3.21 Oil-in-water concentration Testing was carried out at the Statoil Multiphase Flow Loop test facility in Porsgruun [23].

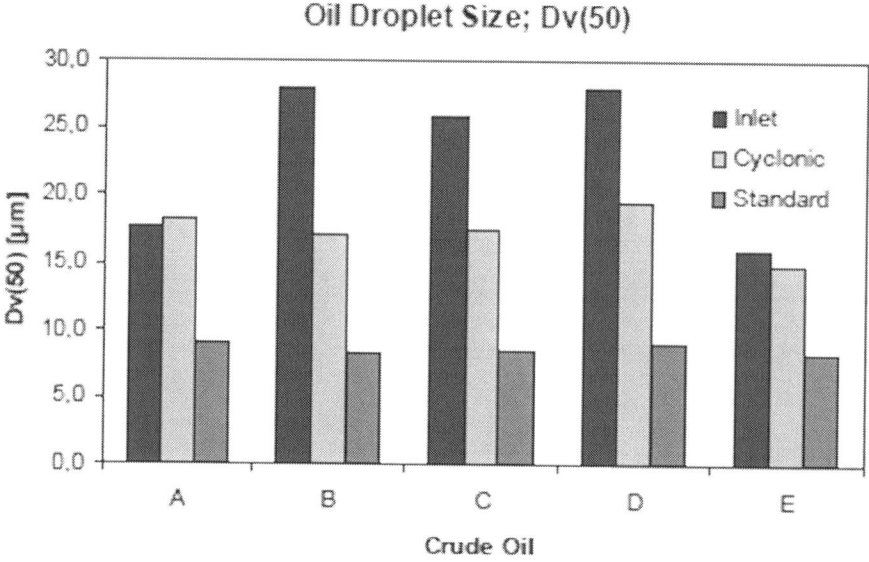

Figure 3.22 Volume-averaged oil droplet diameter upstream and immediately downstream of two test valves as a function of different types of crude oil [23]. Two-phase flow (oil/water) was used with an oil concentration of 500 ppmv. Pressure droplet across the valves was 6.9 bar.

The following table gives the properties of the crude oils given in the figure above.

Crude Oil	API	Density (g/ml)	Viscosity (cP)
A	37	0.84	9
B	38	0.84	7
C	27	0.89	24
D	25	0.91	39
E	41	0.82	3

3.5.3 Shearing Through a Pump

Centrifugal pumps are by far the most widely used type of pump in upstream oil and gas facilities. From a water treating perspective they have benefits and drawbacks. The cost of a centrifugal pump, compared to that of most positive displacement pumps is an obvious attraction. In this section, attention is paid to reducing pump shear and the effect of pump shear. Some background is provided first on how a centrifugal pump works.

Basic Description: Fluid enters the pump through the eye (center) of the impeller. As the impellor spins, the fluid is accelerated and flows from the center toward the impellor edge or tip. There are different designs and shapes of impeller channels but for the most part, all designs are similar in the respect that the width of the channel increases from the eye of the impeller to the tip where the fluid is discharged. This increase in channel width results in a conversion of kinetic energy to pressure energy. In other words, as the fluid traverses the impeller channel it slows down and as it is pushed against the fluid in front of it, its pressure builds up [37].

The blades of the rotating impeller transfer energy to the fluid thereby increasing pressure and velocity. The design of the impeller (type, size) depends on the requirements for pressure and flow rate. The impeller design and the rotation speed are the primary variables determining the pump performance.

In a radial impeller, there is a significant difference between the inlet diameter and the outlet diameter and also between the outlet diameter and the outlet width, which is the channel height at the impeller exit. In this construction, the forces result in high pressure and low flow.

The impeller has a number of impeller blades. The number mainly depends on the desired performance. Impellers with 5 to 10 channels have proven to give the best efficiency and are used for fluid without solid particles. One, two or three channel impellers are used for fluids with particles such as produced water. The leading edge of such impellers is designed to minimize the risk of particles blocking the impeller. One, two and three channel impellers can handle particles of a certain size passing through the impeller.

Pump Shear: It is noted here that whether or not an emulsion forms as a result of pump shear depends as much on the pump as it does on the produced water chemistry. As discussed previously, there are really two properties that come into play. The first is the sensitivity of the fluid to shear. Stated another way, this is the tendency of an oil in water dispersion to form small droplets as a result of shearing. This is primarily related to the interfacial tension of the oil / water mixture, and the viscosity of the oil suspended in the water. The other important property of the fluid is the extent to which mixing and mild turbulence facilitates the coalescing of oil droplets. Examples of how different fluids respond to differing levels of shear was discussed in Section 3.5.1 (Maximum Droplet Size in Turbulent Flow).

In 1988 and 1992 Flanigan et al [27, 28] published measured oil droplet size for a number of different types of pumps over a range of differential head. The results of these measurements are given in the Figure below. All of the pumps were fed the same or similar oil / water emulsion, and were operated at the same flow rate over a range of differential head.

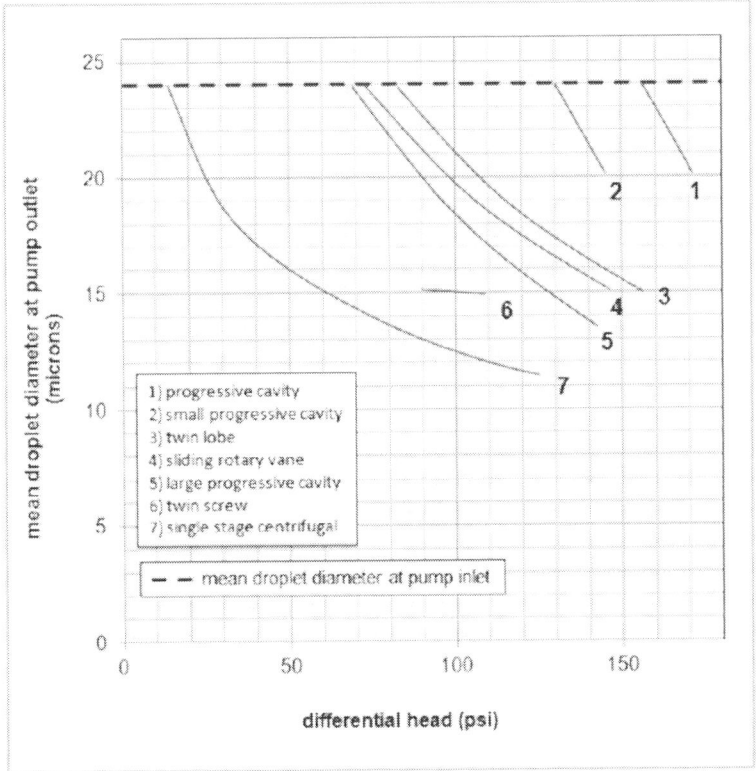

Figure 3.23 The effect of pump shear on oil droplet diameter for centrifugal and other pump types [27]. It is noted that in the original reference [27] the figure legend contained an obvious mistake where lines (6) and (7) had been switched. As shown above, line (7) is clearly that of a centrifugal pump.

Centrifugal Pumps: In 1994, Ditria and Hoyack [29] summarized the benefits and drawbacks of centrifugal pumps versus positive displacement pumps. At that time, centrifugal pumps were not normally designed to minimize shear. Shearing can be reduced by design features that maintain a high hydraulic efficiency (e.g. do not use a recessed impellor). Also a low speed (e.g. large impellor diameter) reduces shear. Schubert [30] has also pointed out the relationship between hydraulic efficiency and shear. Morales et al. have quantified the effect of hydraulic efficiency [1]. The greater the hydraulic efficiency the smaller the turbulent energy dissipation, and the larger the oil droplets in the pump discharge. Multiple stages can be used to reduce the head per stage.

Motor Speed: As discussed, the amount of shear that a centrifugal pump applies to the fluid is related to the rotating speed of the impellor, and a few other variables. One way to reduce the shear is to replace a given impellor / motor combination with a larger impellor that is run at a slower speed. Most centrifugal pumps are run off of an alternating current three phase induction motor. It is straightforward to calculate the so-called synchronous speed of the impellor. This is the speed that the impellor would have when running without a load:

$$\Omega_S = 120 \; f / N_p \qquad\qquad \text{Eqn (3.52)}$$

where:

f = frequency of the alternating current (Hz)
W_S = synchronous speed (rev/min)
N_p = number of poles in the motor

As an example, a motor running off of a 60 Hz alternating current, with four poles, will have an impellor synchronous speed of 1,800 rpm. The actual speed, under load, will be slightly less than the synchronous speed by a factor known as "slip." Slip is calculated as:

$$s = (\Omega_S - \Omega_{FL}) \text{x} 100 / \Omega_S \qquad\qquad \text{Eqn (3.53)}$$

where:

s = slip (percentage)
W_S = synchronous speed (rev/min)
Ω_{FL} = full load speed (rev/min)

For example, a four pole three phase induction motor with a rated full load speed of 1750 has a slip rating of 2.8%. In practice, the synchronous speed and slip are usually specified and the full load speed is calculated. A motor with a synchronous speed of 1,800 rpm and a slip of 3 % would have a full load speed of 1,746 rpm. Since the number of poles of a three phase induction motor is established when it is manufactured, the only way to change the speed of the motor is to change the frequency. Alternatively, the motor and impellor can be changed-out.

Specific Speed: The shear caused by a centrifugal pump can be minimized. A centrifugal pump with a closed impeller design should be selected and operated at a hydraulic efficiency above 70 %, and at a speed of 1800 rpm or less. The Specific Speed (discussed on p. 511 of [9]) is a quantity that can be used to provide guidance on how to minimize shearing in a centrifugal pump. It is used here as a correlation parameter. In other words, the Specific Speed has been calculated for a few centrifugal pump installations where oil droplet size has also been measured on the discharge side of the pump. The values of oil droplet size were then correlated against values of the Specific Speed in an empirical correlation which is given below. As will be demonstrated, the higher the Specific Speed, the lower the oil droplet size discharged from the pump.

The formula for calculating Specific Speed is given here. The formula is based on dimensional analysis. It is presented here without derivation. Page 511 of reference [9] and [38] give the correct formula which can be verified by checking that the Specific Speed is a dimensionless quantity.

$$\Omega_S = \frac{\Omega Q_V^{1/2}}{(gH)^{3/4}}$$

Eqn (3.54)

where:

W_S	=	the Specific Speed, a dimensionless quantity
Ω	=	rotational speed (revolutions / second)
Q	=	volumetric flow rate (m³/second)
g	=	acceleration due to gravity = 9.81 m/second²
H	=	total differential head (m).

A numerical example of the calculation of Specific Speed is given here. The following input information has been given for a particular pump operating in a particular application:

Ω	=	1,750 rpm (29.2 revolutions/second)
Qv	=	3,424 BWPD (0.0063 m³/second)
H	=	33 feet (10 m)

$$\Omega_S = \frac{(29.2 \text{ rev/sec})(0.0063 \text{ m}^3/\text{sec})^{1/2}}{[(9.81 \text{ m/sec}^2)(10 \text{ m})]^{3/4}} = \frac{(29.2)(0.0794)(\text{m}^{3/2}/\text{sec}^{3/2})}{(98.1)^{3/4}(\text{m}^{3/2}/\text{sec}^{3/2})} = \frac{2.318}{31.17} = 0.074$$

Application of this formula, together with a correlation for droplet size suggests that slow speed will result in less shearing. This will always be the case, regardless of the values of the other parameters. In other words, a centrifugal pump operating on a variable speed drive will result in less shear if the rotational speed is reduced. This is physically obvious. As shown below, there is an empirical relationship between the Specific Speed and the droplet size in the discharge of a centrifugal pump.

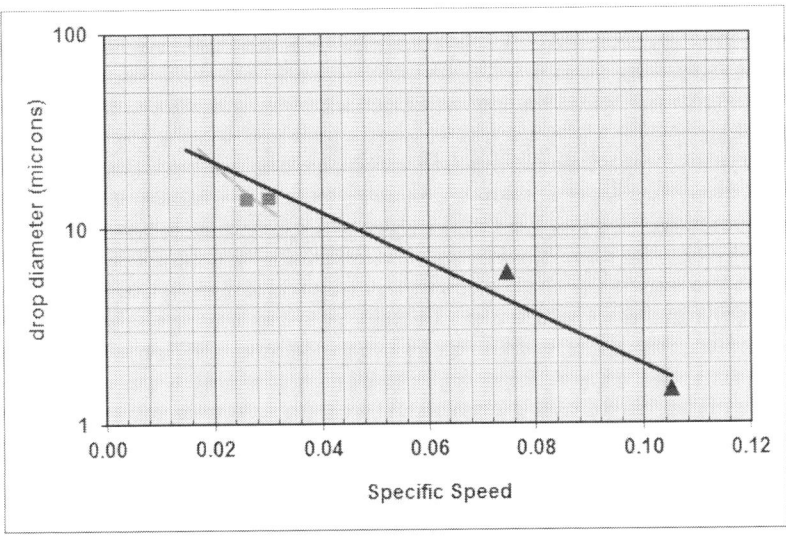

Figure 3.24 The oil droplet diameter in the effluent of the pump as a function of the Specific Speed.

As shown, the Specific Speed increases as rotational speeds increases, as expected. The discharged oil droplet size also increases. This is rather obvious. The figure gives an estimate of the relationship. The few data that are available suggest a straight line but the authors know of no theoretical argument that would suggest that a straight line would be suggested. There is uncertainty and scatter in the data which do not justify anything except a straight line.

Besides rotational speed, the Specific Speed also contains the ratio of $Qv^{0.5} / H^{0.75}$. As is well known, a typical pump curve for a centrifugal pump shows the relationship between total delivered head (H) and volumetric flow rate (Q), at constant rotational speed; as volumetric flow rate increases, the total delivered head decreases. Rotational energy can either be converted into fluid head or volumetric flow. Also well known is the fact that there is an optimal combination of Q and H which gives the Best Efficiency Point (BEP) at a given rotational speed. Lower flow rates (higher head) result in lower efficiency. Likewise, higher flow rate (lower head) results in lower efficiency. The fact that the Specific Speed includes $Qv^{0.5} / H^{0.75}$ gives at least some, albeit not highly accurate, means of taking into account variation in Q and H along lines of constant rotational speed. As this ratio increases, the Specific Speed increases which suggests that droplet size should decrease. This is particularly true for flow rates greater than the BEP, less so for lower flow rates where the head usually does not change much.

Recessed Impellor: A recessed impeller pump causes greater overall shear because it allows fluid to churn and recycle within the pump. This is evident by the low hydraulic efficiency. But this is not correct. Much of the energy being lost is lost to turbulence and friction of the fluid, both of which conflict with gentle handling. The efficiency of a recessed impeller pump is less than the efficiency of a traditional centrifugal pump because of the gap between the impellor and the impellor casing and volute which allows fluids to churn. Efficiency losses result from flow recirculation around the impeller passages, and from the inefficiency of a flow pattern where fluid rotates around the casing numerous times prior to exiting the discharge. Efficiencies in the 40%-50% range are common.

A pump with a full size impeller, and high hydraulic efficiency, prevents internal churning. It does subject a small portion of the fluid to very high shear as the impeller blade passes the volute edge, however, the volume of fluid thus sheared is very much smaller than that in a recessed impellor pump. For a low shear centrifugal pump design, select a high hydraulic efficiency pump, slow speed, large impeller.

Guidelines for design of a low shear centrifugal pump:

For a low shear centrifugal pump design, select a high hydraulic efficiency pump, slow speed, large impeller, as summarized here [31]:

> Slow speed (< 900 rpm)
>
> High hydraulic pump efficiency (> 70 %)
>
> Large impellor diameter (aligns with slow speed for given gpm)
>
> Large discharge nozzle (slow discharge speed)
>
> Limited pressure boost per stage (< 50 psi)
>
> Low specific speed Ns < 0.02

Positive Displacement (PD) Pumps: Another option to using a modified centrifugal pump is to use a positive displacement pump. As shown in the Flanigan figure [27], there are a number of positive displacement pump types that generate less shearing than a centrifugal pump. Several pump manufacturers have improved the reliability of these alternative pump types in the past twenty years making them a viable alternative.

Generally speaking pumps tend to shear liquids as the speed is increased. This makes the PD pump better able to handle shear sensitive liquids. Shear rates in PD pumps vary by design but they are generally low shear devices, especially at low speeds. Internal gear pumps, for example, have been used to pump very shear sensitive liquids.

Not all screw pumps are low shear. Shear rates are available from the manufacturer. Depending on the total flow rate there are several options:

- Progressive cavity pumps - These are very low shear. They can be sensitive to solids in the pump stream depending on the required clearances, which is dependent on liquid viscosity and required differential head.

- Rotating lobe pumps - Also very low shear. Since they are single stage they have somewhat limited differential head capacity. They are positive displacement pumps so the available differential head is much greater than you could expect with a centrifugal.

- Piston or plunger style reciprocating pumps - The shear rate is dependent on the intake and discharge valve design.

- Blowcase - As above, shear rate is dependent on the intake and discharge valve design. Motive pressure issues can make them costly to operate.

- Recessed impeller pumps - Not extremely low shear but sometimes adequate. Differential head is relatively low.

Twin Screw Pump Deepwater GoM: A deep water offshore platform did install a true low shear pump. While the benefits were obtained, the cost and effort to install a low shear pump make the benefit/cost ratio questionable. In 2001, two Flowserve twin screw pumps model NIHP external bearing, VFD-drive, were installed in order to minimize shearing of fluids. Two pumps were installed in a parallel with 2 x 100 % duty, giving 100% sparing. The pumps were designed to deliver 255 psid head, at 200 gpm. The service was to pump oily water from the Wet Oil Tank to the bottom of the FWKO. Water from the FWKO was routed to the hydrocyclones. Twin screw pumps were selected in order to minimize the shearing of oily water from the Wet Oil Tank.

The pumps were manufactured at the Flowserve plant in Canada. Flowserve initially claimed they would have no problem pumping the low viscosity fluid (mostly water), but several problems were encountered in the Factory Acceptance Testing. The pumps did not achieve the rated head at flow rate capacity. Flowserve decreased the clearances between the case and rotors and then had problems with rotor deflection and galling. They tried different hard facing on the rotors and shortened the shaft span between bearings (by changing the seal housing design) and finally demonstrated rated head at rated capacity. Altogether these modifications required months of work and very close engagement by the platform facilities engineer who had many other high priority items to look after.

Once the pumps were installed, they worked well offshore on VFDs, but there was one catastrophic failure in 2002 when the pump was run dry. This required a complete rebuild of the screws and housing. There was a second failure in 2005 which required $62,000 cost to rebuild. These pumps are not very forgiving. If the properties of the fluid change, or if the pump is started up incorrectly even once, damage can occur. A new pump was quoted at $130,000.

The new pump had a larger impeller which rotated at a lower specific speed than the original pump. However, the fact that it was a recessed impellor pump caused it to have greater shear than the original pump. A recessed impeller pump is sometimes promoted for gentle handling, but this is incorrect. Much of the energy being imparted to the fluid is lost to turbulence and friction, both of which conflict with gentle handling.

Pump and Valve Arrangements: In some water treatment process systems, a booster pump is required for recycling fluids into an upstream location. An example would be the case where the water discharge from a Bulk Oil Treater is routed to the Free Water Knockout. This is not necessarily a recommended routing. However, where such routing or similar routing, does occur a pump is required to move the fluid from the downstream lower pressure location into the upstream higher pressure location. Pumps that are imbedded in the process system need to be controlled in such a way that they do not cause level control problems in either the vessel that provides the feed fluid or the part of the system that they are pumping into. Two design scenarios are shown in the figure below.

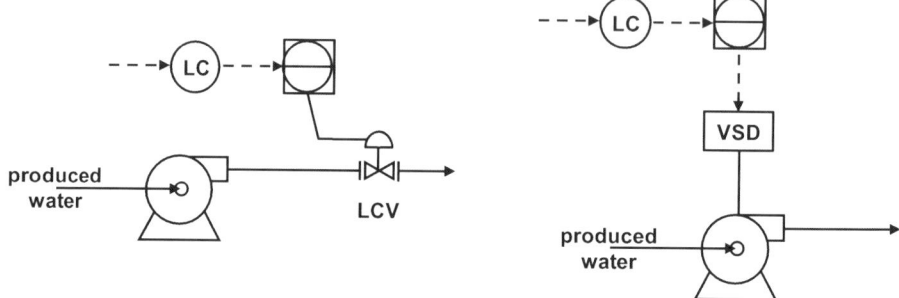

Figure 3.25 Two pump and control schemes. The one on the left involves a constant speed pump and control valve (LCV). The one on the right can be used in the same service but the control is implemented by a variable speed drive (VSD). The scheme on the left provides significantly greater shear of the fluids than the one on the right.

On the Auger platform the scheme on the left of the figure was replaced with the scheme on the right and water quality was improved. In this design, the pump may or may not be controlled. If it is controlled, then it is turned on and off by a controller. The real control is achieved by the valve. If it is normal mode is to run full time, then the Level Control Valve provides all of the control. In this configuration, the produced water is sheared by both the pump and the control valve.

Another control scenario that is sometimes used is to use a constant speed motor and control it (on/off) using the level control system. This has the advantage of eliminating the level control valve but it have a tendency to cause wide swings in flow rate, and level in the vessels that are associated with the system. One way to reduce the shearing is to install a variable speed drive for the pump and eliminate the LCV. This scenario is shown below.

The system in Figure 3.25 will reduce the shear compared to the LCV system. Also, the pump likely run continuously, if the control system is tuned properly. This will reduce wear and tear on the pump and the average impellor speed will be lower than the pump that runs intermittently.

The Effect of ESP Installation: The location of a pump has an enormous effect on the impact that the pump will have on oil / water separation in a facility. A high speed centrifugal pump located within a facility will shear the fluids and create an emulsion that is difficult to resolve. Electrical submersible pumps are typically high speed and high shear. They are installed downhole in order to suck fluids out of the well at a high rate. Oil droplet diameter immediately downstream of an ESP can be in the range of a few microns to 10 microns or so and the emulsion formed can be difficult to resolve. This is extremely small.

There is no doubt that a high speed centrifugal pump causes shearing of the fluids. However, there are instances where the installation of ESP has not had a detrimental effect on water treatment. If the pump is installed far upstream of the facility and the fluids have an opportunity to coalesce and separate in the pipeline, the impact of an ES: on oil/water separation may be minimal. The same could be said of a choke or control valve located far upstream. Also, if the ultimate destination of the produced water is a disposal well, with high permeability, then relatively high concentrations of oil and solids can still be disposed without being noticed. In those cases, the impact of an ESP may not be noticed even if it does promote the formation of a stable emulsion. So there are cases where the impact of ESP installation is either not apparent, or is actually not detrimental due to pipeline coalescence.

On the other hand, the use of production chemicals, particularly corrosion inhibitor can generate small oil droplets for a given shear rate, and can stabilize oil droplets such that they don't coalesce. Heavy oil, and gas condensate in low salinity water are two more examples where coalescence will be hindered and the effect of the upstream shear can be significant, even if it is located kilometers away.

Thus, the impact of ESP's depends on a number of factors. This is part of the reason why there is a lack of consensus in the industry on whether or not ESP are detrimental to water treatment. IF the factors mentioned above were taken into account, there probably would be general consensus.

Lekhwair & Yibal Fields Oman: In the Lekhwair and Yibal fields in Oman, hundreds of electrical submersible pumps were installed in horizontal and vertical wells. The average API across both fields is in the range of 38 to 40 degrees. Water salinity is moderately high. Thus, there is a favorable density difference between the oil and water. The overall impact of these ESP on production of oil and water has been documented [32]. Total fluid production rate significantly increased and for the first few years, oil production increased. But after a few years, the water content started to increase dramatically and oil production decreased. Oil / water separation in these fields is challenging from the standpoint of high water cut (90%) and large volumes of produced water. However, studies of the oil droplet size coming into the facility did not find a significant impact from the ESP. Apparently the flowlines from well head to facility were long enough (one to several kilometers) to allow coalescence of the oil droplets.

Gas Field in Gabon: In the early '90s a major operator in Gabon had chronic pump problems for transfer of an oil / water mixture from various gathering stations to the central Rabi Field station. Twin-screw positive displacement pumps, having four mechanical seals had been selected for their low shear characteristics. Excessive wear of the pumps had reduced the life of each pump to about nine months. Much effort was put into trying to extend the life of the pumps to no avail.

Petroleum Development Oman had recently gone through the same problem with twin-screw pumps but with a different manufacturer. PDO replaced the twin-screw pumps with horizontal, multi-stage centrifugal pumps running 4-pole motors. The synchronous speed of the pumps were 1,500 rpm. In Gabon, the alternating current frequency is 50 Hz (120 x 50 / 4 = 1,500). The experience with the multistage centrifugal pumps was good.

On learning of the PDO experience, the operator in Gabon bought one multistage centrifugal pump on field trial. An emulsion of oil and water was found immediately downstream of the pump discharge. But the flow line to the central station was one km long and no emulsion was found at the Central Station. Eventually all of the twin-screw pumps were replaced with the horizontal multistage centrifugal pumps. The other flowlines were longer distances from the central station. Essentially no adverse effect on the oil / water treatment system was experienced at the gathering station.

Conclusions: There are many sources of shear in upstream oil and gas development. Mild shear is actually a benefit in that it promotes oil droplet coalescence which improves oil/water separation. On the other hand, high intensity shear gives rise to small droplets which can result in poor oil/water separation even with equipment and chemical treatment programs that are working well. Being able to recognize when shearing can and should be reduced, and what are the options for doing so, are useful capabilities. Engineering formulas for calculating the magnitude of shear and relating it to oil droplet size have been provided. Options regarding types of pumps, pumping systems, and valves have been discussed. It must be noted that this is only a brief high level discussion of an interesting and important subject.

3.5.4 The Impact of Droplet Shearing on a Produced Water Process

Most produced water treatment equipment is more effective at oil removal when the oil droplets are relatively large. In order to remove small oil droplets, longer settling time, greater pressure driving force in a hydrocyclone, smaller gas bubbles in a flotation unit or higher concentrations of flocculating agent, or finer filtration are required. While the performance of a water treatment system depends strongly on the inlet oil droplet size distribution, the design of a water treatment system is, in many cases, based on an unknown distribution.

The most critical variable that will determine the performance of a produced water system is the oil droplet size distribution and this is often unknown when the system is designed. Other variables are of course important in some cases such as the presence of natural surfactant crude oil constituents (naphthenic acids), or the presence of surfactant-like production chemicals (corrosion inhibitor and hydrate inhibitor for example), or the presence of fine oily solids which stabilize emulsions. These chemical constituents can often exacerbate the effect of shear.

There are several different ways of estimating the effect of shear on the size distribution of oil droplets, as decribed above. These methods are semi-quantitative in the sense that they give only a relative indication and are best used to compare one situation versus another. Nevertheless, the guidance that these models provide can be valuable in designing a new facility or in identifying the cause of shear in an existing facility.

The purpose of this section is to demonstrate the subtle but nevertheless severely detrimental effect that shear has on the performance of a water treatment system and what can be done to reduce this effect. A model was developed to simulate the effect of upstream shear on the performance of the downstream water treatment equipment. With this model, the relative effect of upstream shear was demonstrated. The starting point for the simulation is the oil droplet distribution of the water discharge from an upstream three phase separator. This distribution is shown below in Figure 3.26. As shown the oil-in-water concentration is 660 mg/L and the Dv50 is 60 micron. These values were chosen based on experience with several deepwater platforms operating in the Gulf of Mexico. Other values could have been chosen. These values are realistic.

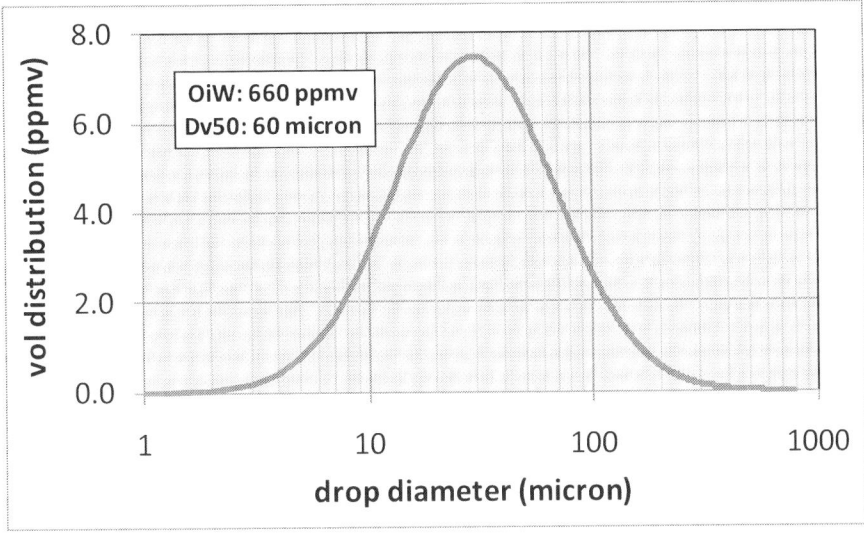

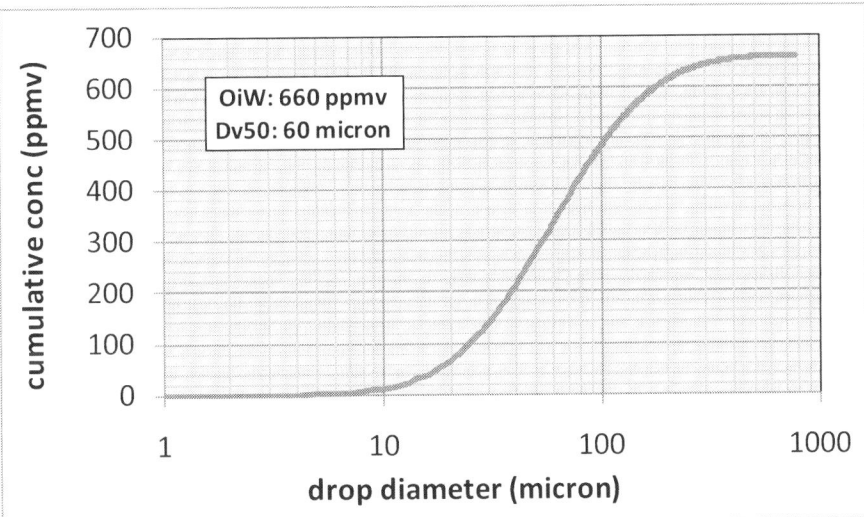

Figure 3.26 An example of an oil droplet distribution for produced water discharged
from a three-phase separator and entering the produced water treatment system.

Two produced water treatment systems shall be studied. In the first system, there is negligible shear of the water that is discharged from the upstream separator. In the second system, there is moderate shearing of the produced water. The result of this shear is shown in Figure 3.27. As shown, the second system has the same oil-in-water concentration as the first system. This validates the fact that the source of shear does not change the oil concentration, only the oil droplet size distribution. As shown, the Dv50 has not shifted to 32 micron, roughly half the size of the first system where there is no shear. Also, as shown, the maximum droplet size is 100 micron. The advantage of using a simulation is apparent in that the effect of shearing can be isolated from all other effects. The only difference between the two sets of figures is the presence of a shearing valve in the second system. In the next several slides the effect of this shear on the performance of the water treatment system shall be demonstrated.

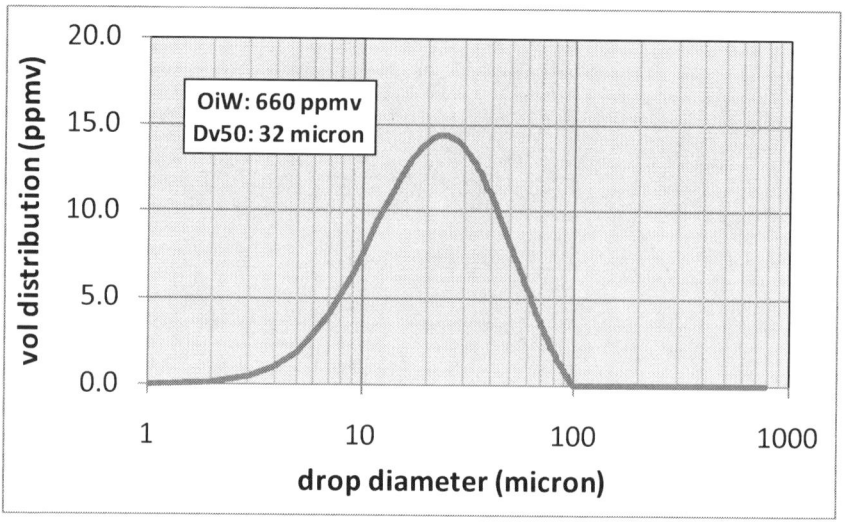

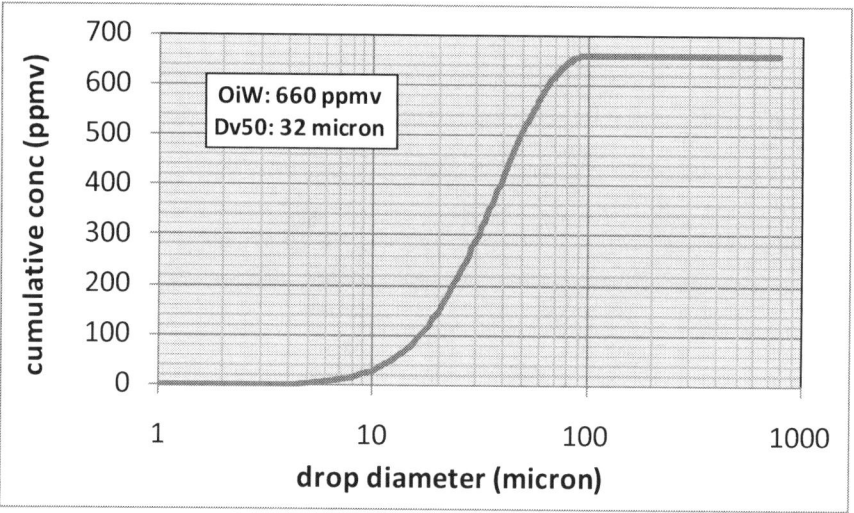

Figure 3.27 Oil droplet distribution discharged from a control valve. The feed to the control valve is the distribution shown in Figure 1. The control valve shears all droplets larger than 100 micron. It reduces the volume average droplet diameter form 60 micron (Figure 1), to a value of 32 micron (this figure). The valve does not change the total oil concentration (660 ppmv in both Figure 1 and 2).

The next step in the simulation is to model the effect of a water treatment system. For the purposes of this demonstration, the details of the water treatment system are not critical. The same water treatment system is modeled for both systems. In other words, the effect of the shearing valve is the only difference between the two systems. The identical water treatment system is applied to the water that is not sheared, as well as the water that has been sheared. The particular system chosen was a hydrocyclone followed by a flotation unit. But a different system could have been chosen. The important feature of the water treatment system is that it has a particular separation efficiency curve and that curve is applied to both systems identically. The results of this water treatment system are shown in Figure 3.29. As shown, all droplets larger than 25 micron are removed. This reduces the oil-in-water concentration to 30 ppmv, and results in a Dv50 of 13 micron. The overall separation efficiency is 95 % (630/ 660).

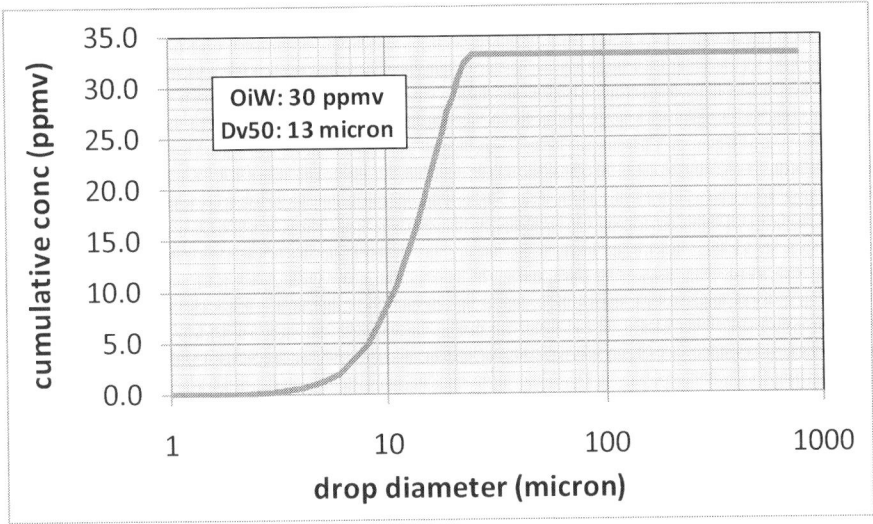

Figure 3.28 Figure 3. An example of an oil droplet distribution for
produced water entering the produced water treatment system.

The next step in the simulation is to model the effect of the same water treatment system on the sheared water. The results of this water treatment system are shown in Figure 3.30. As shown, all droplets larger than 25 micron are removed. This is the maximum droplet size as for the water that is not sheared. This is as expected given that the same water treatment system was applied. The maximum droplet size removed is a function of the equipment and not the droplet distribution. The next result however is dramatically different. The oil-in-water concentration is now 75 ppmv, significantly higher than the water that was not sheared. Now the separation efficiency has dropped to 87 %. The same water treatment system performed poorly and is now inadequate. Obviously, this is not a problem with the water treatment system. It is instead a problem that the system was not designed for the particular droplet size distribution that entered the system.

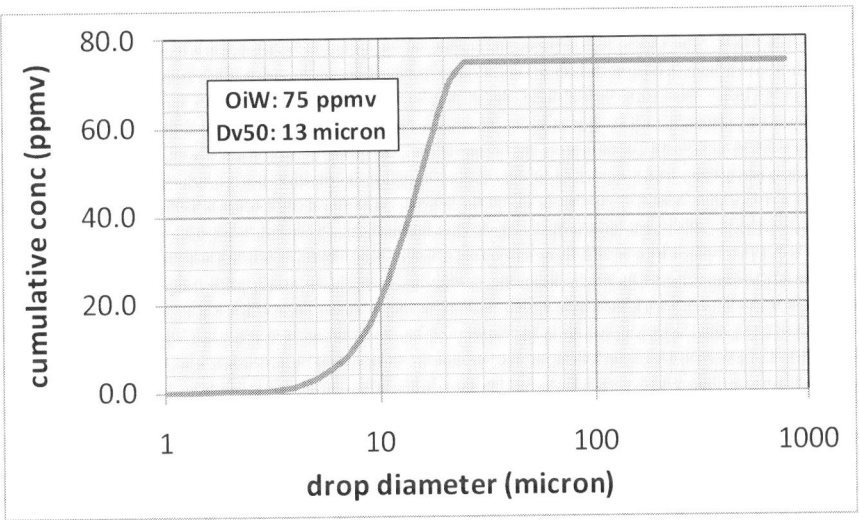

Figure 3.29 An example of an oil droplet distribution for
produced water entering the produced water treatment system.

The pint of the modeling exercise is to demonstrate the detrimental nature of upstream shear on the performance of a typical water treatment system. In both systems the same water treatment system was employed. The second system had an upstream valve that provided shear which generated small droplets in the range of size where the water treatment system was not designed to handle.

3.6 Multi-Phase Flow

The oil and gas production process starts with multiphase fluids flowing from a production well, through a flow line and into a slug catcher, manifold, or primary separator. Fluid flow in these systems is composed of multiple phases (gas, liquid, solid), with stable or unstable interfaces, and phases sometimes mixed together to form various structures (dispersion, foam, bi-liquid foam, etc). The fluids are subjected to shearing and coalescing processes due to artificial lift, pipe flow, riser, choke and valves. Both shearing and coalescence occur to a greater or lesser extent depending on the flow conditions, fluid properties and ultimately on the intensity of the turbulence.

These multiphase fluids are composed of oil droplets dispersed in a continuous water phase, plus water droplets dispersed in a continuous oil phase, plus possibly other fluid phases. The fact that there is a separate oil phase implies that there are physical processes that determine the amount of oil that gets dispersed in the water. From the standpoint of water treatment, the concentration of oil in water as well as the oil droplet size distribution are both important. If there are small droplets of oil dispersed in water and there is a high concentration of these small droplets, then it will be difficult to achieve the desired water quality target.

The oil droplet size distribution in the water phase was discussed extensively above. The models introduced there are accurate for the case of moderate to low (up to a few thousand ppmv) oil concentrations in water. For the case of percentage concentrations of oil and water, the models are less accurate but can still be used to shed light on the turbulence intensity and hence order of magnitude of oil droplet size. The more difficult question to answer is the concentration of oil that gets dispersed into the water phase as a result of the turbulence generating processes.

Unfortunately, the prediction of the oil concentration in the water phase is rather difficult. The scientific study of "Multiphase Flow" does provide some guidance but it is a subject that requires a great deal of depth in order to understand. As with several subjects in this book, the authors have gathered the most useful information that can be described in a reasonable space. Much is left out and can only be referred to in the literature.

Empirical Correlations: Empirical correlations can be used which are derived from extensive laboratory and field analyses. As discussed by Arnold [33] and Juniel [34], such correlations provide a practical means for estimating the inlet fluid condition but they are often not reliable. Small details of piping configuration have a significant effect and cannot be taken into account easily.

Flow Pattern Maps: The material presented here focuses on flow pattern prediction using flow pattern maps. Qualitative description is given to provide an intuitive understanding of what the flow pattern would look like. Quantitative results are presented as empirical measurements. The flow pattern map indicates the transition between one flow pattern and another. These maps are easy to use and for that reason greatly simplify the analysis. However, these transitions depend on fluid properties and so are not universal. In other words, each map represents only one set of fluid characteristics in terms of the viscosity, surface tensions, and oil/water interfacial tensions of the oil and water phases. Theoretical modeling would provide the next level of prediction but this subject is beyond the scope of this book.

Maximum Droplet Size & Separator Cut Size: In this section, flow pattern maps are used to provide a rough idea of the extent of emulsification of oil and water. It should be pointed out that a range of small to large droplets can be generated in the multiphase flow systems leading into a facility. The largest droplets will be separated in the primary separator. Any droplets larger than the separator "cut size" will be separated. The cut size is the droplet diameter for which there is 100 % separation. It can be estimated on the basis of Stokes Law and a knowledge of the vessel size and the configuration of internals. Because of this, it is more important to understand the smaller end of the inlet droplet size distribution than it is to predict the larger end of that distribution.

Gas / Liquid Multiphase Flow Regimes: The Figure below shows the different flow regimes found in multiphase flow.

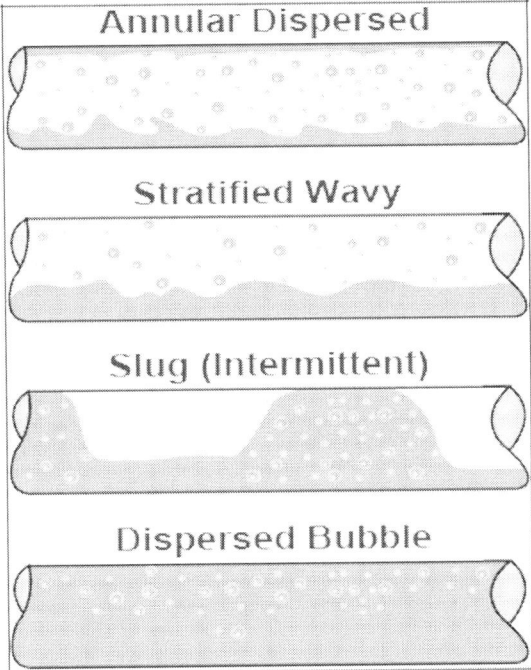

Figure 3.27 Gas-liquid flow regimes (or patterns) in a horizontal pipe [35]

Summary of Technical Terms:

- liquid holdup = liquid volume/pipe volume

- slip velocity = gas velocity – liquid velocity

- no slippage = same gas and liquid velocity

- superficial velocity = u = Q/A where Q is the volumetric flow rate (m^3/sec) and A is the cross-sectional area of the pipe (m^2).

Most of these terms are self-explanatory. The term, "holdup" requires a bit of discussion. "Holdup" for a fluid phase, e.g., water, is the fraction of cross-section occupied by that phase. In the studies that are presented below, all fluids that enter the pipe eventually leave the pipe. In an actual flow line, deposition may occur in which case, the mass balance would not be achieved. Assuming that mass balance is achieved, there can be in some cases very little relation between the flow rate and the volume occupied by the fluid. If the fluid phase has low viscosity, it might travel quickly and occupy only a small volume in the flow line. If the phase has high viscosity, it may travel slowly and occupy a large volume. The volumetric flow rate of a phase will be equal to the linear velocity times the cross-sectional area occupied.

Flow pattern mapping: A typical flow pattern map for horizontal flow is given in the Figure below. Flow pattern maps are ultimately derived from empirical observation with the guidance of dimensionless numbers. They are typically expressed in terms of superficial velocities. Superficial velocities are calculated as: $V_{SG} = Q_G / A$, and $V_{SL} = Q_L / A$, where A is the cross-sectional area of the pipe, V_{SG} is the superficial velocity of the gas, and Q_G is the volumetric flow rate of gas. It should be obvious that the superficial velocities can be quite different from actual velocities. Superficial velocities are used for convenience since volumetric flow and pipe cross-sectional area are usually known, whereas actual velocities are rarely known.

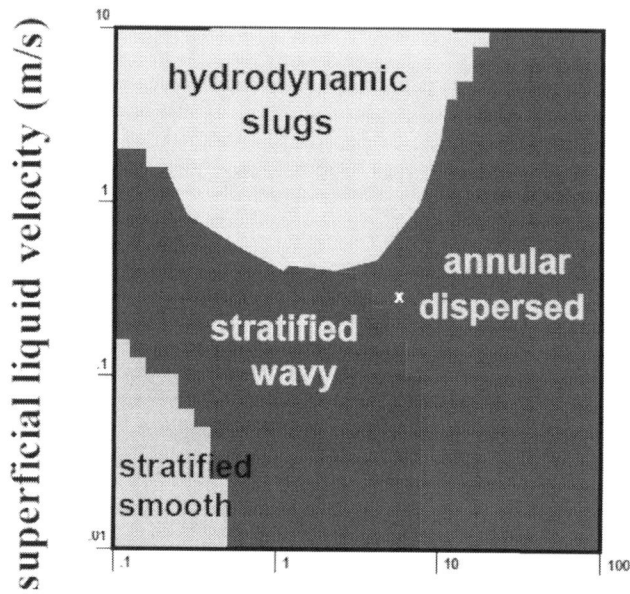

Figure 3.28 Example of a Flow pattern map for a horizontal pipe [35]

There is ample energy available for creating small droplets of gas in the liquid. This situation is described as a wavy instability in a gas/liquid flow. It is referred to as a Kelvin-Helmholtz instability. The amount of liquid entrained in the gas, and the liquid droplets entrained, depends on the gas and liquid densities, surface tensions, and the superficial velocities of the phases.

Most pipelines intended to transport three phase fluids are designed to operate in the stratified flow regime. However, it is not uncommon for the liquid loading, water cut, gas rate, or a number of factors to change the operating conditions significantly from the design assumptions. When this occurs the pipeline will likely generate significant emulsion concentrations of both oil-in-water and water-in-oil. The interface between the oil and water can easily become unstable.

A so-called Kelvin-Helmholtz instability can occur whenever there is a velocity difference across the oil / water interface. The instability that is generated is shown in the figure below. The lower the interfacial tension, the smaller the velocity difference required to generate an instability.

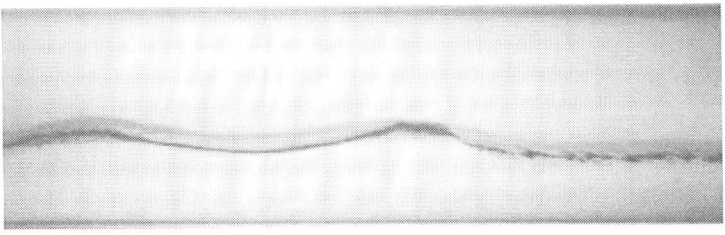

(a) $V_{SG} = 1\ m/s,\ V_{SL} = 0.1\ m/s$

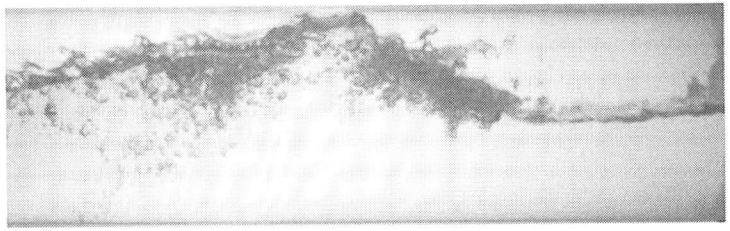

(b) $V_{SG} = 1\ m/s,\ V_{SL} = 0.4\ m/s$

Figure 3.29 Photograph of two flow regimes in a horizontal pipe with gas and liquid. The transition from stratified flow (top) and slug flow (bottom) occurred in this system due to an increase in the superficial liquid velocity [35]. This situation immediately precedes the formation of gas slugs.

<u>**Oil / Water (No Gas) System:**</u> The fluid dynamics of gas / oil / water systems are far more complex than that of oil / water systems. It is currently not possible to provide simple physically reasonable and reasonably accurate correlations of these variables (oil concentration and oil droplet size) as a function of gas / liquid ratio, fluid flow rate, and superficial velocities.

<u>**Oil / Water Multiphase Flow Regimes:**</u> The next three figures summarize the flow patterns that are found in an oil / water mixture, without gas, flowing in a horizontal pipeline. The results shown here are for one set of fluid properties. The flow regimes that were identified are shown in the figures below.

Stratified Flow (ST)

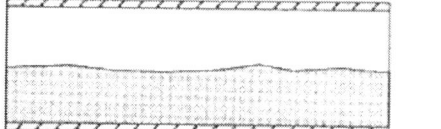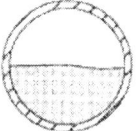

Stratified Flow with mixing at the interface (ST & MI)

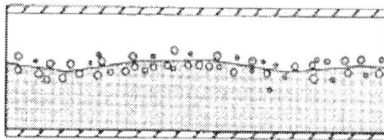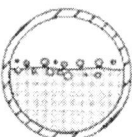

Dispersion of oil in water and water (Do/w & w)

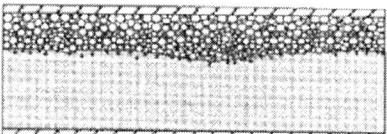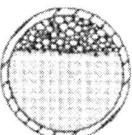

Fig. 1—Horizontal oil/water flow-pattern sketches.

Oil in water emulsion (o/w)

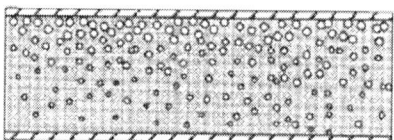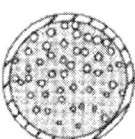

Dispersions of water in oil and oil in water (Dw/o & Do/w)

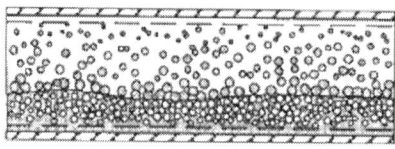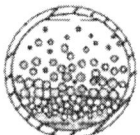

Water in oil emulsion (w/o)

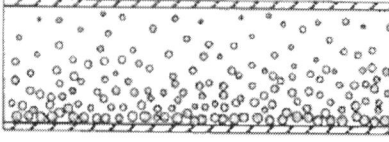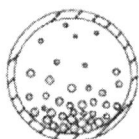

Fig. 2—Horizontal oil/water flow-pattern sketches (continued).

Figure 3.30 Oil / Water Flow regimes [36].

Table 3.6 Flow pattern classification summary for oil / water flow regimes [36].

Flow Pattern Classification Summary		
Segregated Flow		
	Stratified flow (ST)	
	Stratified flow with mixing at the interface (ST&MI)	
Dispersed Flow		
	Water-dominated	Dispersion of oil in water and water (DO/W&W)
		Oil-in-water emulsion (O/W)
	Oil-dominated	Dispersions of water in oil and oil in water (DW/O, DO/W)
		Water-in-oil emulsion (W/O)

Working Fluids: The material that follows is based closely on the work of Brill and co-workers [36]. They used the following fluids in their study:

tap water:
     density: 1.037 gr/mL (at 78 F)

naphthenic crude oil:
     commercial name: Crystex AF-M
     gravity: API 31 degrees
     density: 0.884 gr/mL (at 78 F)
     viscosity: 29 MPa s (cP)
     surface tension: 36 mN/m

Variables Used in Flow Pattern Mapping: In the flow pattern maps presented below, there are two variables that require definition. They are the mean flow rate (u_m), and the water volume fraction (f_w). The previously used variables (superficial velocities) are used to define these two new variables.

$$u_m = u_{so} + u_{sw}$$

Eqn (3.55)

where:

u_m = mixture superficial velocity (m/sec)
u_{so} = superficial velocity of the oil phase (m/sec)
u_{sw} = superficial velocity of the water phase (m/sec)

$$f_w = \frac{u_{sw}}{u_{sw} + u_{so}} = \frac{Q_w}{Q_w + Q_o}$$

Eqn (3.56)

where:

f_w = volume fraction of water (dimensionless)
Q_o = volumetric flow rate of oil (m³/sec)
Q_w = volumetric flow rate of water (m³/sec)

Stratified Flow (ST) - Low Superficial Velocities: For low oil and water superficial velocities, there are essentially no waves at the interface and there are no vortices or disruptions that would cause mixing of the two phases. This flow regime is referred to as stratified flow. The interface forces are dominated by gravity and by interfacial tension. Shear at the interface is relatively small compared to these forces. The upper limit of the stratified flow regime depends on the fluid properties. They measured the upper limit for stratified flow to occur up to a superficial mixture velocity of 0.1 m/sec, with either oil or water (but not both) having a mixture velocity of 0.2 m/sec. Note that if the water superficial velocity is relatively high (but below 0.2 m/sec), then the oil superficial velocity must be correspondingly low in order for the mixture superficial velocity to be less than 0.1. The flow pattern map is shown in Figure 3.31 below.

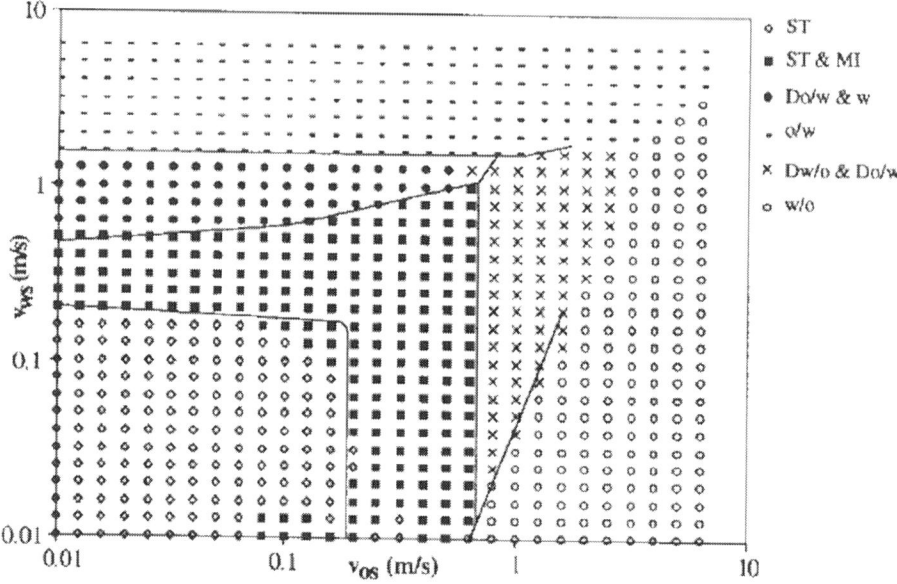

Figure 3.31 Flow pattern map for a crude oil / water mixture (as defined in the text) [36].

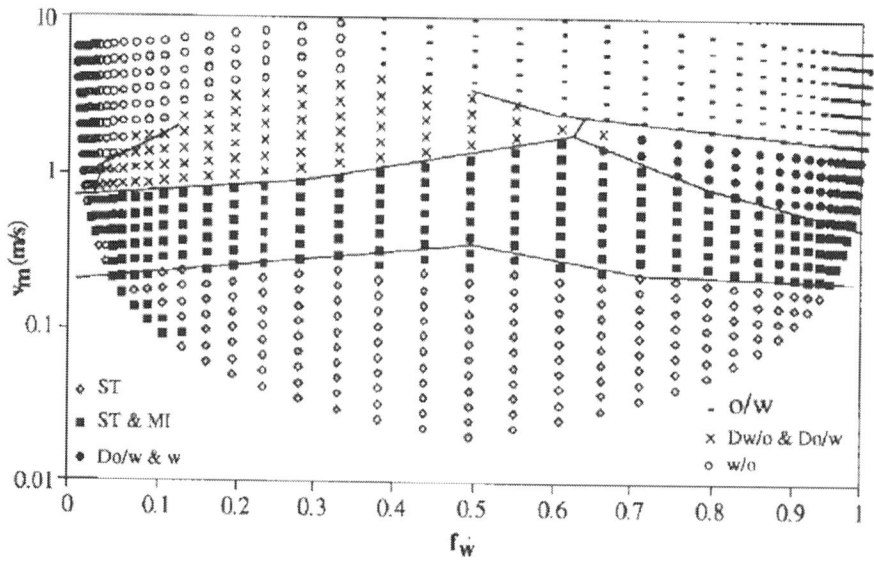

Figure 3.32

Figure 3.31 Flow pattern map for a crude oil / water mixture (as defined in the text) [36]. This map is based on mixture velocity (vm) and water phase volume (fw). This is the same data as plotted in Figure 3.32

Stratified with Mixing at the Interface (ST & MI): This is an intermediate regime between ST (stratified) flow where there is little mixing of the phases to a much more agitated state where the phases become emulsified. As shown in Figure 3.36, the lower velocity limit of this regime spans a very broad range of water volume fraction (f_w), from roughly 0.04 to 0.95. At a low value of water volume fraction (~0.1) the lower limit of this regime occurs at vm = 0.25 m/s. The transition boundary is roughly symmetric such that the lower limit at high water volume fraction is roughly the same (0.25 m/sec). When the water volume fraction is roughly 0.5, the lower limit of this regime occurs at vm = 0.5 m/s.

The upper limits of this regime have a shape that appears similar to that of the lower limits but the values of velocity and water fraction have been shifted. For example, the uppermost mixture velocity occurs at about 1.2 m/sec at a water fraction of 0.62. Oil droplets form in the water phase and water droplets form in the oil phase, as a result of waves and turbulence of the interface.

Dispersion of Oil-in-Water & Bulk Water (Do/w&w): Outside of the two stratified regimes (ST and ST & MI), all of the flow patterns are characterized as dispersions. The lower the velocities, the lower the concentration of oil in water and water in oil. Also, the lower the velocities, the larger the diameter of the dispersed droplets.

For the system studied in [36], the Do/w&w regime can occur when the water volume fraction is about 0.62 and greater. The Do/w&w regime is characterized as two phases. The top phase is a dispersion of oil in water. Essentially all of the oil present is dispersed in the water. Below this dispersion, there is some additional water, depending on the overall water volume fraction. This water has much less oil dispersed in it.

Dispersion of Oil-in-Water & Dispersion of Water-in-Oil (Dw/o & Do/w): At lower water fractions, there is essentially no free water phase. There are still two phases, but they are both dispersions. The upper phase is a dispersion of water in oil and the lower phase is a dispersion of oil in water.

Single phases O/W and W/O: At higher values of the mixture superficial velocity, a single phase forms. Depending on the water volume fraction, and the viscosity of the two pure liquids, the single phase that forms will either be composed of oil droplets dispersed in water or water droplets dispersed in oil. For the system studied in [36], the transition occurs at a water volume fraction of 0.35. Below this value, the mixture is referred to as oil continuous (W/O). Above this value, the system is referred to as water continuous (O/W). The transition from W/O to O/W, as a function of water phase volume fraction, occurs at the so-called phase inversion point.

Inversion Point: As mentioned, the inversion point occurs at mixture velocities above about 4 m/sec and water volume fraction of about 0.35. At water volume fractions below the inversion point, the fluids form a water-in-oil dispersion. At higher water volume fractions, the fluids form an oil-in-water dispersion.

The relative concentration of oil and water at which the inversion point occurs has been studied for a number of different oil and water properties. Generally speaking, the rate of liquid droplet coalescence will determine which liquid forms the continuous phase and which liquid will form the dispersed phase. Coalescence rate is a function of the viscosity of the two pure liquids (oil and water), as well as the interfacial tension, and the interfacial rheology (e.g. shear viscosity). There are a number of mechanisms involved. The most important mechanism appears to be the mobility of the dispersed phase interface and droplet. If the dispersed phase has low viscosity and the interface is mobile, then the dispersed phase will quickly coalesce and the dispersion will be unstable. The ratio of oil droplet and water droplet coalescence rates is the important quantity. In other words, the coalescence rate of oil-in-water must be compared to the coalescence rate of water-in-oil in order to determine which phase will be continuous. If the coalescence rate of water-in-oil is faster than that of oil-in-water, then the continuous phase will be water. This is the case for most of the flow map shown in Figure 3.32. The subject of coalescence is discussed in much more detail in Chapter 4 (Emulsions).

References to Chapter 3

1. R. Morales, E. Pereyra, S. Wang, O. Shoham, "Droplet formation through centrifugal pumps for oil-in-water dispersions," SPE-163055, SPE Journal, p. 172 (2013).

2. W.K. Brown, K.H. Wohletz, "Derivation of the Weibull distribution based on physical principles and its connection to the Rosin-Rammlet and lognormal distributions," J. Appl. Phys., v. 78, p. 2758 (1995).

3. G. Zhou, S.M. Kresta, "Correlation of mean drop size and minimum drop size with the turbulence energy dissipation and the flow in an agitated tank," Chem. Eng. Sci., v. 53, p. 2062 (1998).

4. A.J. Karabelas, "Droplet size spectra generated in turbulent pipe flow of dilute liquid / liquid dispersions," AIChE J., v. 24, p. 170 (1978)

5. N. Vankova, S. Tcholakova, N.D. Denkov, I.B. Ivanov, V.D. Vulchev, T. Danner, "Emulsification in turbulent flow. 1. Mean and maximum droplet diameters in inertial and viscous regimes," Coll. Int. Sci., v. 312, p. 363 (2007)

6. A.M. Kamp, A.K. Chesters, C. Colin, J. Fabre, "Bubble coalescence in turbulent flows: a mechanistic model for turbulence-induced coalescence applied to microgravity bubbly pipe flow," Int. J. Multiphase Flow, v. 27, p. 1363 (2001)

7. R. Kuboi, I. Komasawa, T. Otake, "Collision and coalescence of dispersed droplets in turbulent liquid flow," J. Chem. Eng. Japan, v. 5, p. 423 (1972).

8. K.E. Arnold, "Water treating using a series coalescing flume," U.S. Patent 4,983,287 (1991). See also, U.S. Patent 4,935,154 (1990); U.S. Patent 4,790,947 (1988); U.S. Patent 4,720,341 (1988).

9. V.L. Streeter, E.B. Wylie, K.W. Bedford, Fluid Mechanics, 9th Ed., McGraw-Hill, Boston (1998).

10. G.K. Batchelor, Introduction to Fluid Dynamics, Cambridge University Press, Cambridge, U.K. (1967).

11. R.W. Fox, A.T. McDonald, Introduction to Fluid Mechanics, Second Edition, John Wiley & Sons, New York (1978).

12. J.O. Hinze, "Fundamentals of the hydrodynamic mechanism of splitting in dispersion processes," AIChE J., v. 1, No. 3, p. 289 (1955).

13. M.J. van der Zande, J.H. Muntinga, W.M.G.T. van den Broek, "Emulsification of production fluids in the choke valve," SPE 49173, presented at the SPE ATCE New Orleans (1998).

14. J.T. Davies, "Droplet sizes of emulsions related to turbulent energy dissipation rates," Chem. Eng. Sci., v. 40, p. 839 (1985).

15. G.I. Taylor, Proc. Royal Soc. (London), v. A146, p. 501 (1934).

16. M. van der Zande, "Droplet break-up in turbulent oil-in-water flow through a restriction," Thesis submitted to the Delft University of Technology, the Netherlands (2000).

17. C.A. Coulaloglou and L.L. Tavlarides, "Description of interaction processes in agitated liquid-liquid dispersions," Chem. Eng. Sci., v. 32, p. 1289 – 1297 (1977).

18. V. Alopaeus, J. Koskinen, K.I. Keskinen, "Simulation of the population balances for liquid-liquid systems in a nonideal stirred tank. Part 1. Description and qualitative validation of the model," Chem. Eng. Sci., v. 54, p. 5887 (1999).

19. F. Huchet, A. Line, J. Morchain, "Evaluation of local kinetic energy dissipation rate in the impeller stream of a Ruston turbine by time-resolved PIV," Chem. Eng. Res. Design, v. 87, p. 369 (2009).

20. G.L. Morisson, R.E. DeOtte, G.H. Nail, D.L. Panak, "Mean velocity and turbulence fields inside a beta = 0.5 orifice flow meter," AIChE J., v. 39, p. 744 (1993).

21. D.G.K. Murti, H.R. Al Nuaimi, "Renovate produced water treating facilities to handle increased water cuts," SPE 22831, paper presented at the SPE ATCE Dallas (1991).

22. Typhonix Inc., "Typhoon valve feasibility study," (2006).

23. A.J. Karabelas, "Droplet size spectra generated in turbulent pipe flow of dilute liquid / liquid dispersions," AIChE J., v. 24, p. 170 (1978)

24. B. Knudsen, T. Husveg, "Field test of Typhoon valve at Oseberg C," paper presented at the Tekna produced Water Conference, Stavanger (2013).

25. T. Husveg, T. Bilstad, P.G.A. Guinee, J. Jernsletten, B. Knudsen, H.T. Nordbo, "A cyclone-based low shear valve for enhanced oil-water separation," OTC 20029, paper presented at the 2009 Offshore Technology Conference, Houston (2009).

26. C.A. Fernandes, R.F. Ribeiro, J.B.R. Loureiro, A.P. Silva Freire, "Droplet sizes of emulsions in cyclonic-based valves," ETC14, 14th European Turbulence Conference, Lyon, France (2013)

27. D.A. Flanigan, J.E. Stolhand, M.E. Scribner, E. Shimoda, "Droplet size analysis: A new tool for improving oilfield separations," SPE-18204, presented at the Annual Technical Conference and Exhibition, Houston (1988).

28. D.A. Flanigan, J.E. Stolhand, E. Shimoda, F. Skilbeck, "Use of low-shear pumps and hydrocyclones for improved performance in the cleanup of low-pressure water," SPE – 19743, SPE Production Engineering, August (1992).

29. J.C. Ditria, M.E. Hoyack, "The Separation of Solids and Liquids with Hydrocyclone-Based Technology for Water Treatment and Crude Processing," SPE-28815 (1994).

30. M.F. Schubert, "Advancements in liquid hydrocyclone separation systems," OTC-6869, paper presented at Offshore Technology Conference, Houston, TX (1992).

31. J.M. Walsh, T.C. Frankiewicz, "Treating produced water on deepwater platforms: developing effective practices based upon lessons learned," Paper SPE-134505-MS presented at the SPE Annual Technical Conference and Exhibition, Florence, Italy, September (2010).

32. Z. Khatib, "Produced water management: is it a future legacy or a business opportunity for field development," IPTC-11624, presented at the International Petroleum Technical Conference, Dubai (2007).

33. K. Arnold, "Surface Facilities for Waterflooding and Saltwater Disposal," Chapter 15 of the Petroleum Engineers Handbook.

34. K. Juniel, "Water Treating Facilities in Oil and Gas Operations," Ed. K.E. Arnold, Chapter 4 of Vol III of the Petroleum Engineering Handbook, Ed. in Chief L.W. Lake, SPE, Richardson Tx (2007).

35. O. Bratland, <u>Multiphase Flow Assurance</u> (2010).

36. J.L. Trallero, C. Sarica, J.P. Brill, "A study of oil / water flow patterns in horizontal pipes," SPE Production & Facilities, p. 165, (1997).

37. C.E. Brennen, Hydrodynamics of Pumps, Concepts NREC and Oxford University Press (1994).

38. F.C. Visser, J.J.H. Brouwers, J.B. Jonker, "Fluid flow in a rotating low-specific-speed centrifugal impeller passage," Fluid Dynamics Res., v. 24, p. 275-292 (1999).

39. P.A. Davidson, <u>Turbulence – An Introduction for Scientists and Engineers</u>, Oxford University Press, Oxford U.K. (2009).

CHAPTER FOUR

Emulsions

Chapter Four Table of Contents

4.0 Introduction

The main objective of this chapter is to provide an understanding of:

1. different types of emulsions;

2. the chemical and physical forces that stabilize emulsions;

3. strategies used to break emulsions.

The content of this chapter can be posed in terms of a number of questions:

*   What natural oil and water components stabilize emulsions?

*   What production chemicals stabilize emulsions and by what mechanism?

*   What is the significance of interfacial tension?

*   How low is "low interfacial tension?"

*   What effect do the natural oil and water components have on interfacial tension?

*   What is the significance of viscosity of the fluid or of the oil drop in emulsion stability?

*   What kinds of oil and water mixtures have the greatest tendency to form emulsions?

*   What are the mechanisms by which production chemicals break emulsions?

*   If we know a production chemical will decrease the interfacial tension, what effect will this have on oil/water separation and produced water treatment?

Many of these questions are related to the interaction between the bulk water phase and the oil droplets. Properties such as shear rate, shear stress, viscosity, interfacial tension, drop diameter, drop surface area, density difference between the drop and water, are important. In this chapter, the role of these variables in stabilizing and breaking emulsions and their relative importance are discussed. The important subject of how emulsions are formed was already discussed in relation to turbulence and fluid shear in Section 3.4 (Turbulence), and Section 3.5 (Shearing of Oil Droplets). Factors which stabilize a produced water / crude oil emulsion are discussed in this chapter.

Types of Emulsions: There are several different types of emulsions that are important in produced water treating. The simplest to describe is composed of oil droplets suspended in water. Despite their simplicity, these emulsions can in some cases be difficult to resolve. At the opposite end of complexity would be hydraulic fracture flow back water that has dissolved mineral solids at saturation, oil droplets, solid particles, and oily solids suspended in water that also has dissolved and/or suspended organic polymer and which contains a surface wetting agent (a.k.a., a surfactant).

Oil droplets suspended in water can usually be resolved by simple settling, with perhaps some chemical treatment to accelerate the process. This type of emulsion is most commonly found in conventional oil/gas produced water systems both offshore and onshore (including heavy oil operations). Gravity separation is applied as the first stage. This removes free oil (oil that rises to the top of a sample jar in one minute) and larger droplets of oil (70 to 150 micron and larger). Once this is done, then the water treatment system usually consists of hydrocyclones, flotation, and perhaps some form of media to remove the smaller droplets.

Flow back fluids may contain proppant, reservoir solids (sand stone, clay, shale), hydraulic fracturing polymer, spent acid, and various additives such as biocides, corrosion inhibitor, and mutual solvents. Various types of emulsions can form which have a wide range of separation difficulty depending on the presence of oil, the viscosity of the oil, the viscosity of the polymer, the size and type of solids that are produced, and the type of additives and their surface activity. If there is only gas, then the fluids are composed of mostly the polymer which is slow to settle due to high viscosity. For this reason coagulants and flocculating agents are added. Flow back fluids that have been contaminated with oil from the reservoir may have oil coated solids mixed together with the polymer. These mixtures are difficult to flocculate and have almost no tendency to coalesce. This makes separation more difficult. In some cases, these flow back fluids are not treated and are instead hauled to a disposal facility. In the Gulf of Mexico, temporary water equipment is used to clean the water for overboard discharge. By using temporary equipment contamination of the main oil/water treatment system is avoided.

The type of water treatment process and equipment required will depend on the type of emulsion present. When specifying equipment, selecting water treatment chemicals, or troubleshooting a system, it is important to be able to distinguish the type of emulsion present. The various types of emulsions that are encountered in the industry are discussed in Section 4.1.1 (Types of Emulsions) below.

Contaminants – Dissolved and Suspended: As discussed in detail in Section 2.3 (Composition of Produced Water – General Overview), there are two broad categories of contaminants in produced and hydraulic flow back water: (1) dissolved molecules and ions, and (2) various suspended materials including particles of various shapes and sizes, suspended oil droplets, combinations of oil and solids (oily solids), and various thin film emulsions. The dissolved components include acids, alcohols, BTEX, and ionic species which are polar or ionically charged. They typically have some degree of surface activity and migrate to the oil/water interface as the fluids are being produced. The nonpolar compounds are only slightly soluble in water, if at all, and they form droplets. Formation fines (sandstone, clay, shale, carbonate) are often swept out of the near wellbore region and into the produced fluids. Mineral solids that have a concentration above their solubility limit form particles that can be suspended in the water. Any of these solid particles or liquid droplets will have an interface with the water phase that can attract various polar and surfactant-like compounds making the contaminant even more likely to form a stable emulsion. The physical state (particles, droplets, films) has an effect on the rate of separation and will be discussed in this chapter.

Surfaces and Interfaces: Produced and hydraulic fracture flow back water contains a number of important interfaces, such as oil/water, gas/water, solid/water, solid/oil, etc. All of these are interfaces. Whenever one of the interfaces is gas, it is usually referred to as a surface. In this chapter, the terms "surface" and "interface" are used interchangeably. When a distinction is necessary, the term surface will be used for liquid/gas, or solid/gas. Typically, there is no confusion about which interface is under discussion.

Breaking or Resolving Emulsions: Emulsions must be "resolved" or "broken" in order to achieve the desired quality of treated water. The type of emulsion present will dictate the most promising mechanism for breaking the emulsion. In some cases the separation of oil, solids, and water is accomplished by simple settling. If sufficient residence time for settling is not available, then hydrocyclones or other equipment can be used to enhance the settling and resolution of emulsions. In all cases, emulsion resolution can only be accomplished when oil drops are coalesced, liquid films broken, and solid particles coagulated and settled. These processes, which ultimately lead to emulsion resolution, depend very much on the surface chemistry and characteristics of the emulsion, as well as the equipment being used.

Chapter Content: When reading this chapter, it should become apparent why this chapter on emulsions is preceded by a chapter on produced fluid chemistry, and a chapter on fluid mechanics. All emulsions are formed, stabilized or resolved (broken) by a combination of chemical interactions (Chapter 2) acting in response to fluid mechanical forces (Chapter 3). Of particular importance is the section on Organics (Section 2.6), including the sub-section on Naturally Occurring Surface Active Constituents (Section 2.6.5). Sections 2.9.1 (Corrosion Inhibitors) and Section 2.9.2 (Hydrate Inhibitors) discuss two important classes of production chemicals that contribute, inadvertently, to emulsion stability. During production, oil/water interfacial area is formed as oil and water are sheared into fine droplets (Sections 3.4 and 3.5). As the interface is formed, natural and man-made chemical constituents migrate to the interface where they determine the interfacial properties such as surface tension, surface elasticity, interfacial shape and interfacial area. These surface properties depend on chemical composition and have a impact on the stability of emulsions and whether or not oil droplets will coalesce.

Importance of Mechanism: The material in this chapter provides an understanding of the mechanisms that stabilize emulsions and those that are involved in breaking emulsions. For specialists in the subject, this level of understanding is used routinely. It forms the basis for understanding what is happening inside vessels, pipes, valves and pumps, and what is happening on the molecular surface of an oil drop. It also helps to predict what is likely to occur by changing the conditions in some way, such as by adding heat, or treating with a particular type of chemical or changing piping from one arrangement to another. Various tools are presented for assessing the stability of an emulsion. When these tools are applied in the field, valuable insight is gained which translates into less time to solve field problems, and generally lower production deferment and lower operating costs. True understanding of these systems comes from a combination of field experience and study of emulsion fundamentals.

Having the ability to imagine accurately what multiphase fluids look like inside of a pipe or separator is a powerful skill. It is not natural to most engineers. Indeed, some process and equipment designers have unrealistic ideas about how fluids actually behave in the equipment they design. By forming a more realistic image of the type and stability of emulsions that occur inside water treating equipment, an engineer is much more likely to accurately conceive of a proper design, or solve a problem in the field.

While much of the material in this chapter is of a fundamental nature, it is nevertheless tailored specifically to the oil and gas industry. There are several textbooks on the general subject of emulsions, several of which are referenced in this chapter. However, they tend to focus on rather "clean" systems of only a few components (chemicals) that can be studied systematically in the lab. It is useful to include this material as background. But, this is quite different from the emulsions encountered in the oil and gas industry where hundreds if not thousands of surface active chemicals are found in the produced fluids. Nevertheless, much of the fundamental information does have relevance.

4.1 Emulsions

An emulsion is a liquid mixture of two immiscible phases in which one phase is suspended in the other phase. For our purposes we are mostly concerned with oil drops suspended in water (the continuous phase). In much of the colloid and surface science literature, precise definitions of micro-emulsion, mini-emulsion and macroemulsion are used. The distinction between these emulsion types is important for certain applications such as paint formulation, food, and medicine. These different types of emulsions are distinguished by the particle size distribution, and thermodynamic stability of the discreet phases.

For the most part, the emulsions of importance in the oilfield are either mini-emulsions or macro-emulsions. They are usually thermodynamically unstable, although some are thermodynamically stabilized by high concentration of surfactants. Thermodynamic stability is an incomplete criterion for characterizing the emulsions encountered in the oilfield. It does not consider the time scales of importance in processing oil and water, which is usually in the range of minutes (offshore) or hours (onshore). As a matter of practicality, an emulsion might be unstable thermodynamically, but stable in a "kinetic" sense because there is insufficient time available in the water treatment system for the emulsion be become resolved.

4.1.1 Types of Emulsions

The oil industry is not consistent in the terminology used to describe emulsions. The meaning of words such as emulsion, reverse emulsion, complex emulsion, pad, suspension, and dispersion vary from one location to another, from one sector to another, and from one discipline to another. Since it is our intention to provide a meaningful guide to both onshore and offshore, upstream and downstream sectors, it is necessary to select a single terminology and to use it consistently. That is one of the objectives of this section.

Another important objective of this section is to provide a description of the different types of emulsions, how they can be identified in the field, and to recognize the process systems that cause them. This will help facilitate the discussion between field personnel, and office-based or laboratory-based staff. Many water treatment specialists who have spent years working in the field will, at some point, move into positions of greater responsibility where they must provide advice to many different locations and not spend a lot of time in the field at any one particular location. It becomes necessary for the next generation of field personnel to make the field observations and report back. This can only be done if the field staff understand the basic types of emulsions, how to identify them using simple bottle testing, and if the field and office staff speak the same language. Sending emulsion samples to shore is no substitute for on-site observations because most emulsions will change by the time they are received onshore. Therefore an important objective of this section is to educate field staff so that they can communicate effectively with specialists who need to know what kinds of emulsions are present in a facility.

Once the root cause of a problem emulsion has been determined, the cost-effective solutions can be identified to mitigate the problem on the basis of emulsion type. The process, equipment and chemical treatment employed in design or problem solving should be based on the emulsion type, as well as other that include the origin of the water (primary, IOR, EOR, unconventional), final disposition of the water (reuse, overboard discharge, surface discharge, disposal well), location (onshore, stranded, accessible, offshore deep water), and the oil and gas characteristics. Of these factors, the emulsion type is most closely related to the oil and gas characteristics but other factors must be considered to determine the emulsion type as well.

Overview: An emulsion is not homogeneous. It has distinct phases, the presence of which can be seen by the naked eye. A solution is homogeneous. A solution looks uniform, not turbid, it may be tinted or colored, but it is not cloudy or opaque. An emulsion is composed of different phases of liquids or solids material. In some cases the phases can be seen as particles or droplets. In other cases, the mixture is milky or cloudy looking. It may be uniform or non-uniform. Dairy milk is an example of an emulsion of very small fat droplets suspended in water. Milk is opaque due to the fact that the fat droplets are so small that they scatter light. The droplets cannot be seen by the naked eye but their presence can be confirmed by a microscope. Thus, a solution is identified mostly by what it does not look like (no opacity, non-uniformity, no suspended material). Emulsions have a variety of appearances.

The table below gives the classification that will be used here to distinguish between one type of emulsion and another. These are only applicable to produced and flow back water. At the highest level, a distinction is made between solution and emulsion. Three types of emulsion are identified for produced and flow back water. Four types of suspensions are identified. Most of the emulsions encountered in produced and flow back water are a suspension. However, there are a couple of other emulsion types that can be present and have an important detrimental impact on a facility. It must be kept in mind that any given sample of produced water can have a combination of emulsions or even an emulsion plus a solution phase in the same sample. All of these types of emulsion are discussed below.

Solution			
Micro-Emulsion			
Emulsion	Suspension	1)	oil-in-water
		2)	solids-in-water (not oily)
		3)	organic coated or oily solids in water
		4)	chemically stabilized contaminant in water
	Complex Emulsion		oil-in-water-in-oil
	Dispersion		high internal phase volume

Solution: A solution is a thermodynamically stable homogeneous mixture of one or more substances dissolved in a solvent. It is the counterpart to an emulsion. By thermodynamically stable, we mean that the solution will not separate if allowed to sit over a long period of time. By homogenous we mean that there is only one phase and liquid droplets or solid particles cannot be seen, or inferred by a cloudy appearance. Examples of dissolved contaminants include the mineral ions (Na^{+1}, Cl^{-1}, Ca^{+2}, CO_3^{-2}, etc), organic acids, hydrate inhibitors (methanol, ethylene glycol, etc), dissolved gases, and polar hydrocarbons such as benzene. A more detailed description of dissolved components is given in Section 2.3.1 (The Difference between Dissolved and Suspended Contaminants). A solution is not an emulsion.

Emulsion: For the purposes of this manual, we use the term "emulsion" to mean a dispersion of oil drops in water. Some authors make a distinction between a macroemulsion where the droplets are larger than 10 micron, and a mini-emulsion where the droplets are smaller than 10 micron [1]. In either case, an emulsion is formed by shaking, mixing or otherwise shearing the mixture. For any given oil and water mixture, the intensity of shearing will determine the diameter (and therefore number) of drops in the emulsion. With time, an emulsion might separate into an oil and a water phase. Thus, the word emulsion is used to denote a dispersion of oil drops in water which may or may not be thermodynamically stable.

Reverse Emulsion: In the oil and gas industry, the term "reverse emulsion" is used to refer to an emulsion composed of oil-in-water, where water is the continuous phase. However, the term reverse emulsion is confusing because in the water treatment industry, a reverse emulsion is a dispersion of water-in-oil. Since these definitions are entirely opposite to one another, the term reverse emulsion will not be used. It is simple enough to refer to an oil-in-water (o/w) emulsion as such.

Microemulsion: A microemulsion is a clear, stable, mixture of at least three components: oil, water and a surfactant. The IUPAC definition of a micro-emulsion is a "dispersion made of water, oil, and surfactant(s) that is anisotropic and thermodynamically stable with a suspended discreet phase having diameters varying approximately from 1 to 100 nm, usually 10 to 50 nm". Microemulsions are characterized by their very low interfacial tensions (typically 10^{-2} dynes/cm or less) and they will not settle to form two bulk phases over time. In a microemulsion the Gibbs free energy of the system is at a minimum. To achieve stability a highly effective set of co-surfactants is required. The surfactant molecules may form a monolayer at the interface between the oil and water, with the hydrophobic tails of the surfactant molecules dissolved in the oil phase and the hydrophilic head groups in the aqueous. Microemulsions can form upon simple mixing of the components and do not require the high shear conditions generally used in the formation of ordinary emulsions.

Suspension: A suspension is a two- or three-phase mixture in which droplets of liquid or particles of solid can be seen or inferred by cloudiness. Cloudiness, quantified as "turbidity" is occasionally measured to define the level of contamination present. However, turbidity differences from one oil or water sample to another can be due to the color and opacity of the oil, the droplet size distribution present, or other factors and not just the concentration of contaminants. Therefore in the upstream oil and gas industry, turbidity is only useful for systems that have a consistent composition of oil. This is rarely the case.

In a suspension, the contaminant droplets and particles are dispersed in a continuous or bulk phase. The contaminants are referred to as the suspended or dispersed phase. This two phase system is not thermodynamically stable. In other words, over time, two or three continuous or bulk phases will form. Thus, the mixture may be suspended, but it is not suspended indefinitely. The denser phase (solid or liquid) will separate and sink to the bottom and the lighter phase will rise to the top. This process may take minutes or weeks.

In this book, four types of suspensions are identified and discussed below. They cover most emulsions seen in the field. It can be argued that there are various sub-types that should be discussed. Also, emulsions seen in the field may have two or more of these types of suspensions present. But for now, this short list will keep the discussion as simple as possible. In the description given here, mechanisms, and terminology are used that have not yet been discussed, but which will be discussed in this chapter. The Index can be sued to find discussions of specific terms.

1. **Suspension of Oil Droplets in Water:** Oil droplets suspended in the water phase are characterized by two features. The first is the droplet size which can be estimated by simple desktop settling. This technique was discussed in Chapter 2 (Fluid Mechanics). According to Stokes Law, small droplets take longer to rise to the top. Small droplets, on the order of 1 to 10 micron, are created by a shearing device such as a choke, valve, or pump. Suspensions of hydraulic fracturing polymer fall in this category. Such mixtures often have small droplets (< 10 micron) and high viscosity which is somewhat challenging to clean. The other important feature is the surface chemistry of the oil droplets. This category only refers to oil droplets that have a minimum of surface active components. The intension of this category is to refer to suspensions that have relatively simple chemistry. The main challenge for some of these suspensions is the small droplet size.

2. **Suspension of Non-Coated Solid Particles in Water**: Non-oily solid particles suspended in water are typically not a problem and present to a much greater extent than most operators are aware of. Solids may be composed of reservoir solids (clay, sand, shale), corrosion products (iron compounds), mineral solids (carbonates, sulfates), or combinations (iron sulfide, iron carbonate). Provided that these solids do not become coated by oil or organic material, they do not significantly contribute to Total Oil and Grease. One exception to this rule is that non-oily solids, if small, can stabilize an oil-in-water emulsion without the presence of surface active materials. These emulsions are referred to as Pickering emulsions. Generally, the conditions and chemistry that would give rise to Pickering emulsions have a tendency to form oil-coated solids emulsions instead which are discussed below. Non-coated solids

are characterized as Total Suspended Solids (TSS) with low Total Oil and Grease (TOG). The lack of oil coating can be verified by chemical spot testing of the material collected in the TSS filter measurement. This particular category does not include oily solids. These non-organic or non-oily solids readily sink, depending on the particle size. Larger particles (> 50 micron) can be removed by desanding cyclones. Small particles (< 20 micron) require coagulants and/or flocculating agents followed by filtration or settling. So-called "filter aids" can be used. But in many cases, non-oily solids are not a problem and do not require separation and treatment unless they settle to the bottom of separators and vessels. In that case so-called sand management processes are required. These solids can become a problem if they are vigorously mixed with the oil phase and become oil-coated. Contact of these solids with production chemicals often facilitates the process of becoming oil coated.

3. **Suspension of Organic or Oil Coated Solids in Water**: Solid particles can become coated with a variety of chemicals and contaminants. Once this occurs, they are difficult to remove from the produced and flow back water. These solids may be composed of clay, iron solids, and other minerals combined with asphaltene, wax, and/or resin components from the oil, corrosion inhibitor, etc.. Metallic soaps loosely fall into this category as well. Such materials are formed by the combination of sodium or calcium with ARN and other naphthenic acids. These solids are characterized as Total Suspended Solids (TSS), with finite (non-negligible) TOG. The presence of oil coating can be verified by chemical spot testing of the material collected in the TSS filter measurement. The combination of solids and oil can make some of these particles neutrally buoyant. As such primary separation is not effective. Recycle of these solids should be avoided since they have a tendency to build up in the separation system. The effectiveness of flotation may be reduced if the particles have a cationic charge. Zeta potential should either be measured directly with a zetameter, or should be inferred by bottle testing over a range of pH. Once this is done, then a suitable coagulant or flocculant can be selected. For onshore locations, some form of mild oxidation (hypochlorite, ozone, peroxide, Fenton's reagent) is often effective at releasing the organic material from the solid surface. Wetting agents (also known as dispersing agents) are also effective at separating the solid from its coating. It is also advised to identify the mechanism or source of these solids since their presence often points to other problems in the system such as scale formation.

4. **Suspension of Chemically Stabilized Oil Droplets Water**: If the oil droplets are covered with corrosion inhibitor or some other surface active production chemical, they can be very difficult to remove from the produced water. If the oil droplets are covered with a surface active material, they will form small droplets when subjected to shear. In addition, they will not coalesce. This shear sensitivity combined with a lack of coalescence is referred to as chemical stabilization of an emulsion. Droplets of 1 to 5 micron are typically seen in this scenario. Such droplets are too small for effective separation in vessels, hydrocyclone or flotation without competent chemical treatment. Besides corrosion inhibitor, low dose hydrate inhibitors, and some demulsifiers are problematic. Biodegraded crude oil contains a high concentration of surface active naphthenic acids which can cause small droplets by this mechanism. If the surface active compound is cationic, then gas bubbles may not readily adhere. This will reduce the effectiveness of flotation. The presence of chemical stabilization can often be determined by bottle testing over a range of pH. If the emulsion resolves at high or low pH, then chemical stabilization is likely. Resolution at low pH often indicates stabilization from naphthenic acids. Resolution at high pH often indicates corrosion inhibitor. Such bottle testing, or Zeta potential measurement, can be used to pick a suitable coagulant and flocculant. Such emulsions generally respond well to chemical addition. Often both a coagulant and flocculant are required upstream of a flotation unit for good separation of oil and water.

Figure 4.1 Suspension of polymer particles in water

In a production system, oil-wet solids are commonly produced by several mechanisms. Once formed, they can cause significant water treatment problems with the net result that water quality often fails to meet targets. If a sample is taken and allowed to sit on a desk top, days or weeks will pass before the cloudy water begins to clear up. Over longer time, the preference of water to coat the surface of the solid contaminant finally begins to displace the organic material. Slowly, separation begins to occur and the organic material floats to the top and the solids sink to the bottom. This is shown in the figure below. The time required for this process varies depending on the chemistry of the sample, and the temperature. But it is evident from the two photographs of the same sample that oil-wet solids were suspended initially and over time the oil and solids separated into three phases altogether. This is discussed in greater detail below where the mechanisms of surface wetting are introduced.

Figure 4.29 Desktop settling of an emulsion originally composed of oily solids.
Left frame: sample from produced water system.
Right frame: sample after two months of settling. Over two months, the oil separated from the solids and floated to the top of the sample. The solids sunk to the bottom. This is an example of an emulsion composed mostly of oily solids.

Complex Emulsion: A "complex emulsion" is a system in which very small drops of water are suspended in oil drops, which are themselves suspended in a continuous phase of water [2]. The origin and properties of complex emulsions must be understood since they have a detrimental impact on water treatment and the quality of the effluent water.

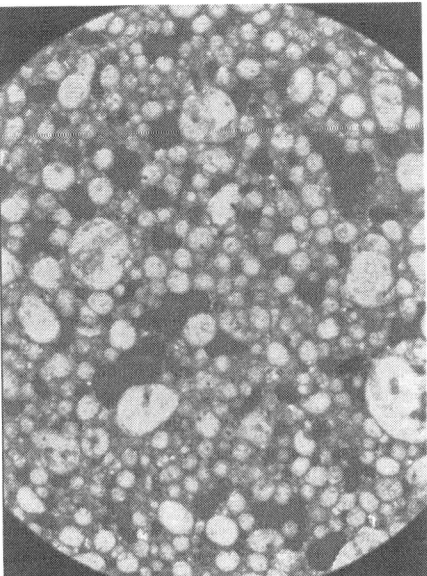

Figure 4.2 Photomicrograph of a complex emulsion. Note the small drops of oil (dark phase) interspersed within water drops (lighter color) which are themselves suspended within an oil phase.

Separator vessels often have a layer of concentrated emulsion between the oil and water phases. These interface layers are referred to as a "rag layer" or "pad layer." There is no single name for these interface layers and the names that are used do not distinguish between one type of emulsion and another. There are several types of emulsion that can be present in these interface layers. The complex emulsion shown in Figure 4.2 is one such emulsion. As shown in the figure, there are drops of oil suspended within water droplets which themselves are suspended within an oil continuous phase.

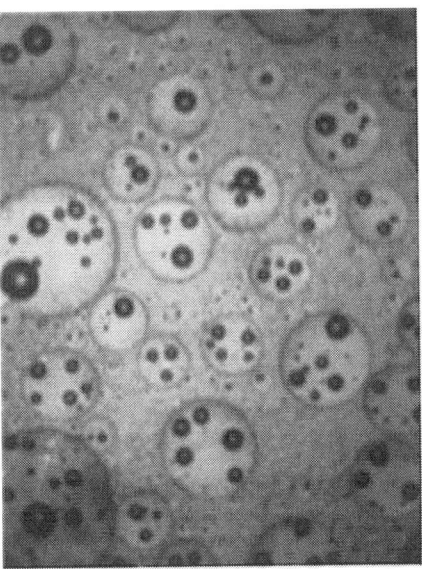

Figure 4.3 Complex emulsion of oil drops in water drops which are suspended in oil.

High Internal Phase Volume Emulsion (HIPV): An HIPV emulsion is a dispersion of water drops in oil – or vice versa - where the continuous phase forms a thin film around the suspended phase. In the oilfield it is most common to encounter a water-in-oil emulsion. In that case, the water is the suspended phase and droplets of water are separated by thin films of oil. Given that this type of emulsion is composed of both oil and water, the density is less than that of pure water and greater than that of pure oil. Thus, this type of emulsion forms at the oil/water interface. Oilfield operators will often refer to material that has collected at the interface as a pad. But a pad is not necessarily an HIPV emulsion. Often it is, but in some cases a pad could be composed of other material such as oil-wet solids. In this book, the term pad is used for an interface material that is of unknown structure and composition. Once the structure and composition is determined, then a more precise and descriptive term is used.

An HIPV emulsion has a structure that is analogous to that of foam. In foam, gas bubbles are surrounded by thin films of liquid. In an HIPV emulsion, there is no gas. Instead, the water droplets take the place of the gas bubbles and they are surrounded by thin films of oil. Breen [3] refers to this type of emulsion as a biliquid foam. Pereyra et al [4], and Henschke et al. [5] refer to this type of emulsion as the Dense-Packed Layer. Both groups provide a model for the rate at which this layer resolves into two continuous phases of oil and water. Polderman et al. [6] refer to this type of emulsion as a Dispersion Band and they provide a practical model with field data for the prediction of the thickness and drainage time. The author worked for several years with Hugo Polderman and learned first-hand from him on this and many other subjects in produced water treatment. Therefore, in this book, this type of emulsion will be referred to as a dispersion.

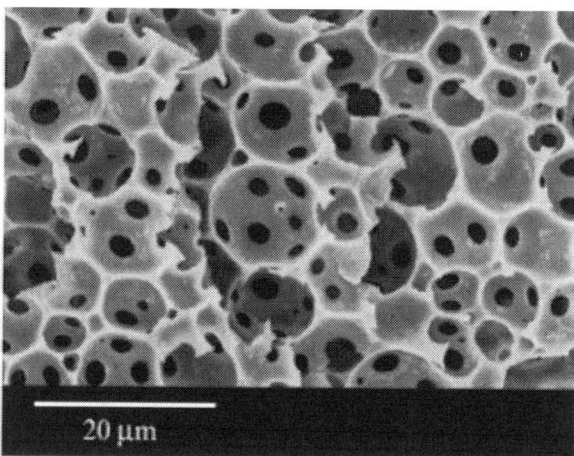

Figure 4.4 A dispersion (also known as a High Internal Phase Volume (HIPV) emulsion). The suspended phase has been evaporated and the continuous phase has been polymerized. Shown here is the structure continuous phase. The structure of the continuous phase is that of thin liquid film between the water droplets. Water occupies much more volume that the oil but the oil is the continuous phase. Hence the name, High Internal Phase Volume (HIPV) emulsion.

A schematic diagram of the different types of emulsions that might be found in a primary separator is shown in Figure 4.5. Starting at the top of this diagram, drops of water are in the process of falling (or settling) due to the density difference between the oil and water as modeled by Stokes Law. The water cut is low and the continuous phase is oil. There may also be some solid particles in the process of settling as well. This emulsion is referred to as a dispersion or suspension of water in oil.

At a certain level in the oil phase, the water drops begin to accumulate. They pack together closely enough that water/water drop coalescence begins to occur. As shown in the figure, large water drops begin to form. These large water drops will fall through the oil phase colliding with other water drops and coalescing along the way. This begins the formation of a dispersion.

A dispersion is formed whenever the oil phase has a relatively high viscosity and drainage of the oil becomes a bottleneck. It must be noted of course that drainage of the oil phase is in the upward direction. There are two water cut levels that are attained. In the upper layer, the water cut is given by the maximum packing density of roughly spherical water droplets. With time, this dispersion packs together even more closely as the sides of the droplets become deformed. The water droplets take on a hexagonal polygon shape as the oil film gets squeezed and becomes thinner. Again, this HIPV phase only forms when the continuous phase viscosity is high enough for film drainage to be a slow process and hence to cause a bottleneck in the process of emulsion resolution. The thickness of this HIPV emulsion phase depends on the oil viscosity and on the cross-sectional area in the separator that is available for emulsion resolution. The importance of cross-sectional area is discussed in Chapter 8 (Primary Separators).

The height of the oil-continuous dispersion depends mostly on the oil viscosity and the drainage rate of the oil film. The percentage of water (water cut) can be relatively high, often 70% or more. The viscosity of this phase can also be orders of magnitude higher than the viscosity of either pure phase. Kokal refers to this as "structural viscosity" [2]. The lifetime (time required for such an emulsion to resolve) depends on the drainage time of the liquid film of oil which depends primarily on the oil viscosity.

There are other chemical factors that stabilize a dispersion as well. Natural surfactants, such as the water soluble carboxylic acids and oil soluble naphthenic acids will act like soap molecules to stabilize the oil films. This is the same mechanism that stabilizes a soapy foam in a shampoo or dish washing detergent. In the upstream oil and gas industry, the height of the dispersion is something that is measured and modeled using empirical formulas. This is discussed in Chapter 8 (Primary Separators). Once an oil film thins to the point of rupture, the water droplets coalesce into their own homogeneous (continuous) phase.

Returning to the figure below, the dispersion is the classic water-continuous suspension of oil drops in water. This is the bottom-most emulsion phase. The level where the oil continuous emulsion inverts to a water continuous emulsion is the theoretical oil/water interface in a vessel or tank. Inversion typically occurs at high water cut, in the range of 70 % water or more. Oil drops suspended in water are in the process of rising (creaming) due to the density difference between the oil and water as modeled by Stokes Law. The distribution of oil drop diameters and the oil-in-water concentration are the main variables that characterize this phase. There may also be some solid particles in the process of settling as well.

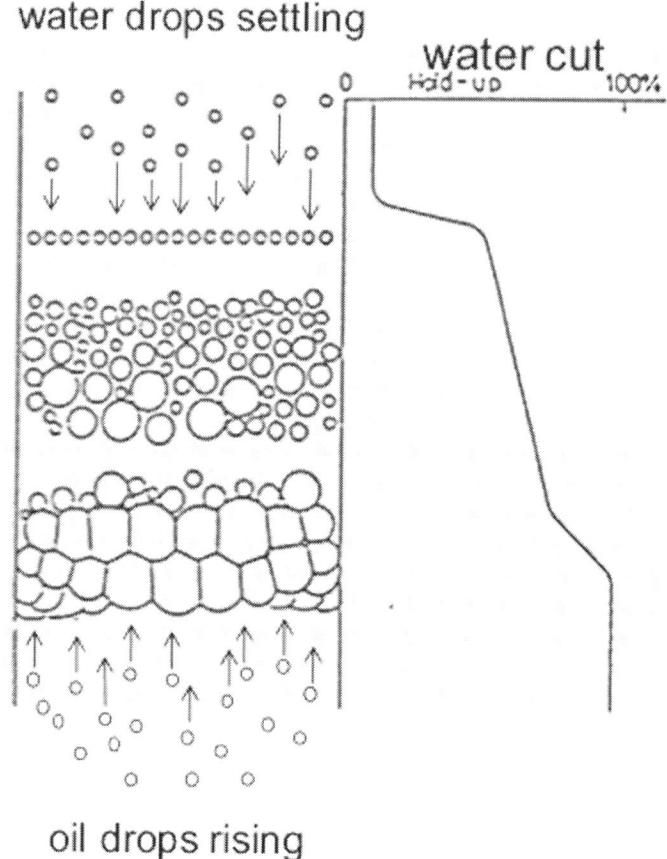

Figure 4.5 Schematic diagram showing a dispersion of water in oil (top), two layers of dispersion in the middle, and a dispersion of oil droplets in water (bottom). A profile of the water cut is shown on the right. In a separator the two dispersion layers would be referred to as a pad.

Not all mixtures of oil and water will demonstrate all of the emulsion types shown in the diagram. The oil/water ratio and the height of the different types of emulsions vary significantly from one fluid to another. Also, as discussed in detail in Sections 3.4 and 3.5, the upstream processing conditions (shearing) have an impact on the kinds of emulsions that are formed. In troubleshooting work, one of the most indicative observations of the process is close visual observation of bottle samples taken throughout the system. Being able to recognize the types of emulsions that form, where they form, and understanding why will provide an understanding of bottlenecks in the process and will lead to strategies to improve the performance.

4.1.2 How and Where Emulsions are Formed

When a mixture of oil and water is pumped from one point to another, or when, for example, it flows through a valve, it experiences shear. Low intensity shear may lead to coalescence of the oil droplets. High intensity shear will cause oil droplets to break apart. Kokal [2] gives a good review of the formation and factors which stabilize water-in-oil emulsions. Many of the same factors are relevant to oil-in-water emulsions.

A dispersion of oil in water is formed when a mixture of oil and water is subjected to shear energy. Typical locations and equipment responsible for shear energy:

- The near well bore formation

- Down hole perforations

- Gravel packs

- Subsurface pumps

- High GOR tubing flow including gas lift

- Chokes

- Manifolds

- Pumps

- Control valves

Typically, the higher the pressure loss and the shorter the time, the greater the shear intensity and hence the smaller the oil drops. Control valves, where the fluid undergoes a large pressure drop over a small distance, generate dispersions of very fine drops. The actual drop size distribution that results depends on a number of other factors including properties of the fluid. A quantitative estimate of shear energy is presented in Section 3.4.2 (Turbulence Energy Dissipation Rate).

Sources of Complex Emulsion: The following figure shows a typical source of complex emulsion formation with this example coming from a deep water platform. In this case, oily water discharged from the Wet Oil Tank is being recycled into an oil stream between the Free Water Knockout and the Bulk Oil Treater. This is not the optimal routing of this recycle stream. The oily water from the bottom of the Wet Oil Tank is mostly composed of water, with anywhere from 1,000 mg/L to 1 % oil in water. This oil is suspended in the water as small droplets. A pump is necessary to boost the pressure of the oily water from the Wet Oil Tank in order to move the oily water to the recycle location. The pump that was used in this particular example was a high speed centrifugal pump (3,600 rpm). The effect of centrifugal pumps on fluid shear was discussed in Section 3.5.3 (Shearing through a Pump). The design of lower shear centrifugal pumps is also discussed in that section. The use of the high shear pump sheared the oil droplets in the recycle water to a much smaller diameter and rendering them much more difficult to separate. At the pump discharge, the oily water was found to contain droplets of oil of between 1 and 10 micron diameter.

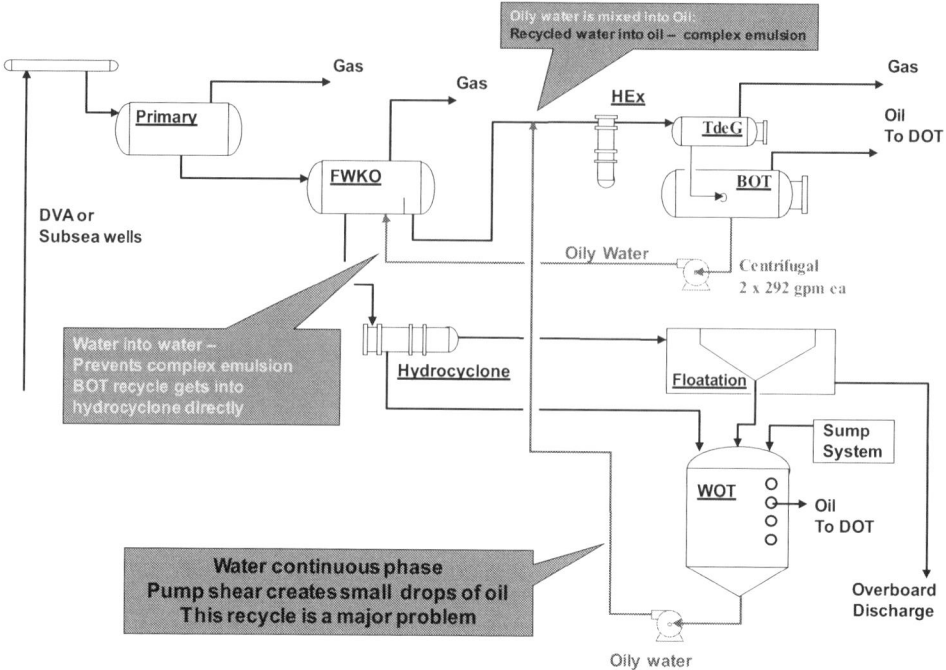

Figure 4.6 System where complex emulsion can form. In this case, oil drops suspended in water drops, suspended in oil drops accumulated in the Bulk Oil Treater

In this example, the oily water from the pump discharge is injected into a stream of oil. During the course of this injection, the oily water was broken up into droplets that were on the order of 10 to 30 micron diameter. But, as described above, the oily water already contains oil droplets that are even smaller (1 to 10 microns). Thus the result of this recycle stream was the formation of a complex emulsion with tiny oil droplets suspended in lager water droplets which were themselves suspended in oil. To make matters worse, the oily water from the Wet Oil Tank contained water treatment chemicals, including a flocculating chemical which had the effect of stabilizing the complex emulsion in the Bulk Oil Treater. This combination of stabilized oil-in-water-in-oil emulsion, together with water treatment chemical resulted in the formation of a stable interface pad in the Bulk Oil Treater that degraded the performance of the treater and necessitated an increase in the use of demulsifier in order to successfully dehydrate the oil.

All things considered, the cause of the complex emulsion in the Bulk Oil Treater is a combination of factors including shearing from the high speed pump, routing the sheared oily water to an oil stream, and the recycling of a fluid that contained water treatment chemicals. This example provides some detail regarding the formation of a complex emulsion. At this point, this is the main objective of the discussion. However, the example also highlights the need for "Best Practices" in process design, which is discussed in Chapter 19 (Applications – Deepwater Best Practices).

4.1.3 Treatment Options for Different Types of Emulsions

In the previous section a number of different emulsion types were discussed. In this section, a high level overview is given which includes how to resolve the various emulsion types. All of these emulsion types are for contaminants in water (oil-in-water and solids-in-water), and does not consider contaminants in the oil phase (e.g. water-in-oil). This information is given in tabular form which makes it easier to access. However, in such a form, great over-simplification is necessary. Much greater detail is given in various sections of this book depending on the emulsion type, how it is formed, and the economical treatment options available.

Table 4.1 Types of Emulsions and the treatment options.

Emulsion Type	Characteristics	How formed	How to Resolve
1) Suspension of oil droplets in water	Oil droplets suspended in water phase. Measured as Total Oil and Grease (TOG). Includes polymer suspended in water. No chemical stabilization.	Shearing of oil / water mixtures.	Settling, hydrocyclones, flotation.
2) Suspension of non-oily solids in water	Non-oily solid particles suspended in water. Solids may be composed of clay, iron solids, carbonate, or other minerals, including silica & silicate minerals. Measured as Total Suspended Solids (TSS), with low TOG.	Seawater flood. Formation solids.	Depends on particle size. Large particles can be removed by de-sanding cyclones. Small particles require coagulants and/or flocculating agents followed by filtration or settling. So-called "filter aids" can be used. Prevent precipitation / generation of the solids at the source.
3) Suspension of oil coated solids in water	Oily solid particles suspended in water. Composed of clay, iron solids, and other minerals combined with asphaltene, wax, and/or resin components from the oil. Measured as Total Suspended Solids (TSS), with finite (non-negligible) TOG. The combination of solids and oil can make some of these particles neutrally buoyant.	Formed from surface active and polar oil components such as asphaltenes adsorbing on clay, iron based, or other solids.	Chemical treatment and flotation. Avoid recycle as much as possible. Wetting agents to separate oil from surface, then settling and/or flotation. Coagulants and flocculating agents also can be used followed by settling. Prevent precipitation / generation of the solids at the source.
4) Suspension of chemically stabilized oil droplets in water	Oil droplets covered with corrosion inhibitor, hydrate inhibitor, or naphthenic acids from a biodegraded crude oil. Small oil droplets. Sensitivity to shear and barrier to coalescence. Typical in gas condensate systems.	Overdose of chemical or biodegraded crude oil plus shearing.	Determine surface chemistry through bottle testing or Zeta potential measurement. Apply suitable coagulant and flocculating agent, apply flotation, followed by media filtration.
High Internal Phase Volume Emulsion (HIPV)	Large drops of water suspended in oil where the oil forms thin films between the water drops. More common when oil has a relatively high viscosity.	Oil / water mixing in pipeline / flow line.	Gravity drainage of oil films – requires adequate cross-sectional area in separator. Use of a diluent (continuous phase solvent) is effective in small recycle / reject streams.
Complex Emulsion	Small oil drops suspended in water drops which are suspended in oil. The final oil phase may by thin liquid films as in HIPV. Sometimes called a pad or rag layer. Might be stabilized by water clarifier.	Shearing of oil water which then gets recycled into an oil stream.	Use interface drain. Send to slops tank. Treat with Heavy Aromatic Naptha. Otherwise, "float" separator to Dry Oil Tank. Or use solvent.

The most important take-away message of this table is that there are distinct emulsion types in the oil and gas industry and it is important to know and understand which type of emulsion is present. Each emulsion type originates from distinctly different mechanisms and the prevention, control, treatment, and management of these emulsions depends on the emulsion type. This is true for both design and troubleshooting.

A design or problem solution that works well for one type of emulsion is unlikely to work well for another type of emulsion. Design work should emphasize the likely characteristics of the produced fluids, and a prediction of various emulsion types being formed. The design of a system will then provide flexibility and options for emulsion treatment based on the likely types. Such design options are discussed in Chapter 15 on Process Design, and in Chapter 19 (Applications – Deepwater Best Practices).

Desktop Settling: Troubleshooting work should begin with the inspection of samples. The various methods that can be used to identify emulsion type are discussed in Chapter 16 (Troubleshooting). The methods include inspection of grab samples, desk-top settling, and filtration for solids, solvent testing, and inspection of samples under a microscope. Multiple samples may be required at different locations in a system. With some experience, emulsion type can be identified quickly with only a visual inspection of a few samples. In this regard, desktop settling is by far the most powerful tool for identifying the type of emulsion. In a desktop settling study, samples are taken throughout the facility and are lined up side-by-side. The sample time is noted so that the time since sampling can be tracked. The initial visual appearance is noted. Taking pictures over time is always helpful. As the emulsion resolves, the settling time is noted. Often it is possible to identify the origin of the emulsion.

4.2 Coagulation, Flocculation, Coalescence Overview

In this section the chemical processes of coagulation, flocculation and coalescence are introduced. A detailed discussion of coagulation is given in Section 4.3 (Charged Surfaces in Ionic Solutions) below and in Section 5.4 (Coagulation Chemistries). Flocculation is discussed in the Section 5.5 (Flocculation Chemistries). That discussion is deferred until later since it involves production chemistry of a complex nature and therefore requires its own section. Coalescence is discussed below in Section 4.5 (Coalescence of Oil Droplets). These processes are, in a sense, the opposite of shearing processes in that they build-up the droplet and particle size and facilitate separation.

Coagulation is the process of reducing the electrical double layer on and near a particle or droplet surface. This reduces the electrostatic repulsion between particles. As will be described, all oil drops and solid particles have an electrical coating, so to speak. This accumulation of charge on the surface is the result of various factors such as ion migration from the bulk to the interface, ionization of chemical groups on the surface of the object, and a natural abundance of charge groups on the surface. In brine water, oppositely charged ions from solution are attracted to the surface. These mobile ions from the solution create what is referred to as an electrical double layer. This double layer reduces the charge-charge repulsion of the original surfaces. If the brine does not have sufficient ability to collapse the charge-charge repulsion, then a coagulating agent (coagulant) can be added to great benefit.

As will be discussed below, at very short distances between particles, an attractive interaction occurs. Thus, the coagulant does not need to eliminate the charge-charge repulsion, but instead it only has to shield it such that the distance over which the electrical repulsion is significant is reduced. Once this distance is reduced, then natural collision of particles will bring them close enough to experience the short-range attractive interaction. This effect is referred to as "collapsing the electrical double layer." It is one of the mechanisms behind the use of a coagulant.

Flocculation is the process of forming clusters of particles or drops. The clusters can be as small as two particles, or as large as many thousands of particles. In conventional municipal water treatment, flocs of particles can be formed that span several centimeters in diameter. In fact, the surface of a water clarifier tank or vessel may be populated with flocs of particles that make the water appear cloudy. These are large flocs of many thousand particles.

Flocs of particles and droplets are much easier to separate from a bulk liquid phase than are individual particles and drops. In gravity based settling, flocs behave like agglomerates with a large Stokes Law settling diameter. In flotation, a floc of oil droplets or oily solids is a much larger object for collision with the gas bubbles. Also, the floc is stickier and facilitates the capture of smaller oil drops. In flotation, small gas bubbles are often trapped on the underside of the floc, giving it buoyancy for rising to the surface where it can be skimmed off.

Flocculation is typically promoted by the addition of polymers. There are several mechanisms by which polymers flocculate particles and drops. These mechanisms and polymers are discussed in Chapter 5 (Chemical Treatment). One aspect of flocculation that must be kept in mind is the fact that flocs are often delicate objects. Mild or moderate shear may break a floc into its component drops and particles which have a substantially reduced tendency to reassemble. Thus, flocculating agents are typically not effective when added upstream of a high shear device such as a control valve. Likewise, they are generally ineffective when added upstream of a hydrocyclone. Chemical agents which are capable of inducing droplet coalescence can be effective at enhancing the performance of a deoiling hydrocyclone.

Coalescence is the process of combining two oil droplets into a single larger droplet. Coalescence does not occur with solid particles because particles are rigid. Solid particles will flocculate under the right conditions. Oil drops are fluid and can combine under the proper conditions. The mechanisms involved in coalescence are complex. However, it is an important process since every facility has multiple devices that shear the oil droplets, and these are often followed by pipelines and separation equipment that promote coalescence. Therefore both shearing and coalescence must be taken into account, if not quantitatively, at least through general guidelines. Coalescence is often a major determining factor for whether or not a facility has a produced water treatment problem or not.

4.3 Charged Surfaces in Ionic Solutions (Brine)

Origin of Surface Charge: The surface of oil droplets and solid particles suspended in produced water are typically electrically charged due to the presence of ions on the surface. Most droplets or particles are negatively charged. The origin of surface charge and the reason for the predominant negative charge as discussed presently. One of the most common exceptions is when a cationic corrosion inhibitor is used and it has an opportunity to adsorb onto the surface of contaminants. In that case, the surface charge of the contaminant can be positive. At low pH, a negative surface can be neutralized by protonation. Also, adsorbed material from the oil or water, as well as the salinity, brine composition, as well as pH will modify the surface charge. This chapter discusses these issues in detail. From a practical standpoint, it is necessary to be aware of the chemical constituents that give rise to surface charge in a particular system.

Electrical charge on the surface of solid particles or oil droplets has a strong effect on the tendency of these contaminants to coagulate, flocculate and coalesce. Charge-charge repulsion occurs at close distances which prevents these particles from naturally flocculating or sticking together. Charge-charge repulsion must be reduced or eliminated in order for contaminant particles to stick together. Water treating chemicals are designed to do this. In order to select, apply, and monitor the performance of water treating chemicals, it is necessary to understand the origin and nature of the electrical charges on the contaminant surface of the particles and how the chemicals interact with these surface charges.

The main mechanisms that result in a charged surface are:

- ionization;
- ion adsorption;
- ion dissolution
- hydrophobic effect.

Ionization: If the surface contains carboxylic acids groups, charge can be formed by the ionization of the acid groups. At relatively high pH (e.g., pH > 8.5), where the proton concentration is low, the acid groups are increasingly ionized. At low pH (e.g., pH < 4.5), where the proton concentration is high, the acid groups are likely to be protonated. The carboxylic groups will still be polar, but not ionized.

Adsorption of ions: Surface active material, either naturally occurring in the produced fluids, or added by way of production chemicals, can adsorb on the surface of suspended material and either neutralize an existing surface charge or add charge to the surface. Various charged ions and non-surfactant small molecules can adsorb onto the surface as well. These entities include carboxyl molecules, quaternary amines, hydroxyl ions, sulfate, sulfide, and/or phosphate ions. These adsorbed surfactants, polar molecules, and ions can bind to the surface by a number of different forces such as hydrogen bonding, charge-transfer complexing, London-van der Waals forces, pi electron resonance, etc.

The net surface charge that results when adsorption occurs depends on a number of solution characteristics such as pH, temperature, presence of co-solvents, salinity, the ionic composition of the solution, the presence of surfactant molecules, etc. At low pH (abundance of protons), most of the acidic groups will be protonated and hence uncharged. Thus, at low pH, surfaces tend to have a more positive charge. At higher pH, as the acid surface groups become charged, the alkaline groups become neutralized, and the surface tends to have a more negative charge.

Ion Dissolution: Most mineral particles suspended in produced water have ionized surfaces. Anionic groups such as carbonate (CO_3^{-2}) or sulfate (SO_4^{-2}) have a net negative surface charge. The counter ions (Na^{+1}, Ca^{+2}) will be present in the mineral. The distribution of these ions will give rise to a local charge density, even if the net charge averaged over a given surface area is zero. Also, there is a tendency for the cations to leave the surface and become hydrated. This leaves a net negative charge on the surface from the remaining anions that have less tendency to solvate. Some inorganic ions may have a strong affinity for certain particle surfaces. Ions that strongly adsorb to a surface are referred to as "charge determining ions".

Hydrophobic Effect: Even if the surface is electrically neutral, the presence of a hydrophobic contaminant in water (e.g. hydrocarbon droplet) will cause the water molecules to align in such a manner that the contaminant together with its closest neighboring water molecules are practically speaking, negatively charged. Since the droplets are composed of hydrocarbon, a nonpolar substance, there is essentially no electronic charge contribution from the droplet. The negative charge comes from the water and the structure of the water molecules surrounding the hydrocarbon droplet. The water molecules surrounding the hydrocarbon droplet are oriented in such a way as to minimize disruption of the hydrogen bonding of the water. This causes an orientation preference which gives rise to a net negative charge [1, 7 – 9].

Indifferent Electrolyte Ions: Ions dissolved in the produced water that do not have an affinity for the surface of suspended material are called "indifferent electrolyte ions". Although they do not directly affect the surface charge, indifferent electrolyte ions do affect the thickness of electrical double layer. They screen the charge of the double layer and reduce the charge-charge repulsion. Indifferent cations that play an important role are sodium (Na^+), and potassium (K^+). In some produced water systems calcium (Ca^{2+}) and magnesium (Mg^{2+}) ions are involved in precipitation and are not therefore indifferent in those solutions.

Oil Droplets in Water: Hydrocarbon oil drops suspended in water have a tendency to acquire a fairly strong negative charge on the surface. This occurs over a range of water salinities, and pH values. It occurs without the presence of any surfactant or other chemicals. The origin of this charge is not completely understood and is the subject of current research. One concept that has some support is that the orientation of the hydrogen bonding network structure around the hydrocarbon as described above. Suffice to say that even this simple system (hydrocarbon drop in water) has significant surface charge.

Surface charge gives rise to attraction and repulsion between contaminants in produced water. Oppositely charged contaminants attract. Contaminants that have similar surface charge will repel each other. The processes of coagulation and flocculation are highly dependent on these molecular interactions. Since the ultimate goal of water treatment is to separate impurities, it is important to understand the origin and effect of surface charge. Controlling and overcoming this surface charge is a primary goal of water clarification chemicals.

Counter-Ion Attraction: Charged surfaces have a tendency to attract counter-ions from the solution. This is particularly true of multivalent surface ions. If surfactant-like molecules are present in the solution, they will have a tendency to migrate toward their counter ions on the surface of a contaminant and adsorb. This will have a dramatic effect on the surface characteristics of the solid particle.

The second layer of charges surrounding a particle is known as the diffuse layer or the "Stern Layer". It is composed of ions attracted to the surface charge via electrostatic attraction. The second layer is more loosely associated with the object than the surface charge, because it is made of free ions which move in the fluid under the influence of electrostatic interactions, thermal motion, and fluid shear, rather than being firmly adsorbed to the surface. The net electric charge of the diffuse layer is equal in magnitude to the net surface charge, but has the opposite polarity. As a result the complete structure is nearly electrically neutral. The residual charge imbalance is the previously discussed quantity known as the Zeta Potential. Zeta Potential measurement instrumentation is commercially available and in fairly wide use in research laboratories.

Dynamic Equilibria of Surface Charge: Most solids that are suspended in produced water brine will dynamically exchange surface ions with water. Some ions will depart from the surface and become solvated in the produced water, and likewise some ions from the produced water will adsorb onto the surface. This dynamic process creates a fluctuating surface charge. The chemical physics of this process is essentially the same as any phase equilibrium process. The rate of dissociation / adsorption is governed by the chemical potential of the ion in the two phases (solid, produced water solution). If the chemical potential of the ion is the same in the two phases, then the rate of salvation and the rate of adsorption are equal, and the net rate of transfer of the ions is zero. Thus, the chemical potential of an ion plays an important role in understanding the surface characteristics of a solid particle in produced water.

Electric Potential: Physically, the electric potential at a particular point in space is the amount of electric potential energy that a single positive unit charge, of essentially zero size (point charge), would have at that point in space. This single positive unit charge is sometimes referred to as a test charge. Like any other form of potential energy, the unit of measure for electric potential is Joules (Newton meter - Nm). However, values of electric potential are most often given in units of volts (V) or millivolts (mV).

Electrical Double Layer: The electric double layer around a particle is illustrated in Figure 4.7 below.

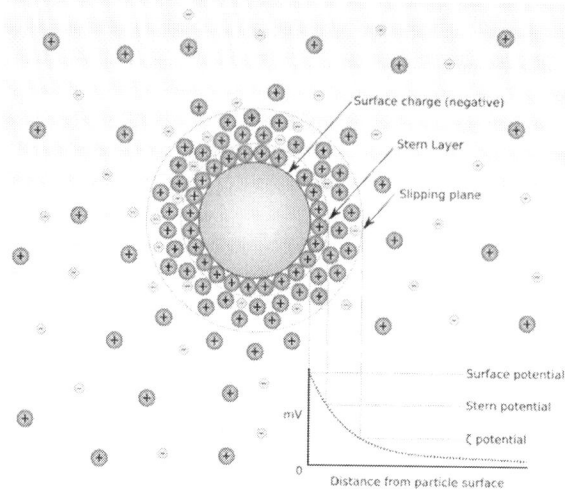

Figure 4.7 The formation of an "electric double layer" around a negatively charged particle is illustrated. The depth of the double layer decreases as the salinity of the water increases.

From a practical point of view, suspended solids may be destabilized by adding to the water a relatively high concentration of salts. The salts increase the ionic strength of the solution. The increase in ionic strength results in higher concentrations of counter ions in the diffuse layer. This balances the surface charge and compresses the diffuse layer of charge around the solid particle. The compression of the electrical double layer reduces the electrostatic repulsion of the particles. This mechanism is referred to as "screening of the surface charge by electrical double layer compression." If the diffuse layer is compressed sufficiently, the other inter-particle forces, such as attractive van der Waals forces or hydrogen bonds will predominate and aggregation will occur.

Poisson's Equation: Without going into details, Poisson's Equation is a fundamental equation in electrolyte solutions. In words it says that the change in electric potential from one point to another depends on the charge density (locations and valence of the ions) in the solution. It is often applied starting at the surface of a contaminant and extending into the solution away from the surface. Conversely, it also says that the locations of the ions in solution depend on the overall distribution of charge (location of all other ions).

The Figure below gives an example of the distribution of electrical charges around a charged surface. The charged surface is in this case a rigid particle with a negative surface charge. The solution is a strong electrolyte (contains a high concentration of ions). The surface charge is negative. The x-axis is the distance between the surface of the charged particle and the center of the ions surrounding it in solution. Note that the ion centers do not come within an ionic radius of the surface (vertical line). As shown, the charge density of negative ions (anions) is very strongly depleted near the negatively charged surface. The charge density of positive ions is enhanced near the surface. The y-axis variable is the distribution of electric charge. This is the same variable in Poisson's equation.

As will be discussed below, the concept of ionic shielding, which is relevant in coagulation, flocculation, and coalescence, can already be discerned at this point. If the positively charged surface represented a calcium carbonate particle and we are interested to know the interaction of this particle with other particles in a brine, this model predicts the charge distribution as shown in the figure below. The effect of the positive surface charge is mitigated by the presence and buildup of negative charge near the surface. This is ionic screening or shielding of surface charge. Thus, another particle that approaches the one in the figure will not experience the full interaction of the surface charge due to the presence of the ions in solution. Enhancing this shielding or screening effect is the role of a coagulating agent. The most effective coagulating agents have multivalent charges which are even more effective at screening the electrical surface charge than the simple NaCl used in the figure.

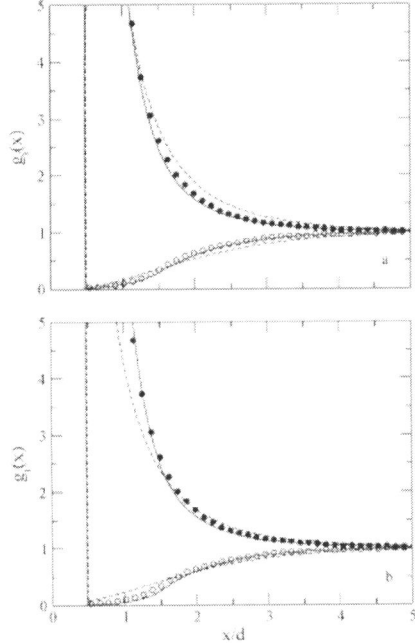

Fig. 3   Density profiles of a 1 : 1 electrolyte ($d = 3$ Å) at 1M and room temperature for the state for which the MC values are $\sigma d^2/e = 0.1685$ and $\beta e \phi(d/2) = 2.6$. The points give the simulation results and the solid and broken curves give the DFT and GCS results, respectively.[33] The comparison is made at the same charge density (part a) and same diffuse layer potential (part b).

Figure 4.8 Distribution of positive and negative ions surrounding a charged surface. Molecular dyanmics was used to simulate this system [10]. The oppositely charged ions (solid circles) are attracted to the surface while the same charged ions (open circles) are repelled. The x-axis is scaled by the diameter of the ions. Note that most of the effect of the charged surface is negligible at a distance of 3 to 4 ion diameters. This is where the concentraion of the positive and negative ions is equal

The material below provides further background, development of the model, and practical calculations that can be carried out to determine the relative magnitude of the effect of electrical shielding in an ionic solution.

4.3.1 The Debye Length

The Guoy-Chapman-Stern (GCS) model [11] provides a way to calculate the electric shielding that occurs as a result of positive ions in solution migrating toward the negatively charged surface of a particle or oil droplet. When this happens, the net charge surrounding the contaminant is reduced. In other words, some of the negative surface charge of the particle is cancelled out by the positive charge of the ions in solution. Of course, it can also be applied to negative ions in solution and a positive contaminant surface. The GCS model provides an understanding of the relative importance of surface charge, and charge polarity of the contaminants, as well as the importance of the pH, salinity, concentration of multivalent ions, and temperature of the water phase. Knowing these factors can help to predict the type of chemicals that will be most effective in water treatment. Coagulants are used when the composition of the produced water does not sufficiently provide this shielding.

The GCS model provides a means to calculate a quantity referred to as the Debye Length. It is a measure of the length or distance over which the electrical double layer of one particle will be felt by that of another particle. When adding a coagulant chemical, the objective is to collapse the electrical double layer (make the Debye Length very small).

A comment regarding the ion concentration is in order here. Typical electrolyte concentrations are given in Molar units (mol/L). If the electrolyte has a monovalent cation and a monovalent anion, then z = 1 electron/ion. Thus 0.1 M is equivalent to 100 mol ions/m³, which is also equivalent to 6.02 x 10²⁵ ions/m³. The conversion to mg/L requires multiplication by the molecular weight.

Table 4.2 Calculations of the Debye length as a function of the ionic strength (I).

c (M = mol ions/L)	C (mg/L)	lk (nm)	3 lk (nm)	3 lk (micron)
3	175,500	0.18	0.53	0.001
2	117,000	0.22	0.66	0.001
1	58,500	0.31	0.92	0.001
0.1	5,850	0.97	2.9	0.003
0.01	585	3.1	9.2	0.009
0.001	58.5	9.7	29.2	0.029

Length or Thickness of the Electrical Double Layer: As shown in the table above, the thickness of the electrical double layer is quite strongly influenced by the concentration of ions dissolved in water. The effect of the charged surface extends to about 3 times the Debye Length, depending on temperature, and ionic strength. For this reason, values of 3 l_κ are tabulated in the Table above.

Collapse of the Electrical Double Layer: As shown in Figure 4.9, the Debye Length decreases significantly as the ionic concentration increases. Although it is not shown here, the addition of a coagulant chemical, with multivalent cations (z=3), will also significantly decrease the Debye Length. In many cases however, electrical repulsions can be shielded naturally in high salinity oilfield brine. As discussed in Chapter 1, most oilfield brines have high salinity. This is why, in many cases in the oilfield a coagulant is not required.

As mentioned previously, at very short distances between particles, an attractive interaction occurs. Thus, the brine or coagulant does not need to eliminate the charge-charge repulsion, but instead it only has to shield it sufficiently to allow particles to approach close enough for the attractive interaction. This is discussed further below.

Condensed Produced Water vs Oilfield Brine: Generally speaking, this repulsive contribution to coagulation and coalescence helps to explain the difference between condensed produced water and more typical high salinity brines. Also, a firm understanding of the electrical double layer is required in order to fully understand "Zeta Potential", which has a number of significant consequences. Also, the process of electrocoagulation can only be understood in the context of the electrical double layer model. These subjects are discussed next.

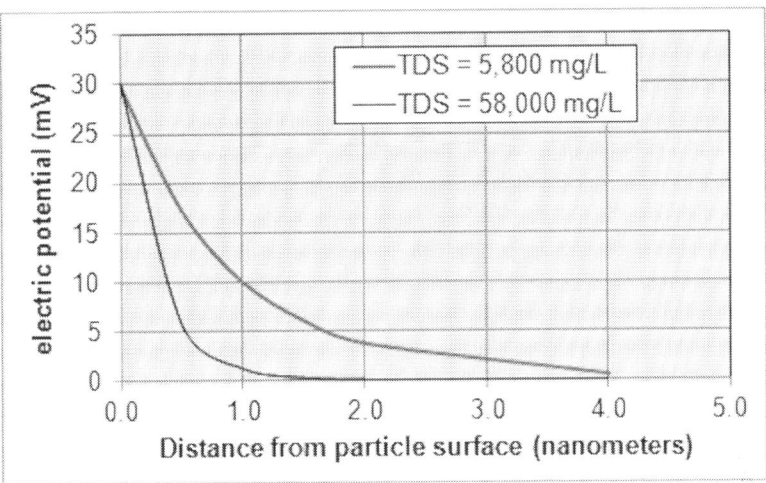

Figure 4.9 Electrical potential as a function of distance from the surface of a charged particle. The surface charge on the partical is 30 mV. As shown, higher concentration of dissolved ions (TDS) reduces the electrical potential more quickly than a dilute solution. This is why condensed water often requires a coagulant whereas produced water nrine does not.

4.3.2 Electrical Double Layer and Zeta Potential

Electrical Double Layer: As discussed, mobile counter ions from the water phase are attracted to the charged surface. As the name implies, these counter ions have a charge opposite to that of the surface of the contaminant. The electrical attraction of the opposite charge leads to an accumulation of these counter ions near the surface. The contaminant can be a solid particle, a gas bubble, or a liquid droplet. The two layers of charge (surface charge and surrounding layer of counter ions) is referred to as the "electrical double layer." These ions are bound tightly to the contaminant.

Helmholtz Layer or Slipping Plane: The outer edge of the electrical double layer is referred to as the slipping plane. The slipping plane differentiates the mobile ions in solution from those that are strongly bound to the surface of the contaminant. The location of this layer is somewhat ill-defined. In actuality there is not a sharp boundary between bound and mobile ions. Nevertheless, various experimental measurements such as electrophoretic mobility can be used to define an electrical potential (zeta potential) for the contaminant and its bound counter-ions. This is discussed in greater detail below.

Stern Layer: There is actually one more layer that is sometimes distinguished. Just outside the surface of the contaminant there is a thin layer that is essentially devoid of charge (or mass) due to the finite diameter of the atoms and molecules involved. This is referred to as the Stern layer. It can be distinguished in Figure 4.8.

Zeta Potential: The zeta potential is defined as the electrical potential at the slipping plane. Within the slipping plane, ions move with the particle. Ions further from the object move in response to fluid shear. Zeta Potential is measured in units of electrical charge per unit surface area (C/m^2). The zeta potential is denoted as ζ-potential, or simply just ζ. Zeta potential can be measured using electrophoresis, as well as a number of other techniques. It is an experimental ramification of the Debye Length.

The magnitude of the zeta potential gives an indication of the tendency of a suspension to coagulate and flocculate. Particles in water are in constant motion due to thermal effects and due to fluid shear, if present. This motion naturally leads to particle-particle collisions. If all the particles in the suspension have a large zeta potential then they will tend to repel each other due to electrostatic repulsion. In that case, there will be no tendency for the particles to stick together when they collide. However, if the particles have a low value of zeta potential then the particles will have a greater tendency to stick together when they collide. Coagulation and flocculation is a bit more complicated than this since there must be an attractive force to overcome the electrostatic repulsion of the two like-charged surfaces. This is discussed later. As a general rule of thumb, suspensions with zeta potential between – 30 mV and + 30 mV have a tendency to coagulate.

The utility of zeta potential measurements is as follows. From observation it has been found that a zeta potential value greater than 30 mV or less than – 30 mV are typical of emulsions that do not readily coagulate or coalesce unless a chemical is added. In such systems, coagulation, flocculation and coalescence are hindered due to electrostatic repulsion of the surface of the solid particles or droplets. When the zeta potential is in the range of - 25 to + 25 mV the object is said to have a low-charged surface. When the zeta potential is low, then other attractive forces (e.g. van der Waals), together with collision momentum can exceed the electrostatic repulsion force. When this occurs, coalescence and / or coagulation will take place and the emulsion will break spontaneously. So, colloids with high zeta potential (negative or positive) are electrically stabilized while colloids with low zeta potentials tend to coagulate or flocculate. This is a rough rule of thumb.

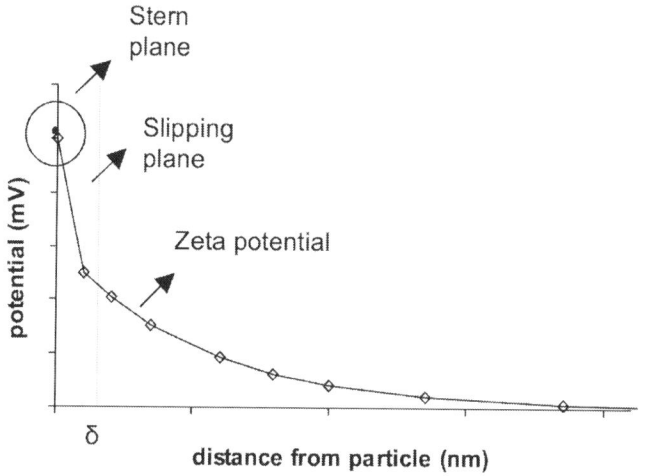

Figure 4.10 Schematic diagram of the potential profile in the double layer (δ = Stern layer thickness)

4.3.3 Electrophoresis and Measurement of Zeta Potential

Electrophoresis is the measurement of the speed at which a particle travels in a fluid as a result of the application of a uniform electric field. That speed is proportional to the strength of the applied field and to the electric potential on the surface and surrounding the particle. Thus, the speed of migration in an electric field provides a measure of the effective surface charge of contaminant particles.

The Smoluchowski equation assumed that there is no effect of the shape of the or size of the particle. It stands to reason that larger particles will travel more slowly as in Stokes Law. But this effect is not present in the Smoluchowski approach. This will normally be the case for suspended solid particles in high salinity brines found in produced water.

$$u_p = \frac{D\varepsilon_0 \zeta E}{\mu_w}$$

Eqn (4.1)

u_p = velocity of the particle, relative to the surrounding liquid at the shear plane (m/s)
D = relative permeability of water (dimensionless, 80.1 at T = 293 K)
e_0 = permittivity of a vacuum = 8.85 x 10^{-12} (C^2 / N m^2 or C / V m)
z = zeta potential (C^2 / N m^2 or C / V m)
E = electric field (V/m or N/C)
m_w = dynamic viscosity of water (Pa sec or N sec / m^2)

It is a worthwhile exercise to verify that the units of the right hand side of the equation are indeed m/sec, as expected.

Analogy with Stokes Law: By analogy with Stokes Law, the zeta potential is the intensive quantity (analogous to density) which interacts with the extensive electric field strength (analogous to gravity).In Stokes Law the variable of interest is the settling rate or speed with which the particle moves through the fluid. The gravitational field of the earth, and the buoyancy of the particle causes the motion. The resistance is the viscosity of the fluid which is found in the denominator of the equation. In the case of electrophoresis, the variable of interest is the speed with which the particle moves through the fluid. The electric field and the charge density of the surface of the particle causes the motion. The resistance is the viscosity of the fluid which appears in the denominator.

The driving force for settling in Stokes Law is $gD\rho$ which has the units of J/m^2. In the above equation, the driving force is $e_0 E_z \zeta$ which also has the units of J/m^2. Both equations have the same dependence on the viscosity which is the measure of resistance to motion in both equations.

4.3.4 Zeta-Potential – Oil Droplets in Water

In this section, measurements of the electrical double layer, zeta potential, and stability of produced water emulsions are reviewed. The objective of this material is to develop a physical understanding of the magnitude of the effect of electrostatic interactions. Hydrocarbon droplets have been found in general to have a relatively large negative zeta potential. The origin of this negative charge is discussed below.

Shaw [11] reports the zeta potential for hydrocarbon oil droplets in 0.05 mol/L salt solution. The particular salt is not reported. If the salt was NaCl, this would correspond to 2,900 mg/L.

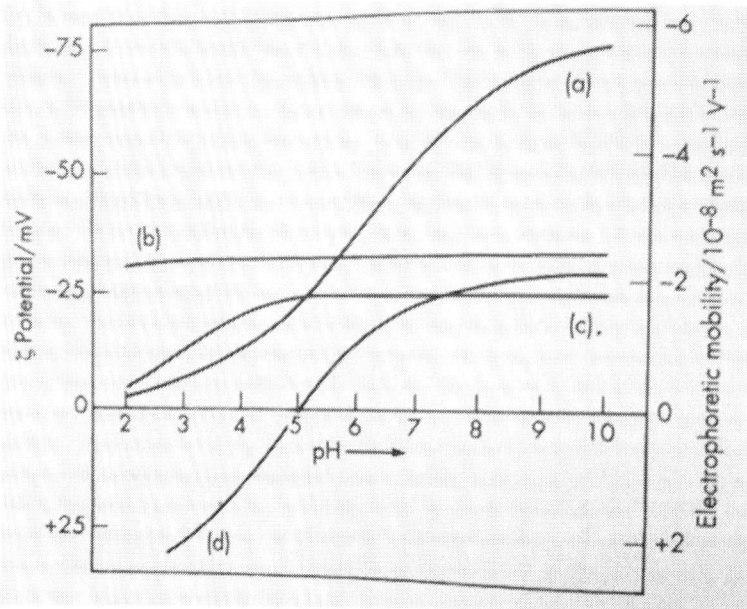

Figure 4.11 The zeta potential of hydrocarbon oil drops (unspecified composition) at approximately 2,900 mg/L salinity. (a) oil drops with no added surfactant; (b) take this one out; (c) carboxylated polymer adsorbed on the oil drops; (d) Serum albumin adsorbed on the oil drops [11].

Wiacek and Chibowski [12] report zeta potential for n-tetradecane in deioinized water and in 1 mM KCl. The concentration of n-tetradecane studied was 0.1 % v/v (1,000 µL/L). The pH of the study was 6.08 to 6.8. The zeta potentials are given below:

Zeta potential:

n-tetradecane in deioized water:	-63.0 mV
n-tetradecane in 1 mM KCl:	-53.8 mV

Table 4.3

Parameter	Dispersing phase	
	Water	KCl (10^{-3} M)
pH	6.75	6.08
Zeta potential (mV)	-63.0	-53.8
Effective diameter (D_{el}, nm)	494	580
$1/k$ (nm)	726.5	9.7
Ka	0.34	29.9

Also reported in [12] is the effect of added salt on the Debye Length. Even with a small concentration of added salt (KCl), the Debye Length ($1k$) is much smaller (0.7 micron) than the case with no added salt (0.01 micron). Measurements of the zeta potential of tetradecane as a function of pH are given in [12]. The n-tetradecane (model hydrocarbon) was found to have a PZC of 9 and a zeta potential of 20 mV at pH 6. The addition of low molecular weight alcohol decreased the zeta potential such that the PTE dropped to 4.5. The zeta potential as a function of alkane chain length, and water pH is given in [13]. The oil was composed of hydrocarbons (C_nH_{2n+2}), obtained from a chemical supply company (Reachim). The concentration of oil was 0.5 %. No mention is made as to whether the concentration is percent by weight or volume. Salt (NaCl) was added at a concentration of 1 x 10-3 mol/L. This corresponds to an ionic strength of 58 mg/L. The oil was sheared in a mixer to produce droplets of 6 to 8 micron diameter. Electrophoresis was used to determine the zeta potential of the oil droplets in water. Zeta potential was calculated using the Smoluchowski formula. As shown in the figure, the pH was adjusted by addition of HCl or NaOH. Obviously, as these reagents are added to the solution, the ionic strength is changed. Measurements were carried out at 23 to 25 C.

In one study [11], emulsion stability was determined by desk-top settling in a beaker where the volume of oil accumulated at the top of the beaker was measured. Results were reported in terms of the percentage of the total oil volume in the beaker that accumulated at the top, as a function of time. The results provide some physical insight. For the mixture studied, the zeta potential is negative, and rather largely negative, over a wide range of pH. This indicates that there is a negative electric potential at the zeta plane surrounding the oil drops in water. At pH 6, the zeta potential was about -80 mV and the settling tendency was non-existent. At pH 5, about half of the hydrocarbon had risen to the top of the beaker and the Zeta potential was about -50 mV. At pH 4, the Zeta potential was -20 mV and 80 % of the hydrocarbon had risen to the top. Lowering the pH to 4 is obviously unrealistic in a production system. But the addition of a cationic coagulating agent or flocculating agent may have the same effect.

Bottle Testing to Estimate Zeta Potential: Bottle testing can be carried out to determine the PZC (point of zero charge - also known as the IEP isoelectric point). The PZC is the pH at which the zeta potential is zero. This is typically done by lining up similar samples and adjusting the pH over a wide range and looking at settling time as a function of pH. The PZC is the pH at which the solution settles fastest. Once the PZC is known, the zeta potential of the emulsion can be estimated by the following formula: ZP = 15 x (PZC - pH). This formula says that the zeta potential is the difference between the emulsion pH and the pH at which the PZC occurs. If, for example, the PZC is 9 and the pH of the emulsion is 5, then the zeta potential is +60 mV. This is a relatively high zeta potential which indicates a relatively high charge density of the electrical double layer and that a coagulant or other counter ion chemistry is probably necessary. Since the estimated zeta potential is positive, this indicates that anionic chemistry would be most effective. If a coagulant be used, then an anionic flocculating agent would be effective. In practical applications, the terminology used is "cationic emulsion."

Zeta Potential and the Hydrophobic Effect: As shown, the zeta potential surrounding hydrocarbon droplets in uncontaminated water is negative. This was discussed already at the beginning of Section 4.3 (Charged Surfaces in Ionic Solution) where the source of negative charges surrounding contaminant particles was referred to as the hydrophobic effect. The measurements given here give the practical measurable result of the hydrophobic effect. The bottom line is that the hydrocarbon droplets show some coalescence at pH of 6, and increase in coalescence as pH is decreased. Thus, as protons are added and the pH goes down, the diameter of the electrical double layer around the hydrocarbon collapses. This allows droplets to come together more effectively such that coalescence efficiency increases. At a pH of 4, the zeta potential is in the range of -30 mV and roughly 30 % of the hydrocarbon drops have creamed in 10 minutes. Given the very small initial diameter (6 to 8 micron) of the droplets, this is a respectable creaming rate.

4.3.5 Zeta-Potential – Solid Particles in Water

In Table 4.4, the Isoelectric Point (IEP) is given [14, 44] for several clean mineral surfaces. The isoelectric point is also known as the point of zero charge (PZC). This is the pH at which there is no net electric charge at the shear plane near the surface of the particle or droplet. At this pH no electrophoretic mobility is measured. The Zeta potential is zero.

The zeta potential will be negative at pH values greater than the IEP. Conversely, the zeta potential will be positive at pH values less than the IEP. This is always the case. Thus, the IEP indicates the pH above which the surface is negatively charged, and below which the surface is positively charged. The table below provides a tabulation of IEP values. A surface will be positively charged if the IEP is greater than the fluid pH.

Table 4.4 Isoelectric points (IEPs) of some common oxides and minerals

Material	Chemical formula	IEP
Quartz	SiO_2	2.0
Sol-gel silica	SiO_2	2.5
Alumina	Al_2O_3	8-9
Titania	TiO_2	5-6
Zirconia	ZrO_2	4-6
Hematite	Fe_2O_3	8-9
Magnesia	MgO	12
Molybdenum oxide	MoO_3	0.5-1
Vanadium oxide	V_2O_5	0.5-1
Zinc oxide	ZnO	9
Chromium oxide	Cr_2CO_3	6-7
Tin oxide	SnO_2	4-5
Calcium carbonate	$CaCO_3$	9-10
Mullite	$3Al_2O_3 . 2SiO_2$	6-8
Kaolin (edge)	$Al_2O_3 . SiO_2 . 2H_2O$	6-7
Apatite	$10CaO . 6PO_2 . 2H_2O$	4-6
Potassium feldspar	$K_2O . Al_2O_3 . 6SiO_2$	3-5

Fe_3O_4 Magnetite 6.5 to 6.8

As shown in the table, amorphous or crystalline silica has a predominantly negative surface charge over a wide range of pH. Conversely, calcium carbonate has a predominantly positive charge over a wide range of pH [14, 15]. It is noted that in produced water the existence of clean mineral surfaces is almost nonexistent. Mineral surfaces will be coated with surface active compounds from the oil and from the water. Nevertheless, the values of IEP for clean surfaces are still of interest since they indicate the tendency for surface active compounds to adsorb. For example, cationic polar groups are likely to adsorb onto surfaces that have a negative zeta potential at the fluid pH. Likewise, anionic polar groups will adsorb onto surfaces that have a positive zeta potential. Thus, the IEP can indicate the pH at which there is a switch from which polarity (positive or negative) of compounds are likely to adsorb.

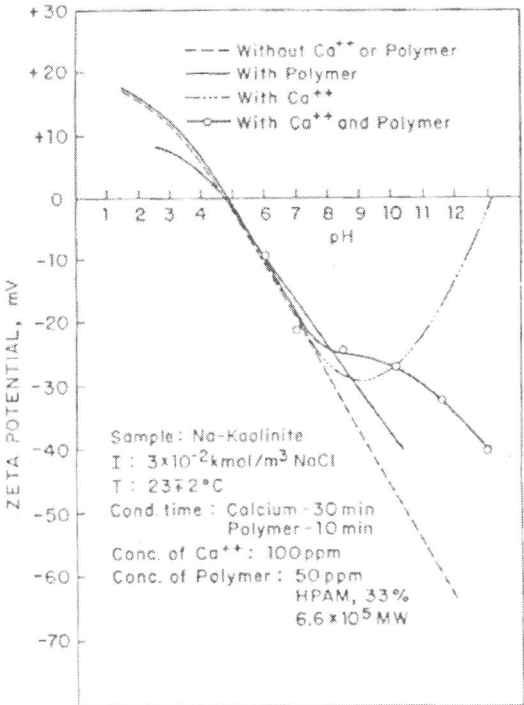

Fig. 10. Effect of calcium, polymer or both on pH response of zeta potential of Na-kaolinite.

Figure 4.12 The zeta potential of Na-Kaolinite (a representative clay) as a function of pH, and additives such as Ca+2 and HPAM polymer [16].

Clay particles are a common contaminant in produced water. As shown in Figure 4.12, Kaolinite has an IEP of roughly 5, which is in the range of many produced water pH. The zeta potential goes from negative values to zero and increases to positive values as pH is lowered. Over a wide range of pH, the absolute value of the zeta potential is relatively small. This means that there is a small barrier to coagulation. When Ca+2 cations are added, the zeta potential of clay becomes more positive in the alkaline range of pH. This is physically sensible. The Ca+2 ions are adsorbed onto negative surface sites of the clay thus reducing the overall negative charge of the particle.

Partially hydrolyzed polyacrylamide (HPAM) was added to the solution with and without added Ca+2 ions. The polymer was 33 % hydrolyzed (33 % of the pendant acrylamide groups were converted to negatively charged carboxyl groups). The effect of added Ca+2 cations is reduced in the presence of polymer.

As discussed in [17], zeta potential has been measured and reported for clay particles in a number of different studies. However, interpretation of these results is complicated by the structure of clay particles. It is well known that clay particles are platelets with a high degree of asymmetry. They have thin edges and broad flat faces. The edges are positively charged. The faces are negatively charged. A suspension of clay particles in water is very highly shear thinning. When left to stand, or under mild shear conditions, the clay platelets stack against each other in a house-of-cards arrangement with positively charged edges facing the negatively charged faces. This creates a very open structure. Ultimately this causes the suspension to have high viscosity. Under more vigorous mixing conditions the house-of-cards structure collapses and the viscosity of the suspension is reduced.

Regarding zeta potential, the structure of clay particles makes it difficult to interpret the results. The mobility of such particles will be affected by both the negatively charged faces and the positively charged edges. Thus, a calculated zeta potential will not reflect either the charge on the surface or the charge on the edge. The true zeta potential will be somewhere in between.

4.3.6 Zeta-Potential – Gas Bubbles in Water

The surface charge at a gas / liquid interface has an effect on the capture of oil drops and solid particles by gas flotation bubbles. The spreading coefficient is discussed in other sections. In this section, the zeta potential is discussed. Graciaa et al. [18] measured the zeta-potential at the air / deionized water interface and obtained a value of – 65 mV at neutral pH. This is a relatively large negative value.

One of the interesting considerations is the effect of surfactants on the zeta potential at the interface between an aqueous solution and a bubble. In this case, surface active compounds will migrate to the gas / water interface with the polar head sticking into the water and the non-polar tail laying along the interface, or at high concentrations of the surfactant, lining up with other surfactant molecules and sticking into the gas.

In [19], two surfactants were studied, sodium hexadecyl sufate (SHS), and hexadecyltrimethylammonium bromide (HTAB). SHS is an anionic surfactant, while HTAB is a cationic surfactant. The molecular weight of SHS is 344 gr/mol. Thus, a solution of 1 x 10-5 mol/L would contain 3.4 mg/L of SHS. This is a small amount of surfactant. Yet, even with this small amount, the effect on the zeta potential is enormous. It must be noted that the solution studied here had low salinity which magnifies the effect of the surfactant. Nevertheless, the zeta potential is very strongly impacted by this type of chemistry (HTAB and other quaternary amines), which is often used in corrosion inhibitors.

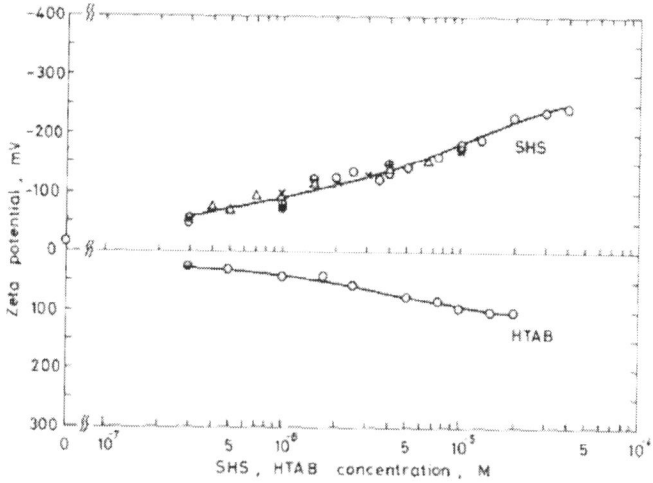

FIG. 5. Zeta potential of argon gas bubbles as a function of SHS or HTAB concentration. ×, Bubbling cell B (20–23°C); ○, 21–23°C; △, use of distilled water B (23–24°C); ●, 24–26°C; ⊕, humidified argon, 23–24°C.

Figure 4.13 Zeta potential of gas bubbles in two different surfactant solutions [19]

Frother chemicals are used in mineral mining flotation operations. They improve the capture rate of valuable minerals. As studied in [20], most frother chemicals do not appear to change the zeta potential of the bubbles significantly particularly in the range of pH (5 to 7) and concentrations of interest (10 to 20 mg/L). Two frother chemicals studied are an exception and do have more than a negligible effect on zeta potential are MIBC (methyl isobutyl carbinol: 4-methyl-2-pentanol) and heptanol. It could be argued that these compounds are surface active. They decrease the zeta potential of the gas bubbles making it more negative. This agrees with the idea that the nonpolar group would lie on the bubble / water interface, and the polar head would stick into the water. In the presence of mineral particles, the polar head would attach to the mineral surface and the non-polar tail would attach to the gas bubble. In this way, a surface active frother chemical would improve attachment between the mineral and gas bubbles.

Summary of this Section: Thus far, a powerful model of the electrical surroundings of particles in water has been developed. The concept of the electrical double layer was developed mathematically which gave the concept of Debye Length and the idea that the electrical double layer must be collapsed in order for particles and droplets to coagulate and flocculate.

The concept of Zeta potential was introduced as an empirical measure of the electric potential at the slipping plane. It was then stated on the basis of empirical observation that zeta potentials inside the range of -30 mV to +30 mV usually permit the facile coagulation of particles. Addition of a coagulant was given as a way to "collapse" the electrical double layer and reduce the zeta potential. Zeta potentials outside of the ±30 mv range favors the stabilization of individual particles. Measured values of zeta potential were then presented as a function of salinity, brine composition, surfactant, and pH for the three types of interfaces of greatest interest: oil/water, gas/water, and solid/water.

4.3.7 Critical Coagulation Concentration & Schultz – Hardy Rule

The Schultz-Hardy rule is an empirical observation about what types of coagulation chemicals are better than others and why. One set of empirical data are given in Table 4.5 below.

Table 4.5 Concentration of coagulant required to coagulate particles of AgI in water [11, p. 211].

Coagulant	Concentration (m-mol/L)
$LiNO_3$	165
$NaNO_3$	140
KNO_3	136
$Ca(NO_3)_2$	2.4
$Mg(NO_3)_2$	2.6
$Al(NO_3)_3$	0.067

The critical coagulant concentration is the minimum concentration of a salt, such as $AlCl_3$, or $Al(NO_3)_3$, that causes a suspension of particles to coagulate. The measurement of coagulation is not discussed in detail here. That discussion is deferred to Chapter 5 where coagulation chemistries and bottle testing are discussed. Suffice for now to say that a common method for determining the critical coagulant concentration is to prepare a set of samples of suspended particles in water. Each sample has the same concentration of contaminant. Add to each sample a different concentration of the coagulant, shake and let sit for a short time, perhaps a few minutes. There will be a difference in the state of aggregation from one sample to another. The solids will settle in those samples where aggregation has occurred. Thus, at low concentrations of coagulant, no clarification will occur. At a particular concentration, and for all higher concentrations, settling will occur. This allows bracketing of the critical coagulation concentration. The precise concentration for which aggregation occurs can be determined by using finer and finer brackets of concentration. Said another way, a narrow range of concentration can be achieved by making up samples with smaller and smaller differences in coagulant concentration until the desired precision is achieved.

The critical coagulant concentration is the concentration at which the electrical double layer becomes compressed. Once the electrical double layer is compressed, electrostatic repulsion is greatly reduced. This results in particles colliding with each other to much closer distances. At close enough distances, the particle / particle attractive forces become significant and particles have a greater tendency to stick together.

In Table 4.5 the solid particles are composed of AgI. As discussed in [21], the surface charge of AgI is negative. The added electrolytes are all salts of NO_3^{-1}. Thus, there is no effect from a change in the anion. The important, and rather obvious point here is that the concentration of the coagulant required decreases dramatically with the valence of the cation.

As discussed in [22], an equation for the critical coagulant concentration can be derived by seeking the minimum energy of interaction. This minimum will be an attractive interaction. The minimum energy and the distance at which it occurs can be calculated by taking the derivative of the potential energy function and setting it equal to zero. The result of this calculation is given by:

$$C_{ccc} = 1 \times 10^{-5} \frac{\varepsilon^3 a k^5 T^5 \gamma^4}{N_A e^6 A^2 z^6}$$

Eqn (4.2)

where:

C_{ccc} = the Critical Coagulant Concentration
V_T = total potential energy of interaction as a function of distance (H) between particles (J)
V_R = repulsive potential energy of interaction (J)
V_A = attractive potential energy of interaction (J)
a = diameter of the spherical particle (m)
H = distance from one Stern layer to the other (m)
A = Hamaker Constant (J)

Without going into details, this formula has the correct behavior. For example, the critical coagulation concentration is predicted to be proportional to $1/z^6$, where z is the electric charge of the cationic coagulant. Thus, the concentration of coagulant required to cause coagulation is predicted to be 1/729 times less for a coagulant that has a +3 charge compared to a coagulatnt that has a +1 charge. Note the critical coagulant concentrations listed in the table above for trivalent ions with (Fe^{+3}, Al^{+3}) compared to a coagulant that has only a +1 charge (e.g. Na^{+1}, K^{+1}).

4.4 Interfacial Tension, Elasticity, Wetting, Capillarity

In this section, the properties of interfaces are discussed. The material starts with a discussion of the properties of interfaces (Section 4.4.1). This material ties together Chapter 2 which discusses chemical composition of produced water, and Chapter 5 which discusses the various types of chemicals used to treat produced water. The properties of interfaces discussed in this section are the result of naturally occurring surface active compounds, and added process chemicals, some of which are highly surface active.

Surface and interfacial tension are discussed in Section 4.4.2. Interfacial tension has a dramatic impact on how produced water responds to shearing. Low interfacial tension results in small oil droplets which are difficult to remove from produced water. Further sections of this chapter discuss the fact that emulsion stability depends on other interfacial properties such as elasticity. Interfacial elasticity can be reduced, thus promoting emulsion breakdown, by adding specific types of surface active chemicals.

Surface wetting is then discussed. Wetting is relevant to not only solid/liquid interfaces, but also gas/solid and gas/liquid interfaces which are central to an understanding of flotation. As discussed in Chapter x, oil-wet solids can be treated by adding a surface active compound that releases the oil from the solid surface.

The concepts of surface and interfacial tension are important for the following reasons:

- produced fluids contain a high concentration of natural and manmade interfacially active compounds;

- there are several different kinds of emulsions encountered in the oil and gas industry; they differ in the composition and behavior of the oil / water interface; these differences can be exploited in order to break the emulsion;

- many of the compounds that are used to resolve emulsions make use of different interfacial mechanisms; these compounds have different characteristics that can are suitable for some emulsion types but not others.

The consequences of interfacial tension occur throughout the water treatment industry. Having a strong background on the subject will allow a water treatment specialist to identify problems as well as opportunities to solve problems.

4.4.1 What is an Interface?

In order to develop an intuitive understanding of a liquid/vapor interface, it is helpful to take a quick look at computerized molecular simulation results as shown in the Figure below. In molecular dynamics simulations the intermolecular interactions are specified and the pressure is calculated. From a molecular perspective, the interface is not abrupt. Rather than having a discontinuity in properties, the properties vary in a gradual manner. On the molecular level, the properties of density and composition fluctuate in time and in location across a dynamic region that spans several molecular diameters. Molecules in the interface region, like the molecules in the bulk fluid, are in constant motion. Molecules from one phase crash into the other phase. Over time, and if the system is at steady state, there is no net movement or exchange of molecules between the interface and the bulk regions. This is sometimes referred to as a state of dynamic equilibrium.

In the figure below [23], the density profile and pressure profile are shown. The y-axis on the right hand side gives the length scale in terms of molecular diameters. The molecules studied here are spherical and non-polar. As shown, the thickness of the interface is about 4 to 6 molecular diameters wide. With the addition of surface active molecules the thickness and tension of the interface will change.

To the right of the density profile is the pressure profile. Away from the interface, the pressure in the liquid equals that in the gas, as expected. But at the interface the pressure in the perpendicular direction goes through a maximum. This pressure is the interfacial tension.

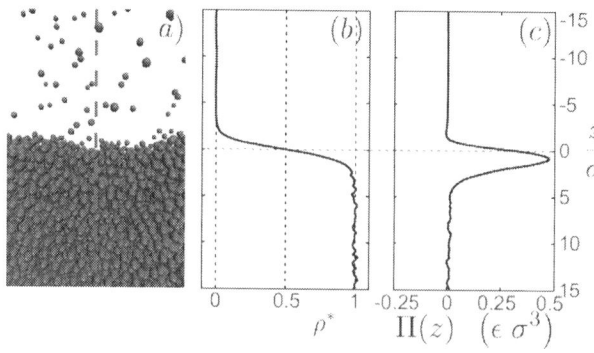

(a)
(b)
(c)
-15
-10
-5
0
5
10
15
z/σ
0
0.5
1
ρ*
-0.25
0
0.25
0.5
Π(z) (ε σ³)
FIG. 5: The liquid-vapor interface. The vertical axis is in units of the molecular scale σ. (a) Snapshot of Lennard-Jones simulation of a liquid-vapor interface. (b) Time-averaged normalized density profile ρ*(z) across the interface. (c) Tangential force per unit area exerted by the left part on the right part of the system. Technically speaking, the plot shows the difference Π = p_NN − p_TT between the normal and tangential components of the stress tensor.

Figure 4.14 Results of molecular dynamics simulations of a liquid/gas interface. The liquid and gas phases are shown on the left. The density profile is shown in the middle. The pressure is shown on the right. The peak in pressure is the surface tension. Results from [23].

The results above show that surface tension is a stress acting parallel to the surface. It is literally a tension along the surface. It is a consequence of the intermolecular forces and the way that molecules pack together at an interface. This is further illustrated in the figure below which shows that in the bulk fluid, the molecules are subject to interactions on all sides. However, near the surface, the interactive forces are only present from one side, thus leaving a net attractive force to be developed at the surface of the liquid. This net attractive force is responsible for the development of the stress or pressure spike described and illustrated above.

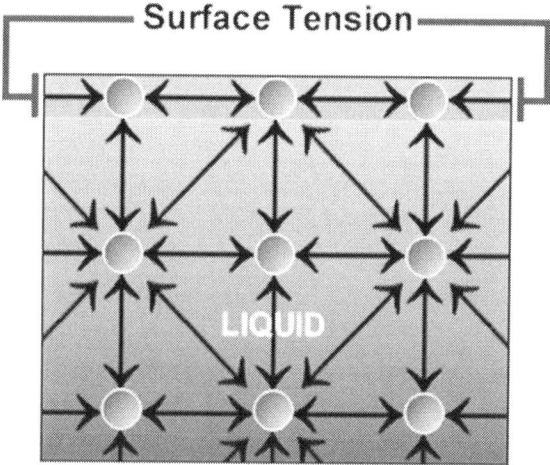

Surface Tension
LIQUID
rame-hart instrument co.

Figure 4.15 The modified intermolecular attractive forces within a liquid and at the surface of a liquid are illustrated. The change in these attractive forces at the surface of the liquid are responsible for the development of a "surface tension" or the liquid

The other concept that is illustrated clearly by the molecular simulation results is the fact that the density and pressure profile do not make a sharp transition from the gas to the liquid phase. They make a gradual transition. In other words, the interface has a finite thickness. It is absolutely essential to keep this in mind as the discussion of interfacial properties is developed below. In the rest of this section, and for that matter, in the rest of the book, the interface will be discussed as if it has essentially no thickness. This provides a very helpful simplification in the mathematics. But it does create a conceptual problem for engineers who are accustomed to thinking about pressure being applied to a finite area. Just to be clear, this is exactly what is happening in reality. So conceptually, this is correct. But for the sake of mathematical simplicity, the interface will be assumed to be infinitesimally thin and the pressure shown in the Figure above will be discussed as if it is a tension acting along this interfacial plane.

4.4.2 Surface and Interfacial Tension

Interfacial Tension – Force/Length or Energy/Area: In order to develop an accurate concept of interfacial tension, it is helpful to think of a particular situation such as an aerosol drop – a small drop of liquid suspended in a gas. Imagine that the gas / liquid interface has mechanical properties similar to those of a membrane. The membrane can be thought of as a thin film, like that of a balloon. A balloon is capable of sustaining a pressure difference between the inside (liquid) and the outside (gas). As will be explained shortly, the pressure inside of the drop is higher than outside the drop due to the curvature of the interface. From a continuum perspective, the higher pressure inside the drop is due to interfacial tension, and the curvature of the interface. If the balloon had a seam, say like a soccer ball, the interfacial tension is the force required to further open the seam. As the seam opens, the length of the opening is extended. *The force per unit length is the interfacial tension. Interfacial tension can also be thought of as the energy required to create a small increment of interfacial area.* If air was blown into the balloon and it expanded slightly then the energy required per unit area would be a measure of interfacial energy.

Surface and Interfacial Tension Units: Interfacial tension is expressed in various units. The most common set of units are dyne/cm. In the SI system, the most common set of units are N/m. There are 100,000 dyne per Newton. Thus, the conversion is:

1 dyne/cm = 1 mN/m (milli-Newton per meter)

In terms of energy per unit area, the appropriate units of interfacial tension are Nm/m^2 (J/m^2). Since a Joule (J) is equivalent to a Newton-meter (Nm), the equivalence of surface tension per unit length (N/m) or energy per unit area (J/m^2) is readily seen.

It would seem that there is a simple equivalence between the concepts of energy per unit area and tension per unit length. On a very simplified level, this is true. However, as will be discussed below, the relation between energy per unit area and tension per unit length is somewhat complicated. It requires a fairly detailed understanding of mechanics and thermodynamics. Suffice for now to appeal to physical concepts.

Surface / Interfacial Tension – Simple Measurement: Shown in the figure below is a schematic diagram of a simple experiment that, in the past, has been used to measure the interfacial tension of a thin liquid film. The force required to extend the surface is measured over a short distance. Actually, the force required to extend the film would be twice the surface tension since there are two sides to the film (two interfaces).

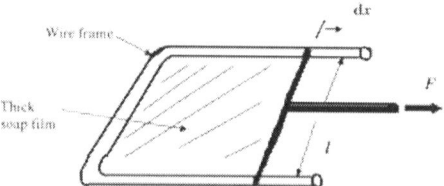

DERIVATION OF THE LAPLACE PRESSURE EQUATION   15

Figure 2.2 Diagram of a soap film stretched on a wire frame.

hence the term 'surface tension'. It is this tension that allows a water boatman insect to travel freely on the surface of a pond, locally deforming the skin-like surface of the water.

This simple experimental system clearly demonstrates the equivalence of surface energy and tension. The dimensions of surface energy, mJm⁻², are equivalent to those of surface tension, mNm⁻¹. For pure water, an energy of about 73 mJ is required to create a 1 m² area of new surface. Assuming that one water molecule occupies an area of roughly 12 Å², the free energy of transfer of one molecule of water from bulk to the surface is about $3kT$ (i.e. 1.2×10^{-20} J), which compares with roughly $8kT$ per hydrogen bond. The energy or work required to create new water–air surface is so crucial to a newborn baby that nature has developed lung surfactants specially to reduce this work by about a factor of three. Premature babies often lack this surfactant and it has to be sprayed into their lungs to help them breathe.

Figure 4.16 Sliding thing to measure surface or interfacial tension.

$$\gamma = 2F / l \qquad\qquad \text{Eqn (4.3)}$$

in which:

g   = interfacial tension (N/m)
F   = force required to extend the film a small amount (N)
l   = length of the film (m)

Alternatively, the energy could be measured and interfacial tension derived from:

$$\gamma = 2F\Delta x / (l\Delta x) = 2F / l \qquad\qquad \text{Eqn (4.4)}$$

where FDx is the energy required to create an interfacial area of 2lDx. The equivalence of force per unit length and energy per unit area is seen directly in this experiment and in the two equations used to calculate interfacial tension.

This is a useful experiment to consider not only because it gives a simple and direct way to think about interfacial tension but also because of the chemical effects involved. If the liquid film is composed of more than one chemical species, say water and a surfactant, then there is a concentration effect which must be accounted for. The composition of the surface of the liquid film will change as the slide is moved. As more surface area is created, the concentration of surfactant molecules per unit area will decrease. Thus, the measured surface tension will increase. A valid interfacial tension can still be measured. However, the measured value is only valid for a specified concentration of surfactant per unit area which must then be reported along with the measured interfacial tension. This will be discussed further below.

Curved Interface: Pressure Inside a Drop or Bubble (Young-Laplace Equation): Having discussed the pressure across a planar interface, the next subject is the pressure across a curved interface. Molecules at a gas / liquid interface experience a net attractive force in the direction of the bulk liquid. This is due to the cohesive forces between molecules in the liquid and is the case up to very high pressures. At very high pressures, and densities, the molecules are squeezed so closely together that their intermolecular forces become repulsive. But in general, at moderate pressures, molecules at a gas/liquid interface experience a net attractive pull into the liquid phase.

While there is always a net increase in cohesive energy across a gas/liquid interface (going from the gas side to the liquid side), there is not always an increase in pressure. The curvature of the interface determines whether the pressure increases, is the same, or decreases across an interface. If the interface is flat (no curvature), there is no pressure change across an interface. In that case, the cohesive force of the liquid occurs uniformly across the plane of the interface. In the case of a convex interface, the convex shape of the interface causes there to be an even higher cohesive force in the liquid than in a planar interface. The opposite is true of a concave interface.

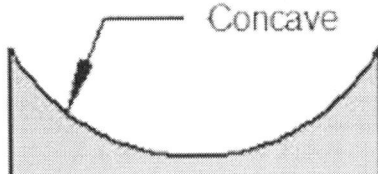 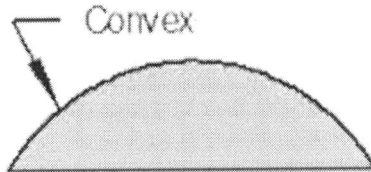

Figure 4.17 Concave and convex curvatures. The pressure inside a convex surface is higher than outside. The gray area is the liquid and the white area is the gas. The pressure inside a concave surface is lower than outside.

The equation that describes this pressure difference is known as the Young-Laplace Equation. For this simple system, it can be written as:

$$P_2 - P_1 = 2\gamma / R$$

Eqn (4.5)

in which:

P_2	= pressure in the liquid drop (N/m²)
P_1	= pressure in the gas vapor surrounding the drop (N/m²)
g	= interfacial tension (N/m)
R	= radius of the drop (m)

The Young-Laplace Equation is discussed in further detail below in terms of the thermodynamics of interfaces, and in terms of capillary rise. It is easily seen that if the interface is planar, R goes to infinity and $P_1 = P_2$.

Hydrocarbon / Water Interface: For the special case of a hydrocarbon liquid suspended as droplets in water, the following table is useful [11].

Table 4.6 Surface tensions and interfacial tensions against water for liquids at 20oC (in mN m⁻¹)

Liquid	γ_o	γ_i	Liquid	γ_o	γ_i
Water	72.8	-	Ethanol	22.3	-
Benzene	28.9	35.0	n-Octanol	27.5	8.5
Acetic Acid	27.6	-	n-Hexane	18.4	51.1
Acetone	23.7	-	n-Octane	21.8	50.8
CCl₄	26.8	45.1	Mercury	485	375

In this case, the molecular interactions are more complex than for a liquid in equilibrium with its vapor. The water molecules at the interface experience an attractive force from the water molecules in the bulk water. But they also experience an attractive force from the hexane molecules in the other side of the interface. Likewise, the hexane molecules at the interface experience an attractive force from the hexane molecules in the bulk hexane. They also experience an attractive force from the water molecules on the other side of the interface. All of these interactions can be taken into account mathematically and in doing so, additional insight is gained about the oil and water interface. As discussed in [11], Fowkes developed a correlation for the cohesive energy density of liquids.

Before the Fowkes correlation can be applied, a short discussion is required about the different types of inter-molecular forces that exist in nature. Water molecules have both dispersion and hydrogen bonding interactions. Hexane molecules have only dispersion interactions. A pair of molecules, one of hexane and one of water, will also have dispersion interactions. The magnitude of this cross-interaction (water-hexane) has been found to be roughly equal to the geometric mean (square root of the product) of the like-like (water-water and hexane-hexane) interactions.

$$\gamma_{ow} = \gamma_o + \gamma_w - 2\sqrt{\gamma_w^d + \gamma_o^d} \qquad \text{Eqn (4.6)}$$

in which:

γ_w = surface tension of water (J/m)
γ_{ow} = interfacial tension between oil and water (J/m)
γ_o = surface tension of oil (J/m)
γ_w^d = dispersion contribution to the surface tension of water (J/m)
γ_o^d = dispersion contribution to the surface tension of oil (J/m)

Substituting values from the table above gives:

$$\gamma_{ow} = 18.4 + 72.8 - 2\sqrt{21.8 + 18.4} = 51.1 \quad mJ/m \qquad \text{Eqn (4.7)}$$

The calculated value of the interfacial tension between hexane and water is 51.1 mJ/m. The units are milli-Joules per meter. This is essentially the same as the measured value reported in the Table above [11]. One of the principles that this calculation is intended to demonstrate is that the interfacial tension between two liquids has its origin in the direction and magnitude of forces between molecules.

Surface & Interfacial Tension Data: In general, the surface tension of a liquid in equilibrium with its vapor, or the interfacial tension between two liquids is a measure of the cohesive energy forces between the molecules. If the interface between two liquids has relatively low interfacial tension, then the molecules have similar cohesive energies. One demonstration of this interpretation is given in the Figure below.

The surface tension of aqueous solutions of acids is shown in Figure 4.19. As shown, the surface tension of the aqueous solution decreases as the concentration of acid increases. Furthermore, the surface tension decreases as the carbon number of the acid increases. This is because the non-polar part of the acid aligns with the gas phase above the solution. The nonpolar part of the organic acid has much less cohesive energy than water itself. Therefore, the more acid that is added, the less polar the interface and the lower the surface tension. The same effect would occur for an acid solution in water in contact with crude oil.

Furthermore, the surface tension decreases as the carbon number of the alcohol increases. The greater the polarity of the head group, or the longer (larger) the nonpolar group, the greater the reduction in interfacial tension. Physically, the acids reduce the disparity in cohesive energy between water and its vapor. The greater the disparity in cohesive energy between water and its vapor, the greater the surface tension. The less disparity, the lower the surface tension. The surface tension of aqueous solutions of alcohols shows a similar trend with respect to carbon chain length [11, p. 78].

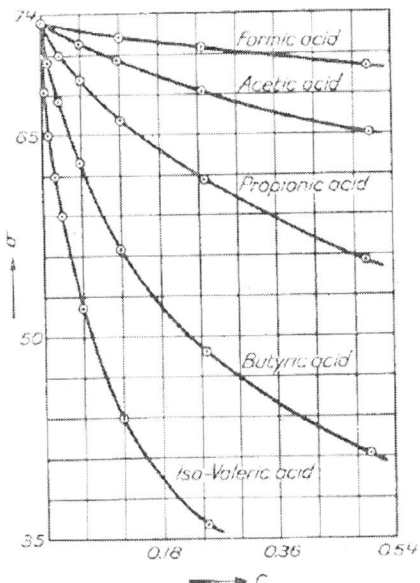

FIG. 18. Surface tension versus concentration (in moles/l) of fatty acids in water (σ–c plot). See TRAUBE's rule.

Figure 4.19 Surface tension as a function of the concentration for a number of organic acids. Data from [24].

Typical amine-based corrosion inhibitors display the same qualitative effect. However, these chemicals display a much greater quantitative effect than the organic acids shown here. Film forming amine corrosion inhibitors are designed to form a strong adherent film on the metal / water interface. This is discussed in the next chapter.

Very few compounds increase the surface tension or interfacial tension. Increasing salinity generally increases the surface tension (or the interfacial tension between water and oil). The ions of the dissolved salt actually increase the cohesive energy of the solution.

As discussed in Section 2.3.2 (Aging and the Thermodynamic State of Produced Water), Sjoblom and co-workers studied the aging of crude oil as indicated by the change in interfacial tension over time [25]. In [26] the oil / water interfacial tension was studied as a function of time and pH.

Pressure in a Thin Film - Disjoining Pressure: As just described, an interface is the boundary between two bulk phases. A thin film is comprised of two interfaces. Thin films are frequently encountered in the oilfield:

1. A thin film of water between two oil droplets. This is the very common case of oil droplet coalescence. In this case, the oil droplets are in the process of colliding with each other. In order for coalescence to occur, the water between the droplets must be squeezed out from between the oil droplets. The driving force is the turbulent energy of the fluid. The resisting force is the viscosity of the water. When the droplets get very close to each other, the surface and van der Waals forces become important.

2. A thin film of oil between two water droplets. This is the case when a High Internal Phase Volume (HIPV, also known as a biliquid foam) emulsion forms. The oil film must drain (upward) in order for the emulsion to resolve.

3. A thin film of water between an oil droplet and a bubble. This is the case in flotation where gas bubbles collide with and associate with oil droplets and capture and remove them.

The Figure below provides a general drawing that can be applied to all three cases. If we neglect the body force due to gravity, and if there is no net movement of the interface, the normal pressure (Pz – in the up and down direction) is the same in all three phases (β above the film, the film of liquid α itself, and the β phase below the film) and it does not change across the interface. If this were not the case, for some moment in time, then the fluid, or the molecules in the fluid (which amounts to the same thing) would move such that the pressure becomes constant in the z direction. In the bulk liquid β, the normal pressure (Pz) is also equal to the tangential pressure (Pz = Px = Py). This is referred to as isotropic pressure. This relation is not true for the thin film.

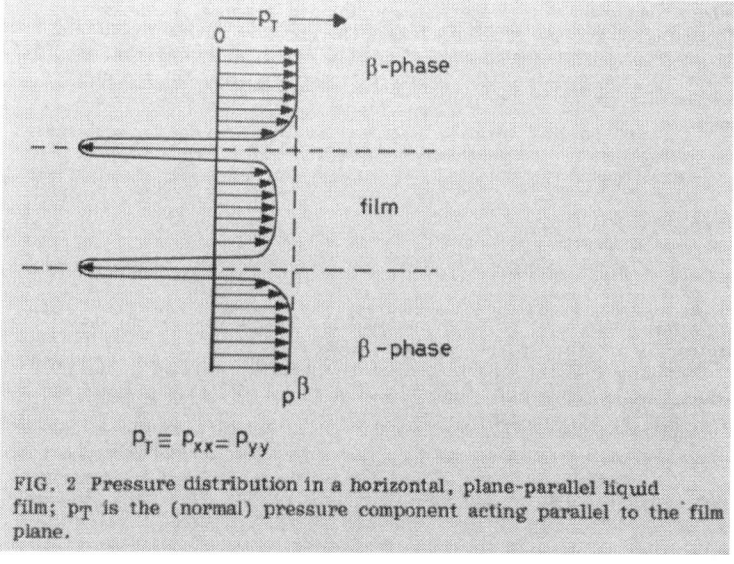

FIG. 2 Pressure distribution in a horizontal, plane-parallel liquid film; p_T is the (normal) pressure component acting parallel to the film plane.

Figure 4.20 Schematic drawing of the film that forms between two droplets or the film that forms in an HIPV (biliquid foam).

The tangential pressure (P_T) in the film is not the same as in the bulk, and the tangential pressure at the interface is also different, as shown in the Figure. Note that P_T it is a function of z. In other words, it varies across the interface and from the interior of the liquid film (α) to the bulk (β) phases. Note also that $P_T\beta$ (tangential pressure in the bulk β phase) is higher than $P_T\alpha$ (tangential pressure in the middle of the α film). This is the Disjoining Pressure which is due to surface and near-surface forces of the interfaces. As will be discussed momentarily, something must hold the film in place otherwise it would shrink to minimize its interfacial area and form a thicker bulk-like phase. This was the case above for the sliding force balance used to measure interfacial intension where a force was applied to stretch the film. Additional discussion is provided by Ivanov [27].

4.4.3 Surface Excess Concentration & Surface Activity

It is important to distinguish between two properties of surface active materials. The first property is the extent to which the compound migrates to the interface. In other words, how much or what fraction of the compound resides in the bulk liquid and what fraction resides at the interface. The second property is the impact of the surface active compound at the interface on other properties such as interfacial tension and the elasticity of the interface. These two properties tend to be related. If a compound has a strong tendency to migrate to the interface, it generally has a large impact on interfacial properties. However, different compounds vary widely in these two properties so it is necessary to distinguish between Surface Excess Concentration and Surface Activity. One strategy for breaking emulsions is to add a compound that has high interfacial concentration but does not cause interfacial elasticity and in fact may actually reduce it. Add what is essentially a surfactant in order to break an emulsion is a clever approach and is used extensively in produced water treatment.

Consider a mixture of two immiscible liquids that form two bulk phases with an interface in between. The mixture is in a closed container. The composition of the entire system is given by:

$$n_T = n_1 + n_2 + n_\sigma = c_1 V_1 + c_2 V_2 + c_\sigma A_\sigma \qquad \text{Eqn (4.8)}$$

in which:

n_T = total number of moles of molecules in the system (mol)
n_1 = number of moles of molecules in bulk phase 1 (mol)
n_2 = number of moles of molecules in bulk phase 2 (mol)
n_s = number of moles of molecules in the interface (mol)
c_1 = concentration of molecules in bulk phase 1 (mol/m³)
c_2 = concentration of molecules in bulk phase 2 (mol/m³)
c_s = concentration of molecules in the interface (mol/m²)
V_1 = volume of bulk phase 1 (m³)
V_2 = volume of bulk phase 2 (m³)
A_s = interfacial area of the system (m²)

This equation is basically a mass balance. It shows that the molecules at the interface are treated in a manner that is analogous to that in the bulk phases. In the case of an interface, the concentration of molecules is given in terms of molecules per unit area. Obviously, there must be some finite volume associated with the interface, otherwise molecules could not reside there. However, the thickness occupied by say two or three layers of molecules is so small that the thickness can be ignored for all practical purposes. This was discussed above in Section 4.4.1 (What is an Interface?).

Surface Excess Concentration: The Surface Excess Concentration is an experimentally measured quantity that indicates the tendency of a compound to migrate from the bulk phase to the interface. A compound with high Surface Excess will be most preferentially located at the interface. This is a useful quantity because it relates directly to chemical strategies for removing detrimental compounds from the interface and replacing them with compounds that are less detrimental. When this is successful, there is a greater tendency for oil droplets to coalesce.

When there is a high concentration of surface active compounds, as in most produced water, adsorption of surface active compounds to the interface is a competitive situation. Those compounds that have the highest Surface Excess are most likely to be located at the interface. Those compounds with lower Surface Excess, will be located in the bulk phase because there is not enough room at the interface to accommodate their presence and they are not by themselves surface active enough to create new interface. Certain chemicals are added to "push" detrimental compounds off of the interface. Low molecular weight organic acids are effective at pushing resins, naphthenic acids, and asphaltenes away from the interface. The later compounds create as elastic energy barrier to coalescence. But they also have a low Surface Excess and can therefore be pushed away from the interface. Low molecular weight organics acids are a good choice to do this because they have high Surface Excess but do not impact the interfacial tension to the extent of other compounds.

Surface Excess provides an explanation of why corrosion inhibitors, which are added at a few tens of ppm cause so much of a change in the interfacial properties compared to resins, naphthenic acids, and asphaltenes which may be present in much higher concentrations. There is no doubt that the later compounds can have a profound effect on water treatment, however, the concentrations of such compounds are orders of magnitude higher than that of corrosion inhibitors. A small amount of a film forming amine corrosion inhibitor can cause havoc due to the very high Surface Excess.

Concentration at the Interface: As discussed on pages 326 – 327 of [28], and in [2], the surface excess concentration of a surfactant can be determined by the following method. The surface tension or interfacial tension is measured and is plotted against the logarithm of the concentration. The slope is then divided by RT (gas constant times temperature in absolute units). The procedure just described can be written in the form of an equation:

$$\Gamma = -\frac{1}{RT}\left(\frac{d\gamma}{d\ln c}\right)_T$$

Eqn (4.9)

where

T	= absolute temperature (Kelvin)
R	= gas constant (8.3145 m^3 Pa / mol K)
c	= concentration of surface active compound (mol/m^3)
g	= the interfacial tension (N/m)
G	= surface excess concentration (mol/m^2)

To understand physically what Γ means, consider the following situation. Consider two samples taken. One sample is taken at the interface. The other sample is taken in the bulk phase away from the interface. In practice this requires a specialized technique. The surface excess concentration Γ is the number of moles of surfactant in the interface sample above and beyond what would be there if the interface was just like the bulk. Hence, the term "excess" in the name. The superficial concentration is the total moles of surfactant added to the solution divided by the total volume of solution. In this way it is not necessary to measure the amount of surfactant that remains in the bulk or the amount that migrated to the interface.

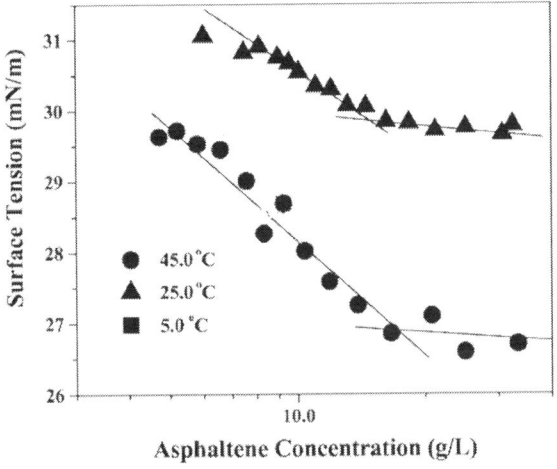

Fig. 2. Surface tension vs. C51 concentration in toluene at 5, 25 and 45 °C.

Figure 4.21 Surface Tension vs C51 concentration in toluene at 5, 25 and 45 °C

The Surface Excess has been measured for asphaltenes. The above data are from [29]. Values of surface and interfacial excess concentrations are given in the Table below. These results are from a number of sources so there may be some differences in the absolute values which will confound comparison. Nevertheless, some interesting comparisons can be made. For example, the commercial surfactants seem to have surface excess concentrations that are about ten times higher than that of the asphaltene and naphthenic acids. This agrees with empirical observations that commercial surfactants can be used to displace asphaltenes from the oil/water interface. When this is done, with moderate concentration of the surfactant, it usually leads to enhanced coalescence.

Table 4.7 Surface / Interfacial Excess at 25C

Substance	Surface or Interfacial Excess at 25 C (mol/m²)	Reference
C51 asphaltene (Fig 4.21)	5.1×10^{-7}	[29] interface with water
sodium naphthenate from bitumen	8.8×10^{-7}	[88] interface with water
asphaltene from bitumen	2.3×10^{-7}	[88] interface with water
sodium dodecyl sulfate	3.0×10^{-6}	page 81 of [30] surface excess
dodecyl ether of hexaethylene oxide	2.9×10^{-6}	page 329 of [28] surface excess
salt of fatty acid amine	13×10^{-5}	[89] interface between kerosene & brine
quaternary amine	7.6×10^{-5}	[89] interface between kerosene & brine
fatty acid amine	4.3×10^{-5}	[89] interface between kerosene & brine
isovaleric acid (i-C5)	8.3×10^{-3}	[24]

It is also interesting to note that the naphthenic acid fraction of bitumen has more surface activity than the asphaltene fraction from the same bitumen. This suggests that asphaltene content alone would not be a good predictor of coalescence barrier. Instead, the asphaltene to resin ratio is likely to be a better predictor with a higher value indicating poorer coalescence. Asphaltene stability is also poorer as this ratio increases. Thus, there are at least two mechanisms (interfacial rigidity and precipitation) that are likely to correlate with coalescence probability.

The surface activity of isovaleric acid is high compared to the interfacial activity of the naphthenic acid. Small alkyl carboxylic acids are indeed highly surface active compared to oil-soluble naphthenic acids. In an oil/water mixture, isovaleric acid partitions to the water phase rather than the oil phase. Conversely, the naphthenic acids partition to the oil phase. The naphthenic acid will help stabilize a water-in-oil emulsion. Conversely, the isovaleric acid, and other short chain acids, will help stabilize an oil-in-water emulsion. In typical crude oil / produced water mixtures, the oil contains more water (usually in percent quantities) than the water contains oil (usually hundreds to a couple thousand ppm). This is despite the higher interfacial activity of the short chain acids and is due to the much higher concentration of naphthenic acids.

The Surface Excess highlights the fact that "surface activity" is a rather broad term that encompasses a number of different characteristics. The surface excess concentration measures the tendency of a surfactant to migrate to the interface. It does not measure the interfacial tension, interfacial viscosity or elasticity that a constituent contributes at the interface. In fact, as mentioned previously, one way to bring about coalescence is to add a surface active compound that has high surface excess concentration but which has low interfacial viscosity and elasticity. Such a compound can be used to displace other compounds that have low surface excess concentration but which have high viscosity and elasticity such as asphaltenes and resins. In this way, it is possible to reduce the effect, for example, of asphaltenes on the interface.

It should also be mentioned that asphaltenes do not necessarily reduce the interfacial tension. Sheu and Storm [31] measured the interfacial tension of various produced water / hydrocarbon systems as a function of time after the addition of asphaltene. What they found is that interfacial tension initially decreased but then increased over time. It was speculated that the increase in interfacial tension was due to structural arrangement of the asphaltenes at the interface and the formation of a network of interlocking / coordinating asphaltene molecules at the interface. This is discussed further below in relation to coalescence.

4.4.4 Interface Concentration versus Bulk Concentration

In an oil/water processing facility, oil is contaminated with water droplets, and the produced water is contaminated with oil droplets. Thus, there are two oil/water interfaces of importance: one in the oil phase and the other in the water phase. One of the questions that can be asked about these two interfaces is the following. Is there a difference in the oil/water interface in the water phase versus the oil/water interface in the oil phase? The quick answer is that yes, there is a significant difference which depends on the concentration and types of the surface active components in the oil versus the water phase.

A related question that can be asked is the following. Given that oil often contains percentage concentrations of resins and asphaltenes, to what extent do these compounds impact the properties of the oil/water interface of oil droplets suspended in the water phase? It is well known that the resins and asphaltenes have a stabilizing effect on the oil/water interface of water droplets in the oil phase. The practical consequence of this is that water in oil emulsions in a heavy crude oil are difficult to break. This is the subject of numerous studies. However, in those studies there is typically no mention of the effect that these compounds have on the oil/water interface in the water phase. Given the subject of this book, it is worthwhile to understand the impact of these compounds (resins and asphaltenes) to the oil/water interface in the water phase.

As mentioned, there are two interfaces of importance: the oil/water interface in the oil phase (which surrounds the water droplets suspended in oil) and the oil/water interface in the water phase (which surrounds the oil droplets suspended in water). In the next three pages, an approximate, but nevertheless useful concept of these interfaces is introduced. In order to do so, certain simplifying assumptions are made. It is assumed that all of the surface active compounds can be grouped into the following categories:

Table 4.8

Category	Example compounds	Range of concentrations
oil soluble surface active	naphthenic acids, resins, asphaltenes	1,000 to 20,000 mg/L
water soluble surface active	short chain acids, phenol compounds, other alcohols	100 to 2,000 mg/L

Water Droplets in Oil: Numerical examples will help to show the relative orders of magnitude. Consider a Suspension of water droplets in water. The concentration of surface active compounds will be calculated for water droplets suspended in oil. For the sake of illustrative calculations, it is assumed that the concentration of water in oil is 2 volume percent. Also, it is assumed that the water droplets are 400 micron in diameter; the water is assumed to have 500 mg/L of surface active compounds such as short chain acids; the oil is assumed to have 2 % of surface active compounds such as resins and asphaltenes.

Table 4.9 Calculated surface concentration of interfacially active compounds from the oil and water phases for water droplets suspended in oil. The bottom two lines are the surface concentration expressed in milli-grams of interfacially active material per unit of interfacial area. These results should be compared with those in Table 4.10 below for oil droplets suspended in water.

Water Droplets in Oil		
Concentration of water in oil	2.0	V%
volume of solution	0.001	m³
volume of water	2.00E-05	m³
volume of oil	0.001	m³
Concentration interfacially active material in water	500	mg/L
Concentration interfacially active material in oil	20000	mg/L
Droplet area and volume		
Water Droplet diameter	400	Micron
Water Droplet diameter	4.00E-04	m
Water Droplet volume	3.35E-04	m³/drop
Water Droplet interfacial area	5.03E-07	m²/drop
Interface area / Volume of water		
Number of water drops	5.97E+05	
Oil / Water interfacial area	300	m2 area
Surface Concentration of Interfacially active materials		
Surface Concentration of Interfacially active materials from water	3.33E+01	mg S/ m²
Surface Concentration of Interfacially active materials from oil	6.53E+04	mg S/ m²

Based on these assumptions, the interfacial concentration of compounds has been calculated as a function of the surface area. This result is expressed as mg of surface active material per square meter of interface. The absolute value of these surface concentrations is not important since slightly different assumptions will give different results.

As shown in Figure 4.21, the interface is populated with material from the oil phase to a much greater extent than it is populated with material from the water phase. Thus, the oil/water interface is populated with asphaltenes and resins to a much greater extent than it is short chain acids. This explains why so many of the oil dehydration studies ignore the water chemistry. This is not to say that water chemistry is completely insignificant in dehydrating oil. But the concentration of surface active compounds from the water phase is small compared to that of the resins and asphaltenes due to concentration and surface area differences.

Oil Droplets in Water: In the next figure, the concentration of surface active compounds will be calculated for oil droplets suspended in water. It is assumed that the concentration of oil in water is 100 ppmv. Also, it is assumed that the oil droplets are 100 micron in diameter. The concentration of surface active compounds in the water and in the oil is the same here as for the above case. This was assumed in order to allow direct comparison.

Table 4.10 Surface concentration of interfacially active compounds from the oil and water phases for oil droplets suspended in water. The bottom two lines are the surface concentration expressed in milli-grams of surfactant per unit meter of interfacial area. These results should be compared with those in Table 4.9 above for water droplets suspended in oil.

Oil Droplets in Water		
Concentration of oil in water	100	ppmv
volume of solution	0.001	m^3
volume of water	0.0010	m^3
volume of oil	1.00E-07	m^3
Concentration interfacially active material in water	500	mg/L
Concentration interfacially active material in oil	20,000	mg/L
Droplet area and volume		
Oil Droplet diameter	100	Micron
Oil Droplet diameter	1.00E-04	m
Oil Droplet volume	5.24E-13	m^2/drop
Oil Droplet interfacial area	3.14E-08	m^3/drop
Interface area / Volume of water		
Number of oil drops	1.91E+05	
Oil / Water interfacial area	0.0060	m2 area
Surface Concentration of Interfacially active materials		
Surface Concentration of Interfacially active materials from water	8.33E+04	mg S/ m^2
Surface Concentration of Interfacially active materials from oil	3.33E+02	mg S/ m^2

Based on these assumptions, the interface is populated with material from the water phase to a greater extent than it is populated with material from the oil phase. Thus, the oil/water interface is populated with short chain acids from the bulk water phase. It is important to note however, that the amount of oil-soluble surface active compounds at the oil/water interface is not insignificant.

These calculations were based on a number of assumptions. Every effort was made to use realistic values. Other values could of source be used but the general conclusions will be similar. These are:

1. **Water-in-Oil:** The oil/water interface for water droplets suspended in oil is dominated by oil-soluble surface active compounds (resins and asphaltenes), and not by water-soluble surface active components.

2. **Oil-in-Water:** The oil/water interface for oil droplets suspended in water is dominated by water-soluble surface active compounds (short chain acids, phenols, etc) but is also influenced by a significant concentration of oil soluble components (asphaltenes and resins).

These calculations should be viewed only as illustrative due to the fact that they are based on assumed values. Nevertheless, they reflect typical oilfield experience as summarized in the four statements listed above. One of the main conclusions from this exercise is that resins and asphaltenes will be discussed throughout this book because they have a significant impact on both water-continuous emulsions and oil-continuous emulsions. Their effect on the oil/water interface will be analyzed, and chemical treatment to counteract their effect will be discussed in some detail.

4.4.5 Bancroft Rule

In this section the Bancroft Rule is presented. The Bancroft Rule allows identification of which type of surface active agents (surfactants) are responsible for the stability of an emulsion. This helps in the selection of water and oil treatment chemicals and in identifying which mechanical equipment is most likely to perform well in the resolution of the emulsion. The practical application of this theory will be presented in Chapter 5 on chemical selection and bottle testing.

A sample comprised of a 50 / 50 mixture of oil and water, with surface active agents will, when shaken, have a certain volume of oil-in-water ($V_{O/W}$) emulsion and a certain volume of water-in-oil ($V_{W/O}$) emulsion. In the former type of emulsion, water is referred to as the continuous phase with oil droplets and lamella suspended in the water. In the later, oil is the continuous phase. The ratio of $V_{O/W}$ / $V_{W/O}$ will have some initial value immediately after shaking, and will evolve with time. If the mixture is allowed to settle, each of the two emulsions will resolve to some extent. Films that have formed between droplets will drain and droplets will coalesce. As they resolve, a bulk water phase may form or a bulk oil phase may form. The rate at which this occurs will, in general, be different for the two emulsions. Over time, one of the emulsions may disappear altogether leaving behind a bulk phase of water or oil. The Bancroft rule helps to predict which of the two emulsions is likely to resolve more quickly than the other.

As discussed by Davies [32], the Bancroft Rule states that the phase in which the surface active compounds are most soluble will be the continuous phase. The Bancroft rule helps to predict the ratio of $V_{O/W}$ / $V_{W/O}$. Another way of saying this is that the Bancroft Rule allows the identification of which surface active compounds will stabilize an oil-in-water emulsion versus a water-in-oil emulsion. This is illustrated schematically in the figure below.

As shown in the figure, most of the surfactants will migrate to the various oil/water interfaces. But some of the surfactants will dissolve in the bulk phases (bulk water, and bulk oil). With time, one of the emulsions is likely to resolve. The remaining emulsion will either be an oil-in-water emulsion or a water-in-oil emulsion. The Bancroft rule states that an oil-in-water emulsion (continuous water phase with dispersed oil droplets) will be more stable if the surface active agents are water soluble. Conversely, a water-in-oil emulsion will be more stable if the surface active agents are oil soluble.

Bancroft Test: The Bancroft Rule has a useful practical application. Whenever an emulsion is encountered in the field it is useful to know which phase is the continuous phase. Once that is known, it is much easier to break the emulsion. Common emulsions that are encountered include flow back fluids from a well stimulation, pads (interface emulsions) in separator vessels, Slop Tank (rejects receiving and handling tanks) fluids. Applying the Bancroft Rule is straightforward. Centrifuge tubes or graduated cylinders are partially filled with the emulsion. Various solvents are added at a concentration of 1 to 5 %. Solvents should include: hexane, xylene, diluted acetic acid, brine, diluted methanol. If one of the oleophilic solvents breaks the emulsion, then oil is likely the continuous phase. If one of the aqueous fluids breaks the emulsion, then produced water is likely the continuous phase. This test is simple and direct. It is referred to as the Bancroft test.

Samples of the emulsion can be further analyzed using microscopy and by determining the phase volume of the oil and water phases after breaking the emulsion. Once the emulsion is broken various analytical methods can be used to determine the presence of scale forming materials, corrosion inhibitor, hydrate inhibitor, mutual solvent and other contaminants that may be responsible for stabilizing the emulsion. In the case where the oil phase is the most contaminated, then the surfactants might include higher molecular weight carboxylic and naphthenic acids, resins, asphaltenes, and some of the higher molecular weight corrosion inhibitors. It is always useful to identify which phase is preferred by the corrosion inhibitor and then carry out the Bancroft test. In practice solvents can and are sometimes used in the field to break the emulsion although such pure solvents are never used for various reasons such as low flash point.

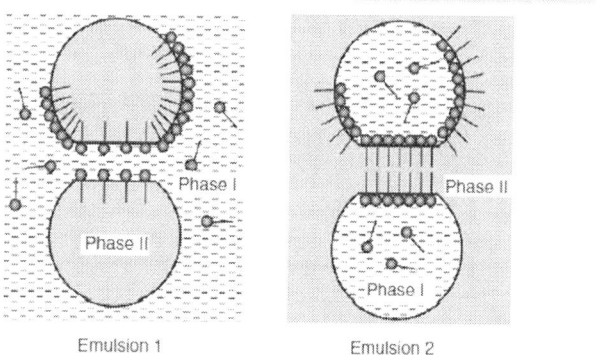

FIGURE 7.54 The two possible types of emulsions obtained just after the homogenization; the surfactant is soluble into phase I.

Figure 4.22 Two types of emulsion: on the left is an oil-in-water emulsion. On the right is a water-in-oil emulsion. The surfactant is the same in both. It is more soluble in the water than in the oil. According to the Bancroft rule, the emulsion on the left (oil-in-water) will be more stable because it is the emulsion where the surfactant is more soluble in the continuous phase.

4.4.6 Hydrophilic – Lipophilic Balance (HLB)

Davies [32] provided a quantitative method for estimating the solubility of various surface active compounds. The approach that he used is referred to as the HPB (Hydrophilic – Lipophilic Balance) method. It is a group contribution method. The net result is an HLB Number. If the HLB Number is low, the compound is more oil soluble (lipophilic) than water soluble. If the HLB Number is high, the compound is more water soluble (hydrophilic) than oil soluble. Once the solubility of the surfactant has been identified, then the Bancroft rule can be used to determine whether the produced fluids will have contaminated oil or contaminated water.

Table 4.11 Classification of emulsifiers according to HLB Values

Range of HLB valves	Application
3.5-6	W/O emulsifier
7-9	Wetting agent
8-18	O/W emulsifier
13-15	Detergent
15-18	Solubilization

Table 4.12 HLB Group Number

Hydrophilic Groups	Group Number
$-SO_4-Na^+$	38.7
$-COO-K^+$	21.1
$-COO-Na^+$	19.1
N (Tertiary amine)	9.4
Ester (Sorbitan ring)	6.8
Ester (free)	2.4
-COOH	2.1
Hydroxyl (free)	1.9
-O-	1.3
Hydroxyl (sorbitan ring)	0.5
Lipophilic Groups	
-CH-	
$-CH_2-$	- 0.475
CH_3-	
=CH-	
Derived Groups	
$-(CH_2-CH_2-O)-$	+ 0.33
$-(CH_2-CH_2-CH_2-O)-$	- 0.15

Table 4.13

Surface Active Agent	HLB from expt. 13, 15, 16, 30-32	HLB from Group numbers of Table III
Sodium lauryl sulphate	40	40
Sodium Oleate	18	18
Tween 80	15	15.8
Methanol	-	8.4
n-Butanol	7.0	7.0
Span 80 (sorbitan 'mono-oleate')	4.3	5.0
Glycerol monostearate	3.8	3.7
Cetyl alcohol	1	1.3
Oleic acid	1	1

4.4.7 Gibbs Elasticity

Gibbs Elasticity is a measure of the elasticity of the interface. Elasticity of an interface may seem like an unexpected property. After all, a liquid / liquid interface should itself be liquid, having been formed between two liquid phases. Generally speaking, our intuitive view of common liquids is that they have viscosity but not elasticity. On the other hand, there are many liquids that do have some elasticity. This is particularly true of solutions of water soluble polymers in water. When a solution of such polymer is sheared, it deforms by some combination of flow and stretch. When the shearing is stopped, the fluid tends to recover some of the deformation caused by the shearing. This is what is meant by elasticity. The same conceptual model can be applied to a liquid / liquid interface. When the interface is sheared by the movement of fluid adjacent to the interface, the interface will deform by a combination of stretching and flow.

When visualizing how this might occur, it is helpful to keep in mind that the interface has finite thickness. While the conceptual model of an infinitesimally thin interface is useful, is only a model and does not fully reflect reality. Thus, it is possible to conceive of interfaces as having elasticity. In fact, as will be discussed later, in produced water systems, the oil/water interface that forms around oil droplets can have a wide range of rheological properties from fully fluid, to highly elastic, to very rigid.

The Gibbs Elasticity is a measure of the increase in interfacial (or surface) tension caused by a small increase in the interfacial (or surface) area [33]. It is determined by measuring the interfacial tension as a function of the logarithm of the area of the interface. The slope of the line is related to the Gibbs Elasticity.

$$E_G = 2\left(\frac{d\gamma}{d\ln A}\right)_T$$

Eqn (4.10)

where

E_G = Gibbs Elasticity (N/m)
A   = area of the interface (m²)
g   = the interfacial tension (N/m)

In the equation, the factor of 2 reflects the case of a thin liquid film surrounded by a second bulk liquid phase. Such a film has two interfacial surfaces. The Surface Dilational Modulus is equal to $E_G/2$ [33]. Measured values of the surface modulus for the system of ethylene oxide surfactant in water are reported [33] in the range of about 15 to 30 mN/m.

The Gibbs-Marangoni effect is a change in surface tension as the concentration of surface active material changes. This change in surface tension can exist from one part of an interface to another. When that situation occurs, the gradient in concentration of surface active compounds gives rise to a gradient in surface tension. This gradient in will create an elastic force on the interface, according to the equation above for Gibbs Elasticity. This effect is commonly seen as the "tears of wine" effect on the side of a wine glass [34]. Oilfield corrosion inhibitors cause this same effect.

4.4.8 Surface Wetting

Speaking in general terms, wetting is the displacement of one fluid on a surface by the spreading of another fluid. The term wetting is a practical term that is easily understood, at least in general terms. However, by providing a precise definition, it is possible to make predictions of the surface coating and thus interfacial properties of solids in produced water. That is the goal of this section.

Surface wetting is important in practically every aspect of produced water treating. The surface properties of solids for example determines if they move to the oil/water interface or stay in the bulk oil phase. Three types of wetting shall be considered:

- spreading

- adhesion

- immersion

Contact Angle: Contact angle provides a simple measurement of the affinity of various solids and various liquids, as well as the affinity of gas bubbles for oil droplets suspended in produced water. Whenever contact angle is reported, a certain convention is followed. The measured angle is always on the inside of the droplet, as shown in the figure below. There are two kinds of contact angle of important in produced water treatment:

- Droplet of oil on a surface of produced water, both of which are surrounded by gas. This is a simple experiment that provides insight into the tendency of gas bubbles to stick to an oil droplet. This has relevance to gas flotation.

- Droplet of liquid on a solid both of which are surrounded by another liquid. The two liquids must be immiscible (oil and water). This experiment provides insight into the tendency of the solid to become wetted by the test liquid. Solids that are partially wetted by crude oil can be surface active and can stabilize oil-in-water emulsions.

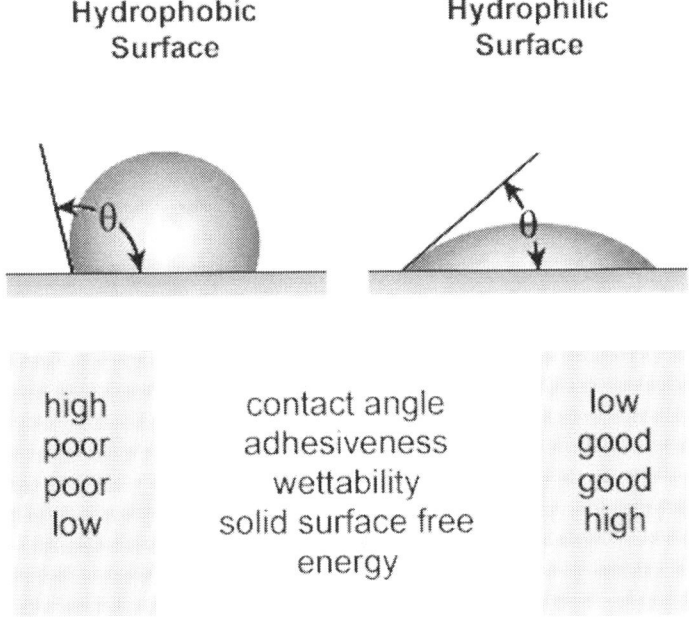

Figure 4.23 Convention used to measure and report contact angle. As shown, contact angle is measured "within" the drop. In this case, the liquid droplet is water hence the use of the terms hydrophilic and hydrophobic.

A further convention is used in describing the surface. When the droplet is water, and the contact angle is less than 90 degrees, the surface is said to be hydrophilic. Likewise, if the contact angle is greater than 90 degrees, the surface is said to be hydrophobic. The contact angle, in the system air / water / solid, was measured for several different solid materials. Note the hydrophobicity for the mineral precipitates $CaCO_3$, $BaSO_4$, and FeS. This property explains why these minerals tend to be oil coated and contribute to the stabilization of oil/water emulsions.

Spreading: A precise definition of spreading is given in the figure below. In this case, a liquid drop is resting on a surface. There are really two fluids in contact with the surface, the liquid drop and the vapor. The system is in phase equilibrium such that the liquid droplet, its vapor, and the solid surface are in equilibrium. A certain amount of vapor has adsorbed on the surface such that the vapor is in equilibrium with this adsorbed layer.

If the drop spreads by a small amount say ΔA, then additional solid / liquid surface is created; liquid / vapor surface is created; and solid / vapor surface is lost. In order for this process to be spontaneous, according to the second law of thermodynamics, the Gibbs Energy must decrease as a result of the spreading. Mathematically this is described by:

$$dG = \gamma_{SL}\, dA + \gamma_{LV} \cos(\theta)\, dA - \gamma_{SV}\, dA \qquad\qquad \text{Eqn (4.11)}$$

in which:

G = Gibbs Energy of the system (J)
dA = small change in the area of liquid / contaminant contact (m^2)
g_{SL} = interfacial tension between the contaminant surface and the liquid (J/m^2)
γ_{LV} = interfacial tension between the liquid and the vapor (J/m^2)
g_{SV} = interfacial tension between the contaminant surface and the vapor (J/m^2)
q = contact angle (degrees)

If the Gibbs Energy is a minimum, as required by the second law, then $dG = 0$, and the following equality results:

$$\gamma_{SV} - (\gamma_{SL} + \gamma_{LV} \cos\theta) = 0 \qquad\qquad \text{Eqn (4.12)}$$

This equation applies at equilibrium where further spreading does not take place because the system has achieved equilibrium. One strategy in water treatment, is to manipulate g_{SL} in order to achieve the desired contact angle and spreading behavior. Translated to field performance, it is common to use production chemicals in an attempt to make solids preferentially oil-wetted or water-wetted in order to improve water quality. These attempts are not always successful because of the variation in the hydrophobicity (contact angles) for the surfaces of the several mineral phases which are likely to be present. It is generally not typical to manipulate g_{SV} and γ_{LV}. They tend to be fixed.

The physical understanding of this equation requires some effort. Recall from the previous discussion that interfacial tension is a measure of the disparity of cohesive energy forces across the interface. If the interfacial tension between a solid and liquid (g_{SL}) is small then there is relatively little disparity in the molecular interactions. In that case, the interactions between liquid molecules are similar to those between solid molecules. Likewise, g_{SV} and γ_{LV} would be expected to be similar in value. After all, if the liquid and solid have similar molecular interactions with each other then it is likely they would have similar interactions with the vapor molecules. The two dominant terms in Eqn 4.12 would be g_{SV} and γ_{LV}. Looking at Eqn 4.12, mathematically, the contact angle would be expected to be small and $\cos(\theta)$ would be expected to be close to unity. A small contact angle implies strong solid/liquid interaction and a tendency for the liquid to spread on the solid. This physical reasoning is consistent with Eqn 4.12.

Considering an alternative case, if the interfacial tension between the solid and the liquid is large, then the opposite circumstance compared to the one above would be true. In this case, g_{SL} is large. It is likely g_{SV} is also large, which requires the contact angle to also be large which means that $\cos(\theta)$ is small and the water wettability of the solid surface will be low.

Flotation Collector Oils: In the flotation of solids, it is desirable that the gas wets the solid surface of the particle. One way to promote this is to add a substance called a collector oil. As will be discussed in the section on flotation, the use of collector oils has a significant benefit on the collection efficiency of solid particles in the flotation process. Most collector oils are amphiphilic, i.e. they have a polar head and a nonpolar tail. But some are non-polar. Collector oils can be anionic, cationic, or non-ionic (e.g. alcohol ethoxylates). Organic xanthenes and thiophosphates are used as collector oils for sulfide ores (e.g. iron sulfide). Long chain fatty acids are used for oxide and carbonate ores. In general, the most successful collector oils for a particular ore are those that have an affinity for a particular mineral of interest. The polar head of the collector oil adsorbs on the surface of the target mineral leaving the nonpolar tail to create a hydrophobic surface. The hydrophobic surface increases the contact angle between the solid surface and water. This increases the tendency of gas bubbles to attach to the target mineral and carry the particles to the surface.

It is worthwhile to apply Eqn 4.12 to understand the mechanism of collector oils. In flotation, it is desirable that the gas / contaminant interaction dominate over the liquid / contaminant interaction. In the best case, the contaminant "sticks" to the gas bubble. Based on physical reasoning, a large contact angle (small value of $\cos\theta$) would be desired. As mentioned previously, it is common in water treatment to manipulate g_{SL} in order to achieve the desired contact angle and spreading behavior. As discussed by Shaw [11, p. 162], this is what a collector oil does. It coats the surface of the solid increasing the disparity of cohesive energy forces between water and the solid. This increases γ_{LV} which, as illustrated above, will decrease $\cos(\theta)$ and increase the contact angle. This makes spreading less favorable and allows greater contact between the gas and the surface.

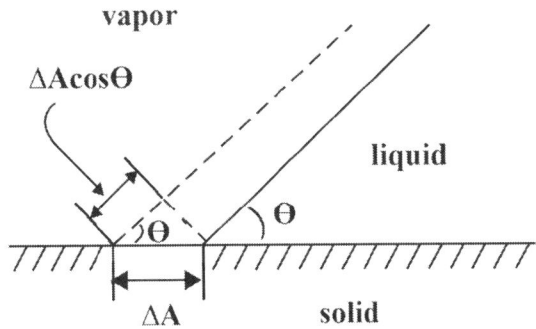

Figure 4.24 Contact angle in the solid / liquid / gas system.

Spreading Coefficient: A quantity referred to as the Spreading Coefficient can be defined from the above analysis.

$$S_c = \gamma_{SV} - \left(\gamma_{SL} + \gamma_{LV} \right)$$

Eqn (4.13)

in which:

$\quad S_c \qquad$ = Spreading Coefficient (dimensionless)

Table 4.15 Initial spreading coefficients (in mN m-1) for liquids on water at 20oC54 (By Courtesy of Academic Press Inc.)

Liquid	$\gamma_{WA} - (\gamma_{oA} + \gamma_{ow}) = S$	Conclusions
n-Hexadecane	$72.8 - (30.0 + 52.1) = -9.3$	will not spread on water
n-Octane	$72.8 - (21.8 + 50.8) = +0.2$	will just spread on pure water
n-Octanol	$72.8 - (27.5 + 8.5) = +36.8$	will spread against contamination

As suggested in the above equation, if the Spreading Coefficient is positive, then the liquid will spontaneously spread over the surface. The Spreading Coefficient will be applied later in flotation. There it is used to determine if an oil drop will spread around a gas bubble. Such contact and spreading greatly increases the capture efficiency of oil drops in flotation, though oil spreading over a gas bubble is not the only mechanism in flotation.

As discussed by Adamson [30], the height of a meniscus above a pool of liquid depends on the contact angle of the liquid on the pore surface, and the radius of the pore. This is particularly relevant to oil/water wetting of the pores in a hydrocarbon reservoir. It is also relevant to the capture of contaminants in media bed filtration.

Wetting of Produced Solids: Much of the previous discussion in this section has provided background for understanding oil and water wetting of produced solids. One of the most common wetting mechanisms is that of asphaltene, wax and organic acid adsorption onto sand, clay, iron compounds (carbonates, sulfides, oxides), and mineral scale particles. Experiments carried out using crude oil, and various crude oil fractions dissolved in aromatic solvents, have shown that asphaltenes can indeed adsorb on the solids just mentioned. Before discussing these wetting studies, it is useful to review the broad subject of solids contaminants in production and water treatment systems.

Types of Solids in Production Systems: Before going further, it is worthwhile to provide an important distinction regarding various types of solids encountered in hydrocarbon production. This is just a brief review since the subject of solids is discussed in Chapter 18 (Applications – Solids).

From a water treatment perspective, there are four basic types of solids. They are categorized based on their interfacial properties:

1. oil-wet solids that primary come from the oil bearing part of the reservoir and are composed of clay, sandstone, carbonate and shale fines; also included in this category are the precipitated wax and asphaltene particles that remain oil wet;

2. water-wet solids that can either come from the water bearing part of the reservoir, or have been formed during the production process; these solids include minerals from the reservoir, minerals that precipitate during production, and corrosion products;

3. water-wet organic solids and interfacially active organic solids which include the lower molecular asphaltene and resin compounds, production chemicals; in hydraulic fracturing, the dominant form of solids are proppant and friction reducing polymer (HPAM, HPAM-AMPS, guar, gelled guar).

4. partially oil-wet solids; these solids are referred to as partially oil-wet but they are also therefore partially water wet; they are surface active, and are formed initially as water-wet solids (formation material or precipitated minerals) which have asphaltene, wax, resin and organic acids, and certain production chemicals (corrosion inhibitor and hydrate in-

hibitor) precipitated onto their surfaces; due to the presence of both oil and water wetted surfaces these particles tend to be surface active and can be a major problem in the water treatment system;

Oil-wet solids that remain oil wet and water-wet solids that remain water wet are both, generally speaking, not detrimental to water quality for overboard discharge. If an oil wet solid remains oil wet it will usually flow with the oil and not cause an oil dehydration problem or a produced water treatment problem. Other problems may result such as erosion or settling in the storage tank or oil pipeline. Also, water-wet solids that remain water-wet will usually be discharged overboard without causing problems in the quality of the discharge water.

Water wet solids that become partially oil wet can be a significant problem because they migrate to the oil / water interface and stabilize oil in water emulsions. Also, they are can be neutrally buoyant (or almost neutrally buoyant) which makes primary separation less effective. This section discusses the chemistry of the solids surfaces and the tendency of asphaltenes, waxes, and organic acids to stick to that surface. Most of the studies discussed here are adsorption studies, carried out in the lab in order to study surface characteristics. As such, they describe the process by which solids surfaces become coated with oil components. This is one of the mechanisms by which water wet solids become transformed to partially oil wet. Another mechanism is the collision of solids particles with precipitated organics particles. Due to the surface characteristics described here, such collisions can result in sticking of the mineral solid to the organic solid. When this occurs, surface active and neutrally buoyant particles can be formed. As stated many times already, this is a difficult problem to deal with. With this in mind, it is now appropriate to discuss adsorption studies of asphaltenes and various solids.

As discussed in Section 2.6.4 (Resins and Asphaltenes), asphaltenes are defined as the fraction of crude oil that precipitates upon addition of excess n-pentane, n-hexane, or n-heptane. Asphaltenes are sometimes referred to as a solubility class. They are typically composed of large molecules with large planar fused aromatic ring structures, with polar and ionizable functional groups imbedded in the molecular structure as part of the fused aromatic ring structure or as pendent groups. This heteroatom functionality promotes the interaction of asphaltenes with charged surfaces such as those just mentioned [35]. Wetting of solid surfaces by asphaltene has several mechanistic effects on the stability of oil-in-water dispersions:

1. Asphaltenes can partially wet the surface giving the solid particle some surfactant-like character. When such a particle migrates to the oil/water interface, it can act as a barrier to oil droplet coalescence.

2. Asphaltenes can combine with solids to form a high viscosity sticky mass. Such conglomerate material (as it is called here) can combine with other solids that would otherwise float through the system. This material can stick to pipes, pump internals, valve internals, and internal devices in water treatment equipment (such as the plate material in a corrugated plate interceptor).

Asphaltene adsorption on clay has been studied [36] with and without the presence of water. Crude oil was provided by the Mobil Oil Company sampled from the South Belridge field, southern California. The asphaltene fraction of the crude oil was precipitated and purified by typical solvent methods (pentane addition). The asphaltene was then dissolved in toluene to the desired concentrations. The asphaltene elemental analysis is given in the table below. The asphaltene molecular weight was measured as 1941 g/mol by vapor pressure osmometry. As shown, the oxygen and nitrogen content are on the high side of typical asphaltenes. Oxygen is usually associated with carbonyl and carboxylic acid groups, which act as Lewis Acids (electron donors). Nitrogen is usually associated with ring groups which also act as Lewis acids. Roughly 50 % of the carbons were determined to be aromatic, and the average number of aromatic rings per molecule was 14. One set of results is shown in the figure below.

It is well-known that clay particles are best described as thin flat plates. They have positive charges on the thin edges, and negative charges on the broad surface of the plates [17]. Thus, they have both positive and negative charged surfaces to interact with asphaltene particles. Asphaltene adsorption on clay particles was studied as a function of the concentration of asphaltene in oil (toluene), and as a function of cation exchange of the clay using multivalent cations. Water was not present in this particular set of data [36]. Prior to the adsorption measurements, the clays were mixed in water containing different cation chloride salts, as shown in the figure. The objective was to populate the surface of the clay with varying cations (Na^{+1}, K^{+1}, Mg^{+2}, Ca^{+2}). The clays were then filtered and oven dried such that essentially no water was present on the clay during the adsorption measurements. There was little difference in adsorption as a function of the cation exchange. Also shown in the Figure, the equilibrium adsorption of asphaltene onto the clay increases and then levels off as the asphaltene content of the oil increases. This is expected. For the particular asphaltene and clays selected, the asymptotic limit is about 50 mg asphaltene / gram of clay.

Kaolin is known to have a high adsorption capacity for asphaltenes compared to other clays and minerals typically found in produced fluids [35]. The poorly crystallized Kaolin had a surface area of 17.5 m2/gram. Using these two values gives a surface coverage of 2.9 mg asphaltene /m^2 clay surface. A further calculation can be carried out to determine the thickness of asphaltene, assuming even coverage. Assuming an asphaltene density of 800 kg/m^3, the thickness of asphaltene coverage is about 3.6 nm, or 36 Angstroms. From a molecular standpoint this is reasonably thick [35] and represents an oil-coated or hydrophobic surface. Experiments like this demonstrate that at least some asphaltenes have a tendency to adsorb onto various solids surfaces that are commonly found in produced water.

The above results are roughly in line with those reported by Sjoblom and co-workers [37, 38] in which the adsorption value of various asphaltenes on Kaolin was found to be in the range of 31 to 55 mg / g. Five different asphaltenes were studied having a range of compositions. However, adsorption was found to be similar among the different asphaltenes. In general, this would not be expected since asphaltenes from different crude oils would be expected to have somewhat different compositions. The kaolin used had a particle size in the range of 0.1 to 4 micron, and a surface area of 19.7 m2/g. Thus, the adsorption was in the range of 1.6 to 2.8 mg/m^2. These experiments establish the affinity of asphaltenes for mineral solids surfaces.

As shown in the table below, the contact angle for asphaltene as the solid phase, water as the liquid drop, and air as the third phase. The surface chemistry is dramatically changed by the adsorption of asphaltenes. All but one of the solids went from being hydrophilic to hydrophobic. Thus, in clean systems without corrosion inhibitor or other production chemicals, and without the presence of water, solids can become hydrophobic by the adsorption of asphaltenes from crude oil.

**Table 4.17 Values of particles contact angles at water/solid/air interface.
The solids were exposed to five different asphaltenes [38]**

Particle	No Coating	Contact angle alteration after asphaltene exposure				
Asphaltene ID No. →		5	7	11	16	30
Kaolin	12±2	106±1	104±1	102±3	108±3	96±3
$CaCO_3$	25±2	74±6	85±1	79±2	84±2	87±2
TiO_2	11±2	90±2	115±1	85±1	117±2	123±2
$BaSO_4$	9±2	64±2	70±3	64±4	60±2	82±2
Fe_3O_4	10±2	94±3	100±3	64±6	65±2	102±2
FeS	16±2	80±1	82±2	57±2	85±2	80±1
SiO_2-i	20±2	93±2	93±2	65±3	102±1	93±2
SiO_2-o	107±2	88±3	100±5	68±1	100±3	98±2

In addition to the dry (no water present) adsorption studies discussed above, a number of measurements were made in the presence of both oil and water [36]. An expected finding was that the presence of water hindered the adsorption of asphaltene on clay. Thus, asphaltenes have an affinity for mineral solids surfaces, but water has a greater affinity.

In general, water-wet particles stabilize oil-in-water emulsions. In other words, water is the continuous phase. This is analogous to Bancroft's Rule. Conversely, oil-wet particles stabilize water-in-oil emulsions, where oil is the continuous phase.

Tambe and Sharma [39, 40] added steric acid to oil/water mixtures containing different types of solid particles. Steric acid is a surfactant-like oil soluble organic acid. According to Bancroft's rule (see Section 4.4.5), steric acid should stabilize a water-in-oil emulsion. The solids, when added to the oil / water mixtures without steric acid, stabilized an oil-in-water emulsion. Without the steric acid, the surfaces of the solids are hydrophilic. As steric acid is added, the surfaces of the solid particles become coated with the steric acid. The polar head of the acid molecule adsorbs onto the solid surface. The nonpolar tail of the molecule protrudes from the surface into the bulk. When this occurs, the solid particles become somewhat hydrophobic and stabilizes a water-in-oil emulsion.

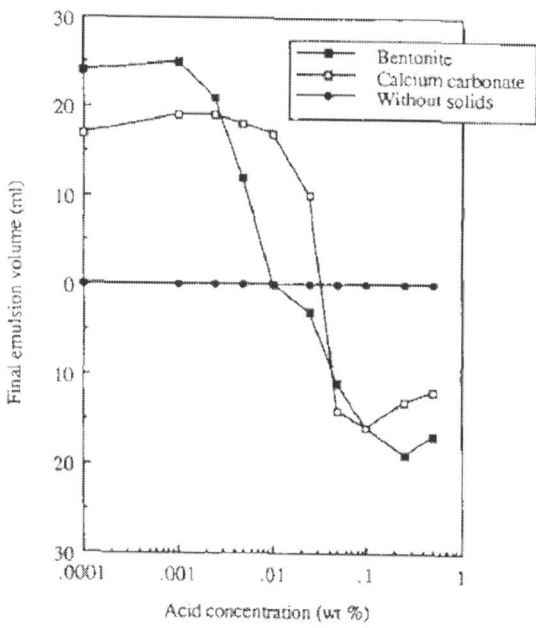

Figure 4.28 Emulsion volume as a function of the addition of steric acid in solids stabilized emulsions. The upper half of the figure represents the emulsion volume for oil-in-water emulsions (water is continuous). The lower half represents that for water-in-oil. Data from [39, 40].

In a study of corrosion inhibitors, Gulbrandsen and Pedersen [41] mixed sand together with CTAB (cetyltrimethylammonium bromide), a quaternary amine corrosion inhibitor, in a water/light oil mixture and evaluated the interaction between the sand and the corrosion inhibitor. The results are given in the Table below. As shown, the sand particles show a strong tendency to aggregate upon addition of the corrosion inhibitor even at concentrations as low as 3 mg/L. Also, the wetted sand became sticky. This is a common occurrence in produced water systems. This occurs with not only sand but a wide range of solids particles including mineral solids, and iron solids.

Table 4.18. Aggregation tests of sand in brine with added corrosion inhibitor (CTAB) [41].

CTAB Conc Aq phase (mg/L)	Tendency to aggregate	Tendency to stick to walls
0	None	None
3	Strong	None
5	Strong	None
10	Strong	Some
15	Strong	Strong
20	Strong	Strong

The above experimental results have the following ramifications. In a production system, oil-wet solids are commonly produced by several mechanisms. Sand, clay, iron solids, and carbonates may be completely water wet at an early stage of production. Addition of production chemicals can change the surface of these solids such that they become partially oil wet. Natural occurring acids in the water or oil phase can cause the same effect.

Once formed, oily solids can cause significant water treatment problems with the net result that water quality often fails to meet targets. If a sample of produced water that contains oily solids is taken and allowed to sit on a desk top, hours, days or weeks may pass before the cloudy water begins to clear up. The time required depends on the surface chemistry, temperature and composition of the various constituents. Over time, the surfactant-like components may detach from the solid surface such that the oil and solids separate. Slowly, separation begins to occur and the organic material floats to the top and the solids sink to the bottom. Figure 4.29, discussed previously, shows the fresh sample with suspended oily solids, and the same sample several weeks later where separation has occurred. The process can be accelerated by addition of high activity surfactants (e.g. short-chain benzylsulfonates), or the addition of oxidizing agents that help break the bond between the surfactant material and the solid. Also, low salinity water (with a low concentration of multivalent cations) has been reported to speed up this process.

4.5 Coalescence of Oil Droplets

Coalescence of oil drops is important in produced water treatment because intense shearing creates small droplets which are difficult to remove. An understanding of the relationship between shearing and the resultant droplet size distribution is important in design and troubleshooting of produced water systems. While intense shearing creates small droplets, mild shearing promotes coalescence which creates larger droplets. It is the give-and-take between shearing and coalescence that determines the oil droplet size that the water treatment system must deal with. Shearing of droplets is primarily a fluid mechanics mechanism, influenced by interfacial tension. It is, to an extent dependent on the surface characteristics of the droplets. Coalescence is a more complex process and involves fluid mechanics as well as the properties of interfaces which are strongly affected by surface active chemicals which can be either natural (in the produced oil and water) or manmade (added as corrosion and surface active hydrate inhibitors).

Practical Examples of Coalescence: Coalescence can occur in pipeline and flow lines under appropriate conditions. It is well known, for example, that an electrical submersible pump is a high speed centrifugal pump and as such it will shear the oil and water into small droplets of both oil in water and water in oil. However, the impact of an electrical submersible pump on the oil and water separation is greatly reduced if there is a long flow line between the well and the facilities. In that case, the effect of shear can be reduced, depending on conditions, due to coalescence.

Mechanisms of Coalescence versus Shearing: The discussion in previous sections provided an understanding of the role of the turbulent energy dissipation rate in shearing oil drops and causing drop breakage. The development concluded with the Hinze and Davies formulas that can be used to estimate the maximum droplet size in a turbulent system. One of the limitations of such models, however, is the inability to predict the actual droplet diameter distribution. The quantity predicted in the Hinze and Davies models is the maximum droplet size, not the average or droplet size distribution.

Computational Fluid Dynamics can be used to study both the details of drop/drop coalescence processes and the effect of turbulence on collision rate. But modeling of turbulent flow including coalescence processes of large numbers of drops is computationally time consuming. Engineering correlations that allow rough estimates are of far greater utility.

Objectives of this Section: The discussion in this section is intended to provide a physical and mechanistic understanding of the processes involved in droplet coalescence. This level of understanding is useful in developing a root cause understanding of the source of problems and potential solutions in a troubleshooting project. Coalescence occurs naturally in most flowlines under appropriate conditions. Thus, most oil/water mixtures will have some degree of shearing and some degree of coalescence by the time they arrive at the facility. When coalescence does not occur, there will likely be problems in oil/water separation. Thus, the material in the following sections is intended to help identify when coalescence might not occur. In addition, the material presented is intended to provide semi-quantitative formulas that can be used to estimate the rate and extent of oil droplet coalescence in a process system. The formulas are discussed in a way that appeals to the mechanisms rather than the detailed mathematics.

Another important reason to discuss coalescence in some detail is the likelihood that new technologies which promote coalescence will be implemented in the next several years. Several technologies are already being developed and presented at conferences. Some of the technologies are not yet practical. The material in this section will provide insight as to why those technologies are not practical. But, some technologies are reasonably practical already and are being tested.

The objectives of this section are to:

1. explain the physical situation and mechanisms in a way that allows water specialists to understand and make intuitively correct judgments on the importance of coalescence as a function of processing conditions and the type of oil and water, and the likelihood that it is significant and would have a positive impact on separation;

2. provide formulas for the calculation of various processes involved in coalescence (collision force, drainage time, etc), and formulas for the estimation of coalescence rate itself;

3. to help water specialists identify promising emerging technologies that are designed to promote coalescence.

4.5.1 The Mechanism and Modeling of Coalescence

When two drops or two bubbles collide a complex physical process takes place. The mechanism of the process is discussed here. In subsequent sections various models are presented that mathematically describe the process. Since these models involve various approximations, it is essential that the mechanism of the process be described first in realistic detail before the models are presented.

Collision between two suspended droplets is usually caused by their relative motion which can have three possible sources:

i) turbulent fluctuations in the water phase;

ii) mean velocity gradient in the flow;

iii) buoyancy arising from rise velocities, wake interactions or helical/zigzag trajectories.

In produced water systems the typical range of suspended droplet size is 1-100 microns. Arnold and Stewart [42] have provided a reasonable formula for the effect of buoyancy on coalescence. Coalescence in a pipe in discussed presently.

Rough Estimate of Coalescence in a Pipe: This is what Arnold and Stewart [42] have to say on the subject of coalescence in a pipe. This is taken from their section on gas flotation.

> "Oil removal is dependent to some extent on oil droplet size. Flotation has very little effect on oil droplets that are smaller than 2 to 5 microns. Thus, it is important to avoid subjecting the influent to large shear forces (e.g. level control valves) immediately upstream of the unit. It is best to separate control devices from the unit by long lengths of piping (at least 300 diameters) to allow pipe coalescence to increase droplet diameter before flotation is attempted. Above 10 to 20 microns, the size of the oil droplet does not appear to affect oil recovery efficiency, and thus elaborate inlet devices are not needed."

As will be shown, the rule of 300 pipe diameters is a reasonable first-order approximation. For a six inch diameter pipe, roughly 150 feet of pipe length would be required to achieve some meaningful degree of coalescence.

However, it must be pointed out that this is only a rule of thumb. It does not take into account the intensity of turbulence, nor the characteristics of the produced water. Some oil in water dispersions may require a much longer pipe length due to coalescence barriers such as corrosion inhibitor, or the presence of asphaltenes or heavy oil constituents at the oil/water interface. Another, more detailed approach is given below.

Drop collision process: when two drops approach each other they have a certain force and energy that was initially provided by their interaction with turbulent eddies. As the two droplets proceed along their collision course, the fluid between the drops is pushed away from the path of the collision. This hydrodynamic effect consumes energy. If there is not enough energy for the collision to continue, then the collision comes to an end and coalescence will not occur for that particular trajectory of droplets. If there is sufficient energy to overcome the hydrodynamic resistance, and other forces, described below, then the droplets approach to a very close distance of less than a tenth of a micron. At this close distance, there may be electrostatic repulsive interactions and van der Waals attractive interactions. If the net interaction is attraction, this will promote further approach. Coalescence occurs if the thin film on the surface of the oil droplets or gas bubbles ruptures and their respective interfaces give way to the combining of liquid from the interior of the droplets. Modeling of this process is envisioned to occur in steps.

Three Step Process of Coalescence: The process of coalescence is envisioned to occur in three steps [43]:

i) Two droplets approach to within a distance of 10 to 100 micron. At this distance the approaching faces of the droplets deform and a disk-shaped film forms. Film drainage begins.

ii) The liquid film between the droplets drains out to a critical thickness of 0.01 to 0.1 micron; at this distance surface forces (electrostatic repulsive and van der Waals attraction) become significant;

iii) The thin liquid film ruptures and results in coalescence.

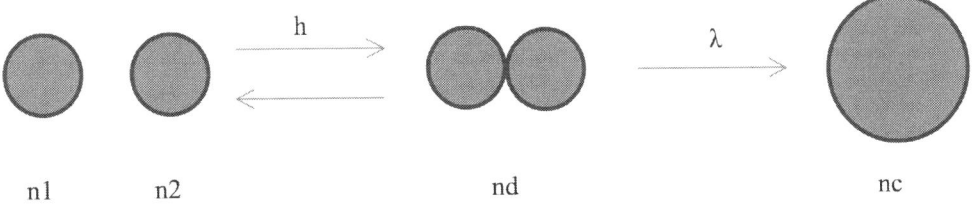

Figure 4.32 Illustration of coalescence and separation of suspended droplets.
h is the collision rate for the formation doublet, and λ is the rate of coalescence of the doublet [43].

Any of the three steps could become the rate liming step. The mechanism of each of these steps is discussed in some detail below. Typically, thin film drainage is the rate limiting step.

Film Drainage: As the two oil droplets approach each other, the pressure in the water between the droplets builds up slightly. As this occurs, there is a tendency for the surface of each droplet to flatten. This creates a disk-shaped film of water between the oil droplets. The extent of flattening depends on how rigid the interface is, and on the inertial energy of the two colliding droplets. If the oil in the droplets is viscous, then the droplets will not flatten so readily, and the radius of the film disk will be small. Likewise, if the oil/water interfaces of the droplets have asphaltene adsorbed onto them, the asphaltene can, in some cases, form a rigid three dimensional structure which will resist deformation, as discussed in [31]. In that case, the film disc radius will be small. If the interfaces are relatively un-contaminated by surface active materials, then they will be readily deformed and the film disk radius will be large. It is worthwhile to keep in mind, during the subsequent discussion that for every pair of colliding droplets, there is one disc-like region of draining fluid, and two oil/water interfaces – one interface for each of the colliding droplets. The situation is illustrated in the Figure below.

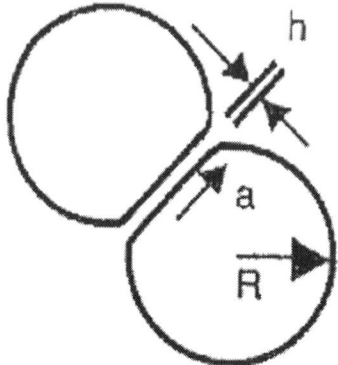

Figure 4.33 Film drainage between two colliding droplets. If the droplets have deformable interfaces then a disc-shaped film (of radius a) of water will form (of thickness h) between them which must drain out in order for the droplets to coalesce.

The larger the deformation of the droplets, the larger the radius of the disk-shaped film. A large film radius will increase the drainage time and reduce the probability of coalescence. Since larger droplets will also form larger film disks, larger droplets require greater time for film drainage, all other things being equal. Thus, if there are no other effects involved, larger droplets take longer for film drainage and have a lower coalescence probability.

In some of the modeling work on coalescence, it is assumed that the flattened surface of the droplets is planar and smooth. In reality this is not typically the case. In reality a dimple forms and contaminants such as solids particles cause the interface to deform. This is discussed below.

Lee et al. [44], and others [22, 43, 45 – 48], envisioned the initiation of the coalescence process by the formation of a liquid film of water between the oil droplets that is between 10 and 100 microns thick. The early mathematical models are reviewed by Chesters [49]. The time consuming step in the coalescence process is the drainage of this film from the initial thickness (h_i) to the final thickness (h_f) at which rupture occurs. Energy is consumed in the film drainage step as well as the final film rupture step. In fact, in order for coalescence to occur, there must be sufficient energy for both film drainage and thin film rupture. If there are surfactants or asphaltenes at the oil/water interface, then film rupture will require a significant amount of energy. The energy required is not typically mea-sured. Instead, the properties of the oil/water interface are measured and correlated to coalescence time. The properties that are measured are interfacial viscosity and interfacial elasticity.

Mobility and Deformability of the Interface: During the collision process, fluid motion in the drop-lets, in the film, and motion of the two interfaces must be considered. Oil droplets experience forces which push against and shear the oil/water interfaces. As mentioned, if the interfaces are deformable, then a film of the continuous fluid will form between the droplets. If the interfaces are mobile, they will be dragged in the direction of fluid flowing out of the thin film. The mobility and deformability of the interfaces has an effect on the shape of the interface and on the fluid flow in the film between the droplets and in the droplets themselves. The mobility and deformability of the interfaces are characterized by parameters that depend on the droplet size and fluid properties such as viscosity, interfacial tension, interfacial viscosity, and interfacial elasticity. Quantitative models of mobility and deformability are discussed below.

Surfactants, naturally occurring in the oil and water, or added as corrosion or hydrate inhibitors, have a dramatic effect on coalescence, as will the presence of small solid particles at the interface. Both surfactant and solid particles will reduce the drainage rate of the film between the two droplets. Also, at close distance the presence of contaminants at the interface will contribute to film stability or rupture, depending on the nature of the contaminant. All of these effects are considered below.

Framework for Coalescence Modeling: Coalescence frequency is defined as the number of droplets that coalesce in a given time and in a given volume of water. For each pair of drops that coalesce, the coalescence frequency is incremented by two. The coalescence frequency can be calculated for all of the drops in the fluid, or it can be calculated for drops of a certain diameter in the fluid. In the former case, a summation can be carried out over all droplet diameters to calculate the total number of coalescence events.

The framework for calculating coalescence frequency is given by the following equation:

Coalescence Frequency = (Collision Frequency) x (Coalescence Efficiency)

Where:

> **Coalescence Frequency:** the number of times that a coalescence event occurs in a given amount of time in a given volume of fluid (e.g. number of coalescence events per second per cubic meter of fluid).

> **Collision Frequency:** number of collisions per droplet per unit time.

> **Coalescence Efficiency:** fraction of collisions that result in coalescence.

Population Balance Modeling: The droplet size distribution that results from transportation of fluids through pipe, through valves and pumps is a result of drop breakage and coalescence processes occurring simultaneously. In some situations breakage dominates (valves and pumps) and the overall distribution is shifted to smaller droplets. In other situations coalescence is dominant (pipeline flow) and the overall distribution is shifted to larger droplets. Modeling such processes is most realistically done using a computational technique known as Population Balance Modeling. In population balance modeling, the birth (coalescence) and death (breakage) of droplets is simulated over time. The rate of drop breakage is modeled using the previously discussed ideas. The rate of drop/drop coalescence is modeled using equations that are given below. A good introduction and historical review of the technique is given by Tsouris and Tavlarides [46]. Comparison with experimental data is given in [46] as well as [50]. It has been applied to produced water systems in [51]. It is the basis for the model presented in Section 4.5.6 (Coalescence Model) below.

4.5.2 Collision Frequency

In this section, a model for the collision frequency of droplets and particles in water is provided. The fundamental framework for this model is based on the kinetic molecular model of ideal gasses. The kinetic model of gases is a billiard ball kind of model. It assumes that a gas is made up of particles which are in constant motion. The moving particles collide with each other at a frequency that can be calculated by assuming that the motion of the individual particles is random. When there is a large number of particles, this is an accurate assumption. It also assumes elastic collisions in which kinetic energy and momentum are conserved. For contaminant droplets and particles suspended in a liquid, this is not the case since the medium, water in our case, is dissipative (i.e., can convert turbulent energy into heat and entropy). Nevertheless, the model is relatively simple and provides useful order of magnitude predictions [52]. It has been tested against direct experimental measurement and has been found to be valid over the range of interest.

The collision model presented here is applied later in the text to the following situations:

- coalescence rate of droplets in shear flow;

- collision rate of gas bubbles with oil droplets in flotation;

- collision rate of oil droplets and solid particles with media particles in fixed bed filtration.

As will be demonstrated, the collision rate depends very strongly on the number of droplets or particles in a given volume. If there are not many particles, then the collision rate will be low. This is important to keep in mind. For example, in the operation of a hydrocyclone it can be demonstrated that coalescence does not play a major role in the separation efficiency although it has been the subject of speculation in the industry for years. Also, in the operation of a flotation unit oil droplet coalescence is rather low due to the low droplet concentrations.

Collision frequency can be modeled in the following way. First, a collision tube is defined for a single moving drop. This is the tube swept out by the moving drop in a unit of time. Second, the number of drops within that tube are calculated from the average number density of drops in the fluid. Visually, consider this to be a large boulder chasing Indiana Jones down a narrow tunnel. It is assumed that any drop within the tube will collide with the single moving drop. This gives the collision rate for a single drop. Then the collision rate for all drops is assumed to be the same as for the single drop. Thus, the collision frequency for the single moving drop is multiplied by the number of drops in a specified volume. This provides an estimate of the total collision frequency for the fluid. It should be mentioned that since all of the drops are moving, the velocity of the single moving drop is the relative velocity (relative to the velocity of the other drops). The step-by-step detail below should help to elucidate the model.

The volume of a cylindrical collision tube is calculated as [47]:

$$V_{12} = \pi d^2 u_r \qquad \text{Eqn (4.14)}$$

where:

V_{12} = the volume of the collision tube per unit time (m³/sec)
d = drop diameter (m)
u_r = relative velocity of two moving droplets (m/sec)

In the formula above, the diameter is used rather than the radius of the particles since the formula requires the excluded volume, and not just the volume of the particles. The formula assumes that the particles are the same size. In practice, the formula is usually applied to systems where the particles are different diameters which R1 + R2 would be used instead of d.

For our purposes, the situation is simplified such that all droplets or particles have the same diameter. This is not a necessary simplification since the formulas for a realistic distribution of sizes are well known. In quantitative calculations and computer modeling, presented in other sections, realistic models that depend of the distribution of sizes are used. For our purposes, we want only a simple formula that illustrates certain important points and which can be used for order of magnitude estimates. The concept of a collision tube is illustrated in the Figure below.

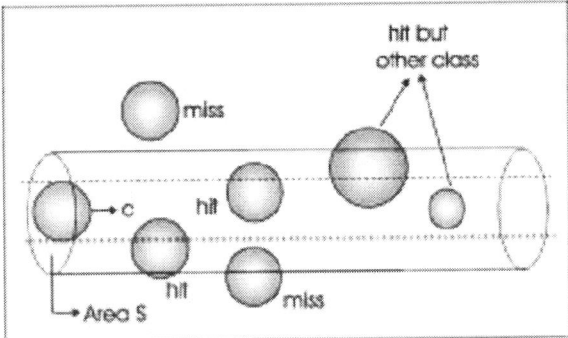

Figure 2. Collision tube of a bubble moving with a relative speed c.

Figure 4.30 Collision tube of a bubble moving with a relative speed c (From Venneker et al., 2002)

The number of drops within the collision tube is calculated as the volume of the tube times the number concentration of drops (n = drops/unit volume). It is assumed that all drops have the same frequency of collision so the frequency of collisions in a unit volume per unit time is just n times the single drop collision frequency.

$$h = V_{12}n^2$$

Eqn (4.15)

where:

h = collision frequency (number of collisions/m³ sec)
n = number density of drops (number of drops/m³)

Collision Frequency: Combining the above equations gives a formula for the collision frequency as a function of the relative velocity of droplets:

$$h = c_1 \pi d^2 n^2 u_r$$

Eqn (4.16)

A numerical constant ($c1$) has been inserted into the formula. The value of this constant depends on certain details of the averaging over trajectories.

Relative Velocity in Turbulent Flow: The relative velocity u_r between the drops is calculated assuming that the velocity of colliding droplets is same as that of an eddy of the same size [44, 46, 53, 54]. Such a relationship was developed in Section 3.4 (Turbulence).

$$u_r = \sqrt{<(u')^2>} = \sqrt{2}(\varepsilon\,d)^{1/3}$$

Eqn (4.17)

where u_t is the velocity of eddy with characteristic size d which is defined by Batchelor (1951) and Kuboi et al. (1972a). Combining the above equations gives:

$$h = c_2 \pi d^{7/3} \varepsilon^{1/3} n^2$$

Eqn (4.18)

where:

h = collision frequency (number of collisions/m³ sec)
e = turbulence energy dissipation rate (m²/sec³ = W/kg)
d = droplet diameter (m)
n = number density of droplets (number of droplets/m³)

The left hand side of this equation is the collision frequency. This is the number of collisions per unit time in a specified volume of fluid. The constant c_2 varies from one form of the model to another. In the discussion below, c2 = sqrt(8π/3) = 2.9.

The right hand side of the equation captures the physics of droplet / droplet interactions. It is based on the assumption that when there is a large number of droplets, the droplets move randomly. This random motion leads to collisions. The speed at which the droplets move (the kinetic energy of the droplets) depends on the intensity of turbulence in the fluid. The contribution of turbulence is given by ε as discussed in Chapter 3 (Fluid Mechanics). The greater the shear rate, the more intense the turbulence, the more frequently droplets will collide. The frequency of collisions of a particular droplet depends on the cross sectional area of the droplet, turbulence intensity and concentration. The bigger and fatter the droplet, the more it collides with other crops. Thus, the size of each member droplet of the pair is included. The final terms represent the number density of droplets. If there are few droplets for a given volume, there will be few collisions.

As originally derived, the above equation gives an accurate prediction of the collision frequency. It does not, per se, predict the frequency of coalescence. Many collisions do not result in coalescence. This is due to the hydrodynamic interactions between colliding droplets, the stability of the liquid film between the droplets, and the presence, in some cases, of chemically and physically adsorbed materials on the surface of the droplets which can give rise to a coalescence barrier. All of these factors are typically lumped into a single term referred to as the coalescence efficiency which is discussed below.

Kuboi, Komasawa and Otake [55, 56] carried out a series of measurements of the collision frequency for liquid / liquid and solid / liquid mixtures. Their measurements are shown in the Figure below. Particle tracking with high speed video was used to measure eddy velocities in mild turbulence. Note that the velocity fluctuation due to turbulence increases with the size of the particle.

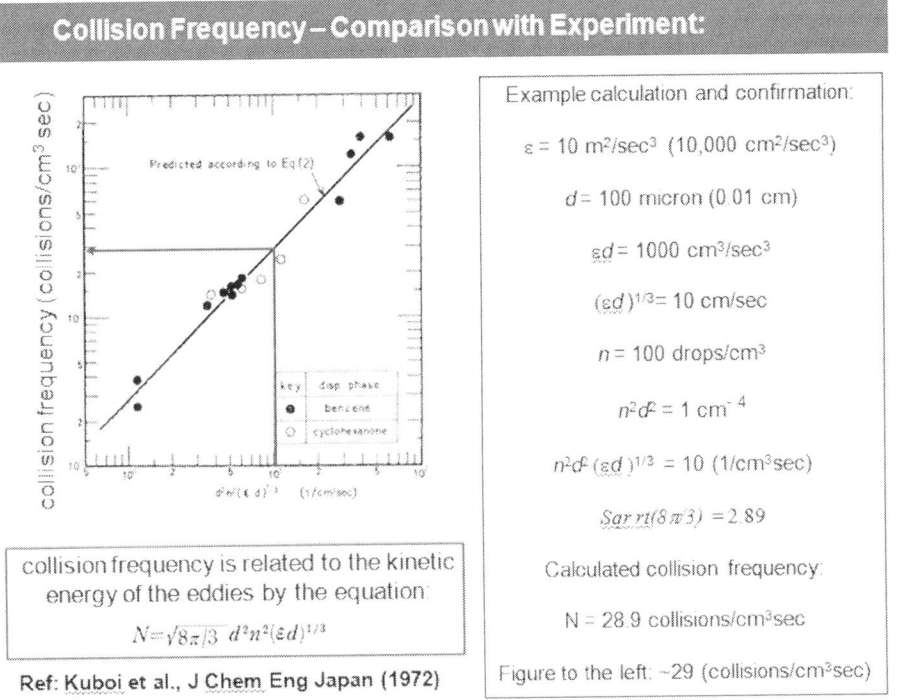

Figure 4.31 Measurements of collision frequency as a function of the turbulence intensity [55, 56]. As shown the collision model combined with the model for velocity fluctuations gives an accurate measure of the collision frequency for oil droplets in mild turbulent flow. The red line highlights the example calculation in the text below.

As shown in the box next to the graph, the results calculated using the model given above agree well with the measured results of Kuboi et al.. This confirms that the model is accurate, at least in the range of conditions studied.

The units of h are the number of collisions per unit time per unit volume. Already an important feature of the collision model is apparent. The frequency of collisions is proportional to the square of the number of droplets in the produced water. If applied to flotation, the collision frequency is proportional to the number of droplets times the number of bubbles per unit volume. The relation between the concentration of oil and the number of droplets is given by:

$$n = \frac{C}{\rho(\pi/6)d^3}$$

Eqn (4.19)

where:

n = the number of droplets per m^3
C = concentration of oil in the form of droplets (mg/L = gr/m^3)
r = density of oil in the droplets (gr/m^3)
d = diameter of droplets (m)

Again, for our purposes here, all droplets or particles have the same diameter. For illustrative purposes, it is useful to carry out some simple calculations using these formulas. For the illustration, the following values are selected:

C = 1,000 mg/L
r = 800 gr/m^3
d = 100 micron = 1 x 10^{-4} m
u_r = 0.01 m/sec

$$n = \frac{1,000 \ gr/m^3}{(800 \ kg/m^3)(1,000 \ gr/kg)(0.52)(1 \times 10^{-4} m)^3} = 2.4 \times 10^9 \, droplets/m^3$$

Eqn (4.20)

Substituting this into the equation for the collision frequency results in:

$$h = 3.1(1 \times 10^{-4} m)^2 (2.4 \times 10^9)^2 (0.01 m/sec) = 1.8 \times 10^9 \, collisions/m^3 sec$$

Eqn (4.21)

This is obviously a large number. In order to put this number into perspective, it is divided by the number of droplets in the unit volume. The resulting number is the number of collisions per droplet per unit time.

$$h/n = 1.8 \times 10^9 / 2.4 \times 10^9 = 0.8 \, collisions/droplet \, sec$$

Eqn (4.22)

where:

h/n = collision frequency per droplet per second (number of collisions for each droplet / sec)

In the above calculations, the relative velocity of the droplets was arbitrarily chosen to be 0.01 m/sec. This is a reasonable value which corresponds to a mild turbulence intensity. However, for the formulas to be most useful, a general equation is required that relates the droplet relative velocity to the turbulence intensity.

Collision Rate – Order of Magnitude: The rate of collisions that was calculated in the example above seems reasonable. It is easy to imagine that in a mildly turbulent flow, with a concentration of 1,000 mg/L and an average droplet diameter of 100 microns, each droplet or particle will experience a collision roughly every second. Only a small fraction of those collisions will result in coalescence. Other values of drop diameter and concentration were used to calculate the collision rates given in the table below.

Table 4.19 Collision frequency of droplets with different size and concentration (collisions per droplet per second).

Drop Diameter (micron)	Concentration (mg/L)	Collisions per droplet/sec	Indicative of
100	1000	0.8	primary separation
50	200	0.06	hydrocyclone
10	100	0.05	flotation

In the Table, the collision rate per droplet per second is given for three sets of droplet diameter and concentrations. These values roughly correspond to the outlet of a primary separator, the inlet of a hydrocyclone, and the inlet of a flotation unit. As shown, the collision rate decreases by a factor of ten to 16 from the primary separator to the hydrocyclone and flotation unit. This is generally observed in most water treatment systems. Not much coalescence occurs in the water treatment system itself, except in the reject streams of the various equipment. This will be discussed in greater detail below. This is an interesting finding since it suggests that droplet / droplet collisions and hence coalescence is a mechanism that mostly occurs upstream in the process.

4.5.3 Collision Force

The compressing force, F, used in all the above equations is the result of turbulent energy. The physical picture is that of turbulent eddies interacting with oil droplets and pushing them in random directions. If the trajectory from these fluid / drop interactions happens to result in two drops colliding with each other, then the coalescence process will start with film drainage between the two drops as described above. Narsimhan [43] provides a discussion of the physical situation and provides a mathematical development of the model.

The force pushing the drops together ultimately derives from velocity fluctuations. This force is assumed constant during the film drainage process. The force F is usually assumed to be proportional to the mean-square velocity difference at either ends of the eddy with a size of equivalent diameter of suspended droplets [43, 46, 53].

$$< F >= \frac{\pi d^2 \rho}{4} < (u')^2 >$$

Eqn (4.23)

where

$<(u')^2>$ = mean square turbulent velocity fluctuation between the centers of the colliding pair of droplets (m²/sec²).

As discussed previously, in isotropic homogeneous turbulence, the average mean square fluctuating velocity is related to the turbulent energy dissipation rate, and the size of the eddies:

$$< (u')^2 >= (\varepsilon d)^{2/3}$$

Eqn (4.24)

Substituting the equation gives:

$$< F >= \frac{\pi}{4} \rho_c \varepsilon^{2/3} d^{8/3}$$

Eqn (4.25)

where

F = compressing force acting on the droplet (N) due to turbulence
d = diameter of droplet (m)
r_c = density of the continuous phase (kg/m³)
e = turbulence intensity (m²/sec³)

Illustrative Calculation of Force: It is instructive to verify the units in the equations and to make an order of magnitude estimate of the force. The force is calculated for the viscous subrange using Eqn (4.25). The following parameter values were chosen for the calculation.

Turbulent Force Calculations		
Water Density	1000	kg/m³
Epsilon	10.0	m²/sec
d	100	micron
d	0.0001	m
F	7.9E-08	N

Using these values, the following value of force is calculated:

$$F = \frac{\pi}{4}(1000 \text{ kg/m}^3)(10 \text{ m}^2/\text{sec}^3)^{2/3}(100 \times 10^{-6} \text{ m})^{8/3} = 7.9 \times 10^{-8} \text{ N}$$

Verifying the units:

$$F \rightarrow (\text{kg/m}^3)(\text{m}^2/\text{sec}^3)^{2/3}(\text{m}^{8/3}) \rightarrow \text{kg m} / \text{sec}^2 \rightarrow N$$

The SI unit of force, the Newton (*N*), is indeed obtained. The following order of magnitude estimate for the force will be used in further discussions below:

$$F \approx 1 \times 10^{-7} \text{ N}$$

This is a fairly significant force considering that the drop is only 100 micron diameter and its mass is therefore only 5 x 10⁻¹⁰ kg. The acceleration of this drop due to the force is 170 m/sec². This is a significant acceleration. In comparison to the gravitational acceleration (9.8 m/sec²), the acceleration of the drop due to turbulence is 17 times greater than that due to gravity. Thus, even mild turbulence has sufficient velocity fluctuations to provide significant collision force.

The force per unit cross-sectional area can also be calculated as an indication of relative magnitude. The cross-sectional area of film that could form between the deforming drops is assumed in this example to be 10 square micron 10 x 10⁻¹² m². Based on this assumption, the force per unit cross-sectional area would be 10,000 N/m² (Pa), which is 1 bar. Again, this is a significant force per unit area for a small drop.

Other Forces – Electrostatic Repulsion and van der Waals Attraction: The net force of interaction experienced by the pair of droplets is the sum of the driving force due to turbulent eddies (just discussed), and the interaction forces between the droplets. The interaction forces have their origin in van der Waals attraction, hydrophobic attraction, electrostatic repulsion, disjoining pressure, surface elasticity, etc.. These forces are collectively referred to as colloidal forces [11, 22, 43].

The repulsive force has been discussed at least qualitatively and is determined by electrostatic repulsion of like-charged surfaces. In order to understand the relative importance of the attractive and repulsive forces, a mathematical model is required for both. Even with a low zeta potential, the potential of the Stern Layer between particles will still lead to a repulsive electrostatic interaction.

The effect of the electrical double layer on colloidal behavior of particles can be reasonably well explained by the DLVO model developed by Deryagin and Landau [57] and independently by Verwey and Overbeek [58]. The model captures the interaction energy between particles suspended in water as a function of the surface charge, and of the valence and concentration of indifferent ions in solution. The model takes into account the electrical double layer charge as a function of distance from the particle, as well as the attractive force that occurs at short distances.

The DLVO model assumes that the stability of a colloidal system (the stability against coagulation), is determined by the sum (V_T) of the Van der Waals attractive (V_A) and electrical double layer repulsive forces (V_R) that exist between particles as they approach each other. If the particles have a sufficiently high repulsion, the dispersion will resist coagulation and the colloidal system will be stable (no coagulation). If the repulsion force is smaller than the attractive force, coagulation will take place.

The forces of attraction between two macroscopic sized particles is an extension of the interaction between say individual atoms of a monoatomic noble gas (e.g. He, Ne, Xe). It is referred to as the van der Waals force. In the case of a noble gas atom, the shape of the electron cloud around the nucleus is constantly fluctuating. In the presence of another noble gas atom, the two electron clouds induce in each other an instantaneous net electronic dipole. Due to the constant fluctuations of these dipoles, a long-lived, but relatively short-ranged, attractive interaction is established. This explains why noble gas atoms condense into a liquid, at low enough temperature to form a liquid phase. Without this attractive interaction, the noble gases would not form a liquid at any temperature. A low temperature is required in order to slow down the kinetic motion of the atoms so that liquid-like clusters will form. Granted the attraction is not particularly strong. But the presence of an attractive interaction is nevertheless proven from the fact that a liquid can form at all.

Solid particles and liquid droplets also have dispersion interactions between each other. For our purposes, the solid particles are much larger than noble gas atoms but the physical effect is the same, at least qualitatively. The atoms within one solid particle will have an attractive interaction with the atoms in another solid particle, when they are very close to each other. The atom – atom attractive interactions are additive. In other words, the net attractive interaction of one particle on another particle is the sum of all of the atom – atom interactions. Adding up all these atomic interactions is the essence of the Hamaker model of attractive interaction between colloid particles. Under a set of simplifying assumptions the following equation is obtained:

There is a range of Zeta potentials for which coagulation can occur. As the zeta potential increases, the potential energy of interaction increases. This potential energy must be compared to the kinetic energy of the approaching particles which depends on thermal motion and fluid turbulence, as previously discussed. This figure gives some validity to the rule of thumb discussed previously that zeta potential within the interval of +/- 30 mV provides a low enough barrier to interaction that coalescence can be expected to occur.

4.5.4 Coalescence Efficiency

In the film drainage model [59], it is assumed that once droplets collide they stay together for some time, called contact time, t_c. The colliding droplets either coalesce or bounce away depending on the approach velocity and the time required for the liquid film between the two droplets to be drained out until a critical film thickness, h_f is reached [53]. The time needed for the film drainage is called drainage time, t_d. Coalescence occurs if $t_d < t_c$ [46]. By assuming the contact time is a random variable and the drainage time is not randomly distributed Coulaloglou (1975) expressed the coalescence efficiency as

$$\lambda = \exp(-t_d / t_c)$$

Eqn (4.26)

where:

t_d = drainage time
t_c = contact time
λ = coalescence efficiency

Contact Time: Contact time (or interaction time) is a very loosely defined concept. Kamp et al. (2001) [60] defined the contact time as the interval between the onset of film formation and the moment at which the drops begin to rebound. The problem with this definition is that it is not possible to define with precision exactly at what point the film starts to form and at what point the drops begin to rebound. This is a drawback of the film drainage mmodel of coalescence.

Nevertheless, there are some benefits to using the concept. First, the definition of coalescence efficiency is simple, as just discussed. Second, the fluid mechanics of film drainage and of various repulsion forces can be incorporated together in a straightforward manner. Levich [61] envisioned the process of coalescence as being synchronized, in a sense, with the turbulent movement of the eddies. He equated the droplet velocity to that of the eddies. Levich derived an expression for the contact time by considering the lifetime of an eddy to be given by:

$$\tau_l = d / v$$

Eqn (4.27)

where:

t_l = eddy lifetime (sec)
d = diameter of the droplets, assuming equal size drops (m)
v = velocity of an eddy (m/sec)

The final result of the energy cascade discussion was that the kinetic energy is related to the size of the eddy (d) and the turbulent energy dissipation rate (ε) by the following equation:

$$v = (\varepsilon\, d)^{1/3}$$

Eqn (4.28)

Substituting this relation gives:

$$t_c = d^{2/3} / \varepsilon^{1/3}$$

Eqn (4.29)

where:

t_c = contact time between two colliding droplets of the same size (sec)

One justification for this equation is that it has the correct units. It can be derived, at least physically if not rigorously, by appealing to the model of the energy cascade which was discussed above in Section 3.4 (Turbulence). Several modifications have been made to the formulas for contact time [44, 53, 60], based on other more complex assumptions. Thus, now that an equation defining the contact time between equally sized droplets is available, the next step is to provide equations for coalescence time. This is done in the following section.

4.5.5 Film Drainage Time

In this section, a model for film drainage is presented. The rate at which film drainage occurs depends on the force exerted on the film by the droplets, on the viscosity of the water in the film, and on the deformability and mobility of the interfaces [18, 46, 62, 63].

Two cases are considered. The first case is that of a pair of droplets with oil/water interface that is completely deformable. Their interfaces flatten out as described above. But these droplets are assumed to have immobile interfaces. In other words, the viscosity of the oil in the droplet, and the interfacial shear viscosity are too high for movement of the interface in the direction of film drainage. This lack of movement of the interface hinders coalescence. The final case is that of partially mobile and deformable interfaces. This latter case allows a variation from light oils to heavy oils and the effect of interfacial properties can be examined. Interfacial shear viscosity and elasticity are the two main properties that impact whether the interface will move in the direction of drainage and thus reduce the energy barrier to film drainage and coalescence.

Deformable Droplets – Reynolds Model: A simple model for the behavior of deformable drops can be derived by focusing on the fluid mechanics of the thin liquid film that forms between the droplets and excluding all other effects. In this approach, the thin liquid film is assumed to have only viscous fluid motion without convection. As the drops approach each other, the fluid is forced out of the film. In this model, this is the only contribution to the fluid resistance. A significant fluid resistance due to flow around the droplets is neglected. As discussed on page 416 of [27], the velocity of approach for this model is given by:

$$u_p = -\frac{dh}{dt} = \frac{2Fh^3}{3\pi\mu_c R_d^4}$$

Eqn (4.30)

where

R_d	= radius of the flat disc, i.e. film (m)
h	= disc (or film) thickness (distance between the surfaces of the two droplets (m)
m	= viscosity of the continuous phase (water) (Pa sec)
F	= force acting on the droplets – usually a result of fluid turbulence (N)
u_p	= rate of film drainage (speed at which the two droplets approach each other) (m/sec)
t	= time (seconds)

The film drainage model is widely used for the case where the Weber Number (see Section 3.4.3) is based on the equivalent diameter of suspended droplets and the relative velocity between droplets is much less than unity [60].

One of the important, though unrealistic, features of this model is that the velocity of the approaching droplets is independent of the droplet diameter (d). This is due to the focus in the model on the thin film between the drops and the exclusion of all other hydrodynamic effects such as the effect of fluid drag due to the surrounding fluid. If the surrounding fluid is taken into account, the drainage rate is observed to depend on the droplet diameter.

Another unrealistic feature is the asymptotic behavior of this model as the radius of the disc decreases (R_d). The radius of the disc is a function of the interfacial tension. For a system with very little surfactant at the interface, and an oil composed of heavy hydrocarbon, the interfacial tension may be high enough that the droplet is rigid and roughly spherical. In that case, the disc radius would be very small. We would expect this model to reproduce the results of two spherical droplets approaching each other, with no deformation of the interface. However, this model does not reproduce that behavior. In the case of high interfacial tension, and roughly spherical droplets, this model deviates significantly from the expected. As will be discussed below, this feature can be eliminated.

In the literature, a force balance is often applied across the interface which, in the simple case of no disjoining pressure, allows use of the Young-Laplace equation:

$$P_2 - P_1 = \frac{F}{\pi R_d^2} = \frac{4\sigma}{d}$$

Eqn (4.31)

where

$\quad$ s $\quad$ = interfacial tension (N/m)
$\quad$ P_1 = the pressure inside a concave curved surface (N/m^2)
$\quad$ P_2 = pressure outside a concave curved surface (N/m2)

For the case where disjoining pressure is not zero, see [64]. Upon substitution:

$$u_p = -\frac{32\pi h^3 \sigma^2}{3\mu_c d^2 F}$$

Eqn (4.32)

In this equation the velocity (u_p) depends on the droplet diameter (d), unlike equation (x.1-1) which does not. But this dependence is only due to the use of the Young-Laplace equation and the introduction of the interfacial tension.

Another important feature of this equation is the dependence of the velocity on the driving force. The velocity goes down as the force goes up. At first thought, this would seem unrealistic. However, what this model says is that the higher the driving force, the more deformation that occurs in the droplet interface. The deformation causes the radius of the disc to go up, thus trapping more fluid in the parallel film between the droplets, and causing more hydrodynamic drag to occur.

Measured Force versus Drainage Rate: Chan and Horn [65] measured the resistance force due to film drainage as a function of the rate of drainage. They used an apparatus that was invented by Israelachvili [66]. They studied the drainage force for three Newtonian, nonpolar liquids. The results below are for OMCTS (octamethylcyclotetrasiloxane). The experimental apparatus consisted of molecularly smooth mica disks. These provide a non-deformable and immobile surface. They found good agreement between the measured results and the theory.

Contact Time: The contact time can be calculated by integrating the approach velocity from the initial contact distance (h_i) to the final contact distance (h_f):

$$t_d = \frac{3\pi\mu_c R_d^4}{4F}\left(\frac{1}{h_f^2} - \frac{1}{h_i^2}\right)$$

Eqn (4.33)

The disk diameter is the diameter of the disk that is formed when the drop deforms. In the literature it is often combined with the Young-Laplace equation which gives:

$$t_d = \frac{3m_c d^2 F}{4 \pi \sigma^2}\left(\frac{1}{h_f^2} - \frac{1}{h_i^2}\right)$$

Eqn (4.34)

Interface Mobility and Deformability: As discussed previously, when two fluid particles approach each other and deform, a disk-like film of liquid may form. The extent of flattening depends on the deformability of the interface. As the liquid drains, it will apply a shear force on the interface, in the direction of liquid flow. If the interface is mobile, then the interface will flow. This flow will create an internal circulation within each droplet. Flow of the interface and internal circulation reduces the drag due to liquid drainage. This will increase the drainage rate and shorten the coalescence time compared to droplets that have immobile interfaces. In produced water, immobile interfaces are often the result of an elastic network formed by asphaltene molecules at the interface.

Deformability for Clean Interfaces – Young-Laplace: The size of the disk that forms upon droplet collision can be readily calculated for clean interfaces where there is essentially no surface active materials. In that case, the Young-Laplace equation applies directly.

$$R_d = \sqrt{Fd/4\pi\sigma}$$

Eqn (4.35)

where:

d = diameter of the droplet (m)
R_d = radius of the disk (m)
s = interfacial tension (N/m)
F = Force acting on the droplets – usually a result of fluid turbulence (N)

This equation says, in effect, that the radius of the disk is larger if the turbulent force that pushed the drops together is large, or if the diameter of the droplet is large, or if the interfacial tension is small. All of these relationships make physical sense. However, this equation is not entirely valid for the case where surface active compounds are present. In that case, the interfacial tension is not the only property involved. The elasticity of the interface must be taken into account.

Mobility Number: For oil/water mixtures that do not have significant concentrations of surface active material at the interface, a dimensionless Mobility Number was defined by Davis et al. [62]. The Mobility Number is based on the ratio of bulk phase viscosities, i.e. the viscosity of the surrounding liquid (water) and the viscosity of the suspended phase (droplets). When m is large, the interface is mobile and it flows in the direction of the film drainage. For example, in the case of bubble coalescence, where the bubble/water interface is clean and not contaminated with surface active materials, the drainage time is at least a factor of ten times faster than the case of oil droplet coalescence. The gas viscosity is low and hence m is high. When m is small, the interface does not flow as much, and instead it imposes a drag on the film drainage process.

The mobility factor assumes that the interfaces are without adsorbed surfactant, particles, and surface active materials from the crude oil or produced water. Thus, interfacial mobility characterized by this approach has limited utility for crude oil/produced water systems. It is nevertheless instructive to note that mobile interfaces have much faster drainage and a much lower film drainage time.

4.5.6 Coalescence Model

A model for coalescence rate is presented here in this section. The discussion and equations that have been presented thus far provide mechanistic insight and practical means of calculating certain effects involved in coalescence. It is beyond the scope of this section to go into the details of this model. Those details are presented elsewhere [50, 51]. Suffice to say that the model is based on more precise versions of the formulas presented in previous sections, together with Population Balance Equations. The model uses the Mobility Number defined by Davis [62] and mentioned above.

The net result of this model is the ability to estimate the time required for coalescence of oil droplets in a flow line, over a range of turbulence intensity (ε – discussed in Section 3.4.2) and over a range of fluid physical properties. One such simulation result is given in the Figure below. In that figure, population balance modeling results are given for a 5 % oil in water mixture, at T = 50 C, and an interfacial tension of 20 dyne / cm. The dimensionless droplet diameter (y-axis) is given as a function of the quantity, time multiplied by epsilon (et) where epsilon is the turbulence intensity, for a range of values of epsilon (from 1 to 20 m²/sec³). The crude oil is assumed to have an API gravity of 27 degrees. When plotted in this way, several curves collapse into a single set of curves. This suggests that the quantity et (x-axis) can be used as a single independent variable. The dimensionless droplet diameter is defined next.

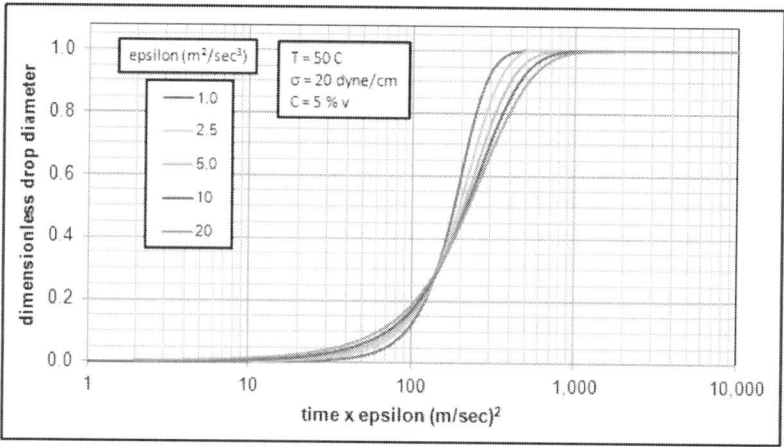

Figure 4.34 Population balance modeling results for the dimensionless droplet diameter versus the quantity: time x epsilon, where epsilon is the turbulence intensity. When plotted in this way, several curves collapse into a single set of curves. This suggests that the quantity time x epsilon can be used as a single independent variable.

Shown in the figure below is the relation between the Hinze droplet maximum diameter (x-axis), and the average droplet diameter (y-axis). This figure is used as follows. First, an estimate is made of the shearing intensity (ε – discussed in Section 3.4.2). The shearing intensity is used in Figure 3.15 to determine the dmax. Once dmax is know, the average droplet diameter (d_{32}) can be determined from the figure below. The final step is to determine if enough time has elapsed for the dispersion to achieve the d_{32} value. This is done by referring to figure 4.34 above. If the dimensionless droplet diameter is 1.0, then sufficient time has elapsed. If not, then the calculated d_{32} can be scaled by the dimensionless diameter: d_{32} actual = dimensionless drop diameter x d_{32} theoretical. The model for coalescence is comprised of these two figures (4.34 and 4.35), plus reference to Figure 3.15 which gives the dmax value.

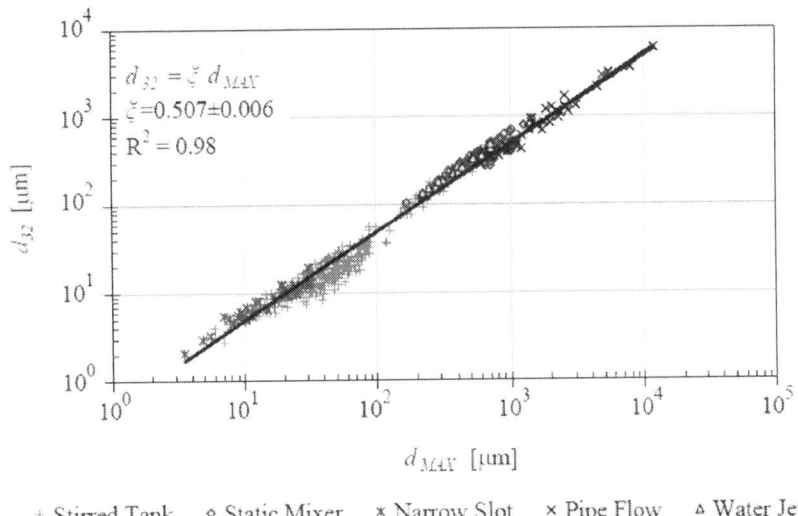

Figure 4.35 Average oil droplet size for flow situations where coalescence can occur such as pipe flow, static mixer, etc. [67] The value of dmax is that obtained from the Hinze correlation (Chapter 3).

Example Calculation: An example calculation is given here based on the information given previously by Arnold and Stewart [42] discussed at the beginning of Section 4.5.1. If we select a system having 4 inch pipe, and an ε value of 20 m^2/sec^3, then dmax (from Figure 2.5.1) is about 100 micron diameter, for a fluid of 5 dyne/cm (mN/m). Using Figure 4.34 above, the average droplet size will be about 60 micron after coalescence has occurred. Using Figure 4.35, the et value is about 800 for complete coalescence. Based on this, and an ε value of 20 m^2/sec^3, the required time is about 40 seconds. Since the fluid is traveling 5 m/sec, the pipeline must be 200 meters (660 ft) long. This is about 2,600 pipeline diameters; twice the time estimated by Arnold and Stewart. It is about 1,000 pipe diameters which is three times longer than Arnold and Stewart. All things considered, this is reasonable agreement.

4.5.7 The Role of Asphaltenes and Resins

Up to this point in the discussion of film drainage and coalescence modeling, the discussion has been restricted to clean interfaces without accumulation of oily solids, surfactants (natural and manmade), resins or asphaltenes. An advantage of discussing clean interfaces is that the role of droplet size, deformability, turbulent energy, and interfacial tension could all be understood rather clearly. Formulas were given that show explicitly the impact of these various properties on film drainage time and therefore emulsion stability. Such discussion lays the groundwork for understanding film drainage and coalescence in real systems of crude oil processing and produced water treatment.

Real produced water systems are far more complex and fascinating. As mentioned several times already, such mixtures have hundreds if not thousands of compounds that have surface active groups (carbonyl, ester, pyridine, etc). Also, the presence of naphthenic acids, resins, and asphaltenes has a significant effect on the interface. All of these features are now discussed.

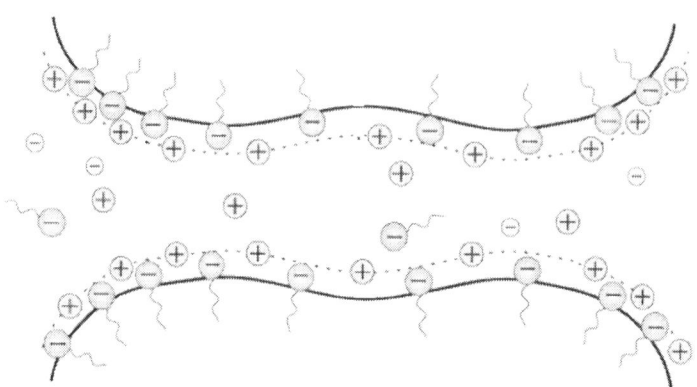

Figure 4.36 A schematic diagram of two droplet surfaces approaching each other in the presence of surfactant at the oil/water interface [68].

Oil-in-Water versus Water-in-Oil: Much of what is known about the oil / water interface in real crude oil / produced water mixtures is a result of the study of water-in-crude oil (WiO) emulsions. Water-in-crude oil emulsions are, to some extent, outside the subject of this book. However, the natural compounds that stabilize WiO emulsions usually also stabilize OiW emulsions. As discussed above in Section 4.4.4 (Interface Concentration versus Bulk Concentrations), interfacially active oil components can accumulate at the oil / water interface even when that interface is formed by oil droplets suspended in water. Examples of interfacially active oil components include naphthenic acids, resins, and asphaltenes. It is noted that naphthenic acids are actually part of the resin fraction. Much of the mechanistic understanding that has been developed for WiO emulsions can be applied to oil-in-water emulsions. In addition, as will be discussed, there are at least a few studies that have focused directly on the oil-in-water emulsions and the role of naphthenic acids, resins and asphaltenes on their stability.

Organic Acids: There are two broad classes of acids in crude oil / produced water mixtures. The first is the lower molecular weight acids that are preferentially soluble in water. These acids are referred to as water soluble acids. They are also known as Volatile Fatty Acids (VFA), or short chain fatty acids. They are not usually part of the resin fraction of oil because they typically partition to the water phase and therefore are not part of an oil sample.

The second class of acids is the so-called naphthenic acids that are preferentially soluble in crude oil. It must be emphasized that the solubility of these acids in water is a function of the pH, temperature, and ionic strength of the produced water. Thus, the definition of these two classes is rather imprecise. Nevertheless, both groups are important for different reasons.

Role of Water Soluble Acids: The water soluble acids lower the interfacial tension which makes the oil / water mixture more likely to form small oil droplets when sheared. This will obviously increase emulsion stability. The presence of these acids will cause emulsion stabilization through the Gibbs-Marangoni effect. The lower surface tension does not, per se, increase emulsion stability. But the presence of the ionized acid group ($RCOO^{-1}$) will cause an electrostatic repulsion between oil droplets and prevent them from coalescing [69]. Both types of acid (water soluble and oil soluble) tend to have a pKa of about 5 [70]. The pKa is the pH at which a mono-acid will have half of its molecules ionized and half protonated.

Role of Naphthenic Acids: There have been several studies of the behavior of naphthenic acids in stabilizing emulsions. As might be expected, these studies are somewhat contradictory. The reason why this is somewhat expected is that naphthenic acid is a very broad category of compounds. Within this broad category are a literally hundreds, if not thousands of compounds and isometrs with different acidities, surface activities, and molecular weights. Some of the acids are fatty acids (long-chain, saturated or unsaturated, single terminal carboxylic group). Other acids have multiple carboxylic groups, aromatic and saturated ring structures, and possibly other functional groups attached. It is thought that the smaller, more hydrophilic acids have high surface activity and displace asphaltenes from the oil / water interface which would reduce emulsion stability. The higher molecular weight acids with more complicated hydrocarbon groups are thought to form a gel-like complex with asphaltenes at the interface and stabilize emulsions [71].

Sjoblum and co-workers [70] have demonstrated that certain naphthenic acid fractions tend to reduce emulsion stability in WiO emulsions. The particular acid compounds studied were found to solubilize asphaltenes in the bulk oil. Thus, the acids hindered the asphaltenes and other resin compounds from forming stable films at the oil / water interface. Naphthenic acids are an important component of resins (SARA analysis) which, for the most part, are responsible for holding asphaltenes in solution.

Arla et al. [69] prepared oil and water mixtures over a range of water cut, from 10 to 50 %. The mixtures varied in the naphthenic acid, resin and asphaltene content. As expected, based on the Bancroft Rule (Section 4.4.5), the mixtures showed a preference to form water-in-oil emulsions, particularly at the lower values of water cut. This was expected because the surface active components were primarily oil soluble. However, OiW emulsions were also formed at higher water cut. The stability of the OiW emulsions increased as the activity of the naphthenic acids were increased (by increasing the pH), and importantly as the concentration of resins and asphaltenes increased. In fact, OiW emulsions were not stable with naphthenic acids alone. Resins and asphaltenes were required in order to stabilize OiW emulsions. The most stable OiW emulsions were formed by highly active naphthenic acids, together with resins and asphaltenes.

Asphaltenes are not highly surface active (Section 4.4.3 Surface Excess Concentration and Surface Activity). This means that the percentage of asphaltene molecules that migrate to the interface is not typically very high compared to simple surfactants like benzyl sulfonate, or the VFA compounds. However, there are sometimes percentage concentrations of asphaltene in the crude oil such that low surface activity can still result in relatively high population at the interface. On the other hand, the naphthenic acid from of the resins are more polar and do have higher surface activity. It has been found that the naphthenic acids tend to displace asphaltene from the interface and tend to break the network structure of the asphaltenes that do not get displaced. Thus, overall the effect of naphthenic acids is to reduce the emulsion forming tendency of asphaltenes, not just by increasing the solvency of asphaltenes, but also by reducing their action at the interface [72].

Thus, at least three important mechanisms for the role of naphthenic acids in emulsion stability have been worked out. The first is the effect of naphthenic acids in solubilizing asphaltenes and making them less prone to precipitation and less prone therefore to partition to the interface. This makes an emulsion less stable. The second is that the more highly surface active acids, with simple low molecular weight structures tend to displace the asphaltenes from the interface. The third is that the more complicated higher molecular weight structures tend to form a gel-like complex with asphaltenes at the interface. This mechanism can greatly increase emulsion stability.

Although these mechanisms have been worked out, the analytical procedures required to determine which, if any, are present in a given emulsion sample are rather involved, as explained in the literature [70 - 72]. Nevertheless, there are important consequences of these mechanisms and at least one simple oilfield test that can be carried out. As discussed several times already, the reduction in pressure that occurs in a facility causes CO_2 to break out of solution (vaporize). This increases the pH. In so doing, the organic acids will become more likely to be ionized rather than protonated. This greatly increases the surface activity which can enhance the second and third mechanisms discussed above. This is an effect of pressure reduction that is worthwhile to keep in mind when trying to understand changes in emulsion stability that may have occurred over time in a facility.

A simple oilfield test to determine if naphthenic acids are a factor in emulsion stability is to carry out bottle (WiO) or jar (OiW) tests at different pH values. Bottle and jar testing is a simple visual way to determine the stability of an emulsion. It is discussed in greater detail in the chapter on Troubleshooting. If there is a discernible difference in emulsion stability as a function of pH, then napthenic acids are likely at play. This can be important information for a chemical vendor to know since there are water clarifiers and emulsion breakers that will not work effectively in this situation and there are others that are specifically designed for it. In most cases, it is not practical to add acid. However, inhibited (with corrosion inhibitor) organic acid (such as glycolic acid) is sometimes used to reduce dissolved oil.

The Role of Resins: It has been determined that asphaltenes stabilize emulsions only if they are near or above their solubility limit in the crude oil (point of incipient flocculation) [72]. In that state, the asphaltene molecules form colloidal particulate aggregates. The molecules have a wide diversity of size, shape and functional groups. Some of the asphaltene molecules are surfactant-like having polar and ionizable pendant groups and other regions of non-polar or moderately polar fused aromatic clusters. The more polar asphaltene molecules migrate to the oil-water interface. They may migrate as individual molecules or as particles or aggregates. It has also been demonstrated that increasing the solvency of the bulk crude oil (by adding chemical solvents) can stabilize the asphaltenes such that they remain solvated in the crude oil. This reduces the tendency to stabilize emulsions.

As described by Wasan and coworkers [3, 73], by Kilpatrick and co-workers [72, 74], and by Sjoblom and co-workers [75], asphaltenes are believed to form a viscoelastic film at the oil / water interface which stabilizes the interface and prevents coalescence. The asphaltene molecules form a cross-linked network by intermolecular association through pi-bonds, hydrogen bonds, and electrostatic interactions. This viscoelastic membrane resists movement and thus the interface is relatively immobile. As discussed in Section 4.5.5 (Film Drainage Time), immobile interfaces require much longer time to drain. In a water-in-oil emulsion, the continuous phase is oil, which means that the film phase is oil. The oil-soluble interfacially active components such as asphaltenes will protrude away from the interface into the film and when the film has drained to the point that it is relatively thin, the protruding asphaltene molecules will exert a steric hindrance between the water drops. This will create a further barrier to coalescence. A similar effect is seen in the case of oil-in-water emulsions except that the steric barrier is caused by chain-like surfactant molecules such as alkyl carboxylic acids.

The resin fraction of crude oil (SARA analysis) helps to solubilize the asphaltenes by forming a resin-solvated aggregate and thus tend to pull the asphaltene away from the interface. The extent to which asphaltenes are solvated is an important determining factor in the effect of asphaltenes on emulsion stability. Asphaltenes that are solvated by resins in the bulk oil phase do not migrate to the interface and so not stabilize emulsions compared to asphaltenes that are precipitated and have formed colloidal suspended agglomerates. Thus, the aromaticity and resin content of the crude oil play a large role in the impact of asphaltene on emulsions [72, 74, 75].

The Role of Production Chemicals: In real crude oil / produced water mixtures, the oil / water interface can become populated with large molecular weight compounds such as asphaltenes and partially oil-wet solids. As will be discussed in Chapter 5 (Chemical Treatment), demulsifiers and other chemicals can be used to soften the interface by displacing the oil soluble interfacially active contaminants into the bulk oil phase. Demulsifiers composed of smaller molecular weight compounds are more interfacially active, thus pushing the larger molecules away from the interface. They tend to reduce film elasticity, and they lower the interfacial tension. This is analogous to the mechanism involved when naphthenic acids have been found to reduce emulsion stability.

Summary of Water-in-Oil Studies: At this point in the discussion, all of the concepts required to understand the role and mechanism of asphaltenes in film stabilization (emulsion stabilization) have been introduced:

1. **asphaltene solubilization:** The solubility of asphaltenes is determined by:

    o   the ratio of weight percent resins to asphaltenes (higher leads to greater solubility of asphaltenes)

    o   the ratio of weight percent aromatic to asphaltene (same as above)

    o   concentration of polar functional groups of resins compared to asphaltenes (CHNOS analysis of SARA fractions, same as above)

2. **surface activity:** asphaltenes are not highly surface active. However, they often do migrate to the oil-water interface and form a tough elastic membrane-like structure. They migrate to the interface rather slowly and the portion of the asphaltenes in the crude oil that do migrate is rather low. However, most crude oils have 1,000 mg/L to a few percent of asphaltenes, so surface activity is not limited by concentration but rather by the polarity of the molecule. It has been shown that naphthenic acids, in particular, and resins in general have greater surface activity and can bump asphaltenes off of the interface.

3. **interfacial elasticity:** when asphaltene solubility decreases, due to production processing (e.g. lower pressure) a fraction of the asphaltenes migrate to the oil / water interface where they form a network structure. This network of asphaltene molecules gives the interface an elastic strength and rigidity. Both properties reduce the film drainage rate and cause a barrier to coalescence.

Crude Oil in Water: While much of the above knowledge was gained in studies of water-in-crude oil emulsions, there have been similar studies carried out for oil-in-water emulsions, with similar conclusions. For example, Oye and co-workers studied the film drainage time for pairs of droplets composed of nine different crude oils [76]. The oils differed in TAN, SARA analysis and in the resin to asphaltene ratio. The strength of the elastic membrane formed at the oil / water interface was measured and found to correlate with the state of asphaltene solvency. As the resin, aromatic and TAN increased, the membrane strength decreased. Further, the drainage time of the film between the droplets was found to vary with the strength of the membrane. The higher the strength, the greater the film drainage time. All of these results are consistent with observations made for water-in-crude oil, as discussed above. The later results are shown in the Figure below.

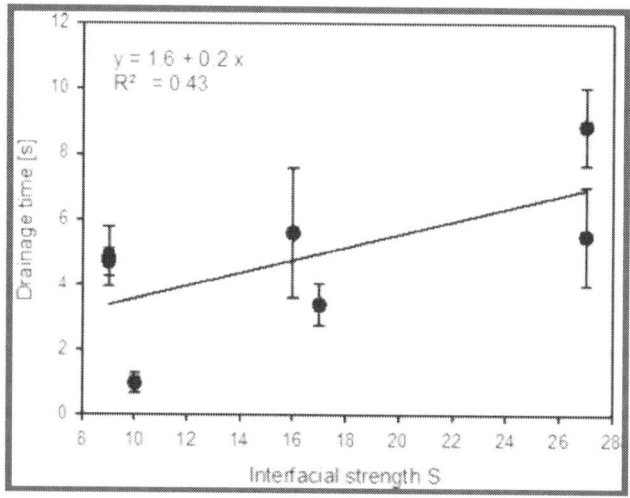

Figure 4.37 Drainage time versus Interfacial Strength (a measure of elasticity) for oil droplets in water measured for a number of different crude oils which differ in a number of different properties [76].

Czarnecki and co-workers have studied the film drainage process and the forces required to overcome the asphaltene interfacial network for oil droplets in produced water [77]. Their results confirm the concepts give here for asphaltene stabilization of oil-in-water emulsions.

Food and Dairy Emulsions: Emulsions in the food industry have been extensively studied [78, 79]. Homogenized milk, for example is a stabilized emulsion of oil (fat)-in-water. The fat droplets in milk are well suspended and are stabilized against coalescence by two proteins: casein and β-lactoglobulin. These proteins have regions of hydrophobicity and other regions of hydrophilicity. They are mildly surface active. They lower the interfacial tension between the fat droplets and the water but not to a large extent. The proteins migrate to the oil / water interface and form a network structure that makes the interface elastic and which provides a barrier to coalescence. This keeps the fat droplets small and well suspended in the solution. The effectiveness of this mechanism in stabilizing the oil-in-water emulsion is evident. The important points for our discussion are that both of these proteins are only mildly surface active, they are preferentially soluble in the fat rather than the water, and they stabilize the emulsion by forming a network structure. The tendency of a molecule to migrate to the interface is not necessarily related to the ability of the molecule to form a tough elastic barrier to coalescence. Thus, the mechanism of emulsion stabilization in milk is much like that of asphaltene stabilization of oil-in-water emulsions formed from crude oil and produced water.

Furthermore, in studies of milk emulsions it has been found [79] that common surfactant compounds are more surface active than the milk proteins casein and β-lactoglobulin. In mixtures of common surfactants with milk proteins, the surfactants displace the milk proteins from the interface. This mechanism is referred to as competitive displacement. Such surfactants have a strong tendency to migrate to the interface but do not provide nearly as much stability to the interface compared to β-lactoglobulin. This suggests that some surfactant molecules can actually behave as emulsion breaking additives. This is entirely consistent with the properties of milk emulsions just discussed. Competitive displacement is a strategy that is used routinely in breaking oilfield emulsions, both water-in-oil and oil-in-water, although different types of compounds are used for these two types of emulsions.

4.5.8 Elasticity and Viscosity of the Oil/Water Interface

Interfacial Viscosity: For oil/water mixtures where significant contamination of the interface is expected to occur, Wasan and co-workers defined the Interfacial Viscosity Number [3, 73, 86] as a measure of the interfacial shear viscosity. The Interfacial Viscosity Number is a measure of the extent to which an interface resists flow. As previously discussed, the drainage of liquid from the space between two oil droplets imposes a stress on the interface which will cause it, and the fluid within the droplet to flow. If the interface has high resistance to flow, then it will flow slowly or not at all. Such an interface would have a high Interfacial Viscosity Number. It is a quantity that can be measured, and which has been measured by Wasan and co-workers. It is not measured in routine oilfield sampling and analysis. It does however provide mechanistic insight as to why some oil-in-water emulsions coalesce rapidly and others do not. Also, it is used by the chemical suppliers to develop demulsifying agents. The Interfacial Viscosity Number is used below to calculate the relative drainage time for oil/water systems.

Presence of Surfactants and Particles at the Interface: Some surfactants and film-forming agents, whether naturally occurring in the oil and water, or added as, for example, corrosion or hydrate inhibitors will have a dramatic effect on coalescence, as will the presence of small solid particles at the interface. Both film-forming agents and particles will reduce the drainage rate of fluid from between the two droplets. Surfactants may either increase or decrease the drainage rate depending upon their surface activity (surface excess concentration) and the impact they have on, for example, interfacial tension, viscosity and elasticity. Also, at close distance the presence of contaminants at the interface will cause a steric repulsive force between the two droplets. The Gibbs-Marangoni effect discussed in Section 4.5.8 (Elasticity and Viscosity of the Oil/Water Interface) and in the next paragraph also reduces the probability of coalescence.

Gibbs-Marangoni Elasticity: The Gibbs Elasticity and the Marangoni effect were discussed in Section 4.4.7 (Gibbs Elasticity). It was pointed out that the interface can take on an elastic quality during the film drainage process. As the film drains, any surfactant adsorbed at the interfaces of the droplets is swept along the interface due to fluid shear. As the disc-like fluid drains from between the approaching droplets, the draining fluid will drag surfactant with it from the center of the disc to the outer edge. This sets up a concentration gradient of the surfactant. The concentration is higher at the edge of the disk and lower at the center. This concentration gradient leads to a fluid tension gradient in the opposite direction. In other words, the interfacial tension is higher in the center of the disc (due to lack of surfactant) and is lower on the edge of the disc (where the surfactant concentration is higher and therefore the interfacial tension is lower). This gradient in surface tension opposes the drainage of fluid from between the approaching droplets. The Gibbs Elasticity provides a relative measure of this effect. It is also noted that Gibbs Elasticity, as the name implies, gives the interface an elastic-like quality which stabilizes it against mechanical disturbances which might otherwise lead to film rupture (see p. 129 of [87]). Gibbs Elasticity is not a typical oilfield measurement. It does however provide mechanistic insight into the effect that surfactants have on the stability of emulsions. While surfactants lower the interfacial tension, they also increase the stability of emulsions, in part through the Gibbs-Marangoni effect.

Interfacial Elasticity Number: The Elasticity Number is essentially the Gibbs Elasticity formula (see Eqn 4.10) divided by the diffusivity of the surfactant, the viscosity of the water phase and multiplied by the radius of the droplet. As the diffusivity of the surfactant goes up, the concentration gradient of surfactant goes down, which lowers the gradient in the interfacial tension. The interface behaves more like a liquid and less like a solid elastic membrane. For calculation purposes, a value of the diffusion coefficient is required. Values of this quantity has been reported for only a few systems. For the case of ethylene oxide surfactant in water a value of $D = 4 \times 10^{-10}$ m^2/s has been reported [33, 80].

Drainage time is given below as a function of the Elasticity Number and as a function of the Interfacial Viscosity Number [73, 78, 81]. The horizontal line (F) corresponds to the Reynolds Model for drainage. In that case, discussed above, the interface deforms to form a disk but the interface is not mobile. Mobility of the interface occurs at low values of the Interfacial Viscosity Number and at low values of the Elasticity Number. Both of these must be moderately low in order for mobility of the interface to occur. When both are very low, then the time required for drainage is only 1/3 that predicted by the Reynolds model.

In the next Figure, the drainage time is given as a function of the Elasticity Number and Adsorption Number. The later quantity is related to the Surface Excess that was discussed in Section 4.4.3 (Surface Excess Concentration and Surface Activity).

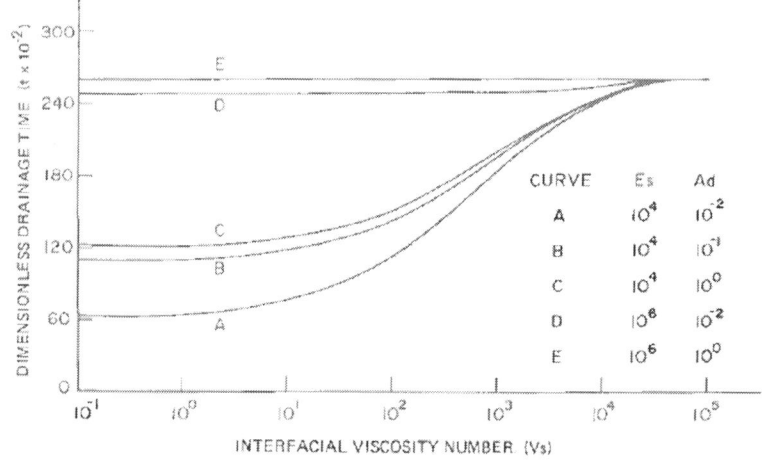

Figure 4.38 Drainage time versus Interfacial Viscosity Number as a function of the Elasticity Number and the Adsorption Number. High interfacial viscosity gives a high drainage. Lower interfacial viscosity gives lower drainage times [78].

As shown in the figure, as the Adsorption Number goes down, the interface becomes more mobile. Physically this says that a surface active compound that is not very surface active will not stiffen or immobilize the interface. This can be a bit misleading. As will be shown, many demulsifiers achieve better coalescence by having a high Surface Excess, or Adsorption Number. They migrate to the interface and displace the asphaltenes and resins. In so doing, they push the rigid network-forming compounds into the bulk and away from the interface. In order to do this of course they require higher Surface Excess than that of the asphaltenes and resins.

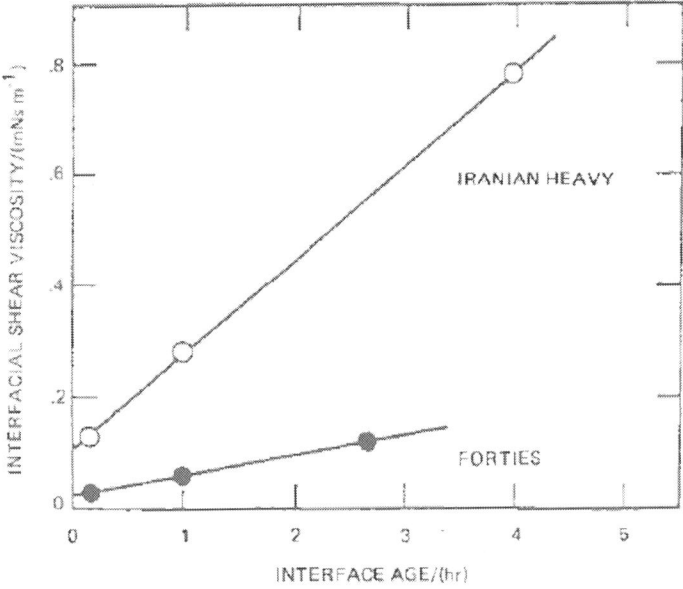

Figure 4.39 Crude oil / water interfacial shear viscosities at 35 C [82]

Based on the values given in Figure 4.39, the Interfacial Viscosity Numbers for the two crude oil/ water systems can be estimated. An aging time of one hour is used. The estimated value for Iranian Heavy crude oil/water is 3,000. The estimated value for Forties for the same aging time is about 500. These values represent a 6/1 ratio in the dimensionless interfacial shear viscosity. A similar ratio is shown in Figure 4.39 where the actual interfacial shear viscosity is shown.

The Elasticity Number will be a relatively large number for crude oil / water mixtures, where asphaltenes and resins are primarily responsible for surface active compounds at the interface. In the case of the two crude oils in the figure, Forties is assumed to have low elasticity and Iranian Heavy is assumed to have high elasticity.

Values for the relevant parameters are collected in Table 4.20 below. The table has an entry for the Dimensionless Drainage Time from Figure 4.38. Referring to that figure, the dimensionless drainage time for these two systems is estimated as 260 and 100 respectively. Thus, the Forties crude/water system drains about 2.6 times faster than the Iranian Heavy system. The next figure demonstrates this with actual measurements of drainage rate. The drainage time was measured for these two crude/water systems. The drainage rate was calculated from the slope of the line at four minutes (1/sqrt(.25) minutes$^{-1/2}$).

There is good agreement between drainage time calculated based on Figure 4.38, and the measured drainage rate shown in Figure 4.40. Based on Figure 4.38 the ratio of drainage times for the two crude oils is 2.6 (260 / 100), as shown in Table 4.20 below. Based on the Figure 4.40 the ratio of drainage rate for the two crude oils is 2.5 (1.47/0.6). Thus, there is good agreement between the theory for film drainage and the measurements of drainage rate.

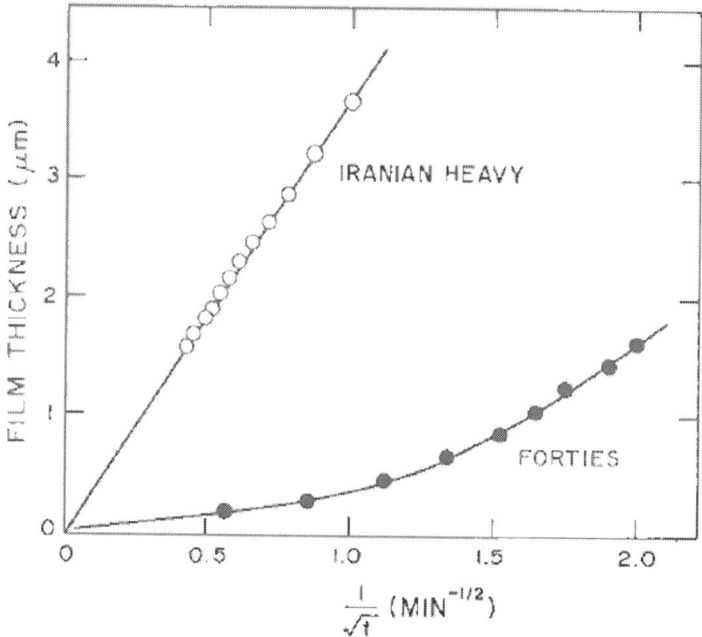

Figure 4.40 Film thickness versus drainage time for two different crude oil/water systems. As shown, there is a dramatic difference in the rate of film drainage for these two systems.

Table 4.20

Parameter	Iranian Heavy	Forties	Units
Interfacial Viscosity Number (estimated)	3,000	500	dimensionless
Elasticity Number (estimated)	1×10^6	1×10^4	dimensionless
Dimensionless Drainage Time (Figure 4.38)	260	100	dimensionless
Measured drainage rate (Figure 4.40.)	0.6	1.47	micron/minute

As demonstrated in the figure below, the presence of a tough network of asphaltene molecules at the oil/water interface can prevent the final coalescence step from occurring. In the experiment depicted in the figure, 1 mL of bitumen was suspended in 10 mL of simulated produced water. The oil droplet diameter ranged between 5 to 30 micron.

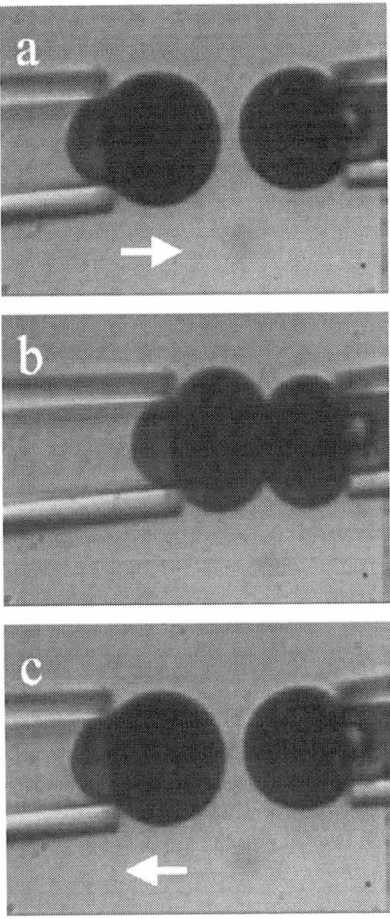

Figure 4.41 Demonstration of the tough oil/water interface that can form when asphaltenes migrate to the interface. Micropipettes were used to push the droplets together. The droplets were pressed together for several minutes during which time no coalescence occurred [77]

4.5.9 Practical Devices

Most of the discussion above has been on the estimate of the effect of pipeline transportation on the oil droplet size. As discussed, sufficient pipeline length can have a significant benefit in promoting coalescence. In addition, over the years there have been a few practical devices developed to maximize oil droplet coalescence over shorter intervals of residence time.

The Hydoflokk [83] process was developed by Norsk Hydro during the conceptual phase of the Troll B Project. The Troll Oil concept is a distributed subsea system assumed to create water treatment problems due to high water cut, long subsea transport and possible flow disturbances in separators. Pipeline slugging and relative low temperature and short retention time in the first stage separator were all factors contributing to the water treatment problem. Several methods were evaluated in the initial conceptual design. The choice was made to install hydrocyclones, but to allow space for future improvements. Studies showed that improved hydrocyclone performance could be achieved by installing an upstream oil droplet flocculation process. Based on water with 1000 ppm oil droplets of an average size of 5μm, pilot scale trials with a flocculation process upstream a hydrocyclone liner were carried out. The pilot study gave the following hydrocyclone separation efficiencies at the different operating conditions.

The dosing rate with one component system was 5 ppm, and with two component systems 1.5 ppm coagulant and 0.5 ppm flocculant, all as 100% active material. The total retention time (flocculation vessel + upstream/downstream pipelines) was 2.5 minutes. Based on the above very promising results, the Hydroflokk process was installed at the Troll B platform.

The Hydrocyclones for treatment of produced water from the 1st stage separator (via the flocculation vessel), had a very good separation efficiency of 92-95% at normal operation (i.e. Hydroflokk was operated without chemical injection).

An increased hydrocyclone efficiency of 96-97% was observed in the tests with flocculant injection (5, 10 and 20 ppm WT-34). A the same time, a reduced discharge of hydrocarbons in the produced water overboard was observed, from 20-23 ppm at normal operation to 10-13 ppm with flocculant injection. Also, the water samples were visually cleaner with flocculant injection.

At the present water production, the flocculation vessel operates as a 1-g separator upstream the hydrocyclones as oil is skimmed off from the vessel. 1-g separation is possible due to larger oil droplets than expected at the inlet of the vessel (dv50 approx. 12μm) together with high oil concentration thereby high coalescence in the vessel.

Free-flow turbulent coalescers: Free-flow turbulent coalescers are a type of device that is installed inside or just upstream of any skim tank or coalesce to promote coalescence. These devices had been marketed and sold under the trade name SP Packers. They are no longer available; but the concept can still be employed in water treating system design. SP Packs force water flow to follow a serpentine pipe-like path sized to create turbulence of sufficient magnitude to promote coalescence, but not so great as to shear the oil droplets below a specified size. SP Packs are designed to coalesce oil droplets to a defined drop size distribution with a dmax of 1,000 Microns [42].

Pre-coalescers: Several new technologies and emerging technologies in the mainstream produced water treatment industry are pre-coalescers. These are often used in treatment processes or pipes; upstream of gravity type or centrifugal separation technologies, and operate by increasing the overall oil droplet size. Increasing sizing makes it easier to remove oil in a subsequent downstream process. These devices are usually composed of numerous fine polyethylene strands packed in tightly with each other. The strands provide high surface area contact within the flowing produced water, and attract small oil particles and aid in their coalescence until they are too large to be held by the media at which point they are released back into the process stream.

PECT –Performance Enhancing Coalescence Technology: PECT is a technology concept unique to Cyclotech, Inc. The first commercialized technology in this range is the PECT-F®, denoting a Fiber based coalesce concept. Cyclotech developed the PECT-F® technology to achieve a significant improvement in separation efficiency of Produced Water Deoiling Hydrocyclone systems. The PECT-F® is a media based coalescer which is installed as a cartridge assembly into either the inlet chamber of the Deoiling Hydrocyclone vessel or into a bespoke vessel located upstream of the PWT system. The inlet chamber of a typical conventional Deoiling Hydrocyclone is the largest chamber in the vessel and has a residence time of up to 20 seconds. The PECT-F® uses this residence time constructively to achieve partial oil droplet coalescence to capture and grow droplets from a size that would not be separated by the Deoiling hydrocyclone to a size that can be separated. The PECT-F® technology is targeted at

- existing system which do not meet current legal or stretch targets of performance, or require excessive chemical dosing to do so

- new building systems where the application of Deoiling Hydrocyclones is potentially marginal due to difficult fluid characteristics [84]

Mare's Tail: Tulloch [85] tested a pre-coalescer device that consists of a bundle of oleophilic polypropylene fibers inside a cartridge positioned along a flow line, just upstream of another separation device (e.g. hydrocyclone, filter). The fibers serve to aggregate small oil droplets for easier downstream removal. The coalescence occurs rapidly (within two seconds). The appearance of the fiber bundle looks somewhat like the tail of a horse, hence the name "Mare's Tail®" Tulloch (2003) [85] reports that oil droplets growth was enhanced, by increasing either the length of the fibers or the number of fibers packed into the cartridge.

4.6 Film Drainage & High Internal Phase Volume Emulsions

As discussed previously, a schematic diagram of the different types of emulsions that might be found in a primary separator is shown in Figure 4.5. The figure shows four different types of emulsions. The top emulsion is referred to as a dispersion of water droplets suspended in oil. The bottom emulsion is a dispersion of oil droplets in water. The middle two emulsions are bi-liquid foams (also known as High Internal Phase Volume emulsions). Bi-liquid foam is also shown schematically in the Figure below where panels (e) and (f) show the bi-liquid foam sitting just under the oil phase. The distinguishing feature of a bi-liquid foam is a high water cut (as seen in Figure 4.5), with large water droplets separated by relatively thin film of oil. The bi-liquid foam is formed whenever the oil phase has a relatively high viscosity and the (upward) drainage of the oil becomes a bottleneck in resolving the emulsion.

The most important aspect of a bi-liquid foam is recognition that it could or does form. Much more will be discussed later about the impact of bi-liquid foams on oil/water separation. Suffice to say for now that they can occupy significant height in a separator and can result in very wet oil being discharged. When this happens, the operators will reduce the height of the oil/water interface in order to give the oil more residence time for dehydration, which will reduce the water residence time and cause dirty water to be discharged from the separator. It is a no-win situation. Bi-liquid foams can be detected by simple visual observation in bottle tests. The Figure below gives the idea. The procedure for bottle testing is given in Chapter 5.

Pereyra et al [4], and Henschke et al. [5] provide a model for the rate at which this layer resolves into two continuous phases of oil on top and water on the bottom. Polderman et al. [6] refer to this type of emulsion as a Dispersion Band and they provide a practical model with field data for the prediction of the thickness and drainage time. The height of the oil-continuous HIPV emulsion depends mostly on the oil viscosity and the drainage rate of the oil film. The percentage of water (water cut) is relatively high, often 70% or more. The lifetime of such a fluid mixture depends on the drainage time of the liquid film of oil which depends primarily on the oil viscosity.

There are two approaches to dealing with bi-liquid foams. In the design phase of a project, the cross-sectional area of the separator can be increased. This will reduce the height of the bi-liquid foam and reduce the tendency to discharge wet oil. Chemicals can be added upstream which will promote film rupture. There are two types of chemicals. One type attacks the emulsion from the oil side. The other type attacks the emulsion from the water side. Such chemicals include short chain acids, or quaternary amine type compounds that pull other surface active compounds away from the interface and into the bulk. On the oil side, where most of the asphaltenes and resins are found, light hydrophobic solvents such as xylene or naphtha are effective at reducing the viscosity of the continuous phase and thereby promoting drainage. Other oil-based chemicals can reduce the elasticity of the film, and various polymers are used to reduce the stability of the film.

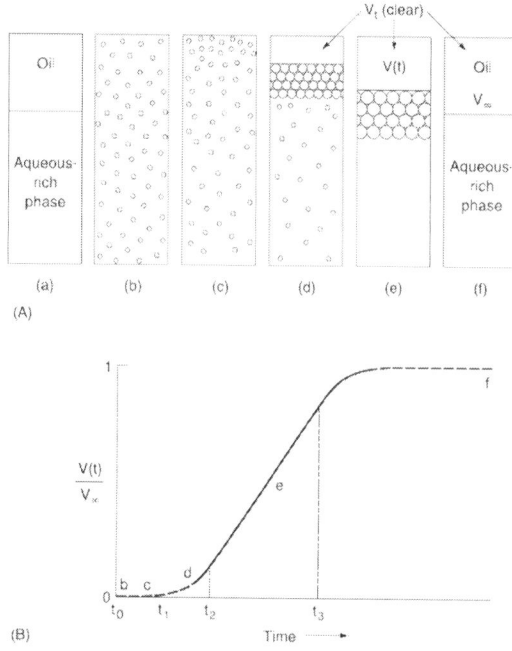

FIGURE 1.25 (A) Commonly observed sequence of events in separation processes. Schematic sequence of direct observations: (a) Before shaking: equilibrated volumes. (b) Just after stopping shaking; $t = 0$. Macroemulsion of oil phase in aqueous phase. (c) Droplets of oil have migrated toward top of aqueous phase, but no appreciable coalescence is detectable. $t_0 < t < t_1$. (d) Appreciable coalescence of oil droplets; V_t increasing. Note larger droplet size just below clear layer, and sedimentation from lower part of aqueous phase still in progress. $t_1 < t < t_2$. (e) Sedimentation effectively complete. Coalescence of droplets continues in layer at top of aqueous phase. $t_2 < t < t_3$. (f) Separation complete. $t \gg t_3$. (B) Schematic kinetic curve of separation, corresponding to conditions in (A). (Parts A and B after I.B. Ivanov, R.K. Jain, P. Somasundaran, and T.T. Traykov, in *Solution Behavior of Surfactants*, Vol 2 (K.L. Mittal, ed.), Plenum Press, New York, 1979, p. 817.)

Figure 4.42 Commonly observed sequence of events in separation processes.

4.7 Concluding Remarks

Significant detail has been presented in this chapter on various subjects related to emulsions in produced water. As discussed in the first section of the Introduction to this book, a guiding principle in writing this book was to provide enough detail to explain field observations from a mechanistic standpoint, and to explain why the certain production chemicals, equipment, and process configurations work well and others do not. In some cases, a simple basic explanation will suffice. In other cases, considerable background and depth is required. In the case of emulsions, considerable depth is required in almost every aspect. This is due to the fact that there is a lot happening on the small size scale of droplets, interfaces, and thin films.

For example, small surfactant-like molecules have high surface activity and have a strong tendency to populate the oil/water or water/gas interface. They lower the interfacial tension which enhances the formation of small droplets due to shear. But, they also soften the interface by pushing the large network-forming asphaltenes and other high molecular weight compounds. Those later compounds on the other hand have less tendency to migrate to the interface. In high enough concentration asphaltenes will migrate to the oil/water interface and cause high interfacial elasticity and prevent coalescence. This is why it is so important in to prevent shearing in the first place. It is also the reason why some so-called demulsifiers are composed of short-chain organic acid. The application of such treatment chemicals must be carried out without causing small droplets due to shearing.

Surface tension, though important, is not the only important variable. Surface activity, interfacial elasticity, ionic concentration, formation of oily solids are also important in understanding how to effectively treat a produced water stream. In most cases, it is not necessary to carry out detailed measurements of these variables. Instead, just knowing the mechanisms can provide enough insight to make the connection between basic measurements such as API gravity, SARA analysis, Total Acid Number, etc., and the various challenges that will be encountered. More will be said about this in Chapter 4 on Characterization and in the various application chapters later in the book.

References to Chapter 4

1. S.S. Dukhin, J. Sjoblom, O. Saether, "An experimental and theoretical approach to the dynamic behavior of emulsions," Chapter 1 from Emulsions and Emulsion Stability, 2nd Ed., Ed. J. Sjoblom, CRC Taylor & Francis(2006)

2. S. Kokal, "Crude-oil emulsions: A state of the art review," SPE – 77497, SPE Production & Facilities, p. 5 (2005).

3. P.J. Breen, D.T. Wasan, Y.-H. Kim, A.D. Nikolov, "Emulsions and emulsion stability," chapter 4 in Emulsions and Emulsion Stability, Ed.: J. Sjoblom, Marcel Dekker, Inc., New York (1996).

4. E. Pereyra, R. Mohan, O. Shoham, "A simplified mechanistic model for an oil/water horizontal pipe separator," Oil & Gas facilities, p. 40 (2013).

5. M. Henschke, L.H. Schlieper, A. Pfennig, "Determination of a coalescence parameter from batch-settling experiments," Chem. Eng. J., v. 85, p. 369 (2002).

6. H.G Polderman, J.S. Bouma, H. van der Poel, "Design rules for dehydration tanks and separator vessels," SPE – 38816, paper presented at the SPE Annual Technical Conference and Exhibition (1997).

7. J. N. Israelachvili and R. M. Pashley, J. Colloid Interface Sci. 98:500 (1984).

8. P. M. Claesson and H. K. Christenson, J. Phys. Chem. 92:1650 (1988).

9. D. Chandler, "Interfaces and the driving force of hydrophobic assembly," Nature, v. 437, p. 640 (2005)

10. D. Henderson, D. Boda, "insights from theory and simulation on the electrical double layer," Phys. Chem. Chem. Phys., v. 11, p. 3822 (2009).

11. D.J. Shaw, Introduction to Colloid & Surface Chemistry, Butterworth Heinemann, 4th Ed. (1992).

12. A. Wiacek, E. Chibowski, "Zeta potential, effective diameter and multimodal size distribution in oil / water emulsion," Coll. Surf. A: Physicochem.Eng. Aspects, v. 159, p. 253 (1999).

13. J. Stachurski, M. Michalek, The effect of the z potential on the stability of a non-polar oil-in-water emulsion, J. Coll. Int. Sci., v. 184, p. 433 – 436 (1996).

14. B.M. Moudgil, P.K. Singh, J.J. Adler, "Chapter 10: Surface Chemistry in Dispersion, Flocculation and Flotation," from Handbook of Applied Surface and Colloid Chemistry, John Wiley & Sons (2001).

15. P. Somasundaran, G.E. Agar, "The zero point of charge of calcite," J. Coll. Interface Sci., v. 24, p. 433 (1967).

16. G. Atesok, P. Somasundaran, L.J. Morgan, "Adsorption properties of Ca++ on Na-Kaolinite and its effect on flocculation using polyacrylamides," Coll. Surf., v. 32, p. 127 (1988).

17. S.L. Swartzen-Allen, E. Matijevic, "Surface and colloid chemistry of clay," Chem. Rev., v. 74, p. 385 (1974).

18. A. Graciaa, G. Morel, P. Saulner, J. Lachaise, R.S. Schechter, "The zeta-potential of gas bubbles," J. Coll. Int. Sci., v. 172, p. 131-136 (1995).

19. S. Usui, H. Sasaki, "Zeta potential measurements of bubbles in aqueous surfactant solutions," J. Coll. Int. Sci., v. 65, p. 36 (1978).

20. A.M. Elmahdy, M. Mirnezami, J.A. Finch, "Zeta potential of air bubbles in presence of frothers," Int. J. Miner. Process., v. 89, p. 40 – 43 (2008).

21. J. Lyklema, T. Golub, "Electrical double layer on silver iodide and overcharging in the presence of hydrolysable cation," Croatica Chem. Acta, v. 80, p. 303 (2007).

22. P.A. Kralchevsky, K.D. Danov, N.D. Denkov, "Chemical physics of colloid systems and interfaces, in: Handbook of Surface and Colloid Chemistry, Ed.: K.S. Birdi, CRC Press, London (1997).

23. A. Marchand, J.H. Weijs, J.H. Snoeijer, B. Andreotti, "Why is surface tension a force parallel to the interface?" Am. J. Phys., v. 79, p. 999 (2011).

24. B. Jirgensons, M.E. Straumanis, A Short Textbook of Colloid Chemistry, Second Ed., The Macmillan Company, N.Y. (1962).

25. M.-H. Ese, J. Sjoblom, H. Fordedal, O. Urdahl, H.P. Ronningsen, "Aging of interfacially active components and its effect on emulsion stability as studied by means of high voltage dielectric spectroscopy measurements," Coll. Surf. A, v. 123, p. 225 (1997).

26. S. Kelesoglu, P. Meakin, J. Sjoblom, "Effect of aqueous phase pH on the dynamic interfacial tension of acidic crude oils and myristic acid in dodecane," J. Disp., Sci. and Tech., v. 32, p. 1682 (2011).

27. I.B. Ivanov, D.S. Dimitrov, "Thin Film Drainage," Chapter 7 in Thin Liquid Films, Ed. I.B. Ivanov, Surfactant Science Series, v. 29, Marcel Dekker, Inc., N.Y. (1988).

28. P.C. Hiemenz, R. Rajagopalan, Principles of Colloid and Surface Chemistry, 3rd Ed., CRC Taylor & Francis, Boca Raton, (1997).

29. A.C. da Silva Ramos, L. Haraguchi, F.R. Notrispe, W. Loh, R.S. Mohamed, "Interfacial and colloidal behavior of asphaltenes obtained from Brazilian crude oils," J. Pet. Sci. & Eng., v. 32, p. 201 (2001).

30. A.W. Adamson, Physical Chemistry of Surfaces, John Wiley and Sons (1990).

31. E.Y. Sheu, D.A. Storm, "Colloidal properties of asphaltenes in organic solvents," Chapter 1 of Asphaltenes – Fundamentals and Applications, Ed.: E.Y. Sheu, O.C. Mullins, Springer (1995).

32. J.T. Davies, "A quantitative kinetic theory of emulsion type. I. Physical chemistry of the emulsifying agent," a chapter in Gas/Liquid and Liquid/Liquid Interface. Proceedings of the 2nd International Congress on Surface Activity, Butterworths, London (1957).

33. E.H. Lucassen-Reynders, A. Cagna, J. Lucassen, "Gibbs elasticity, surface dilational modulus and diffusional relaxation in nonionic surfactant monolayers," Coll. Surf. A, v. 186, p. 63 (2001)

34. J.B. Fournier, A.M. Cazabat, "Tears of wine," Europhys. Lett., v. 20, p. 517 (1992).

35. J.J. Adams, "Asphaltene adsorption, a literature review," Energy & Fuels, v. 28, p. 2831 (2014).

36. K.R. Dean, J.A. McAtee, "Asphaltene adsorption on clay," Appl. Clay Sci., v. 1, p. 313 (1986).

37. D. Dudasova, S. Simon, P.V. Hemmingsen, J. Sjoblom, "Study of asphaltenes adsorption onto different minerals and clays. Part 1. Experimental adsorption with UV depletion detection," Coll. Surf. A, v. 317, p. 1 (2009).

38. D. Dudasova, G.R. Flaten, J. Sjoblom, G. Oye, "Study of asphaltenes adsorption onto different minerals and clays. Part 2. Particle characterization and suspension stability," Coll. Surf. A, v. 335, p. 62 (2009).

39. D.E. Tambe, M.M. Sharma, "Factors controlling the stability of colloid-stabilized emulsions," J. Coll. Interface Sci., v. 157, p. 244 (1993).

40. D.E. Tambe, M.M. Sharma, "The effect of colloidal particles on fluid-fluid interfacial properties and emulsion stability," Adv. Colloid Interfacial Sci., v. 52, p. 1 (1994).

41. E. Gulbrandsen, A. Pedersen, "Alteration of sand wettability by corrosion inhibitors and its effect on formation of sand deposits," SPE – 106492, paper presented at the 2007 Society of Petroleum Engineers International Symposium on Oilfield Chemistry, Houston (2007).

42. K. Arnold, M. Stewart, Surface Production Operations: Design of Oil Handling Systems and Facilities, 3rd Edition, Gulf Professional Publishing, an Elsevier company, Amsterdam (2008).

43. G. Narsimhan, "Model for drop coalescence in a locally isotropic turbulent flow field." J. Colloid and Interface Science, 272, 197-209 (2004).

44. C. Lee, L.E. Erickson, L.A. Glasgow, "Bubble breakup and coalescence in turbulent gas-liquid dispersions," Chem. Eng. Comm. 59, 65-84 (1987).

45. G. Marrucci, L. Nicodemo, D. Acierno, Concurrent gas-liquid flow. Plenum Press, NY (1972).

46. C. Tsouris, L.L. Tavlarides, "Breakage and coalescence models for drops in turbulent dispersions," AIChE J., v. 40, p. 395 (1994).

47. B.C.H. Venneker, J.J. Derksen, H.E.A. Van den Akker, "Population balance modeling of aerated stirred vessels based on CFD," AIChE J. v. 48, p. 673 (2002)

48. K.D. Danov, I.B. Ivanov, Critical film thickness and coalescence in emulsions, in: Proceedings of the 2nd world congress on emulsions, Paper 2-3-154(1997)

49. A.K. Chesters, "The modeling of coalescence processes in fluid-liquid dispersions: a review of current understanding." Trans. IChemE. 69, 260-270 (1991).

50. Z. Chen, J. Pruss, H.-J. Warnecke, "A population balance model for disperse systems: Drop size distribution in emulsion," Chem. Eng. Sci., v. 53, p. 1059 (1998).

51. A. Motin, J.M. Walsh, A. Benard, "Modeling droplets shearing and coalescence using a population balance method in produced water treatment systems," Proc. Int. Mech. Eng. Congress & Expo. (IMECE2015), Paper No.: IMECE2015-53097, Houston (Nov 2015).

52. J. Abrahamson, "Collision rates of small particles in a vigorously turbulent fluid," v. 30, p. 1371 (1975).

53. C.A. Coulaloglou, L.L. Tavlarides, "Description of interaction processes in agitated liquid-liquid dispersions," Chem. Eng. Sci. 32, 1289-1297 (1977).

54. H. Luo, Coalescence, breakup and liquid circulation in bubble column reactors. Ph.D. Dissertation, The Norwegian Institute of Technology, Trondheim (1993).

55. R. Kuboi, I. Komasawa, T. Otake, "Behavior of dispersed particles in turbulent liquid," J. Chem. Eng. Japan, v. 5, p. 349 (1972).

56. R. Kuboi, I. Komasawa, T. Otake, "Collision and coalescence of dispersed drops in turbulent liquid flow," J. Chem. Eng. Japan, v. 5, p. 423 (1972).

57. B.V. Deryagin, L. Landau, Acta Phys. Chim., URSS, v. 14, p. 633 (1941).

58. E.J.W. Verwey, J.Th.G. Overbeek, Theory of the Stability of Lyophobic Colloids, Elsevier (1948).

59. Y. Liao, D. Lucas, "A literature review on mechanism and models for the coalescence process of fluid particles." Chem. Eng. Sci. 65, 2851-2864 (2010).

60. A.M. Kamp, A.K. Chesters, C. Colin, J. Fabre, "Bubble coalescence in turbulent flows: A mechanistic model for turbulence-induced coalescence applied to microgravity bubbly pipe flow," Int. J. Multiphase Flow, 27, 1363-1396, (2001).

61. V.G. Levich, Physicochemical Hydrodynamics, translated from the Russian by Scipta Technica, Inc., Prentice Hall, Inc., Englewood Cliffs, N.J. (1962).

62. R.H. Davis, J.A. Schonberg, J.M. Rallison, "The lubrication force between two viscous drops," Phys. Fluids A 1, 77-81 (1989)

63. G.V. Jeffreys, G.A. Davies, Coalescence of liquid droplets and liquid dispersions: Recent advances in liquid-liquid extraction, 1st ed Pergamon Press, Oxford, UK, 495 (1971).

64. I.B. Ivanov, P.A. Kralchevsky, "Stability of emulsions under equilibrium and dynamic conditions," Coll. & Surf. A, v. 128, p. 155 (1997).

65. D.Y.C. Chan, R.G. Horn, The drainage of thin liquid films between solid surfaces, J. Chem. Phys., v. 83, p. 5311 (1985).

66. J.N. Israelachvili, G.E. Adams, Measurement of forces between two mica surfaces in aqueous electrolyte solutions in the range 0–100 nm, J. Chem. Soc. Faraday Trans., v. 74, p. 975 (1978).

67. E.J. Pereyra, Modeling of Integrated Compact Multiphase Separation Systems, Thesis submitted to the University of Tulsa, (2011).

68. D.S. Valkovska, K.D. Danov, "Influence of ionic surfactants on the drainage velocity of thin liquid films," J. Coll. Interface Sci., v. 241, (2001).

69. D. Arla, A. Sinquin, T. Palermo, C. Hurtevent, A. Graciaa, C. Dicharry, "Influence of pH and water content on the type and stability of acidic crude oil emulsions," Energy & Fuels, v. 21, p. 1337 (2007).

70. P.V. Hemmingsen, S.Kim, H.E. Pettersen, R.P. Rodgers, J. Sjoblom, A.G. Marshall, "Structural characterization and interfacial behavior of acidic compounds extracted from a North Sea Oil," Energy & Fuels, v. 20, p. 1980 (2006).

71. V. Pauchard, J. Sjoblom, S. Kokal, P. Bouriat, C. Dicharry, H. Muller, A. al-Hajji, "Role of naphthenic acids in emulsion tightness for a low-total-number (TAN) / high asphaltene oil," Energy & Fuels, v. 23 (2009).

72. J.D. McLean, P.K. Kilpatrick, "Effects of asphaltene solvency on stability of water-in-crude-oil emulsions," J. Coll. Interface Sci., v. 189, p. 242 (1997).

73. Z. Zapryanov, A.K. Malhotra, N. Aderangi, D.T. Wasan, "Emulsion stability: an analysis of the effects of bulk and interfacial properties on film mobility and drainage rate," Int. J. Multiphase Flow, v. 9, p. 105 (1983).

74. J.D. McLean, P.K. Kilpatrick, "Effects of asphaltene aggregation in model heptane-toluene mixtures on stability of water-in-oil emulsions," J. Coll. Interface Sci., v. 196, p. 23 (1997).

75. J. Sjoblom, N. Aske, I.H. Auflem, O. Brandal, T.E. Havre, O. Saether, A. Westvik, E.E. Johnsen, H. Kallevik, "Our current understanding of water-in-crude oil emulsions. Recent characterization techniques and high pressure performance," Adv. Coll. Interface Sci., v. 100, p. 399 (2003).

76. B. Gawel, C. Lesaint, S. Bandyopadhyay, G. Oye, "Role of physicochemical and interfacial properties on the binary coalescence of crude oil drops in synthetic produced water," Energy & Fuels, v. 29, p. 512 (2015).

77. A. Yeung, K. Moran, J. Masliyah, J. Czarnecki, "Shear-induced coalescence of emulsified oil drops," J. Colloid Int. Sci., v. 265, p. 439 (2003).

78. D.G. Dalgleish, "Food Emulsions" Chapter 1 from Encyclopedic Handbook of Emulsion Technology, Ed. J. Sjoblom, Marcel Dekker, Inc., New York (2001)

79. E. Dickinson, "Adsorbed protein layers in food emulsions," from Emulsions – A Fundamental and Practical Approach, Ed. J. Sjoblom, Kluwer Academic Publishers, Netherlands (1992)

80. J. Lucassen, D. Giles, "Dynamic surface properties of nonionic surfactant solutions," J. Chem. Soc., Faraday Trans. I, v. 71, p. 217 (1975).

81. Y.-H. Kim, D.T. Wasan, P.J. Breen, "A study of dynamic interfacial mechanisms for demulsification of water-in-oil emulsions," Coll. Surf. A, v. 95, p. 235 (1995).

82. J.H. Clint, E.L. Neustadter, T.J. Jones, Proceedings of the Symposium on Enhanced Oil Recovery, p. 135 (1981). Reproduced Malhotra and Wasan.

83. A. Finborud, M. Faucher, E. Sellman, "New Method for Improving Oil Droplet Growth for Separation Enhancement, SPE 56643.

84. Cyclotech catalogue, PECT-F® Pre-coalescer for a Deoiling Hydrocyclone (2009).

85. S.J Tulloch, "Development & Field Use of the Mare's Tail® Pre-Coalescer" presented at the Produced Water Workshop, Aberdeen, Scotland, March 26-27 (2003).

86. A.K. Malhotra, D.T. Wasan, "Interfacial rheological properties of absorbed surfactant films with applications to emulsion and foam stability," Chapter 12 in Thin Liquid Films, Fundamentals and Applications, Ed. I.B. Ivanov, Marcel Dekker, Inc., New York (1988).

87. B. Vincent, "Emulsions," Ch. 6 of T. Cosgrove, Colloid Science – Principles, Methods, and Applications, 2nd Ed., Wiley, West Sussex, UK (2010).

88. Gao, K. Moran, Z. Xu, J. Masliyah, "Role of naphthenic acids in stabilizing water-in-diluted model oil emulsions, J. Phys. Chem. B, v. 114, p. 7710 (2010).

CHAPTER FIVE

Chemical Treatment

Chapter Five Table of Contents

5.0 Introduction:

The main subject of this chapter is chemicals for produced water treatment. Included also are a few sections on chemicals typically used in the oilfield for other purposes such as corrosion inhibition, and dehydration of oil. These chemicals have an important impact on water quality and are therefore included here. There are two main objectives of this chapter:

1. provide information that is helpful in working with the chemical vendors to ensure the selection of the optimal chemicals;

2. provide actual work processes and procedures for bottle testing, field trial, final selection of chemical, and chemical program performance monitoring.

In order to address the first objective, this chapter focuses on the chemistries and chemical products broadly applied in the oilfield for water treatment. Fundamentally, this chemistry is aimed at resolving an emulsion composed of various contaminants in water. There is a variety of emulsion types and characteristics, as discussed in Chapter 4 (Emulsions). Likewise, there is an enormous number of chemical products that could be applied. It is important to be able to identify the different emulsion types and be able to select chemistries that have the greatest probability to resolve them.

In order to address the second objective, various procedures are provided that can be used for chemical performance evaluation and optimization. The procedures are described in this chapter. A bottle test procedure is given at the end of this chapter. Also described in this chapter is the overall process of chemical selection.

The primary focus of chemical treatment for produced water is the removal of oil and oily solids from water. Although the primary interest here is in water treatment, oil dehydration (removing water from oil) is also discussed. This is necessary in order to provide a holistic or system-wide approach. In developing an understanding of oil-in-water (O/W) emulsions it is helpful to compare and contrast them with water-in-oil (W/O) emulsions. These two types of emulsion are vastly different. In describing how they are different, a significant understanding can be gained about both of them. Also, the performance of a water treatment system is dependent to a large extent on the upstream system, particularly the oil dehydration system. If that system is bottlenecked or constrained, it is likely that water quality will suffer as a result of adjustments made by operators to achieve dry oil.

Some of the subjects discussed in this chapter have already been discussed in previous sections of this book. In keeping with the overall approach, some subjects are presented more than once from different viewpoints and perspectives. This chapter was written with a lot of help from production and process chemists.

Production Chemistry: The responsibility for the use and management of chemical applications for a production facility, and its associated pipelines and flowlines, typically resides with the Process or Production Chemistry department. The chemical program is developed by working with various chemical vendors who provide a range of services to compliment their supply of chemicals.

Generally speaking, chemical applications in a facility encompass the following three areas:

- Product / Effluent Quality:
  - oil quality
  - water quality
  - gas quality
- Flow Assurance / Production Enhancement
  - wax prevention, control, monitoring
  - asphaltenes prevention, control, monitoring
  - mineral precipitation (scale) prevention, control, monitoring
  - organic deposition (besides wax & asphaltene) prevention, control, monitoring
  - foam prevention, control, monitoring
  - hydrate prevention in the produced fluids
- Asset Integrity
  - corrosion inhibition, control, monitoring

As demonstrated by this list, in the realm of process chemistry as applied in an oil and gas facility, water treatment falls under the general category of Product and Effluent Quality. Water treatment is only one of several important functions of chemical treatment. All of the applications listed above must be considered in their totality in order to achieve an optimal holistic program for a facility. Compatibility of these chemicals must be tested since one chemical can react with another to make the chemicals inactive or ineffectual, or cause a precipitate or scale to form. This is particularly important in the selection of corrosion inhibitor and hydrate inhibitor since these two classes of chemical have a tendency to stabilize emulsions and therefore can have a detrimental effect on oil dehydration and water treatment. Thus, it is necessary for the water treatment specialist to have an awareness of the entire chemical treatment program.

Product / Effluent Quality: The product / effluent quality of the three phases (oil, gas, water) is achieved through various mechanical equipment together with the injection of chemicals. In the case of product and effluent quality, including emulsion breaking, both chemical and mechanical treatment are required. The chemicals injected span a wide range of properties and are used also for a wide range of objectives. These include oil dehydration, deoiling of the water, and removing various contaminants from the gas such as water, hydrogen sulfide, mercury, and organic materials from carryover.

Oil Dehydration: As discussed below in Section 5.12 (Demulsifiers – Impact on Water Treatment), the use of aggressive demulsifying agents for oil dehydration can result in stable emulsions in the water phase. This is another case where a whole-facility solution is required. Compatibility testing should be carried out to ensure that the chemicals required to carry out all of the functions above do not inadvertently negate each other, or cause an unwanted precipitation.

Flow Assurance: Flow assurance is a critical subject for hydrocarbon producing facilities since fluid flow is what keeps the facility in operation. Any disruption to fluid flow is a major setback. Flow assurance chemicals have a wide range of properties and a wide range of mechanisms for preventing the precipitation, or formation of solids, and the buildup of solids in the flow lines. Some of these chemicals are detrimental to water treatment (hydrate inhibitors), others are generally speaking beneficial (asphaltene inhibitor, scale inhibitor).

Corrosion Inhibition: Corrosion inhibitor can have an enormous effect on water treatment [1]. As discussed in Chapter 4 and in Section 5.13 of this chapter (Corrosion Inhibitors – Use and Impact on Water Treatment), almost all oilfield corrosion inhibitors are surface active. Surface activity is an important part of the mechanism that allows them to prevent or at least reduce corrosion. Molecules that are surface active tend to stabilize O/W (oil-in-water) and W/O (water-in-oil) emulsions according to the mechanisms discussed in Sections 4.4.3 (Surface Excess Concentration and Surface Activity), Section 4.4.4 (Interface Concentration versus Bulk Concentration), and Section 4.4.5 (Bancroft Rule). The tendency to stabilize an emulsion varies significantly from one corrosion inhibitor to another. This suggests that the selection of a corrosion inhibitor should be carried out by considering both the mitigation of corrosion and other factors such as the impact on the water treatment system. The same is true for the other process chemicals.

Holistic and Balanced Approach: In developing a chemical treatment program for a facility, production chemistry staff and chemical vendors will strive to find the greatest overall benefit. For example, as just discussed, the selection of a corrosion inhibitor should be carried out by evaluating both the corrosion and the impact of the chemical on production systems throughout the facility. This is referred to as a holistic approach which has been championed by Horsup and co-workers (see Section 5.13 and reference [1]). In order to do this, competing objectives must be prioritized and compromises must be made regarding gas quality, oil quality, flow assurance, asset integrity, and last but not least, water quality. The particular chemistries selected for these applications can and in many cases do interact with one another. Some of these interactions are synergistic. For example, an asphaltene inhibitor that is used to prevent asphaltene precipitation will reduce the tendency of asphaltenes to migrate to the oil / water interface and stabilize an oil in water emulsion. Often, the use of an asphaltene inhibitor improves water quality. The operating company must participate in the decision making process whereby the chemical program is selected, tested, and optimized. Priorities must be identified and compromises made. In order to do that, it is helpful for the operating company staff to understand the chemical selection options and testing processes. Providing this background and understanding is one of the major objectives of this chapter.

5.1 Terminology:

Across the oil and gas industry the terminology for emulsions and water treatment chemicals varies considerably. The treatment chemicals used to clean produced water are sometimes referred to as reverse emulsion breakers, flocculating agents, clarifiers, etc.. These terms are confusing because they are not universally used in the same way across the oil and gas industry. In most other industries where water treatment is required, such as the paint industry, food and beverage, industrial water, or emulsion polymer industry, the terminology tends to be more consistent.

Terms such as deoiler, clarifier, or reverse emulsion breaker were adopted because many chemicals used in the oil and gas industry are a combination of other chemicals. It is obvious that a deoiler provides chemical assistance to remove oil. But the term deoiler says nothing about whether the chemical is a coagulant which requires extensive mixing, or a flocculant which requires gentle mixing without excessive shear. Furthermore, a deoiler may include a coagulant, a flocculant or it could be a chemical used to adjust the pH or sequester iron or iron compounds.

The chemicals discussed in this chapter will be referred to as coagulant or coagulating agent, flocculant or flocculating agent, coalescing agent, wetting agent, or chelating agent. These terms are preferred since they are descriptive of the mechanism by which the chemical product performs its designated task, they imply certain requirements in application (discussed below), and they are unambiguous. Other chemicals which are discussed include acids, bases, scale inhibitors, and chelating agents. Like their counterparts' (coagulants and flocculants), these are descriptive terms which indicate the mechanisms and imply application requirements and outcome. These terms are used fairly consistently in the scientific literature on water treatment. Actual chemical products are often a combination of these specific chemicals. In those cases, the chemical product name will be used and it will be described as, for example, a combined dispersant and coagulant, or combined acid and flocculant. Also, some coagulants are blended with chelating agents or acids for pH adjustment. As demonstrated below, there is no confusion to simply point this out whenever this is the case. Thus, the terms coagulation, flocculation, and coalescence are descriptive of the mechanism and help guide the application requirements.

In the application of chemical products, understanding the mechanism of treatment is critical to chemical selection and application. Coagulants are a narrow class of compounds that manipulate chemical charge, and which have specific application requirements. Coagulants are not always required. Flocculants are a larger class of compounds with their own requirements and restrictions in application that are quite different from those of coagulants. Flocculants have a wide range of properties including molecular weight, physical structure (linear or branched), and a range of electrical charge polarity, and density. Coalescing agents are most effective when the concentration of oil in water is relatively high and there is moderate turbulence. These conditions result in a high collision rate and a high probability of coalescence.

5.2 Brief Review of Emulsion Chemistry:

Emulsions were extensively discussed in Chapter 4. They are briefly reviewed here from a slightly more general perspective. The review begins with a discussion of the types of compounds found in oil and which can therefore be found in produced water as contaminants.

Oil Phase Hetero-Atomic Compounds: Most crude oil contains a large number of compounds that have polar groups such as carboxylic acid, alcohol, amine, and sulfur. These compounds include the acid-substituted long chain alkyl compounds (e.g. hexanoic acid, decanoic acid), the acid-substituted saturated cyclic compounds (naphthenic acids); the acid-substituted aromatic compounds (benzoic acid); and the asphaltenes and resins which have acidic, alcohol, nitrogen and sulfur groups. Despite having some polar groups, these compounds are predominantly composed of hydrogen and carbon atoms (hydrocarbon groups). It is the presence of the hydrocarbon groups that causes these compounds to be found in the oil phase rather than the produced water. In some crude oils, there is a multitude of these compounds.

Due to the presence of a polar and/or ionizing group together with a hydrocarbon chain or polycyclic structure, some of these compounds have some tendency to migrate to the oil / water interface. But, due to their oil solubility, they have a preference for the oil side of the oil / water interface. As will be discussed below, in connection with the Bancroft Rule, an oil / water interface has two sides, the oil side and the water side. Various molecules have a tendency to migrate to the interface but most compounds have a preference for one side of the interface versus the other. The compounds discussed here have a preference for the oil side of the oil / water interface. As will be explained, they have a tendency to stabilize water drops dispersed in oil, rather than oil droplets dispersed in water.

This group of chemicals is important in the understanding of emulsion stability for water-in-oil emulsions. The stability of water-in-oil emulsions has been correlated to the concentration, composition and stability of these compounds in the crude oil. In heavy crude oils, there is a tendency of these compounds to migrate to the oil/water interface where they form an interlocking network system that has high interfacial elasticity. In practical terms, the interface is tough like rubber. This creates a non-ionic, nonpolar coalescence barrier that requires heat, chemical and an electrostatic field to overcome.

There is some confusion here as to whether these compounds are "polar" or "surfactant-like." Part of this confusion comes from the fact that these compounds are not so easily defined. The smaller molecular weight representatives of this group can be defined. Some examples were given above. But the majority of these compounds are higher molecular weight and have a large diversity of molecular composition and structure. The real definition of these compounds can only be given in terms of how they behave. Whether they are soluble in hexane, or stabilized by the presence of aromatic solvents, for example, or whether they migrate to an oil / water interface provides the functional definition of this class of compounds.

Water Phase Hetero-Atomic Compounds: In order for an organic compound to dissolve in water it must be polar or ionic. Small polar molecules such as formic, acetic acid, or phenol are highly polar and are typically found in the water phase. Other compounds such as benzene are somewhat polar and have a tendency to partition between the oil and water phases. In other words, a fraction of the benzene dissolves in the water phase and a fraction can be found in the oil phase. This is true also of the larger acid compounds such as hexanoic acid, Heptanoic acid, etc.. The impact of these water soluble compounds on water treatment in a facility is somewhat complex and requires a separate discussion.

Surfactant-Like Compounds: Larger compounds that have a hydrocarbon tail, or branch, or which have a set of aromatic rings, can have surfactant-like properties. Such compounds would have both polar and nonpolar (hydrocarbon) groups and would migrate to the oil / water interface. These compounds will stabilize oil drops in water. Examples of these types of surfactant-like compounds are straight chain and branched acids, acid groups attached to aromatic or saturated cyclic groups and their combinations, various substituted phenols, phenolic groups attached to aromatics, phenolic groups on large molecules, asphaltenes, etc.. These are just a few examples of the kinds of chemicals that have surfactant like structures in crude oil. There are literally thousands of such compounds in crude oil.

The chemistry of the crude oil can have a dramatic effect on the emulsion forming tendency in mixtures of oil and water. Crude oils having high Total Acid Number (TAN), and containing a high percentage of organic acids are typical of problematic crude oils. In general such crude oils contain compounds that are surfactant-like and which migrate to the oil / water interface to stabilize both oil-in-water suspensions and water-in-oil suspensions. However, as discussed below, these two types of interfaces behave very differently depending on the nature of the crude oil and water.

Crude oil can be characterized in a number of ways to determine it's relatively tendency to stabilize emulsions. This topic is discussed in Section 4.4 (Interfacial Tension, Elasticity, Wetting, Capillarity). For example, surfactancy can be measured by hydrophilic / lipophilic balance (HLB) and relative solubility numbers. A number of different measurements can be used. The important thing is to measure relative solubility and provide directionality. But the problem with that approach is that it is only really unambiguous to characterize pure substances. If two compounds that have two different solubility numbers are put together in the same test, they are going to influence each other's relative solubility. Extrapolating this to the case of crude oil, where thousands of different compounds are mixed together, the pure compound solubilities mean nothing because each of those compounds will influence each other. This is particularly true when a significant fraction of compounds have acid or base groups in the molecular structure. An acid group from one compound can complex with a base group from another compound to form a complex. The solubility and surface activity of such a complex is significantly different from that of the two individual compounds.

Two Important Characteristics of Interfaces: There are two important characteristics of the interface between oil and water.

1. **Interfaces Have Two Sides:** An oil / water interface has a finite thickness, with two adjacent sides, the oil side and the water side. Surface active molecules migrate to the interface. The polar part of the molecule will protrude into the water side of the interface. The non-polar part will protrude into the oil side of the interface. This is true of all surface active molecules at the oil / water interface. However, surface active molecules also have a solubility preference. They have a slight tendency to partition into one of the bulk phases. In other words, they have a preference for one side of the interface versus the other. Molecules that prefer the oil side of the interface cause a stable water-in-oil emulsion to form. Molecules that prefer the water side of the interface cause a stable oil-in-water emulsion to form.

2. **Ratio of Surface Area to Bulk Volume:** The second important characteristic of the interface is the ratio of the surface area of the interface to the volume of bulk phase. This ratio determines the concentration of the naturally occurring surface active molecules at the interface. When this ratio is low, there is a tendency of these molecules to crowd the interface and stabilize an emulsion. When this ratio is high, the amount of water treatment chemical required may be high as well.

Both of these characteristics are discussed in further detail below.

Interfaces Have Two Sides: The first important characteristic of the oil / water interface is that it has a finite thickness, with two adjacent sides. In Section 4.4 (Interfacial Tension, Elasticity, Wetting, Capillarity) the chemical and physical nature of an interface was discussed. It was mentioned that the concept of a thin interface is really just a mathematical convenience. Real interfaces have finite thickness. As discussed in Section 4.4.5, the Bancroft rule basically points out that the oil/water interface in the oil phase (where there are water droplets dispersed in oil) is different from the oil/water interface in the water phase (where there are oil droplets dispersed in water). The main difference, as far as the Bancroft rule is concerned, is the type of surfactants at the interface. In the case of produced water treatment, the surfactants can be either the natural surfactants from the produced fluids, or they can be the man made surfactants added for flow assurance, or corrosion control. The Bancroft rule predicts which of the two emulsions will be more stable on the basis of the type of surfactants present in the mixture.

This has immediate application to produced water treatment. The Bancroft rule helps explain why produced water from gas condensate fields is so notoriously difficult to treat. Besides the shearing associated with large pressure drop, and the expanded electrical double layer from the low salinity, we can now understand that the light acids play an important, if detrimental, role. As discussed in Section 2.1.3 (Gas Production – Interstitial and Condensed Produced Water), produced water from gas fields can contain high concentrations of short chain carboxylic acids, low molecular weight aromatic compounds and phenol compounds. The carboxylic acids are preferentially soluble in the water phase. Thus, they tend to stabilize the oil/water interface in the water phase, i.e. oil-in-water emulsions, as predicted by the Bancroft rule (Section 4.4.5).

The Bancroft rule also helps explain why some corrosion inhibitors are more detrimental to water treatment than others. A corrosion inhibitor that has a preference for the water phase will be more detrimental to water treatment than a corrosion inhibitor that has a preference for the oil phase because it will stabilize oil-in-water emulsions.

Ratio of Surface Area to Bulk Phase Volume: The second important characteristic of the interface is the ratio of the surface area of the interface to the volume of the two bulk phases. This was discussed in some detail in Section 4.4.4 (Interface Concentration versus Bulk Concentration). In an oil-in-water emulsion, the continuous or bulk phase is water. The dispersed phase is oil. Both the oil and water phase share the same interfacial area but the two phases have significantly different bulk volumes. In a produced water stream where there is no bulk phase of oil, the ratio of interfacial area to water volume is relatively low. The ratio of interfacial area to oil volume is high. This means that there is a large reservoir of bulk water from which surface active compounds can migrate to populate the interface. Such surface active compounds include the short chain acids. However, there is a much smaller reservoir of bulk oil from which surface active compounds can migrate.

The dispersed oil droplets have a much smaller volume than the water phase. On the other hand, the dispersed oil droplets contain surface active components such as resins and asphaltenes. Typically the concentration of these compounds are in the ballpark of a few percent, at least. Although the dispersed phase (oil) volume is relatively small compared to the continuous phase volume (water), the high concentration of surface active compounds typically found in the oil gives rise to surface active compounds migrating from the oil phase to the interface, as well. Therefore when considering the population of surface active compounds at the oil/water interface for the case of oil droplets suspended in water, it is important to understand both the oil composition and the water composition.

Water Droplets in Oil: The previous discussions should help to explain why a water drop in oil is fundamentally different than an oil drop in water. While this may seem like a rather obvious statement, it has profound consequences that should not be overlooked. The composition of the oil/water interface, the structure of the interface, and the processes of coagulation, flocculation and coalescence are complex subjects. In almost all situations, these subjects do not require a rigorous mathematical description. Most often, a simple qualitative view will suffice. Nevertheless, it is helpful to have a rigorous understanding of these subjects in order to make accurate practical guesses about how to treat a particular type of emulsion.

The stability of a suspension of water drops in oil is dominated by the chemistry of the oil side of the oil / water interface. If given enough time, molecules in the oil containing polar groups will migrate to the oil / water interface. The molecules will orient such that the polar groups face the water. Such molecules include the asphaltenes, resins, alkyl phenols, and alkyl acids. The variety of these compounds is almost unlimited.

If the oil contains molecules that have carboxylic acid groups, the acids will both orient toward the water and may protrude into the water and become ionized. This typically occurs at moderate to high pH, which for produced water systems is in the 6 to 8 pH range. This range is considered high for produced water.

Oil in water is composed of nonpolar materials and water is very polar. There is a functional charge difference between the oil drop and the water drop. Functional charge difference. Helmholtz layer. Coagulate oil drops in water by using that effect.

Coagulation of water droplets in oil is possible. Typically, high molecular weight alcohol ethoxylates are used. These molecules are designed to have large sequences of nonpolar groups which mostly reside in the oil phase together with blocks of polar groups which prefer the water phase. The mechanisms by which these molecules promote coagulation are all related in one way or another to binding, or bridging between water drops. One polar end of the molecule may encounter a water drop and stick to that drop. As other water drops move around, brush by and collide with that drop, the other end of the molecule may by chance bind to a second drop of water. Thus, a long chain high molecular weight polymer molecule can form a bridge between two water drops and bind them together. In reality, several molecules may actually be involved in binding two water drops, and a kind of network or mesh may be formed. In any case, the mechanism is one of forming a bridge across the oil film that separates two water drops.

As described, it is easy to see why such chemistry is capable of promoting coagulation, but incapable of promoting coalescence. In fact, such chemistries often make coalescence worse than if there were no production chemical involved. They create a barrier to coalescence.

Oil Droplets in Water: An oil drop in water typically has a polar surface. If given enough time, the molecules containing polar groups will migrate to the oil /water interface. The molecules will orient such that the polar groups face the water. These are naturally occurring surfactant-like molecules that were discussed previously. Some high acid containing crude oils also have ionizing groups (mostly acids) that move the compound to the interface. Compared to water, the oil contains a low concentration of ionizing groups. Typically the oil contains enough acids such that the surface of the oil drop is slightly negatively charged. For various other reasons, the surface of solid particles typically found in water is also negatively charged.

Produced water contains fully mobile mineral ions, and acids. These chemical species can migrate to the interface and accumulate there to shield or neutralize the surface of an oil drop or solid particle. This shielding effect lowers the repulsive charge-charge barrier between the contaminant particles and promotes coagulation. As mentioned already, the produced water also contains surfactant-like compounds which hinder coagulation by adsorbing to the oil / water interface and creating a physical barrier.

The Effect of Contaminant Size on Chemical Demand: Chemical treatment of emulsions often involves the injection of polymers which adsorb onto the contaminant surface and promote coagulation, flocculation or coalescence. In addition to the concentration of contaminant, the total surface area of the contaminant determines the amount of chemical required.

The total surface is calculated as the surface area of each particle times the number of particles. Thus, it is worthwhile to have a rough understanding of the magnitude of surface area created when droplets are sheared. The formula for the volume of a sphere is $V = 4\pi R^3/3$ where R is the radius of the sphere. If a spherical 30 micron diameter droplet is sheared into 10 micron droplets, then the total volume of the 10 micron droplets must be equal to the volume of the single 30 micron droplet:

$$1.33\pi(R_0)^3 = N \times 1.33\pi(R_f)^3$$

Which becomes $\qquad (15)^3 = N\,(5)^3 \qquad$ and $\; N = 27$

The equation for calculating N is $\quad N = (R_0/R_f)^3$

Thus each 30 micron droplet has been sheared into a total of 27 droplets with a diameter of 10 microns each.

Because the amount of chemical required to induce droplet coalescence is proportional to the total surface area of the droplets being treated with the coagulant, it is of interest to also calculate the increase in the total surface area of the droplets when a single 30-micron droplet is sheared into 27 10-micron droplets.

The area of the initial 30-micron droplet is $\qquad A_{30} = 4\pi(R_0)^2 = 4\pi(15)^2$

The area of the final 27 10-micron droplets is $\qquad A_{10} = N \times 4\pi(R_f)^2 = 27 \times 4\pi(5)^2$

Using the equation for N derived above, one can now calculate the ratio of areas for the final droplets to the area for the initial single droplet:

$$A_f/A_o = N \times 4\pi(R_f)^2 / 4\pi(R_o)^2 = (R_o/R_f)^3 \times (R_f/R_o)^2 = R_o/R_f$$

Thus the total surface area of the 27 final 10-micron droplets will be 3X the surface area of the initial 30-micron droplet.

If a certain quantity of chemical was required to flocculate the 30 micron droplets, three times more chemical would be required to flocculate the 10 micron droplets.

In the case of a shearing process that converts 100 micron droplets to 10 micron droplets, the number of droplets formed and the amount of surface area formed are calculated as:

$$N/N_0 = (100/10)^3 = 1,000$$
$$A/A_0 = (100/10) = 10$$

Thus, the total surface area has increased by a factor of 10. If a certain quantity of chemical was required to flocculate the 100 micron droplets, ten times more chemical would be required to flocculate the 10 micron droplets due to the increase in surface area.

This simple calculation demonstrates the dramatic effect of shearing has on increasing the amount of chemical required to break emulsions with smaller droplets. The most detrimental effect of shearing is the breaking of the large drops into a much larger number of smaller drops. Effective water treating focuses on separating the large drops, and getting them out of the water, before they get sheared. The dose of water clarifier required depends on the surface area of the oil/water interface. Small drops have much greater interface area. Thus, the required chemical dosage increases dramatically due to the shearing of oil drops.

5.3 The Mechanisms of Coagulation, Flocculation and Coalescence:

Three key processes for the separation of contaminants from water are coagulation, flocculation, and coalescence:

Coagulation   = neutralization of surface charge

Flocculation   = particle bridging and building of macro-floc

Coalescence   = two or more liquid droplets combine to form a larger single droplet

In order to understand the mechanisms and chemistry involved in water treating, it is important to have a definition of these terms. These terms have already been discussed in various sections. The basic background was given in Section 4.2 (Coagulation, Flocculation, Coalescence Overview). An in-depth treatment of electrical double layer was given in Section 4.3.2 (Electrical Double Layer and Zeta Potential). Coalescence was discussed in Section 4.5 (Coalescence of Oil Droplets). In keeping with the overall approach in this text, the same subjects are presented and discussed throughout from different perspectives. The perspective presented in this chapter is that of a chemist. Much of the rest of this text has an engineering perspective. This section was written with help from chemists for the benefit of other chemists.

Historical Starting Point – Solids Removal from Water: Historically, the removal of suspended solids from water in order to provide municipal water supply predates water treatment in the oil and gas industry by hundreds if not thousands of years. The municipal water treatment industry, which was and currently is today, a very large industry, provided the drive for the development of solids coagulating chemistries.

Solids Coagulation in Municipal Water Treatment: It is instructive to start with solids coagulation both from an historical and scientific perspective. Solid surfaces do not deform. This makes them easier to study than oil droplets. Well defined suspensions of solids are easier to generate in the lab than well defined suspensions of oil droplets. At the same time, the chemistry of the solid surfaces can be complex with a range of charge density and a wide variety of adsorbed species. This makes them relevant to other complex contaminants but easier to control and manipulate in the laboratory. Thus, there was both a practical need and a scientific accessibility for solids coagulation to develop long ago. Many of the concepts that are used today in the coagulation and flocculation of oil have their origins in the coagulation and flocculation of solids.

In order to provide potable water, municipal water systems are supplied by large volumes of readily available surface water from rivers, lakes and reservoirs. The main contaminants are fine suspended solids such as small particles of clay, sand, silt, (calcium/magnesium/aluminum silicates, other minerals) and suspended particles of organic material such as humic acid and dissolved organic acids such as fulvic acid. Humic acids are insoluble in water at acid pH. These are analogous to naphthenic acids in produced water samples. Fulvic acids are similar to humic acids but are lower molecular weight and are water soluble. Municipal water typically has a pH in the range of 6 to 8.5, which means that the organic acids are mostly ionized. Essentially all of these suspended solids have negative surface charge. A coagulant chemical will be used to shrink the electrical double layer and reduce charge-charge repulsion between contaminant particles. Coagulants are based on a tri-valent cationic ion such as iron or aluminum or any from the synthetic organic polymers and copolymers such as poly-aluminum chloride (PACL). This will reduce charge-charge repulsion. When thermal motion causes the particles to collide, they will flocculate. In some cases, a small concentration of flocculating agent is added. A floc of suspended particles has a larger effective diameter which makes it settle more quickly than the individual suspended particles. Coagulation followed by flocculation and settling works very well for surface water which, as mentioned is composed of small solids particles dispersed in low salinity water. If the surface of the solid particles are relatively free of adsorbed polymers or cationic corrosion inhibitor, then the dominant interaction between particles is electrostatic and the mechanisms involved in using multivalent cationic coagulating agents are effective.

Humic and FulvicAcid – Organic Acids in Municipal Water Treatment: The primary organic acid constituents in "dirt" are humic and fulvic acids. They are the biological degradation products of plant matter [2]. Fulvic acids are actually a subset of humic acid. These organic acid compounds have a negative surface charge over the typical pH range of municipal water (6 to 8.5). While their chemical structure is highly complex, coagulation, flocculation and settling works well for most of these contaminants.

In addition to the suspended particles, some of the humic and fulvic acid are low molecular weight and are therefore dissolved in the water. For centuries, chlorination (oxidation) was a standard method of treatment for these soluble components. However, as a result of chlorination, these naturally occurring organics would become halogenated forming a class of compounds known as Trihalomethanes (THMs), which were identified as carcinogens in the latter part of the last century. Halogen use for drinking water production was replaced with chloramination, chlorine dioxide or other alternative oxidizers to eliminate the formation of THMs. These natural, soluble organics that are not removed in the raw water plant will foul ion exchange resins and create additional process challenges depending on the nature of the industry. Chlorination is obviously also used for eliminating biological contaminants as well.

As discussed by Faust and Ali (page 236 of [2]), dissolved humic material can be precipitated, flocculated and allowed to settle by the application of cationic polyelectrolytes. The negatively charged or polarized phenolate, alcohol, ether, and carboxylate groups on the humic molecules are electrically attracted to the positively charged pendant groups of the cationic polyelectrolye. This charge-neutralization causes the humic molecules to drop out of solution. The presence of the polymer then promotes flocculation. The large flocs can then be treated by flotation or sedimentation. This approach will be discussed in greater detail below and in Chapter 21 in relation to the treatment of Water Soluble Organics (WSO).

In many municipal water treatment systems, there is a high Biological Oxygen Demand (BOD), and comparatively low Chemical Oxidation Demand (COD), lending itself to effective biological digestion of the soluble organics. As increasing amounts of pharmaceuticals and other industrial complex organics, much of which are known as "refractory" organics reach the municipal wastewater treatment plants, additional treatment, post-biological digestion is necessary (advanced oxidation). The additional treatment typically consists of oxidation of these complex organics into a biodegradable form for further digestion. Biological treatment of produced water is obviously not possible offshore because of the volume of water and the long residence time required for biological processing. However, being able to recognize when biological processing should be applied onshore is very important in developing the optimal process concept. Also, biological processing is used in a few offshore locations for specialty applications and it is used extensively offshore for processing gray and black water that is generated.

Similarities / Differences – Municipal vs Oilfield Water Treatment: Oilfield oil and water chemistry has many similarities with municipal water, but it also has some important differences. The suspended solids size and composition (clay, silt, sandstone fines, carbonate fines, etc) found in municipal and in produced water have a wide range of values. Both sources of water can contain the so-called colloidal solids which are very small, in the range of 10 microns or less. Suspended solids whether from surface waters or produced water all have negative surface charges and exhibit charge-charge repulsion. Municipal water is typically low salinity, and relatively low hardness compared to most produced water. In such systems, coagulating agents are effective. Produced water varies in salinity from moderately fresh (condensed water) to highly saline. Hardness can be extremely high in produced water. Coagulants are less effective in treatment of highly saline produced water because the electrical double layer is already collapsed due to the high concentration of ions and presence of divalent and trivalent ions.

Produced water contains various acids (volatile fatty acids, short-chain fatty acids, naphthenic acids, resins, asphaltenes), which are similar to humic acids found in municipal water treatment, and other dissolved organics such as phenols. These compounds contribute to BOD. Biological water treatment processes are common in municipal water treatment but not used offshore for produced water except in rare situations due to the weight and space requirements of such systems. As of 2012, consideration was being given to the installation of a Membrane Bioreactor for the Shtokman platform in the Barents Sea [3] to treat 250 m3/day of produced water containing mono- and tri-ethylene glycol. Almost all platforms utilize a bioreactor of some sort for gray and black water treatment but the water volumes are very small compared to that of produced water. Biological reactors are seldom used onshore for produced water treatment because BOD reduction is seldom a requirement for disposal well injection.

The diversity of emulsion types is much greater in produced water than in municipal water. Contaminants in municipal water are most often based on suspended solids. Such emulsions are rather simple in comparison to the variety and complexity of emulsions that can form when two liquid phases are involved, and in produced water. In the case of oil and water, emulsions can range from a simple suspension of oil drops in water to complex emulsions (oil-in-water-in-oil), to that of a High

Internal Phase Volume (HIPV) emulsion where large water drops are separated by thin films of oil. Between these extremes there is a wide range of emulsion oil/water ratio.

Carbon dioxide can play an important role in produced water systems but is almost never a factor in municipal systems. By themselves, at neutral pH solids and acids would have negative surface charge. However, if there is CO_2 present in the gas, then the water will have a low pH, depending on the CO_2 concentration and gas pressure. Low pH will protonate the solid and acids thus increasing the solubility of the organic acids in the oil phase. The pH of produced water and the stability of an emulsion can vary significantly in a facility depending on the CO_2 content and resulting pH. This is not typical in municipal water systems.

Produced water chemistry can be very complex. Most solids that are suspended in produced water have some degree of oil coating. Natural surfactants tend to migrate to the oil / water interface. These include asphaltenes, resins, short chain fatty acids, naphthenates, and any other molecules containing S, N, or O functional groups. Also, produced water may have surfactant-like compounds of corrosion inhibitor, and/or hydrate inhibitor. These are not seen in municipal water. These chemicals will adsorb onto the oil/water interface and can absorb onto solid surfaces and flip the surface charge from anionic to cationic. Also, in oilfield produced water there typically is a few or several tens of mg/L of iron. This is sufficient to saturate any remaining anionic surface sites. Thus, the most effective coagulants and flocculating agents can be either anionic or cationic. Both must be tested.

While there are important differences between municipal and produced water systems, there is a great deal of knowledge gained from the study of municipal water that can be applied directly to oil/water systems. This is another reason why the terminology of coagulation and flocculation, used so extensively in water treatment of solids suspensions, is also used here.

Coagulation: Another way of defining coagulation is to say that it is the process whereby a suspension of solid particles is destabilized. This is obviously one of the first and primary steps in separation. The extent of coagulation is fundamentally measured in terms of surface characteristics and forces which contribute to the particle-particle potential energy of interaction. However, a practical consequence of coagulation is the formation of a floc, which is in some cases actually visible, or the formation of a precipitate, or the dewatering of a sludge, etc.

In order to promote the separation of oil droplets and oily solids particles at the smaller end of the size scale the like surface charges have to be neutralized. Many naturally occurring particles in oilfield waters as well as organic contaminants have a net negative surface charge – the resulting repulsive forces, then, stabilize these contaminants in suspension. Destabilization colloidal silica particles and oil droplets is then accomplished through addition of chemicals that either generate a cationic charge or carry that charge. Oxidation of dissolved iron in the produced water will produce Fe^{+3} which works to neutralize the surface charge of the anionic droplets and particles. The neutralization of the surface charges eliminates the repulsive forces from the like charges allowing the particles or droplets to collide, stick together (agglomerate) and make a bigger particle. In most cases, the cations are added directly as iron or aluminum salts or cationic inorganic or organic polymers or a combination of the two. An abundance of positively charged species can "cationize" a portion of a droplet or particle surface creating an attractive force for other particles and other droplets. If the solid contaminants are predominantly clays then anionic polymers might perform better at coagulating these particles.

According to the DLVO model, charge-charge repulsion can be reduced by increasing the salinity, which is not practical, or by adding a relatively small amount of trivalent cations. The Schultz-Hardy Rule indicates that trivalent cations are over 700 times more effective than monovalent cations, on a molar basis, at reducing the electrostatic repulsion of suspended solids. The simplest such coagulants are salts of iron (iron chloride) and aluminum (alum). By adding a coagulant, the electrical double-layer becomes compressed and charge-charge repulsion is reduced to the point where the short-range van der Waals attractive forces can bind particles together.

The term, "coagulation" can be a bit misleading since it is a common and broadly used term outside of water treating. In biology it describes blood clotting. In cheese manufacture it describes the process of making curd from milk. Both of these processes imply aggregation of suspended material. In water treating, coagulation does indeed lead to aggregation. But, the important aspect is the prior step of chemically modifying the surface of the suspended particles to reduce the inherent electrostatic repulsion between particles. Without this step, flocculation and aggregation in water treatment would be limited. In the water treatment industry, coagulation refers to surface modification to allow subsequent agglomeration and flocculation.

Coagulation is enhanced by intense mixing, applied immediately after chemical addition. Intense mixing promotes the migration of the coagulant chemical to the surfaces of the suspended contaminant material. However, it must not be so intense to shear the particles and make them smaller since that would defeat the overall purpose. Through the process of coagulation, aggregates of particles begin to form since the electrostatic repulsive barrier has been reduced in size and intensity.

Flocculation: As more collisions occur and particles and/or droplets are bridged, the particles form very large charge neutral "flocs." A large floc is visible to the unassisted eye. In order to accelerate the rate of floc formation, an anionic polymer, a flocculant, can be used. The advantage of a large floc is that it settles (or rises) as if it is a single sphere (with some equivalent large sphere diameter. As discussed in Section 3.2.3 (Stokes Law Settling Rate), Stokes Law indicates that the rate of settling is proportional to the diameter squared.

Flocculants are typically, high molecular weight, low charge density, anionic polymers, sufficiently large so as to create bridges between particles or droplets – dramatically increasing the particle size to obtain significant improvements in separation velocities. The bridge is created with the anionic polymer attaching to the cationized sites (cationized by the cationic coagulant) on the particle or oil droplet. It is important to remember that there are exceptions to the process where some solids are positively charged and coagulation requires an anionic.

In a typical water treating system designed to remove solids, an initial step would be to add a chemical coagulant to destabilize the solid suspension. This would be followed by the addition of a flocculating agent to promote the formation of agglomerates of solid particles. After coagulation and flocculation, the solution is treated in some form of solid / liquid separation process such as gravity settling or filtration. In both cases, the formation of flocs enhances the removal process. In the case of gravity settling, the floc is larger than the individual particles and therefore has a higher volume (mass) to surface area ratio. This enhances gravity settling as predicted by Stokes Law. In the case of filtration, the floc has a greater probability of being adsorbed onto a surface or being caught in a pore of the filtration media.

Flocculation, takes the aggregation process a step further. Flocculation is the process whereby a flocculating agent, usually a polyelectrolyte polymer, is added to promote the formation of macroscopic aggregates or flocs. The polymer binds suspended particles together by forming bridges, strands or by forming a mesh. Once aggregated, the suspended material will migrate either up or down, depending on its density relative to water. The larger the flocs and the more tightly aggregated the suspended solids, the faster the migration rate.

Gentle mixing is important for uniform distribution of the chemical. But flocs are delicate objects and too much mixing will break the flocs apart. Flocs are typically settled in a quiescent tank in order for separation to occur. As mentioned, filtration or flotation can also be used.

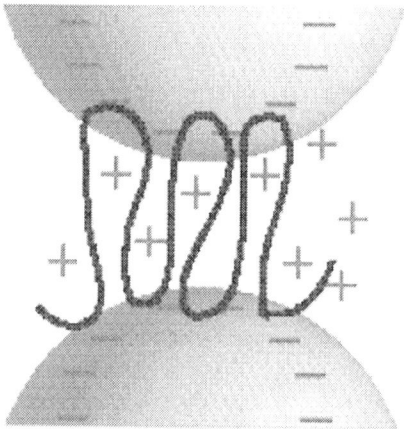

Figure 5.1 Flocculation of two particles by the use of a bridging polymer

Floc-n-Drop (Flocculation followed by Sedimentation): Flocculation, followed by sedimentation (also known as "floc-n-drop") is a classic water treatment processes. It is used in the treatment of surface and ground water for drinking. In those applications, the suspended solids loadings are relatively low. As discussed in the previous section, they are also used to treat industrial waste water containing high concentrations of suspended solids, such as from a pulp and paper mill. Paper mill effluent can contain a high concentration of starch and cellulose components which is analogous to hydraulic fracture flow back fluids where guar is used.

The main mechanism of coagulation / flocculation / sedimentation is the aggregation of suspended solids in order to promote separation. Once the solids have formed aggregates, separation can be carried out by settling, filtration or by flotation. In the case where flotation is used, flocculation of the suspended solids improves separation by creating a larger target for flotation bubbles to collide with. This is the conventional mechanism used in offshore water treatment.

Coagulation and flocculation are considered to be separate processes in large part because they involve separate chemical treatment, and separate unit operations. As described, the two processes differ in the intensity of mixing and in processing time. They differ somewhat in the chemicals used as well but there is also quite a bit of overlap.

In conventional coagulation / flocculation chemical selection, dosage and treatment time are important. Overall, there are three types of chemicals used in coagulation / flocculation. The trivalent salts (iron chloride, alum) were historically developed first. They are mostly used for coagulation. In industrial and municipal water treatment where there is a steady water quality and flow, they can be optimized. They are effective at breaking down the electrical double layer but are a bit tricky to use in the field. They require a precise knowledge of the water chemistry, and pH adjustment in some cases in order to avoid side reactions, unwanted precipitates, and to assure their effectiveness. Under optimum conditions they form polymer-like poly-hydroxide structures which reduce the electrical double layer and help bind suspended solids particles together.

Coalescence: Coalescence is the process whereby two or more droplets combine to form a single larger drop. In order for this process to occur, the drops must come together with sufficient kinetic energy to overcome the coalescence energy barrier. This energy barrier is discussed in Section 4.5 (Coalescence of Oil Droplets). It involves a combination of electrical forces caused by surface charge, steric forces caused by the adsorption of surfactant-like molecules, and intermolecular forces as measured by surface tension. Chemical addition can reduce the magnitude of this energy barrier.

At least one of the mechanisms is similar to that discussed in solids coagulation, i.e. the destabilization of surface forces. Thus, coagulants are an important class of chemicals which aid coalescence. Another mechanism involves the substitution of an interfacially active compound for another, or several others. In this case, a small molecule such as acetic acid will be added which migrates to the interface and pushes away larger molecules that will get in the way of the coalescence process. A mechanisms that has been shown to be effective in treatment of Alkali Surfactant Polymer systems is to add a small molecule cationic surfactant which binds with the sulfonate used in ASP and makes it less sterically repulsive. Once coalescence has occurred the larger drop is more easily separated. The extent of coalescence is measured by the drop size distribution.

References for Further Reading: Both coagulation and flocculation are more complex than discussed above. Only the dominant mechanisms, and most common chemicals have been described. An excellent in-depth discussion of all aspects is given in the MWH Water Treatment book [4]. The literature of scientific studies and a wealth of data is provided in the Faust and Ali book [2]. Bratby's book is somewhat dated but provides practical knowledge on chemical treatment and subsequent separation [5]. When the right chemistry is applied, solid particles coagulate and flocculate. In the case of oil drops, when the right chemistry is applied, the droplets may actually coalesce (combine into a single drop). For coalescence to occur, collision must occur with sufficient velocity (energy) to overcome the chemical forces which stabilize the oil / water interface.

Hydraulic Fracturing Flow Back Fluid: As discussed in Section 4.1.1 (Types of Emulsions), there are several components that contaminate flow back water. The main contaminants are:

1. Suspended organic solids. These organic solids are mostly composed of the polymer used for viscosity control and proppant placement. The most commonly used polymers are HPAM (partially hydrolyzed polyacrylamide – slickwater polymer), guar and cross-linked guar (gel).

2. Depending on the location, this water also contains suspended oil droplets. Shale hydrocarbons tend to be light and easily removed from the flow back water. Offshore flow back will contain a wide range of oil API. This oil is much harder to separate particularly if it co-mingles with the polymer.

3. Proppant. Several different materials can be used as proppant. The most common is sand. Typical sand sizes used are rather large and easy to separate provided that the sand particles are not crushed. Crushed sand particles belong in a separate class of material due to the very fine particle size (see next category).

4. Suspended shale fines, crushed proppant, iron oxides, and iron sulfide. These contaminants constitute a separate category due to the very small size (sub-micron to less than 10 microns). They are very difficult to remove.

5. High concentration of dissolved salts.

Coagulation and flocculation is very effective at promoting the separation of the suspended organic solids. The trivalent cationic charge of coagulant chemicals is effective at forming gel-like aggregates of the polymer which can be separated by settling or flotation. In some cases, the HF polymers adsorb onto solid particles. When that occurs, it is helpful to apply a mild oxidation step prior to coagulation. Mild oxidation can be achieved using bleach (hypochlorite), ozone, or peroxide. Oxidation helps to release the polymer from the solid. It also helps to destabilize the polymer.

Electrocoagulation: Electrocoagulation technology makes use of coagulation, flotation and oxidation together in one treatment step. Some shales have a higher clay content than others. Depending on the water chemistry the clay can be positively charged. In that case, an anionic flocculating agent is usually found to be effective.

All things considered, coagulation and flocculation are well understood and there is precedent and good reason why they should be successful in treatment of HF flow back water. The significant challenges that have been overcome by the service companies are in the areas of logistics, cost, universality, robustness, and all of the challenges associated with bringing such technology to remote locations and applying it to unknown fluids with highly variable characteristics.

5.4 Coagulation Chemistries:

The first step in chemical treatment is often to prepare the oil drop surface, or solid particle surface so that flocculation can occur rapidly and extensively. This step is known as coagulation. Coagulation, if done properly will collapse the electrical double layer that surrounds the contaminant particles and allow aggregation and facilitate subsequent flocculation with or without flocculating agents. The mechanism of coagulation has been discussed extensively already (see Sections 4.2 and 4.3). In this section, the chemistries used are discussed. A number of coagulating agents are listed in the table below.

Table 5.2 Summary of the main chemicals used as cationic coagulating agents.

Acronym	Structure	Polymer M. Wt. (gr/mol)	Also used as
ferric chloride	cationic salt	--	
ferric sulfate	cationic salt	--	
aluminum chloride	cationic salt	--	
aluminum sulfate	cationic salt	--	
PAC (poly aluminum chloride)	metallic polymer	180	
DADMAC (di-allyl di-methyl ammonium chloride)	homopolymer quaternary salt	500,000 to 2,000,000	flocculating agent
epi (epichlorohydrin)	homopolymer	100,000 to 500,000	flocculating agent

As discussed previously, dispersed oil droplets and suspended solids in produced water typically have a negative surface charge. By adding the Lewis acid ions, which then form large association polymers, two mechanisms occur. The first important mechanism is the so-called collapse of the electrical double layer. As described in Section 4.3.2, the electrical double layer of the oil drops and solid particles acts as a repulsive potential energy barrier that prevents close association of oil drops with other oil drops and of solid particles with other solids particles. The effect of this electrical energy barrier can be reduced both in magnitude and in length by the addition of cations. Even simple cations such as Na^{+1} and K^{+1} are somewhat effective. However, multivalent cations such as Al^{+3} and Fe^{+3} are significantly more effective. This is known as the Shultz-Hardy rule.

Coagulating Agents Composed of Lewis Acid Metal Salts: Coagulation of oil droplets in water is promoted by metal salts, and self-associating metal salts. These are usually Lewis acids. A Lewis acid is an ion that is an electron pair acceptor. A Lewis Base is an electron pair donor. When a Lewis acid and base interact, it usually results in the formation of an adduct, a stable combination of the acid and base that shares the electron pair. Typical Lewis acids that are effective as coagulating agents are

Al^{+3}, and Fe^{+3}. These Lewis acids can be injected into the produced water in the form of $AlCl_3$, $FeCl_3$ and the analogous sulfates, or as polyaluminum chloride (PACl). These Lewis acid compounds react with the lone pair of water molecules. These association complexes react further with other Lewis acid ions and associated water molecules to form self-association polymers, as shown in Figure xx. These reactions are a strong function of pH.

The second important mechanism that occurs is a site-site binding effect that is similar to that which occurs with long chain flocculating agents. When the polymeric association complexes form, they act as bridging and enmeshing agents like those discussed already for water drop flocculation in oil.

Coagulant Application and Zeta Potential: The concept of zeta potential is highly applicable in the use of coagulants. Zeta potential is a measure of the surface charge of contaminant particles. By neutralizing the surface charge, as measured by zeta potential, it is more likely that contaminant particles will stick together. The figure below illustrates this idea.

High Salinity Brines: It must be noted that coagulants are not always effective in oilfield water treatment. While flocculating agents are almost always used, typical oilfield brine does not always require the use of a coagulant. As discussed in Section 4.3 (Charged Surfaces in Ionic Solutions), common produced water salinities from oil fields and from shale are usually high enough that coagulant has little beneficial effect. The electrical double layer is already collapsed by the high salinity and presence of divalent cations. Nevertheless, for lower salinity produced water systems, such as in gas fields, the use of a coagulant can have an enormous beneficial impact and cost saving.

5.4.1 Metallic Salt Coagulants:

As discussed previously, the simple tri-valent salts of iron and aluminum are the simplest of the coagulants. When added to water these salts form a number of complex hydroxyl by-products then polymerize. These polymeric hydroxides of the trivalent cations neutralize the negatively charged colloids (small suspended oil drops and solid particulate). Their polymeric structure also gives them an ability to act a bit like flocculating agents. This mechanism is discussed later in relation to electrocoagulation. As shown in the table, the polymeric coagulating agents can also act as flocculating agents. In fact, the mechanism of interaction (coagulation or flocculation) becomes a bit indistinguishable for the polymer electrolytes since they have both charge and polymer bridging properties. Generally speaking the chemicals used for coagulation (charge neutralization/collapse of the electrical double layer) are lower in molecular weight and charge density. The same chemical structure may also be used as a flocculating agent but for that purpose the chemical will be formulated as a much higher molecular weight and charge density version, thus giving it the ability to bridging and hold together suspended particles in the water.

Aluminum Cation: One of the chemistries that is used typically in municipal waste water treating is the aluminum cation (Al^{+3}). This cation can be applied in the form of aluminum sulfate (AlSO3), aluminum chloride (AlCl3), or other forms of aluminum salt. It is effective at lower concentrations than required by the addition of NaCl, as described above for double layer compression. However, as will be discussed, the concentration required is still moderately high compared to modern polymer chemistries.

Typically, two mechanisms are involved in Al^{+3} coagulant chemistry. The first involves surface adsorption at anionic sites such as carboxyl, sulfate, and silanol, which results in partial charge neutralization. The second involves the formation of a colloidal floc of aluminum hydroxide which results in precipitate enmeshment.

Charge Saturation and Reversal: It has been found that Al^{+3} is effective over a wide range of concentrations. The required concentration for a particular application depends of course on the concentration and composition of contaminant. Finding the most effective concentration generally requires some initial bottle testing followed by field optimization. In detailed studies of surface chemistry it has been found that Al^{+3} does more than simply compress the electrical double layer. It has been found to adsorb at anionic sites on the solid particle surface. This adsorption results in neutralization of some of the solid particle surface charge. Neutralization reduces the repulsion between particles. However, neutralization can become saturated. Once neutralization has become saturated, additional Al^{+3} starts rebuilding the electrically repulsive double layer, albeit with opposite charge. While charge neutralization has its limits, it is an important mechanism in coagulating solid particles.

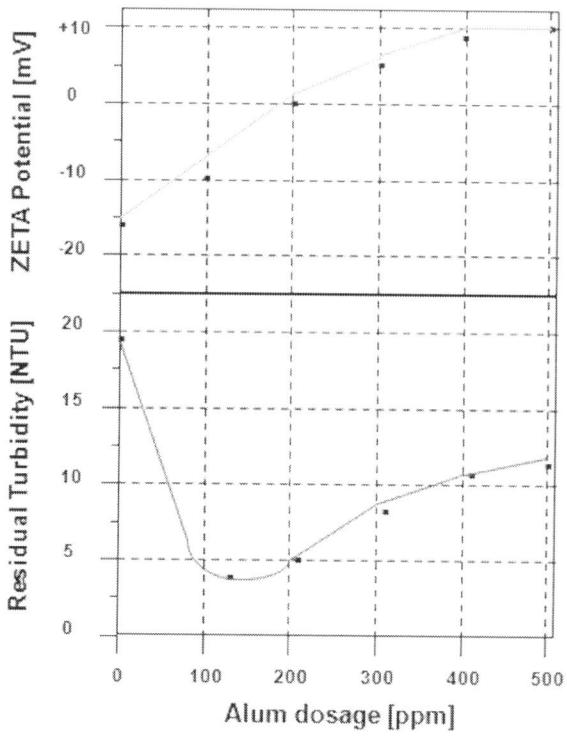

Figure 5.2 Determination of the optimal coagulant dosage (minimum residual turbidity). Alum was added and the solution was allowed to settle. Zeta potential was also measured. At zero coagulant dosage, the zeta potential was negative indicating a net negative surface charge. As Alum was added, the negative surface charge was neutralized, at the optimal coagulant dosage. Higher dosage of coagulant causes the zeta potential to increase indicating that the surface has become positively charged due to excess coagulant.

Zeta Potential: As shown in Figure 5.2, the optimum concentration corresponds to a zeta potential near zero. Zeta potential is a practical measure of net surface charge at the slip plane (the plane that divides the bound ions from the mobile or fluid ions). When the zeta potential is near zero, then contaminant particles can approach each other with minimal charge-charge repulsion. In the nest case, the particles can approach so closely that attractive van der Waals and other interactions can occur. This allows aggregates to form and it provides the ideal surface charge environment for application of flocculating agent. It is not necessary to measure the zeta potential to determine the optimal concentration. A simple visual assessment using a jar test procedure or a turbidity measurement after a short settling time is sufficient.

Aluminum Chloride: When the Al^{+3} ion is added in the form of $AlCl_3$ aluminum chloride, hydrolysis occurs. Hydrolysis produces protons which lower the pH, and hydroxyl ions which become bound to the aluminum cation.

Aluminum Hydroxide: When the Al^{+3} ion is added in the form of $Al(OH)_3$ aluminum hydroxide, a precipitate is formed. This precipitate has a three dimensional polymeric chain structure with potentially thousands of repeating aluminum hydroxide units. The precipitate grows by the entanglement of other precipitated polymeric chains. At sufficiently high concentration, a solution will appear white and cloudy. This white floc sticks to solid particles and entraps the solid particles. This mechanism is referred to as "precipitate enmeshment." Since the polymer chain contains Al^{+3} ions, it will also have a degree of surface adsorption onto the solids particles thus strengthening the attachment of the mesh with the particles.

Other metal salts besides Aluminum: Other chemicals that function in this way are ferric chloride, ferrous sulfate, sodium aluminate, and lime. The required concentration of these chemicals and their effectiveness typically depends on the specific surface sites available for adsorption, and on the solution ionic strength, pH, and alkalinity.

5.4.2 Polyaluminum Chloride (PACl) Coagulants:

In the oilfield, aluminum chloride (or its related mineral coagulants) is not often used, for various reasons. It is corrosive. The polymer form (polyaluminum chloride) is much better for handling and is more reactive over a wider range of pH, temperature, and salinity. In effect, the polymer form is "pre-hydrolyzed" which improves its performance over a wide range of water properties. The third class of chemicals is the pre-hydrolyzed alum salts. Generically they are known as PAC (polyaluminum chloride). But there are many varieties. They are pH stabilized such that there is essentially no pH shift when applied. They are formulated with an optimum hydroxide, sulfate, and alumina ratio to promote the formation of large cationic aluminum hydroxide polymer chains which are effective as a flocculating agent. They are more expensive than the simple metal salts but they have essentially none of the compatibility and performance problems.

PACl is the pH stabilized polymeric form of the salt, aluminum chloride. When used as the salt, the aluminum chloride hydrolyzes in water to form aluminum hydroxide, which polymerizes to form an amorphous gel-like polymer structure. Even at low concentrations, solutions of aluminum chloride can be seen to form white flocs of polymerized aluminum hydroxide. The flocs can be several centimeters in diameter which makes for a very effective flocculant. The aluminum hydroxide has dispersed cationic adsorption sites which bind to negatively charged sites on the solid surface or to negative charges in the inner layer of the electrical double layer. The hydrolysis lowers the pH somewhat which helps this binding.

In the oilfield, aluminum or iron chloride (or its related mineral coagulants) are not commonly used for various reasons. They are corrosive. The polymer form (poly aluminum chloride) is much better for handling and is more reactive over a wider range of pH, temperature, and salinity. In effect, the polymer form is "pre-hydrolyzed" which improves its performance over a wide range of water properties.

5.4.3 Polyelectrolyte Coagulants:

Polymers: As a class of chemicals, the polyelectrolyte coagulants include synthetic polymers, natural compounds, and derivatized natural polymers. The polyelectrolytes are more expensive than the simple salts but are far more robust in terms of compatibility with other ions, and are more effective over a wider pH range. They too can have limitations in terms of compatibility with the waste water. But pre-mixing is often effective at overcoming compatibility problems. Given the large number of products on the market, on-site bottle testing is often necessary. When a proper coagulating step is included upstream, the effectiveness of a polyelectrolyte flocculating agent added downstream can be exceptional. This reduces the cost of the polyelectrolyte significantly.

Like the inorganic clarifiers described above, modern polymer water clarifiers utilize both surface adsorption and precipitate enmeshment. Such polymers are sometimes referred to as polyelectrolytes. Clarifiers based on organic polymers are synthesized with much higher molecular weights than achieved by flocs of inorganic clarifiers. If the right chemistry is synthesized, then polymer chains will both adsorb onto surface sites, and entangle each other to form large flocs. A floc of polymer molecules will form a three dimensional bridging structure between solid particles. There are many such polymer water clarifiers on the market. Typically they are branchy molecules in order to maximize the probability of enmeshment. They also have positively charged sites in order to maximize surface adsorption.

Effect of pH: pH has an important effect on the action of these chemicals. The vendor may say that his PAC is "pH stabilized." This means that the polymer will not lower the pH as much as the salt, and that its activity will be higher over a larger range of pH. That is true to some extent but it may still be sensitive to pH.

As a practical matter, bottle testing is carried out using vented samples. This means that the samples contain less CO_2, and higher pH than the water in the flow line. Venting of the sample can be minimized by carefully capping the sample as soon as the sample jar is full. However, there may not be anything more that can be done except to add a small amount of acetic or mineral acid (HCl) to compensate for the loss of CO_2. But it is difficult to know how much acid to add. In some cases, where the CO_2 partial pressure is low, the loss of CO_2 and the shift in pH is minor.

Reaction Rate of Polymer versus Salt: One good thing about the polymer compared to the salt is that it is faster acting and is pound for pound more effective. Seems counter-intuitive that a polymer would act faster than a simple salt but it does. The flexible water soluble polymers have a strong tendency to adsorb on surfaces.

Dosage Sensitivity: As mentioned in the preceding paragraph, some polymeric coagulants form a three dimensional bridging structure between solid particles. These coagulant polymers have particular sensitivity to dosage rate (concentration). For example, as the initial polymer dose is increased beyond that required for initial aggregates to form, any additional polymer will participate in bridging between the aggregates to produce a network of aggregated particles or floc. This is perhaps the optimum concentration in that particles are captured in the network and flocs are bridged to form effectively large bodies with high separation velocity and high potential for sweeping additional particles into the floc. Slightly higher dose will increase the polymer concentration of the flocs and may actually impart strength to the floc thus allowing it to survive some degree of shearing without breaking up. This is not necessarily detrimental but is potentially a waste of chemical. Further increase in dose beyond his point is however detrimental. At high polymer concentrations, solid particles can become covered with polymer before bridging and network formation occurs. What results are individual particles with a high concentration of polymer and very little floc formation. From a macroscopic standpoint, a sticky gummy mess is generated which usually requires vessel cleanout.

Although organic polymers are highly effective in most applications, there are situations where they are not sufficient. One of the causes is the presence of a high number density of very small particles, on the order of a micron or submicron in diameter. In such cases, the performance of the organic polymer can be aided by the addition of inorganic salts. For some of the organic polymers it has been found that the combination of an organic polymer and an inorganic coagulant has a synergistic effect. Presumably, the inorganic coagulants help to bind a large number of small particles into short chains and small networks which are then enmeshed into larger flocs by the polymer materials. Those polymers which are most effective in bridging with the inorganic coagulants, are referred to as coagulant aids. It is typically the low and medium molecular weight cationic polymers that are most effective in this manner.

Blends: These processes are not as well defined nor as precise as presented here; varying in nature and degree. For this reason physical blends of the differing molecular weights of the same polymer often provide significant performance advantages over a product consisting of a single molecular weight polymer with a reasonably tight molecular weight profile.

DADMAC (di-allyl di-methyl ammonium chloride): DADMAC (or DMDAAC as it was originally known) is a quaternized ammonium compound (quaternary amine). This means that the pendant group has a positive or cationic charge. Acrylamide monomer is frequently used to produce a copolymer with reduced polymer charge density. Acrylamide pendant group is nonionic.

The basic DADMAC polymers, such as the homopolymer polydiallyldimethylammonium chloride (poly-DADMAC), are known to perform well as flocculants. High molecular weight polymers (>1,000,000 Da) are typically used. The polymers contain manly five-ring but also some six-ring cationic groups. Various methods of making the polymers, as aqueous solutions, suspensions, or emulsions have been claimed. An improvement on the use of the homoploymer, poly-DADMAC, is to add 5-10% by weight of residual DADMAC monomer to enhance the formulation. DADMAC monomer can be copolymers with acrylamide or other nonionic monomeric hydrophilic monomers to vary the charge density in the polymer.

A small amount of branching in some of the polymers is claimed to give improved flocculant performance. Typical cross-linking agents are vinyltrimethooxysilane and methylenebisacrylamide. Hydrophobically modified DADMAC polymers can give improved flocculant performance. A preferred hydrophobic monomer is ethylhexylacrylateester. The hydrophobic groups are attracted to hydrophobic oil droplets, enhancing their flocculation. A combination of aluminum chlorohydrate and a polyamine such as polydiallyldimethyl ammonium chloride, is claimed to give superior flocculant performance.

Polyacrylamide (PAM): Polyacrylamides are so-called "coagulation aid polymers" because they are so effective at coagulation [6]. Besides polyacrylamide itself, there are a number of varieties. There is a wide variation in functionality (number of functional groups per chain length), molecular weight, and ionicity (charge density). This gives rise to many PAM produces on the market, with vast differences between them. Charge density ranges from neutral to nearly complete ionization of the amide group. Also, most vendors have a wide variety of copolymer polyacrylamide / polyacrylate which vary in adsorption tendency, and speed of reaction. If a particular polymer can be found that works well, knowledge of the polymer chemical structure can give insight about the contaminant.

HPAM (partially hydrolysed polyacrylamide): polyacrylamide is a nonionic homopolymer of acrylamide groups. The acrylamide groups can be hydrolyzed to form an anionically charged group as in HPAM (partially hydrolyzed polyacrylamide). HPAM has a fraction of hydrolyzed groups meaning that a fraction of the pendant groups are anionic. The percent hydrolyzed refers to the mole fraction of groups that have been hydrolyzed to form the carboxylate anion.

AMPAM (amino-methylated poly acrylamide): polyacrylamide can be reacted such that the pendant groups are cationic as in AMPAM (cationic PAM – amino-methylated polyacrylamide). Polyacrylamide (PAM), is a Cationic Poly-Electrolyte (CPE).

5.5 Flocculation Chemistries:

The discussion below is intended to provide a mechanistic understanding of flocculant chemicals, and to provide an introduction to their chemistry. The practice of chemical selection and application is discussed in previous chapters and generally involves onsite competitive evaluation of products. As discussed, the process of chemical selection is typically carried out by the chemical vendor who performs onsite bottle testing. This is typically followed by field testing of a few products in the actual system. The selection process does not typically allow the user to understand the mechanism and chemistry of the products being tested. Chemical vendors will claim that information about their products is proprietary. Thus, to a great extent it is difficult for a user to learn much about the chemistry of flocculants. However, some knowledge is necessary in guiding the chemical selection process, and in recognizing what may or may not be reasonable strategies for chemical selection.

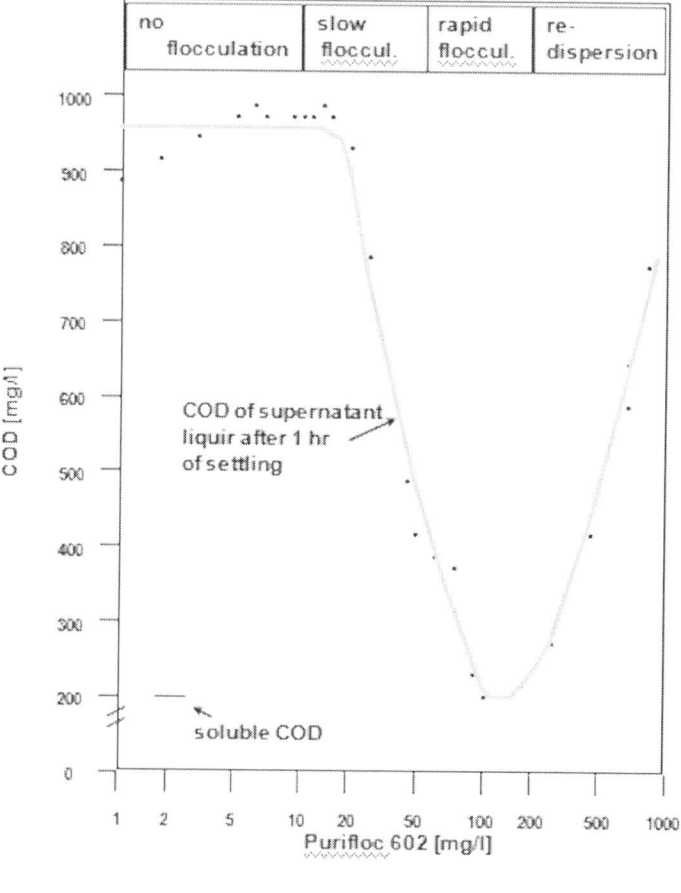

Figure 5.3

Two of the critical functionalities of flocculants are "bridging" and "entrapment." These require that the polymer be fully active (fully extended in the case of linear polymers with the exception of amphoterics) and high molecular weight. Shearing the polymer/produced water mixture will reduce their ability of the polymer to bridge and entrap. At very high hear rates, the polymer molecular weight can actually decrease. But even mild shearing can have a detrimental effect. Contact makes coagulation work faster and better but over-shearing can destroy the effectiveness of a polymeric flocculating agent. Shearing the floc is more of a problem than shearing the polymer.

Table 5.3 Summary of some of the chemicals used as flocculating agents

Acronym	Scientific Name	Structure	Polymer M. Wt. (gr/mol)	Electric Charge Polarity	Typical Function
DADMAC	di-allyl di-meth-yl ammonium chloride	homopolymer	500,000 to 2,000,000	cationic	coagulant or flocculant
epi	epichlorohydrin	homopolymer	100,000 to 500,000	cationic	coagulant or flocculant
PAM	polyacrylamide	homopolymer	400,000 to 1,000,000	nonionic	flocculant
HPAM (20 to 40 % hydrolyzed)	partially hydrolyzed poly acrylamide	homopolymer	1,000,000 to 5,000,000	anionic	flocculant
AMPAM	amino meth-ylated poly acrylamide	homopolymer	2,000,000 to 5,000,000	cationic	flocculant

Flocs are Delicate: Make sure that there are no valves or pumps between the chemical injection point and the separator or flotation unit. Injecting these chemicals upstream of a valve or pump in order to achieve better mixing is a mistake. Better mixing will occur but the turbulence will break the floc and cause the polymer to collapse around individual particles, saturating the available adsorption sites and render the chemical useless and contribute to a sticky precipitate with high erosion properties.

5.5.1 Chemistry and Characteristics of Polymeric Flocculating Agents:

Essentially all flocculating agents are polymers. There are two basic types of flocculating agents, the natural products, and the synthetic polyelectrolytes.

The synthetic polyelectrolytes are characterized by the following features:

1. chemistry

2. structure (linear or branched)

3. molecular weight

4. charge polarity, and charge density

5. physical form (dry powder, aqueous solution, emulsion, brine, gel polymer)

In addition to all of the different products that can be derived by varying one or more of these characteristics, an even greater number of products can be derived by blending these products together in various ratios and concentrations in order to achieve synergy in their water treating effectiveness.

Chemistry: most polymeric flocculating agents are water soluble. This narrows the chemistry considerably since most polymers manufactured by the chemical industry and used by consumers are not water soluble (e.g. polyethylene, polypropylene). Most, but not all, of these water soluble polymers have what is referred to as a polyethylene backbone. This backbone is essentially a long sequence of carbon-carbon linkages made up of CH and CH$_2$ groups. Since the backbone is essentially hydrocarbon, most water soluble polymers achieve their solubility by having polar or ionic pendant side groups. The pendant groups provide water solubility and functionality.

An important exception to the ethylene (hydrocarbon) backbone structure are the polyethylene oxide polymers, and derivatives. Although polyethylene is not particularly polar, the spacing of the ether $(CH_2 - O - CH_2)$ groups is such that water solubility is achieved.

Polymer Structure or Shape: There are two basic polymer structures used in the water treatment industry: linear and branched. Long liner polymer molecules are most effective at making contact and forming a bridge between two oil droplets, or oily solids particles. However, the bridge may not be stable, relatively speaking, due to the fact that a linear polymer forms only a limited contact on the surface of the particle. Such polymers dominate the municipal water treatment industry where the particle surfaces are clean and attachment between the polymer and the particle surface is electrostatically favored. At the other end of the spectrum, a branchy polymer may form a very strong attachment due to multiple contacts made between the branchy pendant groups and the particle surface. These polymers tend to be required in upstream oil and gas application where the particle and droplet surface is heterogeneous, brine salinity is high and thus the electrostatic interaction is muted, and the surfaces are generally contaminated with various surface active compounds either from the oil or from production chemicals.

Branchy polymers tend to have lower molecular weight. The branching improves the polymer shear resistance. These polymers tend to not degrade when subjected to shear. Also, the flocs formed with contaminant particles can withstand higher shear without falling apart.

Whether the polymer is linear or branched, it can also be classified as a homopolymer or a co-polymer. A homopolymer is manufactured as a polymer of a single type of mer (monomer or repeat) group. A copolymer is manufactured as a polymer composed of two or more mer groups. Copolymers can be block copolymers, regular alternating copolymers, or random copolymers. In a block copolymer, there are strands of one mer group followed by strands of another mer group. The category of regular alternating copolymers is self-explanatory. Most co-polymers are however, random copolymers. As the name implies, a random colpoymer is a random assembly of mer groups.

Molecular Weight: Molecular weight is also an important property of a flocculation polymer. One of the features that distinguishes a coagulation chemical from a flocculation chemical is the molecular weight. Coagulation chemicals are generally small molecules. High molecular weight is an important characteristic for a flocculation polymer because the longer the polymer, the greater the probability that it will collide with and form a bridge between two oil droplets or oily solid particles. Some of the building blocks for flocculation polymers are chosen on the basis of the ease with which they polymerize into long chain polymers.

low molecular weight do not need to haveLong flexible polymer molecules are able to bridge between contaminant droplets or particles and start the buildup of a floc.

Molecular weight of these polymers ranges from:

- Low   10,000 to 100,000 gr/mol

- Medium: 100,000 to 1,000,000 gr/mol

- High: 1,000,000 to 5,000,000 and higher gr/mol

In the case of the high molecular weight polymers, the chain length, when fully extended can be as long as one micron or more. In produced water the chain length will be considerably shorter due to the random folding of polymer chain. Nevertheless, flocculating polymers that are formulated to bridge between contaminant particles are quite long compared to the thickness of the electrical double layer which can be less than 0.01 microns in thickness.

Electronic Charge Polarity and Density: There are four basic charge types of polymer: anionic, non-ionic, cationic and amphoteric. Within each charge group there are differing densities of charged sites along the polymer. Charge density is related to molecular structure in the sense that a homopolymer will have a uniform distribution of charges. Every repeat group in a homopolymer is the same so the charge distribution will be constant along the polymer length. In contrast to this, a random copolymer will have a random distribution of repeat groups. Depending on the charge of the repeat groups, the random copolymer will have a random distribution of charged groups. Some polymers are manufactured with a certain charge distribution which is then modified by subsequent reaction such as the hydrolysis of polyacrylamindes.

It is worthwhile to point out that high charge density is not always most desirable. The reason for this is best understood by reference to the concept of a Flory theta solvent. In this concept, the average end-to-end distance of a polymer is theoretically estimated and is used as a measure of the extent to which a polymer molecule uncoils (stretches out) in solution. In general, an uncoiled (stretched out) configuration is the most desirable situation for a polyelectrolyte flocculating agent since it maximizes the probability of attachment to other polymer molecules and to the contaminant. If the solvent environment is favorable, as in a theta solvent, then the polymer has both high entropy and high (negative) enthalpy of mixing. If the solvent environment is not favorable, the polymer will coil in upon itself, and thus have a short end-to-end distance. This coiled state has lower entropy of mixing. The coiled state of a polyelectrolyte can be caused by a high concentration of divalent and tri-valent counterions (such as calcium in the case of an anionic polymer). If a polymer has so many charged groups that it becomes internally cross-linked with its counterions, then it has too many charged groups to be useful. Also, the polymer may precipitate from solution or form a gel. Thus, the optimum charge density of the polyelectrolyte is one which gives a theta solvent state, and maximizes interaction with contaminant particles and droplets.

Electronic charge polarity (anionic or cationic) is selected, at least initially, on the basis of the charge polarity of the contaminant surfaces. If the polymer has an opposite charge to the particle surface, there is a good chance that it will be attracted to that surface. Since most contaminant surfaces are negatively charged, cationic polymers are typically evaluated initially. However, opposite charge polarity is not always most effective. Anionic polymers are sometimes most effective particularly when the contaminant surface has a layer of strongly bound cationic counter-ions. Such ions form what is referred to as as the Stern layer. As discussed previously, this layer of counter-ions is essentially immobile. Anionic polymers can bind to these immobile counter charges in essentially the same manner that a cationic polymer might bind to the negative surface site itself.

Toxicity is sometimes a problem. These polymers "go with" the solids, meaning that, for example, when polymer is fed, upstream of a solids contact clarifier, the polymer will be found in the solids bed. To demonstrate the point, cationic polymers typically exhibit acute toxicity to gill breathing organisms yet fish can be found living happily in the clear water above the solids bed in clarifiers employing cationic coagulants.

Physical Form of Polymer: All polyelectrolyte flocculating agents are relatively long chain polymers. Each polymer type has certain limitations in solubility and stability. The form in which they are delivered will depend on stability in transport and ease of application. This varies from one polymer to another and sometimes within a polymer type depending on the molecular weight. The three physical forms of polymer are:

1. dry powder

2. aqueous solution

3. emulsion polymer

Since a dry powder does not involve water, the shipping mass and volume are minimized. The main drawback is of course that water must be added and good mixing must be achieved in the field.

Latex or Emulsion Polymer: High molecular weight cationic polymers have been found to be effective flocculating agents for a variety of produced water types, particularly those that have contaminants with an anionic surface charge. The chemistry of these cationic polymers is discussed in the next section. Suffice to say that high molecular weight cationic polyacrylamide polymers and copolymers are often used. If these high molecular weight cationic polymers were simply mixed in water, the solutions would have very high viscosity. In fact, the viscosity would be too high for practical application. To avoid this problem, a water-in-oil emulsion is formed [7]. The water-in-oil emulsion is often referred to as a latex. The oil phase is the continuous phase and it is typically composed of low molecular weight solvents that have relatively low viscosity such as kerosene. The water phase, which is suspended in the oil phase, contains the polymer. The overall viscosity of the chemical product is relatively low while still containing an aqueous solution of the high molecular weight polymer.

Polymers that are packaged this way are typically manufactured in a process referred to as emulsion polymerization. In this process surfactants are used to control molecular weight, and latex particle size. The use of surfactants also contributes to the stability of the latex. Polymerization inhibitors are also used to control molecular weight. When an emulsion polymer is added to the produced water, the latex particles de-emulsify. This allows the polymer molecules in the core of the latex particle to become solvated by the produced water. In the case where the chemistry is favorable, the polymer elongates in the produced water. This is referred to as "unwinding" of the polymer. As the word implies, the polymer stretches out which is an important characteristic for a flocculating agent. This improves its ability to attach to multiple contaminant particles and form a floc. Emulsion polymers do require special handling since they are somewhat less stable than a simple polymer in water solution. They are sensitive to temperature and water must be kept out of the product. Application problems relate to the unwinding of the polymer and the tendency of a poorly unwound polymer to agglomerate and form organic scale.

5.5.2 Cationic Polyelectrolytes:

Most cationic polymers are based on amine group chemistry. Secondary, tertiary or quaternary nitrogen make up the main amine group chemistries in use. In the case of a quaternary amine (with four groups attached) the nitrogen then has a positive charge. When there is less than four side groups, the nitrogen can still develop charge in solution depending on the side group chemistry, the pH, and the other constituents in solution. Amine groups tend to be positively charged particularly at lower pH values. Thus, the activity and effectiveness of polymers based on amines is dependent on solution characteristics.

Most of the cationic flocculants are composed of the following chemistries:

- derivatized polyacrylates and acrylimides

- poly(alkanolamines)

- poly(amines)

- blends of the above

Blends may be composed strictly of the polymers listed above or they may be blends of polymers together with inorganic salts. There are two points to keep in mind. Each chemistry has certain specific properties and performance as a coagulant or flocculating agent. The properties and performance are a result of the number and type of functional groups in the molecule. But another aspect is not so obvious, at least to the end user. That is, each of these chemistries of polymer has properties and behavior in chemical synthesis and reaction chemistry. In other words, these chemistries provide different reaction pathways which can be manipulated to form certain types of molecules. For example, forming copolymers from poly(aminoacrylates) is far easier than from poly(amines). Therefore if multifunctionality is desired, then the poly(aminoacrylates) would be the category of choice. Reaction chemistry impacts the density of functional groups, the types of groups employed, and the price of the product.

The most commonly used water treatment polymers are based on acrylamide monomers or polyacrylamide. Polyacrylamide is nonionic yet it is polar enough to be water soluble. Polyacrylamide is referred to as a homopolymer having a simple linear-chain structure.

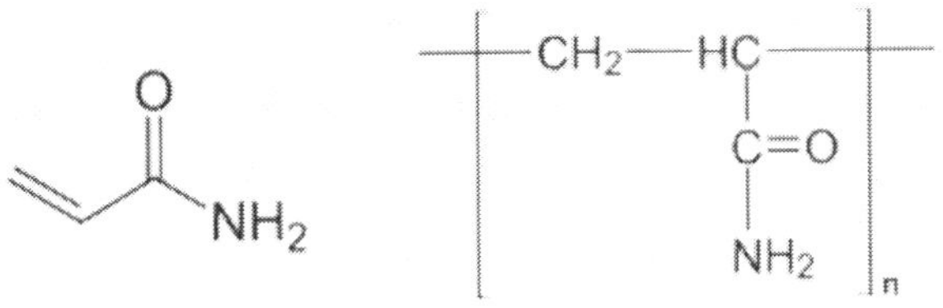

Figure 5.4. Acrylamide monomer (left), and the repeat group of polyacrylamide (right).

Following polymerization, the pendant group of the polymer can be derivatized to form cationic and anionic polymers. An example of this is HPAM (partially hydrolysed polyacrylamide), discussed below. To manufacture HPAM, polyacrylamide is hydrolyzed in a controlled manner such that some of the pendant groups are converted from the amide to a carboxylate anion. When this is done, the polymer has some non-ionic side groups (amide) and some anionic side groups (carboxylate).

Alternatively, the acrylamide monomer can be derivatized (reacted) in various ways to form cationic and anionic monomers which then can be polymerized. Prior to polymerization, other monomers can be added to form random or block co-polymers. An example of this is the addition of the acrylic acid monomer which introduces an anionic carboxylic pendant group to the polymer. Cationic polymer can be formed by introducing a cationic monomer prior to polymerization as in AMPAM (cationic PAM – amino-methylated polyacrylamide). Likewise, other monomeric groups can be added which will form cross-linked polymers upon polymerization.

At these high molecular weights, the viscosity of the polymer solution in water would be very high. Instead of mixing these high molecular weight polymers in water, they can be suspended as a water-in-oil emulsion [7]. The active ingredient(s) are mixed in the internal water phase of suspended micelles. These micelles are kept suspended by surfactant. The external or continuous phase is oil. One method to reduce viscosity is to polymerize neutral, hydrophobic monomers such as dimethylaminoethylmethacrylate and thyl acrylate in a water-external latex emulsion. The polymer becomes cationic and water-soluble on addition to saline solutions. Suspensions of hydrophilic cationic copolymers of acrylamide in a salt media have also been claimed as easier ways of handling these otherwise high viscosity polymer solutions. Di-quaternary acrylic monomers can make useful cationic polymers, can be prepared from a vinylic tertiary amine, such as dimethylaminopropylmethacryamide, by reacting it with (3-chloro-2-hydroxypropyl)trialkylammonium chloride.

Cationic polymers with a percentage of hydrophobic monomers have been claimed to give improved flocculation compared with the homopolymer. While conventional polymers can attach themselves to oil droplet by coulombic attraction, hydrogen bonding, and other undefined or not clearly understood mechanisms, the hydrophobic groups of these terpolymers can also be attached by a hydrophobic group-hydrophobic oil droplet association.

Others claim the use of lipophilic alkylacrylate comonomers with cationic acrylate or acrylamide monomers. A water-dispersible terpolymer formed by polymerization of an acrylamide monomer, a water-soluble cationic monomer, and a water insoluble, hydrophobic monomer such as an al-kyl(meth)acrylamide or alhyl(meth)acrylate is claimed to be a superior flocculant.

A variety of acrylamide or acrylate-based cationic monomers are used to make cationic polymeric flocculants. The most common cationic monomers in this class of polymers are quaternary salts of dimethylaminoethyl (meth)acrylate and methylacrylamidopropyxxx chloride. Another cationic polymer is based on quaternary ammonium salts of 1-acryloyl-4-methylpiperazine. As with DAD-MAC polymers, a small amount of branching obtained by using a cross-linker can improve the performance.

The cation density and effectiveness as a flocculant can be varied by copolymerization with neutral hydrophilic monomers such as acrylamide, for example, 20:80 mol% copolymer of acryloxyethyl-trimethyl ammonium chloride and acrylamide with molecular weight above about 2,000,000 Da

It may be also possible that hydrophobic groups on different polymer molecules interact to form a bridge or network which could aid in floc formation and oil flotation. While coulombic attraction still appears to be the strongest type of attraction, the hydrophobic association, or hydrophobic effect seems to add significant strengthening to this attraction, as evidenced by improved emulsion breaking and wastewater cleanup.

AMPAM (amino-methylated poly acrylamide): polyacrylamide can be reacted such that the pendant groups are cationic as in AMPAM (cationic PAM – amino-methylated polyacrylamide). AMPAM is referred to as a cationic polyacrylamide (CPAM).

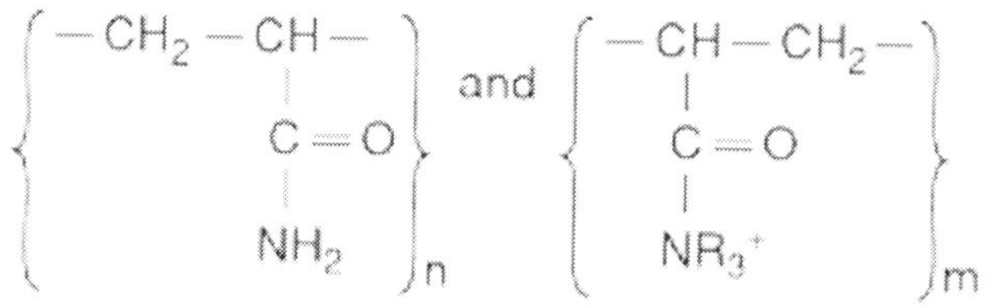

Figure 5.5 Acrylamide monomer plus quaternary ammonium monomer results in a cationic polymer.

Cationic polacrylamides are often formulated using the Mannich reaction. Such polymers are often referred to as Mannich acrylamide polymers. The Mannich reaction, as applied to polyacrylamide chemistry, involves the acrylamide monomer, plus formaldehyde and one of several possible amine compounds, including several quaternary amines. The result is a co-polymer with a combination of acrylamide side groups and acrylonitrile, methacrylamide, di-methyl, tri-methyl, or other amine structure, in amounts up to 50% of the resultant copolymer. The polymers have molecular weights from 10,000 to 3,000,000. Alternatively, amide pendant groups can be modified by quaternization; for example with dimethyl sulfate, to provide aqueous cationic polymers. Mannich acrylamide polymers, in the form of inverse microemulsions, can give superior performance. It also allows the polymers to be prepared at high solids content while maintaining a very low bulk viscosity.

Poly(alkanolamines): One of the earliest products developed for produced water were the poly(al-kanolamines). These compounds are derived by reaction polymerization of compounds such as triethanolamine, or triisopropanolamine, for example. Both straight chain and branched chain polymers can be formulated in this way. The branches can be composed of a wide variety of molecular structures including hydrophobic chains, hydrophilic chains, and cyclic structures as well.

Thus, these polymers are characterized by the amount of branching present, the proportion of ring structures, the number and type of reactive sites on the branches (pendant groups), and the molecular weight of the product. Cross-linked structures are also synthesized. A wide variety of polymers can be synthesized by varying these characteristics. These characteristics in turn affect the performance of the product.

Poly(amines): Polymers of small amine units such as ammonia, methyl amine, dimethyl amine are referred to as poly(amines). They differ from the poly(alkanolamines) in that they do not have oxygen in the backbone. But like the poly(alkanolamines) they can be synthesized with a wide variety of different pendant groups. By reaction with fatty acid, an amide functionality can be formed which imparts a high degree of hydrophilicity. Typically, poly(amines) are synthesized as homopolymers. In general, the poly(amines) are widely used as coagulant aids.

polyDADMAC: a homopolymer of di-allyl di-methyl ammonium chloride. It is a quaternary amine. Thus it is cationic. Every monomer group is charged. Thus, it has high charge density. It is very soluble in water and can be delivered in concentrations up to 50 w%. The range of molecular weights of the polymer for use as a flocculant is hundreds of thousands of grams/mol. It is a water-soluble polymer used world-wide for flocculation of surface water to make drinking water. It is also used in produced water treatment.

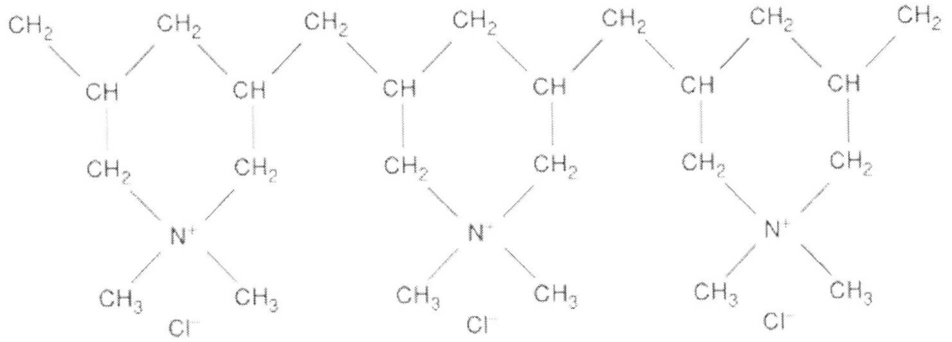

Figure 5.6 Structure of polyDADMAC.

As mentioned above, the introduction of a hydrophobic monomer improves the performance of the cationic polymer. DADMAC or acrylate cationic polymers with vinyl trimethoxysilane are examples. The alkyl silane is hydrophobic and provide improved attachment to oil droplets.

The charge density can be reduced by copolymerizing the DADMAC with non-ionic acrylamide monomer to produce AM/DADMAC. The AM/DADMAC has a lower charge density. In produced fluids that have low or mixed charge density on the contaminant particles, then lower charge density polymers often perform better than high charge density polymers. The determination of charge density varies from one supplier to another. One method is the charge units as a percent of the molecular weight. A common method of calculating charge density is as follows: If the ratio of the number of AM units to that of DADMAC units is 75/25 then the polymer has a 25% charge density (25% charged monomer based on number of monomer units). AM/DADMAC copolymers are commonly used in oil-in-water emulsion breaking while clarification of fresh surface waters is more often accomplished with DADMAC homopolymers. In the former case, the physical structure of the polymer is the important polymer property, while in the latter, the polymer charge is critical in destabilizing the suspended solids.

Blends of DADMAC together with aluminum polymers such as poly-aluminum hydroxychloride, poly-aluminum chloride, or poly-aluminum siloxane sulfate have been used in removing organic and inorganic contaminants from waste water. It is reported [8] that these blends are significantly more effective than any of the individual components used separately. The aluminum polymers can be replaced with aluminum salts such as alum or aluminum chloride. The quaternized polymers have a high molecular weight, in the range of 1,000,000 gr/mol. The polymer is supplied at a concentration of 20 w% in water. This solution has a viscosity of of about 1,000 cP.

epi-DMA: epichlorohydrin-dimethylamine. It is one of the most widely used polyelectrolytes for drinking water clarification. As such it is found to be effective at removing suspended solids.

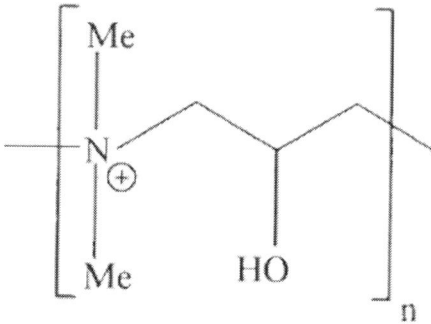

Figure 5.7

In this case the cationic group is in the backbone of the polymer rather than the pendant group.

Poly(aminoacrylates): These are polymers with an ethylene backbone and an acrylate group attached to every other carbon. The end of the acrylate group can be modified with a wide variety of functionality. Synthesis usually involves the formation of copolymers which imparts multifunctional pendant groups on the same polymer.

5.5.3 Anionic Polyelectrolytes:

Anionic polymers are used for flocculating particles that have a net cationic charge on or near the surface. There is a variety of mechanisms that can give rise to cationic charge. A negatively charged surfaces will attract counter ions (cations) from the solution. As discussed previously, adsorption of cations will form a Stern layer of relatively tightly bound ions. Anionic polymers can bind to these immobile counter charges in essentially the same manner that a cationic polymer might bind to the negative surface site itself. Another mechanism is caused by the use of high concentrations of cationic corrosion inhibitor. Most anionic polymers are based on the presence of carboxylic acid or sulfonic acid pendant groups. Just as with other polyelectrolytes, the extent to which these anionic polymers take on charge depends on the characteristics of the solution.

Most anionic or nonionic flocculants are derived from poly(acrylamides). These polymers typically have an ethylene backbone with pendant acrylamide and carboxylate groups. The amide groups are nonionic but can be converted to anionic carboxylate groups by reaction of the polymer with caustic (hydrolysis). The proportion of negative charges can be varied depending on the degree of hydrolysis. Such anionically charged polymers are usually converted into the sodium salt of the acid.

Apart from the above, there are several other types of commercially available anionic polyelectrolytes. The important ones are poly(styrenesulfonicacids) and 2-acrylamido2methyl propane sulfonic acids.

In addition, copolymers of acrylamide and acrylic acid are available. The main difference between these polymers is the charge density, molecular weight, and of course ease of manufacture and cost.

5.5.4 Chemistry of Carbamates and Chelating Agents:

Dithiocarbamates (DTCs) are a somewhat unique class of flocculant. There are multiple mechanisms by which they help remove contaminants from produced water [9]. Each mechanism will be discussed. Use of DTC for produced water clarification was initially introduced by the mineral mining industry. In that industry, DTCs are added to process water just prior to flotation. The DTC dramatically boosts the efficiency of the flotation process. Dithiocarbamates are employed and registered as pesticides. The Vinnings company used to market potassium dimethyl dithiocarbamate. In the E&P industry both flotation and dithiocarbamate chemistry have been adapted from the mineral mining industry. The general formula for a DTC molecule is given by:

$$R_A NHCS_2 Na$$

In which the sodium salt of a DTC is shown. R_A is a hydrocarbon functional group attached. The chemistry of the R_A group has a strong impact on the oil and water solubility of the DTC, its flocculating tendency and its effectiveness to clarify produced water in various brine chemistries.

DTCs act as classic flocculating agents. The DTC molecule has a hydrophobic part that attached to oil droplets or oil patches on solid particles. This hydrophobic part also entangles with other DTC molecules to form a network or bridging structure. This is the classic flocculating mechanism. They are rapid acting and quickly form a floc. If the right DTC is selected for a given produced water chemistry, the floc that is formed can be very effective at removing contaminants.

Another mechanisms is the following. It has been shown, in controlled laboratory experiments, that DTC is much more effective as a flocculating agent in the presence of dissolved iron. One or more DTC molecules will form a complex with multivalent metal ions such as iron. Once the complex forms, the iron has characteristics of iron sulfide. Iron sulfide is surface active but relatively hydrophobic. This combination of surface activity and relative hydrophobicity gives it a very high affinity for the surface of oil drops. The very high affinity is manifest in rapid movement to the oil/water interface. This is a general characteristic of iron sulfide which is shared by the DTC/iron complex, once it has formed. The R_A functional group (R_A) provides flocculating action with other oil droplets and with other R_A groups through an entanglement mechanism. Thus, the iron sulfide-like character of the DTC/iron complex, together with its flocculating functional group, results in a high activity and fast acting flocculating agent.

In Section 2.7.1 (Dissolved Inorganic Constituents), the chemical treatment of produced water to remove heavy metals is mentioned. Dithiocarbamates (DTC) are particularly effective. This class of chemical compounds functions as both chelating agent and as flocculating agent. As a chelating agent, the DTC forms an electrostatic bond with the dissolved or dispersed metal. If the metal ion is divalent, then two DTC molecules will attach. The metal / DTC complex will then form a network with other metal / DTC complexes. As DTC molecules proceed to bind metal ions, the concentration of the complexes increase until a floc is formed. The floc can then be removed by settling or flotation.

A wide variety of DTC products, which differ in the R_A group, are available. The molecular weight of the branch, and its functional groups are design parameters which affect the solubility of the DTC, the extent to which it flocculates, its sensitivity to salinity, and its tendency to form what is known as a DTC floc. In the early application of DTC chemistry to the oilfield, there was a tendency of the products to form a sticky precipitate which was difficult to remove from treating equipment. This was particularly true in systems that relied on extensive recycling of treating water rejects. In those systems, the sticky DTC precipitate would accumulate throughout the system. Such problems have been mostly overcome by better design of the R_A group. But performance testing is highly recommended to ensure long term operability of the DTC.

The performance of a DTC water clarifier varies depending on the produced water total dissolved solids, the concentration of multivalent cations, temperature, mixing, and the presence of other chemicals – both naturally occurring and injected. The table below gives solubility data for three DTC water clarifiers.

Table 5.4

DTC Water Clarifier	Solubility in fresh water	Solubility in 10 % TDS	Solubility in 20 % TDS
A	soluble	soluble	soluble
B	soluble	dispersible	dispersible
C	soluble	dispersible	floc forming
D	dispersible	dispersible	dispersible

As shown in the equation below, in controlled experiments, DTC was found to have a higher affinity for calcium ions than the carboxylic acid.

$$2\,R_A NHCS_2 Na + (R_B CO_2)_2 Ca \rightarrow 2\,R_B CO_2 Na + (R_A NHCS_2)_2 Ca$$

This is an important finding since carboxylic acids have a high affinity for calcium. The finding suggests that DTC has an even higher affinity for calcium than carboxylic acids. The experiments have been carried out with other multivalent cations besides calcium, with similar results. In the case of iron, cation exchange occurs to an even greater extent.

The affinity of carboxylic acids for multivalent cations is used as the basis for an effective class of water clarifiers. An explanation of the significance of this finding is a bit involved but it does serve to further elucidate the chemistry of water clarifiers. A typical produced water will contain organic acids that have an affinity for the oil/water interface. They tend to stabilize the formation of small drops of oil. In extreme cases, a "chemically stabilized emulsion" can form. This occurs when the concentration of the acids is so high that small drops are stable for hours, or days, or longer. Since the oil drops are stabilized by these acid components, it makes practical sense that the only way to resolve such an emulsion is by chemical means.

Some water clarifiers push these organic acids away from the interface and into the oil phase. Once removed from the interface, the oil drops are less stable and more likely to coalesce. It has been observed that water clarifiers that are effective in this way tend to have something in common. They tend to be less water soluble than the acids that they displace from the interface. In other words, the water clarifiers are still water soluble but less so than the acids that they displace.

The mechanism can be illustrated by the following chemical equilibrium equation.

$$2\,R_C CO_2^{-1} + R_B CO_2 - Ca - O_2 CR_B \rightleftharpoons R_B CO_2^{-1} + R_C CO_2^{-1} + R_B CO_2 - Ca - O_2 CR_C$$

in which the water clarifier is denoted by $R_C CO_2^{-1}$, the naturally occurring acid is denoted by $R_B CO_2^{-1}$. In solution, the species on the left of the equation are in equilibrium with the species on the right. In this mechanism, the surface active complex is denoted by $R_B CO_2 - Ca - O_2 CR_B$. By cation exchange, the surface active complex is transformed into an oil soluble complex denoted by $R_B CO_2 - Ca - O_2 C$-

R_C. By forming the oil soluble complex, the acid is pushed away from the interface and the oil drop is no longer chemically stabilized.

For this mechanism to occur, the water solubility of $R_BCO_2 - Ca - O_2CR_C$ must be less than $R_BCO_2 - Ca - O_2CR_B$. Therefore, the water solubility of $R_CCO_2^{-1}$ must be less than $R_BCO_2^{-1}$. This fine-tuning of solubility is one of the reasons why so much field work is required to determine the best DTC compound for a given platform. In reality, the identity of the naturally occurring acids will not be known. In fact, there may be a number of different acids, and therefore a blend of DTC molecules may be required. This is best determined by trial and error. Water solubility alone is not the important characteristic. Instead, the water solubility relative to the naturally occurring acids is the important parameter. In general, the less water soluble, while still being somewhat water soluble is the most likely strategy to follow. However, precaution must be exercised since gunking tendency increases as water solubility decreases.

5.6 Coalescence Chemistries:

In order for coalescence to occur, there must be sufficient turbulence so that the droplets crash into each other frequently and with sufficient kinetic energy to overcome the energy barrier that exists between droplets. The intensity of such turbulence must be high enough for this to happen but low enough so that shearing of droplets does not occur. Chemical addition can reduce the magnitude of the energy barrier. That is the objective of adding a chemical that enhances coalescence. Once coalescence has occurred, the larger droplets are more easily separated. The extent of coalescence is measured by the drop size distribution.

Chemicals that promote coalescence can employ a wide range of mechanisms. For example, one mechanism involves the displacement of naturally occurring interfacially active compounds at the interface. In this case, a small molecule such as glycolic acid will be added which migrates to the interface and pushes away larger molecules that will get in the way of the coalescence process. The glycolic acid is very interfacially active (migrates to the interface) but it does not hinder coalescence. The larger molecules that it pushes away might be asphaltenes or resins that form an elastic membrane. By pushing these compounds away from the interface, the elasticity of the interface is reduced. In fact, it lowers the energy barrier. As discussed previously, any of the coagulating agents might be effective at reducing the electrical energy barrier between droplets.

Another mechanism that has been shown to be effective in treatment of Alkali Surfactant Polymer systems (Chemical Enhanced Oil Recovery) is to add a small molecule cationic surfactant which binds with the sulfonate used in CEOR [10]. Typically the anionic sulfonate, which is a strong surfactant, has strong surface activity. It migrates to the oil / water interface and creates very stable emulsions. When the cationic surfactant is added, it forms a chemical complex with the anionic sulfonate making it much less surface active. In effect the cationic surfactant pulls the sulfonate away from the interface. Without the sulfonate at the interface, the interface becomes much less electrically and sterically repulsive thus, coalescence is promoted.

Another class of coalescing agents is based on polymers. The chemistry of these coalescing agents is not well documented in the literature. There is a wide range in the performance of these products some of which can greatly improve oil dehydration and water treatment. As such, there is significant economic advantage for companies that supply the high performing products. For most specialty chemical supply companies the chemistry of these products is a closely guarded secret. As discussed previously for the case of flocculating agents, in order to select and apply chemical products it is usually not necessary to have detailed chemical information. The process of bottle testing, if carried out properly, will allow selection and application of the optimal products. In the case of coalescing chemistries, it may be necessary to bottle test with some degree of added shear since mild shearing promotes coalescence. Also, the injection location can have a significant effect on the performance of these products. In some cases, injecting the chemical upstream of a valve or pump can be effective. In that case, the chemical promotes coalescence in the region where shear has mostly dissipated and the shear intensity is optimum for drop-drop collisions. This is roughly in the region starting at about 4 to 5 pipe diameters and extending to about 10 to 20 pipe diameters.

While most companies will not divulge the chemistry being used, there is some information that is interesting and which provides some mechanistic insight. Breen, Wasan and co-workers [11, 12] have studied various chemistries and determined which properties correlate with oil/water demulsification. They found a correlation between chemicals that lower the interfacial film elasticity and coalescence. This in turn was found to have an optimum molecular weight suggesting two competing mechanisms. High molecular weight polymers generally lead to a softer interface since the molecules are flexible and tend to displace the elastic forming compounds such as resins and asphaltenes. On the other hand, low molecular weight provides high diffusivity and high migration rate to the interface. Thus, the optimum molecular weight must be investigated for each crude oil / water system.

5.7 Dispersants:

n-Alkylbenzenesulfonic acid: Dodecylbenzenesulfonic acid (DDBSA) is known to efficiently disperse asphaltene deposits and form stable suspensions [13, 14]. The molecule has an anionic sulfonic acid group and an alkylbenzene hydrocarbon group. It is included as a component in many commercial asphaltene dispersant products. The mechanism behind its effectiveness is thought to be a strong acid-base interaction between the anionic sulfonic acid of the dispersant and cationic groups and sites in the asphaltene molecule. There are several variations of the chemistry all of which have the sulfonic acid group but differ in the structure of the hydrocarbon group.

Typically the alkyl group is formulated to be long enough to disperse the asphaltene molecules in oil and to give steric stabilization (n-C10-13) of the asphaltene particles. Often the alkylbenzene sulfonic acid is blended with compounds that enhance or supplement its effect. One of the commercial products on the market contains the following ingredients.

Table 5.5 Component Listing

Chemical Name	CAS Number
Xylene	1330-20-7
Toluene	108-88-3
Hexane	110-54-3
Cyclohexane	110-82-7
Isopropanol	67-63-0
Ethylene Glycol Monobutyl Ether	111-76-2
Glacial Acetic Acid	64-19-7
Isopropylamine	75-31-0
Dodecylbenzenesulfonic Acid	27176-87-0
Methanol	67-56-1

The aromatic compounds (toluene and xylene) are very effective asphaltene solvents. The dodecylbenzene sulfonic acid is an effective asphaltene dispersant. The acetic acid also acts as an asphaltene dispersant by attacking the nitrogen groups (amine, pyrrol, pyridine) on the asphaltene molecule.

The glycol (EGMBE) and alcohols (methanol and isopropanal) are mutual solvents so that the hydrocarbon compounds and the polar compounds will mix together without forming two phases. The mutual solvents also help to suspend the dispersed asphaltenes in with water or oil.

5.8 Oxidation:

Oxidation as an Aid to Coagulation: Oxidants have been found to aid coagulation and flocculation in several ways [reference 4, page 464]. In produced water many solid particles will adsorb organic materials, acids, asphaltenes, resins, etc.. In the municipal water treatment industry, these compounds are referred to as NOM (Normally Occurring Organic Matter). This material imparts an organic steric repulsion between the solid particles which reduces the tendency to flocculate. Oxidation is thought to break the bond between the surface of the particle and the NOM and thus release the organic material. This allows a flocculating polymer to adsorb onto the solid surface. A second mechanism is thought to be the oxidation of the NOM to make it more susceptible to coagulation and flocculation via aluminum and iron coagulants. A third mechanism is to oxidize NOM to form carboxylic acid groups which sequester calcium and magnesium. The net effect of this reaction is to create a network of multivalent cations that bind to the acidic NOM to precipitate these compounds from the water.

Common oxidants that are used in the oil and gas industry:

1. sodium or calcium hypochlorite

2. chlorine dioxide

3. hydrogen peroxide

4. Fenton's reagent

5. ozone

Oxidation to Remove Dissolved Organics: In a field application, wastewater containing about 160 mg/liter total phenols (including phenol itself, various methyl-phenols (a.k.a. cresols), some chloro-phenols, and other unspecified phenols) was successfully treated using ClO2. The dosage was about 3,000 mg/L of ClO2. It was possible to drop the concentration of the phenol compounds to below detection limits using ClO2. It was unclear what the phenols were being converted to, however, it was discovered that the major byproduct identified was benzoic acid, but that only represented about 10% of the initial concentration of total phenols.

Depending on the source of the wastewater, efficient BOD removal may also reduce the COD to acceptable levels. (BOD and COD were originally intended to be two measures of the same thing, but some compounds that contribute COD aren't very biodegradable. The relationship depends on the specific composition of the wastewater.)

When considering this application, it is necessary to have specific information on wastewater (e.g., concentrations of total and specific individual phenols, source of wastewater, BOD / COD ratio, other important constituents of wastewater, overall flow rate) and the current treatment operations and processes that are being used, but ClO2 shows some evidence of being effective.

5.9 Chemical Treatment of Naphthenates, Dissolved Organics and Water Soluble Organics:

A class of compounds of major importance are the tetra-acid naphthenates [15, 16]. These compounds are also known as Arn acids. Arn (Ørn) is the Norwegian word for eagle. Under certain conditions various soap-like compounds precipitate. In general, naphthenate precipitation occurs because the naphthenate is a multifunctional acid (four acid groups in complex organic) and combines with multivalent ions such as calcium, magnesium, and iron. The multivalent ion acts like a crosslinking agent binding two or more segments together from different molecules. This results in a chain or network of naphthenate molecules. Calcium is often the most active multivalent cation because it fits into the carboxylic acid structure better than other cations. Thus, it has a high binding affinity. This is the essential mechanism of calcium naphthenate structure.

The Bleo Holm processing facility with Blake and Ross production is a facility that is famous for naphthenate problems dating back at least ten years, maybe 15. Years ago, the operator settled on the use of glacial acetic acid together with an oil-based naphthenate inhibitor. In this strategy, 1 to 5 % (wt % basis produced water flow rate) of acetic or citric acid is added to the produced fluids. Delivering lots of acid creates problems with corrosion. This is essentially what Chevron ended up using on the Kuito facility in Angola. It has been reported that PAC has been applied but has not worked well in the presence of so much added acid. Regarding the treatment chemicals, PAC is a bit pH sensitive. It is better than the iron and aluminum salts which are very pH sensitive, but it still requires pH above about 5.

DTC was evaluated and found not to work well. Not all DTC are alike. It is possible that the molecular weight was not optimized. Also, DTC generally requires iron, preferably an iron precipitate, in the water for best performance. It complexes the iron and forms a bridge between the iron and suspended oil droplets. Almost all produced water has some iron. However, at low pH the iron would not be accessible for the DTC to complex. Also, the naphthenate molecule can bind to iron. Thus, DTC would compete for the same multivalent cations that the naphthenate does.

Naphthenate inhibitors are formulated with the idea that they have the right surface activity and oil / water solubility to close-pack at the oil / water interface. The objective is to prevent the interface from getting crowded with the naphthenic acids. The naphthenic acids are primarily oil soluble. But like all amphoteric molecules, they will also partition to the oil / water interface. At the interface they will bind with multivalent ions dissolved in the water phase. The acetic acid is a small molecule and thus can pack very well at the interface and bind to the calcium. This prevents the naphthenic acid molecules from binding to the calcium.

When production fluids are degassed, CO_2 comes out of solution and the pH increases. This tends to deprotonate the acid which drives it to the interface and it pushes the carbonic acid / calcium carbonate equilibria in the direction of lower solubility. The activity of the calcium ion increases which makes it more likely to precipitate as either calcium carbonate or calcium naphthenate.

The inhibitor should be added before the cross-linking occurs. This will limit the molecular weight of the calcium naphthenate. The inhibitor can act as a plastic modifier as well. Once the naphthenate has formed is very stable. The cross-linking reaction leads to a cementing effect. Once the cross-linked complex is formed, it is not possible to dissolve the complex.

Phosphorate ester has been designed for the interface demulsifier. It works by a similar mechanism to the naphthenate inhibitor (i.e. it pushes the naphthenate molecule away from the interface). It is less corrosive than the organic acids (acetic and citric) but it is more expensive than acetic. Also, when the oil goes to the refinery, the phosphate causes fouling and foam partition the WSO components to the water gives water quality problems.

Electrocoagulation has a decent chance of working if placed upstream of the flotation unit. But EC will have a limitation in the flow capacity that can be used offshore. There is a process patent that uses EC to promote flocculation of the naphthenic acids (US 8658014).

5.10 Water Treatment Chemical Selection:

There are thousands of water treatment chemical products available from chemical suppliers for application in the E&P industry. Most of these products fall into roughly a dozen chemical categories. Each category is most effective for a one or a couple of characteristics of the fluid. No single chemical is effective for all types of fluids in all applications. If the fluid has a particular characteristic, it makes sense to include a large number of candidates that are effective on that particular characteristic. Because characterization is an imprecise science, it also makes sense to include a few candidate chemistries that would not necessarily be expected to be effective. This is done in order to verify the understanding of the fluids and treating problem.

It is unlikely that the vendor will share this information since it is considered proprietary. One way to circumvent this is to include some known compounds in the bottle or jar testing kit. To do this, the operating company would source samples of water treating chemicals which would be included in the testing. A small number of these known treatment chemicals can indicate a lot about the chemistry of the oil and water.

Application Strategy: Understanding and successfully managing an application requires some pre-screening of the available polymer products and a disciplined policing of the overall mix of chemicals being added to the system. Key elements in selecting and managing a polymer application include:

1. **Electronic charge compatibility** - Mismatching charges can negate the effect of the added polymer and other functional chemical additives like scale, corrosion and deposit inhibitors as well as non-oxidizing biocides.

2. **Bench testing** – from simple jar testing to compare various polymers and varying dosage rates to process simulations using bench top flotation cells or other equipment is strongly recommended

3. **Select the Polymer Structure** – There are two basic polymer structures used in the water treatment industry; linear and branched. Linear polymers dominate the landscape but some degree of branching is sometimes incorporated. The branching improves the polymer shear resistance and offers some performance benefits in physical interaction with particles.

4. **Select the Charge Density** – There are anionic, non-ionic, cationic and amphoteric polymers. Within each of these groups there are differing amounts of charge sites. For example, a homopolymer of DADMAC has 100% charge density – each monomer unit carries the cationic charge. The charge can be diluted copolymerizing the DADMAC with nonionic acrylamide, AM/DADMAC. In the former case, the physical structure of the polymer is doing more to work at the oil/water/particle interface while in the latter, the polymer charge is critical in destabilizing the suspended solids.

The high molecular weight polymers (coagulants and flocculants) can be purchased in a variety of physical forms, each with its own advantages and disadvantages as described in the following table:

Table 5.6

Polymers are available in several physical forms	
Physical Form	**Comments**
Solution Polymers	High viscosity, low solids. High viscosity makes handling the products difficult. With low active solids (often around 5%) not appropriate for high volume applications.
Emulsion Polymers	High solids but with low viscosity. Takes long time to get polymer full active. Large volume applications are easier to manage. Typical active solids levels are 20 to 40%.
Dry Polymers	100% active polymer is desirable when shipping all over the world. However, most polymers are hydrophillic so storing the polymers in humid parts of the world can be challenging. Difficult to make up into a feed solution without forming "fish eyes," undissolved globs of polymer that will adhere to system surfaces and can cause equipment performance failures
Brine Polymers	High strength, lower viscosity. Short activation time. Active solids concentrations of 40+%.

Typical oilfield applications require large volumes of polymer. The preference is for dry products. Feeding dry products without creating a litany of other problems is a particular challenge. Generally, dry polymers are "made down" prior to introduction into the system to be treated. "Made down" means the dry polymer is converted to a solution of (typically) 1% active polymer concentration. A preferred method of making down the polymer is to stir the water sufficiently quickly so as to create a vortex or "whirlpool" in the center, into which the dry polymer is slowly and evenly introduced. Poor mixing will create "fish eyes" – large blobs consisting of wetted polymer on the outside and dry polymer on the inside. These blobs can clog valves and the pores on separation equipment. The 1% solution should be allowed to stand for some time before it is added to the system to be treated.

Solution polymers can be fed neat to the system. The polymer is generally ready to do the desired work. Managing the polymer can be a challenge as many of these products can have neat viscosities of several thousand centipoise (cps). Emulsion polymers carry the active polymer in concentrated water drops suspended in a mineral oil (or other) continuous phase. Feeding the polymer requires inverting the emulsion and then aging the polymer to be fully active. Brine polymers can have high solids concentrations, lower viscosities and be readily quick and easy to feed.

The first stage which is often overlooked is the proper characterization of the produced fluids. This characterization can often point to substances or characteristics of the fluids that must be taken into account in selecting candidate chemistries for field testing. Keep in mind that the chemical vendors each have 100 or more products. It would be impossible to field test each product. So initial screening must be carried out before taking candidates to the field.

Characterization can be carried out by the operating company or by the chemical vendor. The best case is when both groups work together. In either case, information should be shared between the two companies in order to facilitate the selection of deoiler chemicals. Chapter 1 on Chemistry, Chapter 4 on Sampling and Analysis, and Chapter 5 on Characterization can all provide some guidance.

The overall the process involves initial sampling and analysis of produced fluids (oil, water, gas). Also, there may be some major challenges that must be identified and overcome before water treatment chemistries can be selected. For example, if iron sulfide is present, then this must be addressed in combination with the selection of the water treatment chemistries. Iron sulfide typically forms as very small particles that can quickly consume general water treatment chemicals.

A rough knowledge of the droplet and particle size distribution should be obtained. As a first pass, simple desktop settling can be used to get a rough idea of droplet size and coalescence tendency. This will help to determine the overall challenge involved in cleaning the water. Also, if there is a high concentration of small droplets or particles, then the surface area will be large and the chemical concentration required may be higher than usual.

Knowledge of the upstream production chemicals is required in order to determine if there are any chemistries that will compete at the water / contaminant interface, lower the interfacial tension, or contribute to the stabilization of small particles or droplets.

For this purpose, a produced water sample is taken under minimal shear conditions, the droplet rise (creaming) rate, and the solids settling rate is visually assessed. The rate of rise of oil drops can be related to the average drop size using Stokes Law. If the suspension is stable for days, then the dispersed phase is a true emulsion. This would indicate that the dispersed drops or particles are small and thus suspended and kept in thermal motion by Brownian motion. It would also indicate that there is a stabilizing mechanism that prevents coalescence. These mechanisms were discussed above and include natural surfactants, surfactant-like production chemicals, excessive polymer used for water treating or dehydration, or excessive shearing together with some combination of the above.

The electrochemistry of the surface of the produced water contaminants does not necessarily need to be measured precisely, but it should be known to a reasonable degree. In the initial selection of candidate treatment chemicals, it is important to have at least a basic knowledge of the surface charge of the contaminants. Mismatching charges can negate the effect of the added polymer and other functional chemical additives like scale, corrosion and deposit inhibitors as well as non-oxidizing biocides.

Bottle testing can be carried out to determine if the contaminants have cationic or anionic surfaces. This is typically done by lining up similar samples and adjusting the pH over a wide range and looking at settling time as a function of pH. The PZC (Point of Zero Charge) is the pH at which the solution settles fastest. Once the PZC is known, the zeta potential of the emulsion can be estimated by the following formula: ZP = 13 x (PZC - pH). This formula says that the zeta potential is the difference between the emulsion pH and the pH at which the PZC occurs. If, for example, the PZC is 9 and the pH of the emulsion is 5, then the zeta potential is +60 mV. This is a relatively high zeta potential which indicates a relatively high charge density of the electrical double layer and that a coagulant or other counter ion chemistry is probably necessary. Since the estimated zeta potential is positive, this indicates that anionic chemistry would be most effective. If a coagulant be used, then an anionic flocculating agent would be effective. In practical applications, the terminology used is "cationic emulsion."

5.11 Bottle and Jar Testing:

It is important to view bottle testing as both a selection process and a diagnostic tool. However, in order for it to be effective as a diagnostic tool, the water specialist must have a very good and open relationship with each of the chemical vendors. Each vendor will likely learn details about the system which gives them potentially a competitive advantage over their competition. They may not tell the operator those details unless there is an established a relationship of trust. It is important to work closely with each vendor and establish that trust.

The bottle test consists of taking a representative sample, injecting various test chemicals into the sample, providing mixing and reaction time, and observation of the settling rate, and ultimate oil-in-water concentration. The details of this test are given at the end of this chapter. There are differences between the bottle test and the system. Bottle test is static, run at lower temperature than the actual system, there is no way to pressurize practically. Also, at higher temperatures, convective currents, driven by temperature gradients, will form.

Once the chemicals are injected and mixed, the sample jars are set on the bench top without further mixing. The objective at this point is to achieve quiescent fluid dynamics such that oil drops rise to the surface and water drops and solids particles settle, without disruption from secondary flow. In this stage of the test, the rise velocity of settling is measured, or at least compared from one bottle to another in an effort to characterize the size of the flocs and agglomerations that have been formed by the addition of the chemical.

Bottle testing is usually run at temperatures up to but not higher than 190 F. Usually, the temperature of the bottle test is controlled to match the temperature of the stage in the system where the chemical will be injected. For moderate temperatures, this is not a problem. But there are a number of process systems that operate at greater than 200 F. Bottle test temperatures higher than about 190 F would require a pressurized sample bottle. This is not practical since volatile organics and hydrocarbons will flash off and make the test lab somewhat noxious.

Also, if the more volatile components flash off then the composition of the bottle test mixture will not match that of the system. This may not seem as if it is a significant issue but for young, viscous, high API crude oils, that have a high content of volatile acids, BTEX, and other volatile compounds, the composition can indeed change dramatically with high temperature.

One further complication can occur at high temperatures. If the sample temperature is raised too close to boiling, convection currents will form. Convection currents are density driven flows that result from thermal gradients. This will cause unwanted mixing, and secondary flows in the bottle which will not mimic the quiescent flow that is the basis of bottle testing. The bottle tester must be careful to make sure that any flashing of oil in the bottle matches that in the system.

Once the chemical has been injected into the bottles containing produced fluid, an attempt is made to simulate the level of physical mixing that will occur in the process system. Mixing energy can be characterized by turbulence dissipation, or turbulence kinetic energy, as discussed in Section 3.4 (Turbulence). In the process system, two levels or kinds of mixing will occur. The first is a bulk liquid mixing. This is the mixing that disperses the chemical in each of the bulk phases. The chemical will likely have a higher solubility in one phase versus the other phase. It will therefore mix into that preferential phase, and will diffuse into the other phase and then become mixed. This is one of the reasons why it is important to have accurate phase volumes and accurate drop sizes of the oil and water phases.

The second kind of mixing that occurs is more of an interfacial displacement than a typical mixing process. Almost all demulsifiers and water clarifiers have surface activity. They have a preference to reside at the interface rather than the two bulk phases. The speed with which they migrate to the oil/water interface is a function of the surface activity of the molecule, the molecular weight and the bulk diffusion coefficient, the amount of oil/water interface, the character of the chemicals already at the interface, and the amount and character of the mixing energy put into the system by the bottle tester. High intensity mixing over a short period of time will not be equivalent to low intensity mixing over a long period of time. They may have the same total energy of mixing but the high intensity mixing will result in a large number of small eddies that will generate a large number of small drops. This will help to even out the performance differences between molecules that would otherwise have differences on the basis of surface activity.

In developing a bottle test procedure, a good place to start is to match the conditions or temperature, mixing duration and intensity, settling time, injection concentration, etc.. However, the bottle test is a significant departure from the process conditions. So too it may be necessary to depart from process conditions in order to have a test that gives the correct relative ranking.

It should be kept in mind that the objective of developing an accurate bottle test procedure is to mimic the effect seen in the field. The most important effect seen in the field is the relative performance of a few chemicals versus each other. If the rank order of performance in bottle testing matches the rank order seen in the field, then the bottle test procedure is useful. Such a test can be used then to determine if any other chemicals can be found that perform better than the current set of chemicals. This is all that can be expected.

It is implied, and perhaps should be emphasized that to develop an accurate bottle test, field testing of several chemicals is required. An initial bottle test should be carried out to determine suitable candidates for field testing. These chemicals are then tested in the actual process. Their performance is compared, one against the other. Then the precise bottle test procedure is developed to mimic the performance of these chemicals.

A further implication of this requirement is that it is better to work with a small number of chemical companies, each of whom is allowed to field test several chemicals, than it is to work with several companies and restrict the number of chemicals from each company to just one or two. Each company should be given the opportunity to establish a relationship between their bottle test methods and actual field performance of at least a few chemical products. This can only be done by bottle testing and field trial. Since this is an involved process, requiring significant efforts, it may not be worthwhile to repeat the effort with several chemical vendors. Focusing on just a couple of vendors, and allowing them to do a thorough job of correlating their bottle tests to field performance is far more valuable than working with several vendors each of whom is not given the chance to establish such correlations.

5.12 Demulsifiers – Impact on Water Treatment:

Strategies for Oil Dehydration: There are four strategies that are used to dehydrate oil [17]. They are typically applied in combination. They are:

1. provide residence time for settling;

2. raise the temperature in order to lower the oil viscosity and promote setting;

3. use electrostatic treatment to promote water droplet migration;

4. add chemicals to promote various separation mechanisms.

In some cases, the strategy used for oil dehydration has a direct effect on the water treatment system. This effect can beneficial or detrimental. If the oil dehydration system is bottlenecked or constrained, it is likely that water quality will suffer as a result of adjustments made by operators to achieve dry oil. Dry oil production usually takes precedence over clean water for various reasons. One of the adjustments that operators will make in such circumstances is to increase the residence time of the oil in the separators and the oil dehydration equipment. This results in reduced residence time for the water. Another adjustment is in the chemicals used to promote oil dehydration. Whatever chemistries are used to remove water droplets from oil, there typically is some residual chemical which can carry oil droplets into the bulk water phase. When oil dehydration is a bottleneck, chemical treating for dehydration is applied more aggressively. This is generally to the detriment of water treating. It is worthwhile to keep in mind these relationships when troubleshooting a produced water system. It can sometimes be the case that water quality can be improved by improving the oil dehydration process.

In some systems achieving dry oil requires aggressive chemistry which leads to the discharge of dirty water from the Oil Treater. This is particularly true of the flocculating demulsifiers. In that case, several water droplets arrive at the oil / water interface in a floc or cluster carrying a pocket of oil between them. With time, some of this oil is released into the oil phase. But typically, not all of the oil is released from the water droplets due to the nature of the demulsifying chemical. This can be a problem, to varying degrees, for the water treatment system. This is one of the ways that the strategy for oil dehydration has an effect on the challenges involved in water treatment. If the oil dehydration strategy were to rely more heavily on raising the temperature, increasing the residence time in the Oil Treater, and increasing the intensity of the electrostatic treatment, then less chemical demulsifier is required and there is less potential consequence on the water treatment system. Another effective strategy to reduce this effect is to dehydrate the oil as much as possible upstream of the Oil Treater. The less water that is taken out of the oil by way of chemical treatment, the less impact the oil dehydration chemicals will have due to over all material balance of the water streams.

One of the ways that demulsifiers can benefit the water treatment system is, of course, the converse of the situation above. If the demulsifier is effective, then the oil residence time can be reduced leaving more residence time for water in the separators. Another impact that is less well known and not commonly applied is the use of demulsifiers to promote oil droplet coalescence. Many of the demulsifiers used soften the oil/water interface, particularly on the oil side where the asphaltenes can otherwise form a tough, elastic or rigid film. All of these solid-like mechanical properties are a result of network formation on the part of the asphaltenes. Many of the oil dehydration chemicals migrate to the oil/water interface and soften the interface making it less solid-like and more fluid. This promotes coalescence. Another obvious beneficial effect of oil dehydration is the use of heat to reduce the oil viscosity and promote water settling. Higher temperature will of course improve oil droplet migration in the water treatment equipment.

Onshore versus Offshore: Many of the oil dehydration chemicals are surface active long chain polymers that are added to the oil stream and which must then migrate to the oil/water interface. This migration time can be on the order of tens of minutes. Typical offshore residence times are on the order of several minutes per vessel. Such long migration time can only be provided if the chemical injection is carried out rather far upstream of the separator vessel. In some cases this is possible. In that case, mixing by way of mild turbulence is a benefit. In the offshore settling, such long residence times are quite common. Very large skim tanks are commonly used. In those cases, the main challenge is to ensure adequate mixing upstream of the tank. In any case, it is important to keep in mind the time requirements for many of these chemicals. Such data will be provided wherever it is available.

Residence Time: residence time is important for both oil droplet creaming from a produced water stream, and for water droplet settling in an oil dehydration vessel or tank. While the mechanism is the same, as described by Stokes Law, there is a substantial difference in the actual values. Stokes Law is discussed in Section 3.2.3 (Stokes Law Settling). First, it is generally the case that oil droplets in the produced water are somewhat smaller than water droplets in oil. According to Stokes Law, this leads to faster settling of water droplets in oil than oil droplets rise in water. But the other factor to contend with is the viscosity of the bulk phases. The viscosity of water is generally less than 1 cP (0.001 Pa sec), whereas the viscosity of oil can range from 1 cP up to several hundred, or higher for heavy oil. This can significantly impede the settling rate of water droplets.

Importance of Coalescence: Given the relatively high viscosity of most crude oils, there is a great effort put into chemistries that promote water droplet coalescence in order to promote oil dehydration. As discussed in Section 4.5 (Coalescence of Oil Droplets) coalescence of oil droplets in water and of water droplets in oil is impeded by the presence of asphaltene accumulation at the oil / water interface. As discussed in Section 4.4.4 (Interface Concentration versus Bulk Concentration) and Section 5.2 (Brief Review of Emulsion Chemistry) the oil / water interface is different depending on whether the bulk phase is oil or it is water. In the oil, the oil / water interface tends to have a greater accumulation of oil-soluble surface active material. This is described by Bancrofts Rule. In any case, many demulsifiers are formulated in order to push the asphaltenes away from the interface and allow coalescence to occur more rapidly.

Electrostatic Treatment for Oil Dehydration: The use of electrostatic treatment for oil dehydration is discussed in [17] and the references given in that paper. Arnold and Stewart [18] provide an excellent discussion of the subject of electrostatic treatment and the entire subject of oil dehydration. Electrostatic treatment for oil dehydration is outside of the scope of this book.

Chemical Mechanisms: Yang et al. [19], and Kokal [20] discuss the use of demulsifiers and the mechanism of how they work. Chemical demulsifiers have three mechanisms. Some chemicals used employ a combination of these mechanisms. Yang gives the following description of the three mechanisms [19]:

1. The demulsifier adsorbs at the oil / water interface to displace surface active components residing there and push those components into the oil phase. Once this occurs, the interface is more flexible and more fluid. It is less solid-like (less elastic and less rigid). When an oil film forms between two water droplets, that film ruptures faster and with less collision energy. This promotes water droplet coalescence.

2. The demulsifier acts as a flocculating agent which joins water droplets together. This increases the effective diameter of the water droplets which accelerates separation by settling of the flocculated water droplets. It also increases the likelihood that the water droplets will collide and coalesce.

3. The demulsifier acts like a wetting agent for oily solids. Depending on the composition of these solids, they may be surface active and migrate to the interface where they stabilize the emulsion. In this case, the demulsifier will adsorb on the solid surface and release the oil from the solid particle. This makes the particle less surface active and promotes migration of the particle to the water phase. It also raises the effective particle density which promotes settling.

In order for oil dehydration to be effective, it must remove water droplets from the oil. As the above mechanisms suggest, this is done by chemical treatment followed by settling of the water drops from the oil into the water phase. In most cases, and as described above, the action of the chemicals helps to promote this settling. Electrostatic coalescence and raising the temperature promote this migration as well. Ultimately, separation occurs by settling.

Settling causes the water droplets to migrate to the oil / water interface. At that interface, the water droplets have demulsifier attached to their surface which typically has carried with it smaller droplets of oil. Ultimately, this water will be discharged from the Oil Treater. This can create a slight problem from the standpoint of water treatment. This "dehydration water" coming from the Oil Treater can contain oil entrained in the water as small chemically stabilized oil droplets.

At the very least, some bottle testing should be carried using this water of dehydration. Much of the bottle testing will be based on water sampled from the Free Water Knockout, or Water Skimming Vessel, upstream of the Oil Treater. This makes sense due to the fact that this will be the major water stream. However, when selecting the optimal water treatment chemistry, the selection of the demulsifier must be taken into account as well.

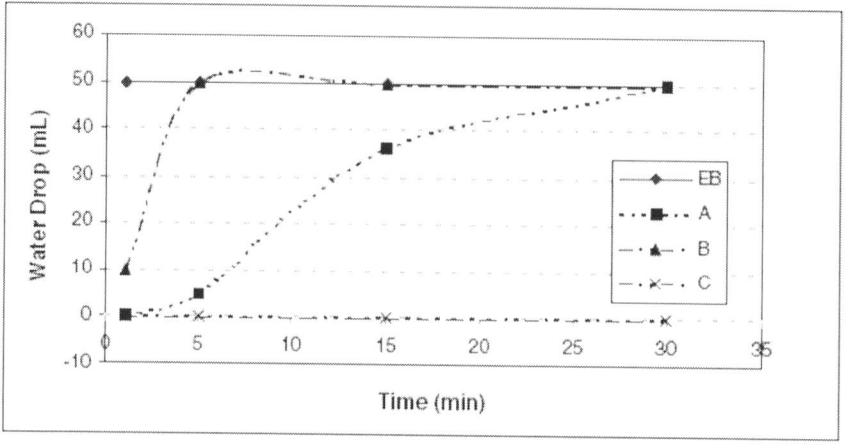

Fig.4: Water Drop Profile for Crude Oil Emulsion at Demulsifier Concentraton of 20 ppm

Figure 5.8 Water drop versus time for four demulsifiers.
Water drop is the volume of water that appears under the oil phase in a bottle test. Results from [21]

As discussed by Chen and Towner [21] there are four characteristics of effective demulsifiers:

1. Extent to which the demulsifier reduces the interfacial film elasticity;

2. high interfacial excess;

3. the rate at which the demulsifier migrates to the interface;

4. good BS&W grind-out.

The first characteristic is measured by interfacial film elasticity. This was discussed in Section 4.5.8 (Elasticity and Viscosity of the Oil / Water Interface). Asphaltenes create an elastic film around the water droplets. An effective demulsifier is able to displace the asphaltenes at the interface and reduce the film elasticity. The second characteristic is a measure of the extent to which the emulsion breaker partitions to the interface versus the bulk phases (oil or water). This is measured by the interfacial excess concentration, as discussed in Section 4.4.3 (Surface Excess Concentration & Surface Activity). The higher this value, the less emulsion breaker required for application. The third characteristic is the rate at which the emulsion breaker migrates to the oil / water interface. The faster the emulsion breaker migrates, the more effective the chemical. Residence time is always a limitation offshore. The final characteristic is related to the quality of the oil, i.e. how much water is left in the oil after demulsification. If this value is high, then the crude oil quality is said to be poor.

Table 5.8 Results of on-site bottle testing using an Alaska crude oil. Results from [21]

Demulsifier	Concentration Demulsifier (ppmv)	Water in Crude after 30 min (ppmv)
EB - 1	5	0.8
	10	0.3
	15	0.1
	20	0.1
EB – A	5	1.0
	10	0.5
	15	0.2
	20	0.0
EB - B	5	5.8
	10	3.1
	15	2.6
	20	2.0
EB - C	5	x
	10	x
	15	x
	20	x

Breen [11] provided an excellent summary of the detailed mechanisms involved when a demulsifier is used to promote coalescence. The physical picture of two colliding water droplets is shown in the figure below.

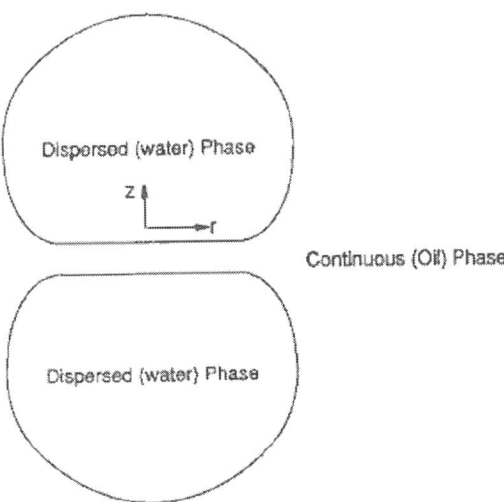

Figure 5.9 Approach of two water droplets in oil. As the droplets approach, the oil between them is pushed out. This creates a pressure gradient that flattens the droplets forming a configuration of two parallel flat discs with a film of liquid between them. Depending on the momentum of the two droplets and a number of other factors, the film may eventually get squeezed out and the droplets will coalesce.

In many crude oils a thin film of asphaltene and resin forms at the oil / water interface. This film prevents coalescence by several mechanisms most of which are related to migration to the interface, pushing the asphaltene and resins off of the interface, changing the interface mechanical properties to be less solid-like and more fluid-like. Each of these mechanisms have been studied and various interfacial properties have been measured to help quantify the effect [21, 11]. This is really a remarkable achievement in surface and colloid chemistry.

In Table 5.9 below, the separation performance (1 = good) is compared to measured values of the surface activity of the demulsifiers. Surface activity is a measure of the partitioning of a compound between the bulk and the surface. The higher the surface activity, the greater the tendency of the compound to reside at the surface. This has the consequence that it will displace other compounds at the surface that do not have as high of a surface activity. In the case of crude oil containing asphaltenes, this means that the demulsifier will push the asphaltenes and resins away from the interface. As discussed below, this will soften the interface and promote coalescence. Thus, as shown in the Table, the compound with the highest surface activity was the most effective at water removal from oil.

Table 5.9 Coalescence Rate Constants and Interfacial Activities

Coalescence Rate Constants and Interfacial Activities for Various Demulsifiers in Water-in-Crude-Oil Emulsions				
	Demulsifier			
	RE-1748	**RE-1753**	**RE-1868**	**RE-1756**
δi	1.97	1.05	0.63	0.15
Coalescence Rate Constant (K)	0.75	0.56	0.32	0.01
Performance	1	2	3	4

In Table 5.11 below we see another example of the effect of high surface activity with a different crude oil and a different set of demulsifiers. Again, the demulsifiers with higher surface activity are more effective at oil dehydration.

Table 5.10 Partition Coefficients and Interfacial Activities for Various Demulsifiers in Water-in-Crude-Oil Emulsions

	Demulsifier				
	RP-4011	**R-77**	**RP-2327**	**RP-484**	**No EB**
K_p	0.40	0.19	0.19	0.16	0.0
δi	2.11	1.03	0.65	0.55	0.05
Performance	1 (0.5 min)	2 (2 min)	3 (10 min)	3 (days)	5

As shown in the Table below, separation efficiency also correlates with a property referred to as film modulus. It is a measure of how tough an interface is. The lower the modulus the more the interface acts like a liquid and flows. the higher the modulus the more the interface acts like a rubber or elastic membrane. The displacement of resins and asphaltene from the interface by the demulsifier compound reduces the interfacial film modulus and allows coalescence to occur.

Table 5.11 Data Summary for Demulsifier Performance, Film Modulus and Initial Slope in Film Stress-Relaxation

Demulsifier	Efficiency* (%)	Film modulus (dyn cm^1)	Initial slope+ (dyn cm^{-1})
RE-2306	86	14.4	0.89
RE-2307	42	16.8	0.38
RE-2308	4	45.8	0.15
RE-2309	15	20.4	0.28
* Water separation as volume percent after 6 hours			

5.13 Corrosion Inhibitors – Use and Impact on Water Treatment:

The most common type of corrosion inhibitor used in the oilfield is referred to as a film forming amine (FFA). There are many varieties of FFA, all of which are surface active with a polar or ionic head group and a hydrophobic nonpolar tail group. It is this combined hydrophilic / hydrophobic nature that causes an inhibitor to migrate to the metal surface and adsorb, forming a surface film that reduces the wetting of the surface by water, and that retards the electrochemical corrosion process. In the ideal case, when the concentration is sufficient, the corrosion inhibitor molecules align perpendicular to the surface and parallel to one another forming a consistent film, without defects, of relatively even thickness. The tendency of the hydrophobic tails to align and exclude water molecules is, in part, a manifestation of the hydrophobic effect, which was discussed in Section 4.3 (Charged Surfaces in Ionic Solutions). It is important that the corrosion inhibitor film not have any breaks, holes, or edges since imperfections in the film would become anodic to the covered surface causing accelerated pitting corrosion. The hydrophobic effect helps to ensure that the film is continuous provided that sufficient corrosion inhibitor is present.

The basic corrosion cell is shown below. As shown, some material has shielded a part of the metal surface. The metal beneath the deposit becomes anodic to the exposed metal surface around the deposit – electrons begin to flow. This is the basic mechanism of corrosion when corrosion inhibitor is absent or not effective.

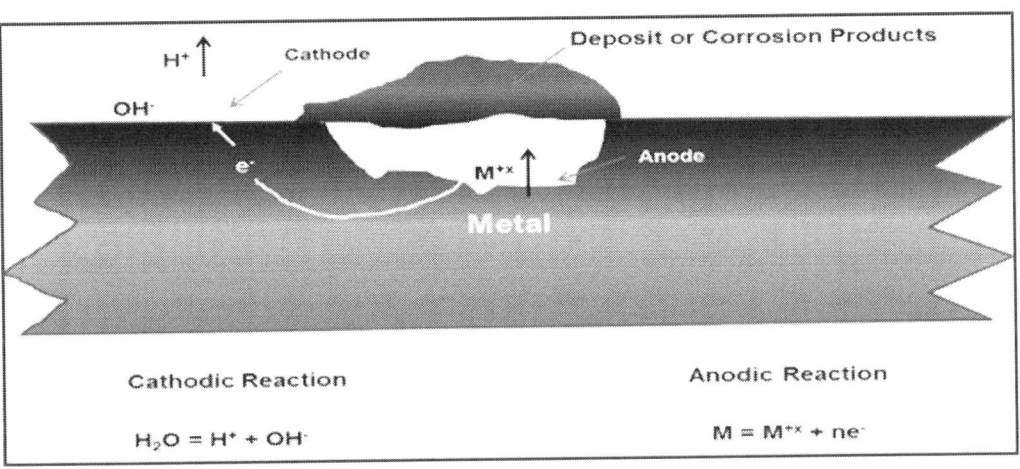

Figure 5.10 Basic corrosion cell showing the anodic and cathodic reactions

Other Types of Corrosion Inhibitors: There are other categories of corrosion inhibitors, including neutralizing amines that protect against carbon dioxide induced corrosion, sulfide scavengers that remove the sulfide to prevent sulfide-induced corrosion, oxygen scavengers, and precipitating inhibitors like phosphates that react with the metal surface to form a protective barrier.

Corrosion inhibitors that function by creating a barrier between the electrolyte or corrodant and the metal surface require certain environmental conditions that will drive the creation of the barrier and that will hold the barrier on the metal surface. The performance of the barrier is a function of the uniformity of thickness and the lack of discontinuities in the film. Areas of the metal surface where there are holes in the film (discontinuities) or even thinner areas of the film become anodic to surrounding areas where the barrier is intact or thicker driving pitting corrosion.

Inorganic polyphosphates are examples of inhibitors that form corrosion inhibiting barriers. Polyphosphates control corrosion of steel surfaces by precipitating with calcium and iron in the high pH region of the cathode, stifling the cathodic reaction. Orthophosphates react with iron going into solution at the anode stifling the anodic reaction. Polyphosphates are then classified as "cathodic inhibitors" while orthophosphates are classified as "anodic inhibitors." In order for the required reactions to proceed the bulk water pH has to be sufficiently elevated to drive the precipitation reactions. Water dissociation reactions at the cathode evolve hydrogen which moves away from the surface and hydroxyl ions that remain. The hydroxyl ions will drive an increase in the pH of the water at the metal surface by 1.0 to 1.5 pH points.

Consequence of Surface Activity: Surface activity is an important part of the mechanism by which corrosion inhibitors prevent or at least reduce corrosion rate. From a water treating standpoint, a major consequence of the surface activity of corrosion inhibitors is that they also stabilize O/W and W/O emulsions. There are two mechanisms involved. One mechanism is the lowering of interfacial tension. As discussed in Section 3.5 (Shearing of Oil Droplets), lower interfacial tension leads to greater susceptibility to forming small oil droplets due to shear. For a given shear intensity, the lower the interfacial tension, the smaller the droplets that are formed. The other mechanism involved in stabilizing emulsions is the creation of an interfacial barrier to coalescence. The presence of any surface active compound at the oil / water interface will create a barrier to coalescence. Asphaltenes at the interface create a stiff and immobile barrier that prevents coalescence. Corrosion inhibitors are chain-like amphoteric molecules like fatty acids, fatty acid amines, quaternary alkyl compounds. These all have a high affinity for the oil / water interface and create an elastic barrier in part due to the Gibbs-Marangoni effect.

Holistic Selection Process for Corrosion Inhibitors: The stability of the emulsion varies depending on the corrosion inhibitor and does not necessarily correlate with effectiveness of corrosion inhibitor. In other words, the most effective corrosion inhibitor from a corrosion inhibition standpoint is not necessarily the most surface active compound, nor it is necessarily the worst compound from an emulsion stability and water treatment standpoint. Thus, there is often an opportunity to find a "best" corrosion inhibitor that both provides good corrosion inhibition and which minimizes the impact on the water treatment system.

A good example of a holistic selection process has been documented by Moon and Horsup [1]. They tested the corrosion inhibition effectiveness together with emulsion stability and found one chemical gave good results for both criteria. They also measured phase partitioning of the inhibitor between the oil and brine phases, interfacial tension and determined the surface activity (as measured by Gibbs Excess). This later three sets of measurements are not required in the selection process but they do add mechanistic insight and verification of the emulsion stability effect. All of the compounds studied had a preference for the water phase. Surface Excess was discussed in Section 4.4.3 (Surface Excess Concentration and Surface Activity). It is worthwhile to summarize the results of this inhibitor selection process here. However, it should be noted that the particular results obtained are specific to the conditions and should not be interpreted as a recommendation for a particular type of corrosion inhibitor. Instead, the testing methodology should be understood and used for any particular location.

Table 5.12 Corrosion Inhibitors and Solvents used for Study

Chemical	Description
FA Am	Fatty acid amine
FA Am Salt	Salt of fatty acid amine
Quat 1	Quaternary amine
Quat 2	Quaternary amine
FAD	Fatty acid derivative
Solvent	Description
LVT-200	Kerosene
ASTM Brine	Laboratory-prepared sea salt solution

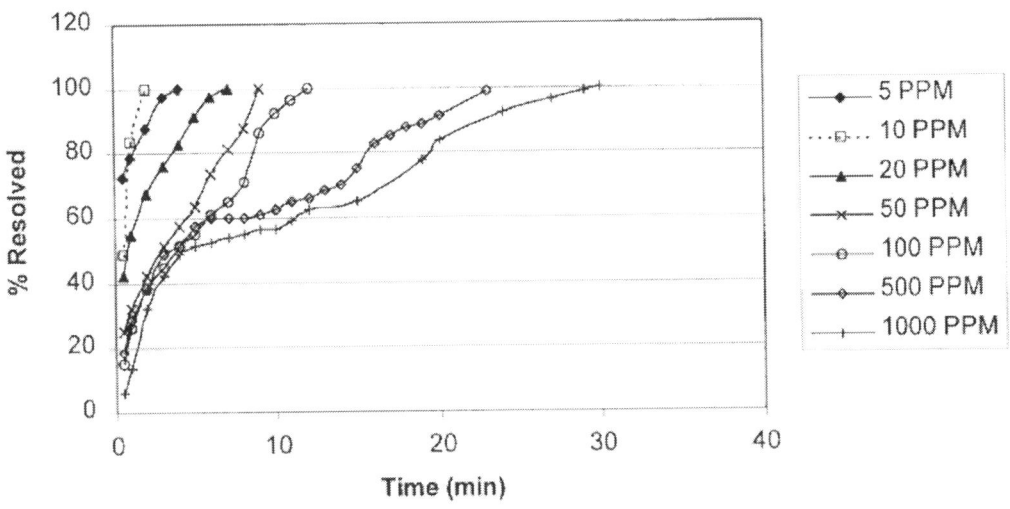

Figure 5.11 Percentage of resolved volume of sample containing 50/50 kerosene / brine with corrosion inhibitor (salt of fatty acid amine) added. Concentration of corrosion inhibitor is shown on the right. Without corrosion inhibitor the mixture resolved completely (100%) in 10 seconds. Data from [1].

In the study, LVT-200 was used as the oil phase. The brine was prepared in accordance with ASTM Method D-1141-52. The pH of the brine was 7.1 at 25 C. Mixtures of 50 / 50 volume percent oil / brine were made up. Corrosion inhibitor was added at various concentrations. Emulsion stability testing was carried out by hand shaking the mixtures and then letting the emulsion sit without further agitation. The percent of the emulsion that had resolved versus time was observed. The most stable emulsion were formed by FA Am, Salt of FA Am, and Quat-1. These emulsions were oil-in-water, as would be expected by application of the Bancroft Rule, and the observation that all of the compounds studied had a solubility preference for the aqueous phase.

Representative results of this testing are given in the table below. As indicated in the title of the table, these results represent the type of testing that should be carried out more so than the class of corrosion inhibitor that should be used. The class of corrosion inhibitor that should be used will vary from one location to another. In this study, for this particular set of undisclosed corrosion inhibitor compounds, there is an unambiguous best compound which demonstrated good corrosion reduction and good emulsion resolution (FAD). Under another set of conditions, another compound or compound class might be superior.

Table 5.13 Results of corrosion inhibitor evaluation for a group of candidate corrosion inhibitors. Note that the results are specific to a particular oil / water mixture chemistry, set of testing conditions, and cannot be applied to other conditions without testing under those conditions. These results are shown in order to demonstrate the type of testing that should be carried out in order to select a corrosion inhibitor that provides good asset integrity protection as well as environmental compliance (water quality). Data from [1].

Compound	IFT at 50 ppmv (mN/m)	Emulsion Resolution Rate	Corrosion Reduction
Fatty acid amine (FA Am)	5 x 10-3	X	X
Salt of fatty acid amine FA Am Salt)	9 x 10-4	X	X
Quaternary amine – 1 (Quat1)	1.65	X	√
Quaternary amine – 2 (Quat2)	27	√	X
Fatty acid derivative (FAD)	15	√	√

5.14 Scale Inhibitors – Impact on Water Treatment:

Before the discussion of scale, it is important to point out the difference between scale and precipitated solids. From a flow assurance standpoint, the objective of scale control is to prevent the build-up and eventual blockage of flow in a pipeline, flow line, or vessel. The mechanism of scale build-up has several steps which begins with precipitation of some compounds or material. Precipitation can occur in the bulk fluid, at the interface between fluids, and on the walls of the system. If precipitation occurs in the bulk or at the interface of the fluids, then scale can only form if there is subsequent migration of the precipitated material to the wall followed by adherence of the material to the wall. Precipitation can occur directly on the wall as in the case where the wall is slightly cooler than the bulk fluid and the precipitate is driven by temperature as in the case of wax deposition. From a mechanistic standpoint, scale deposition can be more complex than precipitation. However, scale deposition must always involve precipitation.

Generally speaking, from a water treatment perspective, precipitation, whether it occurs in the bulk or at the interface is the main concern. Scale deposition can also be a concern, in some cases, but precipitation is almost always associated with the generation of oily solids which are always a major challenge in water treatment.

In this section, the main subject is the inhibition of mineral precipitation. As just mentioned this is only one part of the scale deposition mechanism. However, from a water treatment perspective, it is an important part. The other parts that are associated with deposition, are not discussed in any detail. There are two main classes of precipitates in the oil field, mineral and organic.

The control of mineral scale precipitates can be accomplished in several ways:

1. Remove the scale forming ions – there are numerous means of reducing or fully removing the scale forming ions including chemical precipitation softening, ion exchange and distillation. This approach, removing the scale forming ions, is commonly practiced in oilfield applications in Steam Assisted Gravity Drainage (SAGD). Applications of this approach outside of SAGD are less common.

2. Adjusting system conditions to improve salt solubility and avoid precipitation. Adjusting the system conditions includes:

   a. Changing the temperature – many calcium scales exhibit a retrograde solubility with temperature: calcium solubility $\propto 1/T$.

   b. Adjusting the pH – many sparingly soluble salts are much less soluble under alkaline conditions than acid. Field operations often involve "acidizing" which is increasing the acidity of the environment to the extent where precipitation reactions of the acid soluble salts are reversed and scales are solubilized.

   c. Reducing "time" at adverse condition – irrespective of the combination of conditions driving precipitation, time or reaction kinetics, are part of the process. Time "at elevated temperature" or "elevated pH" are often adjustable elements in controlling scale formation. For example, water passing over a hot surface can potentially reach conditions where precipitation occurs at the water/hot surface contact. By moving the water away from the surface, the time required for the precipitation reactions to take place is is short circuited and often the precipitation process that might have been initiated begins to reverse as the water temperature is reduced.

3. Addition of chemical scale inhibitors. Products employed in controlling the formation of scale in water systems include a variety of cationic and anionic, polymeric and single-molecule chemicals.

Threshold versus Sequestration / Chelation Inhibitors: Several inhibitor products have emerged and are commonly applied in industrial water treatment, including oil and gas applications. These inhibitors are anionic and function as threshold inhibitors. The chemical inhibition of scale formation is approached by one of two pathways: Threshold Inhibition or Sequestration/Chelation. Threshold inhibitors work at concentrations a fraction of the concentration of sparingly soluble salts they are selected to control. Sequesterants work according to the stoichiometric relationship, binding the scale forming ions and preventing their precipitation.

The advantage of using threshold inhibitors is the limited quantity required of chemical required to protect the system. Sequesterants or chelants, because they work on a stoichiometric basis, typically require a lot more chemical to get the job done. However, the latter tend to have other functions that increase their overall respective performance profile.

Below is a chart showing the common scales with common scale inhibitors:

Polymeric Inhibitors: The polymeric scale inhibitors are predominantly anionic and contain some of the same monomers that comprise the coagulants and flocculants but have much lower molecular weight. Liquids/solids separation, (coagulation and flocculation), is achieved using polymers of a minimum molecular weight in the range of 200,000 to 500,000 while the scale inhibiting polymer's molecular weight rarely exceeds 20,000 and most often is less than 5,000. These low molecular weight, polymeric scale inhibitors react stoichiometrically with divalent cations like calcium and magnesium preventing the formation of the carbonates and sulfates. In addition to stabilizing soluble divalent cations the polymers act to disperse suspended solids and oil droplets. Coagulants are (nearly without exception) high molecular weight, cationic polymers that neutralize the net-negative surface charge on suspended particles and oil droplets. The idea of suspension, rather than neutralize the surface charges and allow particles to combine, is to increase the anionic charge density of the water – dramatically increasing the repulsive forces that stabilize materials in suspension. Low molecular weight means more molecules per unit mass of polymer yielding a lot more opportunity for contact with solids and polymer to polymer contact – all governed by the nature and degree of the surface charges.

Given the often elevated TDS of produced water, the use of sequesterants to control scale formation is typically cost prohibitive. The use of threshold inhibitors, like the diphosphonate, HEDP, is common and effective. Oxidation is a common practice in controlling iron and other sulfides. Selecting the right inhibitor is not difficult – most companies utilize the same raw materials. But most suppliers do maintain a solubility program to predict what scales are likely to form. When evaluating the potential for scale precipitation, the identification of sparingly soluble salts, and the conditions that affect their respective solubility is quite important. The prevention of scale is guided by the following questions.

1. What scale forming ions are in abundance? Understanding which scale or scales are likely to form (e.g., Calcium carbonate v calcium sulfate) is the first step in designing a control regimen. There are several factors that are common to most scales encountered in the oil-field: Solubility is reduced across a pressure drop – this is why nozzles often plug while the system might show no other evidence of precipitation. Solubility is reduced within a biological matrix – accumulations of microorganisms will generally reduce the solubility of the sparingly soluble salts. A table is provided below that shows common scales and offers comments on their respective solubilities.

2. Where is protection required? Where is scale formation likely to cause the most problems? Scale can form in places that do not impact operability, equipment life or performance. Precipitation of calcium carbonate in a tank is not necessarily going to cause problems and might not be worth making operating changes or adding inhibitor to prevent. However, precipitation in the formation can reduce well productivity. Understanding where a problem might occur is also important. High silica concentrations at elevated downhole temperatures are fine but when the water is brought the surface and cools, entire transfer piping can become blocked with silica (glass) – which requires mechanical energy and/or HF for removal.

3. What are the conditions that exist where scale formation would cause the most problems? (Temperature, ion concentration, suspended solids, presence of sulfides, pH, TDS, etc.). When evaluating scale forming potential it is important to consider the actual conditions where precipitation would be of most concern – not "generic" conditions.

Table 5.16 Chemical Scale Inhibitors and Solubility Comments

Mineral	Formula	Inhibitor(s)	Key Solubility Factors
Calcite	$CaCO_3$	HEDP Sodium polyphosphate DTPMP PAA ATMP CMI	Decreases with increasing temperature Decreases at pressure drop Decreases with increasing pH
Aragonite	$CaCO_3$	HEDP Sodium polyphosphate	
Vaterite	$CaCO_3$	HEDP Sodium polyphosphate	

Mineral	Formula	Inhibitor(s)	Key Solubility Factors
Anhydrite	$CaCO_3$	PAA Polyacrylic Acid/2-Acylamido methyl propane sulfonic acid	pH independent Function of Ca and SO4 concentrations Solubility increases between 20°C and 80°C
Gypsum	$CaSO_3$	PAA HDTMP BHPMP Polyacrylic Acid/2-Acylamido methyl propane sulfonic acid	
Barite	$BaSO_4$	PAA Acrylic Acid/2-Acylamido methyl propane sulfonic acid	
Celestite	$SrSO_4$	PAA Acrylic Acid/2-Acylamido methyl propane sulfonic acid	
Mackinawite	FeS	Oxidation PAA Acrylic Acid/2-Acylamido methyl propane sulfonic acid	Formation strong kinetics Not typically a hard crystalline deposit-dispersible
Pyrite	FeS_2	Oxidation PAA Polyacrylic Acid/2-Acylamido methyl propane sulfonic acid	
Halite	NaCl	Dilution – no chemical inhibitor	Increases with increasing temperature
Fluorite	CaF_2	Hydroxyehtylidiene disphosphonic acid	pH independent
Sphaerlite	ZnS	Oxidation PAA Polyacrylic Acid/2-Acylamido methyl propane sulfonic acid	Formation strong kinetics Not typically a hard crystalline deposit – dispersible
Galena	PbS	Oxidation PAA Polyacrylic Acid/2-Acylamido methyl propane sulfonic acid	Formation strong kinetics Not typically a hard crystalline deposit – dispersible
Silica, amorphous	SiO_2	Acumer 5000, DOW Chemical Co. (Propritary)	pH determines form Increases with increasing temperature

HEDP: Hydroxyehtylidiene disphosphonic acid
DTPMP: Diethylenetriamine-penta methylene phosphonic acid
HDTMP: Hexamethylenediamine tetra(methylene phosphonic acid)
BHPMP: Bishexamethylenediamine penta(methylene phosphonic acid)
PAA: Polycarboxylic acid
ATMP: Amino tri(methylene phosphonic) acid
CMI: Carboxymethyl inulin

Chemical Compatibility: A high concentration of iron can interfere with the performance of polymeric scale inhibitors. This has been observed in treatment of shale produced water in the Marcellus shale formation of Pennsylvania and West Virginia [22]. Water samples contained 10,000 to 25,000 mg/L of calcium, and dissolved iron up to 200 mg/L. In order to inhibit calcium mineral precipitation, a high concentration of inhibitor was required. However, at high concentrations, most of the inhibitors tested formed a complex with the dissolved iron which significantly lowered their ability to inhibit the precipitation of calcium minerals. Shen at al. [22] found that the addition of iron chelating agents (e.g. citric acid and the sodium salt of citric acid) greatly improved the performance of the scale inhibitors. Also, they developed new inhibitors with greater iron tolerance.

Scale Inhibitor Testing and Evaluation: Much of the work that is typically carried out to test scale inhibitors is done in a laboratory setting. Initially, samples are collected in the field. Care is taken in sample preservation and handling. The composition of the dissolved solids (TDS) is then analyzed by various methods (see Chapter 5). Synthetic brines are then made up to mimic the composition of the samples and testing is carried out to determine the precipitation tendency, which minerals precipitate, etc.. In many cases, particular species are varied in order to test the sensitivity of the system to those particular species. Static bottle tests and dynamic tube-blocking tests are the most common tests carried out. Shen et al. [22]; Graham and co-workers [23], and Tomson and co-workers [24] discuss the methods.

Table 5.17 Summary Table of Polymeric Coagulants, Flocculants and Scale Inhibitors / Dispersants

	Homopolymer Copolymer Terpolymer	Molecular Weight Range (typical)	Primary Function	Cationic	Anionic	Nonionic	Amphoteric
Polyacrylamide (PAM)	Homopolymer	400k to 1,000,000	Flocculent / Viscosity Build			X	
Hydrolyzed poly-acrylamide	Homopolymer	1,000,000 to 5,000,000	Flocculent				
Acrylic Acid / Acrylamide (AA/AM)	Homopolymer	200,000 to 15,000,000	Viscosity Build				
Aminomethylated polyacrylamide (AMPAM)	Homopolymer	2,000,000 to 5,000,000+	Flocculent	X			
Polydymethyldi-allylammonium chloride (DM-DAAC or DAD-MAC)	Homopolymer	500,000 to 2,000,000	Coagulant Reverse Breaker	XX			
Acrylamide DADMAC	Copolymer	500,000 to 2,000,000	Coagulant Reverse Breaker	X			
Poly Aluminum Chloride (PAC)	Inorganic homopolymer	~175	Coagulant Reverse Breaker	XX			
Epichlorohydrin	Homopolymer	<100,000 to 500,000	Coagulant Reverse Breaker	XX			
Other Quaternized Polyamines	Homopolymer	NA	Coagulant Reverse Breaker	XX			

	Homopolymer Copolymer Terpolymer	Molecular Weight Range (typical)	Primary Function	Cationic	Anionic	Nonionic	Amphoteric
Acrylic Acid / DADMAC (AA/DADMAC)	Copolymer	NA	Coagulant Reverse Breaker				X
Ferric Chloride	N/A	NR	Coagulant	X			
Ferric Sulfate	N/A	NR	Coagulant	X			
Aluminum Sulfate	N/A	NR	Coagulant	X			
Aluminum Chloride	N/A	NR	Coagulant	X			
Polyacrylic acid	Homopolymer	<10,000	Dispersant Threshold scale inhibitor		X		
Acrylic Acid / 2-Acrylamido methylpropane sulfonic acid	Copolymer	<10,000	Dispersant Threshold scale inhibitor		X		
Polymaleic Anhydride (PMA)	Homopolymer	<10,000	Dispersant Threshold scale inhibitor		X		
Polymethacrylate (PMA)	Homopolymer	<10,000	Dispersant Threshold scale inhibitor		X		
Sulfonated Styrene/Maleic Acid (SS/MA)	Copolymer	<10,000	Dispersant Threshold scale inhibitor		X		
Acrylic Acid / Hydroxypropyl acrylate (AA/HPA)	Copolymer	<10,000	Dispersant Threshold scale inhibitor		X		

HMW (High Molecular Weight) polymers (MW > 50,000 gr/mol) are used for separating suspended particles and oil droplets. LMW (Low Molecular Weight) polymers (MW <10,000 gr/mol) are used as dispersants for suspended particles and scale inhibitors. A tradeoff occurs with increasing MW as the number of molecules decreases. This reduces the number of strands of polymer available for forming engagements and meshes. Generally, for coagulation and flocculation it is preferable to have a small numbers of large molecules. For scale inhibition it is preferable to have a large numbers of relatively small polymer molecules.

Change density refers to the amount of charge per unit weight of polymer, e.g. adding AM to DADMAC dilutes the density of cation charge sites. Bridging occurs with very HMW polymers, 5,000,000 to 15,000,000. Bridging requires physically large polymers: HYPAM at 1 million to 5 million MW estimated at 0.2 microns to 0.8 microns. Physical blends of different molecular weights of polymers have often shown improved results versus single polymer compositions

Amphoterics are often employed when other contaminants interrupt the function of the homopolymers. For example, in building gels for a frack using polyacrylamide, in high iron containing waters, the dissolved iron will react with PAM wrapping the polymer up in a ball and prevent any viscosity build until all of the iron is reacted.

5.15 H$_2$S Control – Impact on Water Treatment:

One of the options for removing H$_2$S from produced hydrocarbon gas is to inject an amine-based or triazine chemical scavenger. While such scavengers can be effective at removing H$_2$S, scaling from carbonates is a common problem [25] due to the increase in pH which typically occurs with the use of triazine scavengers. Both the unreacted scavenger and its reaction products contain alkaline amines. If either the scavenger or its reaction products are mixed with produced water containing carbonates, the increase in the pH will increase the probability of carbonate scale3-5.

Amine-based chemicals used for scavenging of hydrogen sulphide will cause precipitation of carbonate scales, under a wide range of conditions. They are effective as scavengers, and have health, safety and environmental advantages compared to other scavenger chemicals. Unfortunately, they significantly raise the pH of all process waters that they are mixed with, and thus exacerbate the scaling tendencies of dissolved carbonate minerals. Calcium carbonate, magnesium carbonate and iron carbonate scales can occur over a wide range of temperatures, pressures, and carbonate concentrations. In many process configurations there is no alternative but to allow the mixing of scavenger with produced water. In those cases, extensive laboratory work is typically required to select an appropriate scale inhibitor. It is usually necessary in such studies to understand in detail the chemistry of the solution and its pH and alkalinity characteristics.

A modeling methodology has been developed which can approximate the effect of triazine type H$_2$S scavengers using an analogue amine as an effective weak base. The effective weak base has a base dissociation constant of the right order compared to monoethanol amine. It therefore simulates the pH effects, and the titration curves of the amine / CO2 / carbonate buffer system that were measured in the lab. Using this methodology, we show how the production process has a significant impact on the in situ scaling potentials and how improved modelling and laboratory tests are able to better simulate the field conditions and aid understanding of the process chemistry issues and solutions.

However although the figure shows the complete reaction, treatments would not normally be designed to go to completion since the final trithiane product is relatively insoluble. Scavenger is therefore normally applied on a basis of a minimum dose rate of 1 mol of triazine to 2 moles of sulphide. One of the strategies used to prevent scaling is to inject the scavenger directly into the gas stream. This can be accomplished downstream of any one of several gas/liquid separation vessels in a facility. By adding the scavenger to the gas, the scavenger does not immediately contact the produced water. This strategy works well to prevent scaling in the injection system. It does reduce the scavenger reaction rate.

However, transporting the reacted scavenger through a production facility, and ultimately disposing of reaction products often poses a challenge. Many questions arise regarding the ultimate fate of the reacted scavenger. Fouling and corrosion of downstream equipment is a major concern among operators. When considering disposal into a subsurface reservoir, the possibility of scaling in the reservoir and injectivity impairment are issues that must be addressed. If the reservoir is a producing reservoir, then souring from a nitrogen food source must also be considered. When considering overboard disposal, the Environmental Impact must be assessed.

Often it is not possible to completely segregate the scavenger reaction products from the produced water in a facility. The process configuration may not provide two parallel water-based processing systems. At some point, there may not be any alternative than to mix the produced water and the scavenger reaction products together.

Process constraints may leave no alternative than to mix the scavenger reaction products with the produced water and to mitigate the consequences with scale inhibitor. Laboratory tests are usually required to select an appropriate scale inhibitor and to determine the required minimum dosage.

Laboratory conditions must mimic the conditions in the field, at least with respect to the important variables such as temperature, and solution composition. It becomes critical to understand the pH and alkalinity effects of the fluids, before and after the scavenger has been added. The mitigation of scale formation becomes in this case a complex problem involving solution chemistry. The extent to which the chemical vendors are willing to help solve such problems, or are even aware of the issues varies significantly from one vendor to another.

Commonly used H_2S scavengers contain a triazine-type chemical as the active ingredient. Examples are tri(hydroxyethyl) triazine and trimethyltriazine (see Figure 1). The tri(hydroxyethyl)triazine is the more water soluble of the two, and is generally favored when it is likely that the scavenger will come in contact with produced water. The trimethyltriazine is the more oil soluble and is often used when the scavenger will contact predominantly oil-based produced fluids. This later compound however, does have finite partitioning into both oil and brine. Both of these triazine examples contain a similar amine structure and therefore undergo similar hydrolysis and scavenging reactions1, 2. The product of these reactions include primary amines. The scavenging reactions for tri(hydroxyethyl)triazine are shown in Figure 2. As shown, electrophilic attack of the nitrogen groups by hydrogen sulfide results in sulfur substitution into the cyclic structure. Also, a primary amine (ethanolamine in the example) is produced in each step of the sequential reaction.

As indicated in Figure 2, both the unreacted triazine, and the reaction products are amine-type compounds. Thus, the pH of the solution will be alkaline along the entire reaction path. In a typical application, only a fraction of the scavenger will react. Therefore the product of the scavenging treatment will contain both reacted and unreacted scavenger. In fact, complete reaction of the scavenger (Figure 2, part D) results in a cyclic sulfur compound that is insoluble in either hydrocarbon or brine. If that compound is allowed to form, it will precipitate on pipe walls and vessel internals. Scavenger is therefore normally applied in excess of the stoichiometric ratio, and the contact time between scavenger and H_2S containing gas is controlled within certain limits. Both of these actions help prevent the formation of the insoluble reaction product, while still achieving the desired scavenging target. However, a consequence is the presence of unreacted scavenger in the product. Thus, the product of typical scavenger applications in the oil and gas industry is a complex mixture of unreacted triazine, partially reacted triazine, and primary amines.

As discussed above, the chemistry of the H_2S reaction with the amine scavenger is such that both unreacted scavenger and reacted scavenger both have high pH. Thus, the pH is expected to remain elevated over the entire course of the reaction. This conclusion does not take into account the presence of CO2. In addition to the amine chemistry discussed above, it is necessary to consider the CO2 / bicarbonate equilibrium. The CO2 / bicarbonate equilibrium tends to drive the pH to acidic values. That equilibrium also has a direct impact on scaling tendency since it determines the concentration of carbonate anions.

5.16 Overall Chemical Optimization Strategy:

Selecting the optimal chemicals for a given process is a multi-stage process:

1. characterization of produced fluids and selection of compatible candidates;

2. sampling of produced fluids for onsite bottle testing;

3. bottle / bench testing of candidate chemistries;

4. field trialing of candidates;

5. repeat steps (3) and (4) as necessary to find the optimum chemistries.

Initial Characterization and Compatibility Tests: The overall the process involves initial sampling and analysis of produced fluids (oil, water, gas). This stage is required so that the chemical vendor has at least a rough idea of the fluid properties. Without this basic information, the list of candidate chemicals is impractically long. Also, there may be some major challenges that must be identified and overcome before water treatment chemistries can be selected. For example, if iron sulfide is present, then this must be addressed in combination with the selection of the water treatment chemistries.

Produced Water Sampling: The next stage involves sampling the fluids for bottle testing. This is a critical stage in the process. Poor sampling, which may not be the fault of the chemical vendor, can be very costly because it can lead to poor selection of chemicals.

Bottle / Bench Testing: The actual onsite bottle testing is the next stage. A bottle testing procedure is given below in Appendix 13.A. A description of the strategy is given in Section 5.10 (Water Treatment Chemical Selection). Bottle testing may be combined with bench testing such as flotation bench testing, testing with a bench electrocoagulation tester, or with a type of filtration media.

To the uninitiated, the process of chemical selection seems odd since most of the work is carried out by a chemical supply company and you, the engineer working in the operating company, will not be given much information about the chemical product that your company will pay for and which may have a huge impact on the profitability of a facility. You will be aware that there are at least a half dozen vendors who can supply chemicals, and between them literally hundreds of chemicals to choose from. The chemicals have strange names like Cleartron PZ-20000, which sounds bizarre and tells you nothing. The MSDS and Product Bulletin (if one exists) will give you no significant application information. Even if you do learn the actual organic chemistry name, that really tells you nothing. Essentially you will be given no chemical information upon which to make your selection.

To appreciate how odd this is, consider a facilities engineer who must select a level control valve. If he followed the process used in chemical selection, he would invite several manufacturers to test their recommended products in the field without any technical specifications of valve type (globe, ball, gate, etc), seal type (packing vs mechanical), trim size, materials of construction, and control parameters. In the actual process of selecting an LCV, all of these parameters are clearly spelled out in the specification sheet for the valve. The engineer would calculate flow rate, pressure drop, determine the corrosion resistance required, and select an appropriate valve.

In selecting a chemical, characterization is usually not carried out adequately, and very little information about the chemicals is available. The chemical class, the molecular weight, the charge type and density, the co-solvents, sensitivity to dissolved solids, are almost never known to the customer. The chemical vendors consider this level of detail to be proprietary.

What is even more interesting about the chemical selection process is that basic chemistry and performance are usually known by the vendor. The larger vendors have relatively sophisticated laboratories and highly qualified research and development staff. Such vendors do carry out the relevant surface science and physical chemistry studies. They do carry out controlled evaluations of the performance of their products. The experiments, either laboratory or field based, demonstrate the performance of one product versus another in a variety of applications. Vendors use such studies to decide whether to bring a new chemicals to market, or to expand an existing product line into new applications. Such information is important to the vendors in making business decisions, but will almost never be shared with the customer.

As odd as it may seem, it is a given fact that chemical selection is not based on chemical information. In such an environment, proper chemical selection is based on what is known as competitive chemical evaluation. Two or more vendors are invited to test their products in the field. Vendors typically provide highly experienced specialists to do this type of work. The operating company sets up the parameters and guidelines for the evaluation. In the best run evaluations, some third party testing is involved, and the operating company is directly involved in the gathering of field performance data, ensuring the integrity of the evaluation, and ensuring that every vendor has the best possible opportunity to perform well. Details on how to conduct a competitive chemical evaluation are available.

Overview of Competitive Chemical Evaluation Process:

1. Invite chemical vendor company(ies) for Competitive Chemical Evaluation.

2. Review the commercial proposals from the vendors.

3. Establish an agreed evaluation method with the vendors on the basis of a Performance Scorecard that provides numerical, specific and well-defined criteria and that includes both process criteria and the Total Cost of Operation (TCO).

4. Give the incumbent supplier a period of time, free of major planned process modifications, to optimize the system. The incumbent supplier should already have optimized the system and the degree to which optimization is achieved upon the threat of a Competitive Chemical Evaluation must be taken into account in awarding the contract.

5. Define the stages of the evaluation. It may include an initial (optimization) period and a final (test/evaluation) period. During the initial period, the invited supplier will optimize their treatment system without evaluation by Shell. The supplier will be given some time for optimization.

6. Define the reporting requirements such as process data, chemical injection rates, gauging of chemical for reconciliation of rate versus inventory. Agree that a company representative will witness the gauging of all chemical totes, pumps, tanks, and drums.

7. After all tests are completed, the Chemical Engineer will work with Operations, the Chem Team, and the Asset Team Leadership to evaluate the Performance Scorecard.

8. Award the new contract.

Example Score Card:

Table 5.7

Evaluation Criteria	Wt %	Score (scale: 0-1.0)		Weighted Score	
		Vendor-1	Vendor-2	Vendor-1	Vendor-2
1. Safety - Region	5				
2. Safety - Corporate	5				
3. Quality Control	5				
4. Service Performance					
4.1 Asset experience	20				
4.2 Competence	20				
4.3 Staff plan	10				
5. Cost Performance	30				
6. R&D Location Support	5				
	100		Weighted Totals:		

Weighted score is the Score x Weight % for each evaluation category.

Explanation of items in Scare Card:

1. Safety score based on the evaluation of the 2000 TRIR of each company for recent past year in the region.

2. Safety score based on the evaluation of the 2000 TRIR of each company for recent past year for the corporation.

3. Quality control score based on Operations experience across the Gulf of Mexico.

4. Service Performance score

4.1   Asset experience

- Time in location which has lead to extensive understanding on the physical hardware, fluid characteristics and system dynamics. Score will be given on how this experience will apply to the particular field location.

4.2   Competence

- Technical capabilities and demonstration of service from representatives in the areas of: chemical treating system, separation processes, and overall subsea and facilities processes (include historical GoM experience - Chem Team input).

- Quality and completeness of documentation of treating program, surveillance and monitoring programs. Documentation for operators on what to watch, how to watch, how often to watch, and criteria for acceptable chemical treating system. Documentation on how to change and troubleshoot the chemical treating system in response to process changes, planned and unplanned (include Chem Team input).

- Competence in achieving hydrocarbon and water quality targets, e.g.:

- Capable of BS&W to 0.6% and free oil in water below 15 ppm as needed.

- Reduce chemical so BS&W trends steady at 0.7-0.8%, as appropriate.

- Identify likely problems ahead of time and have contingency chemicals on-board to resolve likely problems and prevent upsets (include Chem Team input).

- Chemical inventory and logistics practices set, checked, and ordering practice defined and documented.

- Demonstration of good communication and coordination: both on-site and with off-site support teams.

- Training for operators, w/operations leadership participation (include Chem Team input).

- TCO implemented ideas (include Chem Team input).

4.3   Staff plan:

- Percentage of time that a service representative is on location.

- Ability of organization to provide rapid and experienced support on-site to respond to process upsets.

- Support response based on historical GoM experience (Chem Team input)

5. Cost performance:

- Cost will be based on the estimated cost for one year. The lowest cost company will be given a 1.0 and other companies will be given a value based on their ratio to the lowest cost company. If all else equal, to award business to the competitor, cost savings must be greater than 5% decrease from current program costs.

- Demonstrated continuous cost improvement (include Chem Team regarding other locations in Gulf of Mexico).

6. R&D location support:

- Demonstrated ability to anticipate problems and have tested answers ready to implement as they are needed. This requires coordination between field development, operations, and the R&D labs (include Chem Team input).

- Look at opportunities and bottlenecks and work with Shell to get the benefit.

- R&D activities responsive to BPC, WTC, and SOI input.

5.17 Chemical Injection Systems:

Regarding location of chemical injection, there are several issues to consider. In order for water treatment chemicals to be effective, they must be well mixed into the produced water, and they must be given a certain amount of reaction time without high shear prior to mechanical separation. Enough time must be provided for coagulation, flocculation, or coalescence to occur, prior to oil/water separation.

While good mixing is required, high intensity shearing must be avoided once a flocculating agent has been added. Shear will destroy the flocs, and reduce the oil drop size, both of which reduce separation efficiency. Flocculant injection must occur downstream of valves and pumps, and upstream of a quiet zone which is either part of, or is followed by an oil/water separation device or piece of equipment. Sometimes the optimum location is a tradeoff between reduced shear and reaction time.

Wemco Quiet Cell: One of the better locations for injecting a flocculating agent is upstream of the so-called "quiet cell" of a Wemco Depurator (a type of horizontal flotation unit). The quiet cell is the first cell. No bubbles are introduced or generated in this cell. This cell only provides roughly one minute of retention time which is actually shorter than usually required. Therefore fast acting flocculating agents are usually selected for this injection location. Another obvious disadvantage of this location is that it is "end of pipe." In other words, it is usually the last treatment stage in the system leaving little room for error. However, another advantage of this location is that once the floc is built, it is immediately subjected to flotation. Flocculating agents optimized specifically for this location can also have flotation aids which improve adhesion of oil drops to the bubbles and which help stabilize the foam (froth). As it turns out, in the tradeoff between reaction time and reduced shear, this is an excellent location.

Free Water Knockout (FWKO): Another favorable location is the feed stream to the Free Water Knockout vessel. From a retention time standpoint this is an excellent location since most FKWO water buckets have three to four or more minutes of retention time. Once the floc is formed it will have a higher rise velocity and may in fact reach the oil/water interface. This is an ideal situation because the floc then combines with the export crude oil, and is removed from the produced water stream. This is a major advantage which distinguishes a separation vessel from a water treating unit. In a water treating unit, separated oil is rejected and must be further concentrated in a wet oil tank, or recycled upstream into one of the separators in order for the oil to make its way into the crude oil for export. Upstream recycle requires the use of a pump which is a significant source of shear.

What makes this location less than optimal is the sensitivity toward over-treatment of the chemical. Over-treatment of the chemical does not necessarily lead to better separation. It can in fact make separation more difficult. One of the ways this can happen is due to the presence of an interface control valve on the water discharge of the FWKO. If a floc of oil drops is formed but does not have time to reach the oil/water interface, then the floc is discharged with the produced water. The shear action that occurs in the valve will destroy the floc. Once the floc is broken, the polymer has a tendency to coat the drop or solid particle. It will no longer act effectively like a flocculating agent. In fact, when this occurs, coalescence is actually hindered and subsequent addition of flocculating agent is also less effective. Since typical produced water from a FWKO may contain 100 to 1000 ppmv oil in water, this means that careful control of chemical addition is required at this location to prevent over-treating.

Because of the shearing action of the FWKO Interface Level Control Valve (ILCV), any deoiler added upstream of the FKWO does little benefit to promote separation in the vessels or equipment (such as hydrocyclones) downstream of the FWKO itself. This is generally true of chemical injection upstream of any equipment where an ILCV exists. Due to the presence of these control valves, multiple application of deoiler is sometimes necessary. However, in that case, care must be exercised not to over-dose with chemical. Such overdosing will generate a sticky deposit in vessels.

Recycle Streams: Recycle streams always involve shear which breaks the floc and leaves the oil and solid particles coated with polymer. This polymer coating makes subsequent flocculation more difficult. Also, flocculating agents have a tendency to stick to vessel internals. They form a sticky mass that is difficult to clean.

On Bullwinkle (deepwater US), the mode of chemical injection was critical and dramatically impacted oil removal efficiency. Initially the chemical was being injected in its "neat" form (in liquid state as provided to the platform). Performance of the flotation unit was poor. After modification of the injection system, by a 20:1 dilution with fresh water, oil removal greatly increased while overall chemical injection concentration was greatly reduced.

Hydroflokk Process: A straightforward but nevertheless underutilized technology is the Hydroflokk Process [26]. It was developed by Norsk Hydro for the Troll B project. In this process, produced water is first treated with a coagulating agent injected into the feed pipe. This is followed by a flocculation vessel. This vessel consists of two chambers connected in series and stacked one atop the other, in which the oil droplets are first allowed to coalesce. The flocculating agent is injected into the second chamber. A mixer extends the length of the two chambers to provide gentle mixing to promote coalescence and to ensure adequate mixing of the flocculating agent.

This improvement occurred for two reasons. The original chemical injection (neat form) caused the chemical to encapsulate into small "blobs." This is known as the Flory theta solvent effect. For this particular polymer the produced water brine was significantly removed from a Flory theta solvent. When the polymer was added after pre-dilution with fresh water, the polymer did not form blobs and achieved significantly greater blending with the produced water. In this case, the fresh water was more of a Flory theta solvent allowing the polymer to unwind and have maximum contact with the oil drops. Pre-mixing resulted in greater overall chemical performance and much improved performance of the flotation unit.

Dosage Control – in oilfield and downstream water systems (slop tanks especially), gooey, gummy accumulations of overfed, high molecular weight polymer are unfortunately quite common and are often a key contributor to performance deficiencies in water process equipment. Accurate dosing is required to achieve the desired results while not creating additional problems. The proper dosage is best determined empirically in the lab. Underfeed of polymer results in the predictable consequence that the effect desired by the addition of the polymer is not achieved.

Overfeed of high molecular weight polymers can be devastating. When using a cationic or anionic polymer to remove suspended solids, overfeed will go past charge neutralization and partial cationization of particles and droplets and recreate repulsive forces. Instead of a suspension of particles stabilized by net anionic surface charges, overfeed of a cationic polymer will replace the abundant anionic driven repulsive forces with cationic ones – the result is that the particles are re-stabilized with the cationic polymer now acting as a dispersant.

Overfeed of high molecular weight polymers typically presents as "pin floc;" tiny, high density floc particles rather than large "fuzzy" particles. In addition to re-stabilizing particulate and oil droplets, the unreacted polymer will travel through the system in globs – collecting in the flotation cell and impairing performance or even going downhole and substantially impairing formation rheology.

Overfeed of low molecular weight polymers being used as scale inhibitors can be equally as devastating as overfeed of the HMW products. Low molecular weight polymers are generally aggressive to mild steel and when overfeed can result in a corrosion pattern resembling "measles." Equipment failure typically results from other corrosion accelerators working on the metal surface areas damaged by the overfeed of the scale inhibiting polymer.

Many tanks have a layer of "goo" which is an accumulation of polymer that has been fed (overfed) for whatever purpose and found its way to the tank. The polymer goo is difficult to remove and handle and represents a lot of money spent on polymer that was purely wasted. Certain specialty chemical companies were known for overfeeding polymer to refinery desalters. Inevitably, some tank in the refinery would wind up being the collection point for all of the wasted polymer – often a 1 to 2 meter or more layer.

Dosage location – Polymers often require time for the desired reactions to occur. First the polymer needs to hydrate, to unwind in order to become fully effective. The time required to reach full activity is greatly affected by the physical form of the polymer product. The polymer should be fed to the system as far from the mechanical separation device (clarifier, flotation cell or filter) as is typically possible in the limited space in most oilfield well sites. Strong mixing, preferably from a low shear source is recommended. Low shear, in-line, "static mixers" are an excellent choice to ensure the majority of the polymer fed is activated and the desired contact between the polymer and particles/droplets occurs. High or low molecular weight polymers should not added to a system in close proximity to where oxidizing materials, like hydrogen peroxide or bleach, are being dosed. Most of these polymers are readily oxidized and thereby rendered ineffective.

Shear – polymer chains can be broken when shear energy is applied. Passing the polymer treated water through pumps and mixers or through the formation, such as in a polymer flood, will degrade the polymer. Once sheared, broken into smaller pieces, the polymer is not typically going to be able to perform the desired task.

Depending on the particular product, there is a specified temperature range outside of which the product will de-emulsify. When this happens, the polymer latex particles come out of suspension. The material becomes very viscous and essentially unusable. Also, the emulsion polymer must not contact water until it is injected into the produced water system. If water comes in contact with the polymer it will de-emulsify.

One other challenge is in the final application of the material. The process of unwinding the polymer usually takes time, on the order of a few minutes or less. Thus, it is not recommended to inject these material directly into the produced water. When this is done, the polymer will not be effective in attaching to contaminants and will have a tendency to stick to other polymer molecules. This results in agglomerates of polymer and so-called fish eyes. These agglomerates will stick to the piping, valves, and surfaces of water treatment equipment.

The application problems relate to the incomplete unwinding of the polymer after injection, and the tendency to agglomerate solids at the interface in an oilfield application. The only way to increase the efficiency of application in a water system is to ensure the latex polymer is completely hydrated before injection. This is typically attempted with a make-down unit. A make-down unit is a device that is intended to mix the latex as delivered with a large amount of fresh water before injection. The efficiency at which a make-down unit operates is inversely proportional to the amount of neat latex in the fresh water. In other words, the lower the concentration of latex in the fresh water, the more efficiently the chemical is applied. This creates more handling problems simply because more equipment is involved. The ideal concentration for complete hydration of a latex polymer is 100 ppm in fresh water. In practice, this would require very large tanks. Therefore, a compromise is made in efficiency, and most of these systems are operated at ~ 1%. The alternative is to inject the latex neat. However, this is very inefficient, and results in injection of more latex than is ideal for the system. While this is not unusual in many different types of chemical injection, the problems when using latex polymers are worse because of the high cost per gallon and the operational problems associated with the latex concentrating at the interface of oil systems. The effect of latex concentration at interfaces is probably the most expensive problem associated with its use, although it is very difficult to quantify this expense. Generally, the amount of slop generated in heavy oil systems using latex polymers for reverse emulsion breaking is orders of magnitude higher than in systems using lower molecular weight solution polymers. This results in very high rework costs.

References to Chapter 5

1. T. Moon, D. Horsup, "Relating corrosion inhibitor surface active properties to field performance requirements," NACE – 02298, paper presented at the Corrosion 2002 meeting (2002).

2. S.D. Faust, O.M. Aly, Chemistry of Water Treatment, Second Ed., Ann Arbor Press, MI (1998).

3. B. Aunan, X. Paloumet, S. McBride, "Produced water treatment design development for the offshore Shtokman field," presentation to the Tekna Produced Water Management Conference, Stavanger, Norway (2012).

4. J.C. Crittenden, R.R. Trussell, D.W. Hand, K.J. Howe, G. Tchobanoglous, Water Treatment Principles and Design, Publsihed by MWH and Wiley Interscience, 3rd Ed. (2012).

5. J. Bratby, Coagulation and Flocculation, Uplands Press Ltd, Croydon, England (1980).

6. F.J. Mangravite, "Synthesis and properties of polymers used in water treatment." Proc. AWWA Sunday Seminar, Use of Organic Polyelectrolytes in Water Treatment, AWWA, Denver, Colo. 1983.

7. R. Bosch, E. Axcell, V. Little, R. Cleary, S. Wang, R. Gabel, B. Moreland, "A novel approach for resolving reverse emulsions in SAGD production systems," Can. J. Chem. Eng., v. 82, p. 836 (2004).

8. US patent 6,120,690 assigned to R.A. Haase, "Clarification of water and wastewater," (2000).

9. D. Poelker, "Chemical control of produced water solids," paper presented at the 11th Annual Seminar of the Produced Water Society, League City, TX (2001).

10. D. Nguyen, N. Sadeghi, "Stable emulsion and demulsification in chemical EOR flooding: challenges and best practices," SPE – 154044, paper presented at the SPE EOR Conference, Muscat, Oman (2012).

11. P.J. Breen, D.T. Wasan, Y.-H. Kim, A.D. Nikolov, "Emulsions and emulsion stability," chapter 4 in Emulsions and Emulsion Stability, Ed. J. Sjoblom, Marcel Dekker, New York (1996).

12. Y.-H. Kim, D.T. Wasan, P.J. Breen, "A study of dynamic interfacial mechanisms for demulsification of water-in-oil emulsions," Coll. Surf. A, v. 95, p. 235 (1995).

13. J. Sjoblom, N. Aske, I.H. Auflem, O. Brandal, T.E. Havre, O. Saether, A. Westvik, E.E. Johnsen, H. Kallevik, "Our current understanding of water-in-crude oil emulsions. Recent characterization techniques and high pressure performance," Adv. Colloid Int. Sci., v. 100, p. 399 (2003).

14. M.E. Newberry, K.M. Barker, U.S. Patent No.: 4414035 (1983).

15. A.G. Shepherd, A mechanistic analysis of naphthenate and carboxylate soap-forming systems in oilfield exploration and production, thesis submitted to Heriot-Watt University (2008).

16. "Naphthenate Deposits, Emulsions Highlighted in Technology Workshop," Journal of Petroleum Technology (July 2008).

17. J.M. Walsh, G. Sams, J. Lee, "Field implementation of new electrostatic treating technology," OTC-23200, paper presented at the Offshore Technology Conference, Houston (2012).

18. K.E. Arnold, M. Stewart, "Surface production operations, Volume 1: Design of oil handling systems and facilities," Gulf Professional Publishing, an Elsevier Company, 3rd Edition, Amsterdam (2008).

19. M. Yang, A.C. Stewart, G.A. Davies, "Interactions between chemical additives and their effect on emulsion," SPE-36617, paper presented at the Annual technical Conference and Exhibition, Denver (1996).

20. S. Kokal, "Crude-oil emulsions: A state of the art review," SPE Production & Facilities, p. 5 (2005).

21. G. Chen, J.W. Towner, "Study of dynamic interfacial tension for demulsification of crude oil emulsions," SPE – 65012, paper presented at the 2001 SPE International Symposium on Oilfield Chemistry, Houston, TX (2001).

22. D. Shen, D. Shcolnik, R. Perkins, G. Taylor, M. Brown, "Evaluation of scale inhibitors in Marcellus high-iron waters," SPE Oil and Gas Facilities, p. 34 (2012).

23. S.J. Dyer, G.M. Graham, "The influence of iron on scale inhibitor performance and calcium carbonate scale formation," paper presented at the 11th NIF International Oilfield Chemical Symposium, Geilo, Norway (2000).

24. C. Fan, A. Kan, G. Fu, M. Tomson, D. Shen, "Quantitative evaluation of calcium sulfate precipitation kinetics in the presence and absence of scale inhibitors," SPE – 121563, SPE J. v. 15, p. 977 (2010).

25. N. Goodwin, J.M. Walsh, R. Wright, S. Dyer, G.M. Graham, "Modeling of the effect of triazine based sulfide scavengers on the in-situ pH and scaling tendency," SPE 141583 (2011).

26. A. Finborud, M. Faucher, E. Sellman, "New method for improving oil droplet growth for separation equipment," SPE – 56643, paper presented at the SPE ATCE, Houston, TX (1999).

CHAPTER SIX

Petroleum Microbiology

Chapter Six Table of Contents

6.1 Introduction – The Microbial World:

Historically, all life on earth has been classified into two super-kingdoms: Prokaryota, those organisms whose cells do not have a nuclear membrane, such as bacteria, and Eukaryota, those organisms whose cells have a nuclear membrane such as plants and animals. The rapid advancement in the biological sciences led, in about 1990, to a proposed reclassification of life forms into three super-kingdoms, shown in the figure below [1]. The new classification comprises Eukaryota, Bacteria, and Archaea. Archaea was recognized as a new life form in 1977 by researchers from the University of Illinois, which they called Archaeabacteria [2]. These organisms had formerly been classed with bacteria in the Prokaryota super-kingdom. As advances in technology allowed the cellular structure to be examined in greater detail, it became appropriate to recognize Archaeabacteria as a separate super-kingdom. Archaea, as they were called, are so fundamentally different from bacteria that they merit a name change to help distinguish them from bacteria. The life forms of interest in petroleum production are primarily bacteria and archaea.

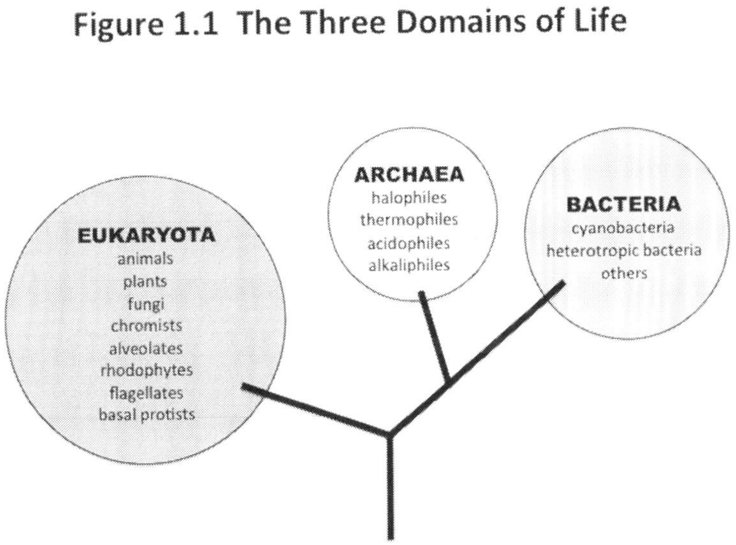

Figure 1.1 The Three Domains of Life

Figure 6.1 The Three Domains of Life

6.1.2 Bacteria:

Bacteria are usually very small, on the order of 1/50,000 inch (0.6 micron) in diameter. As a general rule, the smaller a species, the greater its overall metabolic activity. For example, given the relatively harsh conditions to which many bacteria are exposed, these bacteria can double their population in about 20 minutes. This high rate of metabolism means they can produce enzymes which are able to metabolize tremendous amounts of substrate per unit time. In general, the surface area to volume ratio is a good indication of metabolic activity. For example, a bacterium of 0.5 micron size has a surface area to volume ratio of about 120,000. For an amoeba, which is somewhat larger, the ratio is about 400. For a 200 lb. man, the ratio is about 0.3 [3]. To contrast the relative metabolic activity, a lactose-fermenting bacterium may degrade 1,000 to 10,000 times its own weight of lactose in an hour.

It would take about 250,000 hours (28.5 years) for a man to metabolize 1000 times his own weight of sugar. Many types of bacteria are found in petroleum production zones, and their rapid growth explains why microbiological problems can manifest in very short time frames [3].

Classification of Bacteria: There are several ways to classify bacteria. These include, but are not limited to their shape, how they respire, differences in the composition of the cell wall, how they get carbon, and nutritional requirements. In terms of shape, bacteria can be classified most generally as round or spherically shaped (coccus), rod shaped (bacillus), and curved or spiral shaped (spirillum). In terms of respiration, there are three types of bacteria: aerobic, anaerobic, and facultative. Aerobic bacteria grow in the presence of oxygen and may go dormant in the absence of oxygen, while obligate or strict aerobes have an absolute requirement for oxygen and will die if it is not present. Anaerobic bacteria grow in the absence of oxygen and may go dormant in the presence of oxygen, while obligate or strict anaerobes are killed in the presence of oxygen. Facultative bacteria can adapt to either an oxygen rich or oxygen depleted environment.

An early developed test used to identify the presence of bacteria, used mostly in human pathology, is the Gram staining method. Certain bacteria react to crystal violet dye by turning blue. This makes them easier to spot under an optical microscope. More recently the test is used to differentiate between bacteria that turn blue and those that do not stain (Gram-positive or Gram-negative). This method, called the Gram staining method, is rarely used in the oilfield.

Bacteria are further classified in terms of how they obtain carbon. Autotrophs are organisms which can obtain their energy from inorganic compounds and utilize CO_2 directly for their carbon requirements. Heterotrophs generally require an organic carbon source.

Nutritional requirements of bacteria include, in addition to hydrogen and oxygen (which they obtain from water), carbon, nitrogen, phosphorous, and sulfur. In addition, trace amounts of the elements potassium, magnesium, calcium, iron, manganese, cobalt, copper, molybdenum, and zinc are required by the vast majority of organisms, principally as cofactors for certain enzymes.

For surface equipment and fluids that are being injected downhole and which contain oxygen, heterotrophic aerobic bacteria are highly competitive. Some aerobic bacteria are prolific slime-formers; others can produce enzymes that degrade hydrocarbons. As the water proceeds downhole, the oxygen is rapidly depleted. For petroleum reservoirs where dissolved oxygen has been consumed, anaerobes and facultative organisms become dominant. These include organisms that can hydrolyze organics found in oil and gas, including organisms that produce volatile fatty acids (VFA), fermentative bacteria, methanogens, and sulfate-reducing bacteria [4].

Bacteria utilize nutrients and at the same time affect their immediate environment. Figure 1-3 shows some of the types of nutrients that bacteria utilize, and what happens to other compounds that must be present in order to accept the electron produced by the oxidation of nutrient.

Aerobic Bacteria: Aerobic bacteria are bacteria which utilize oxygen for respiration. Consequently, they are found primarily in surface equipment, any part of the facility that involves sea water, or in fluids that have been exposed to air. A number of aerobic bacteria are notorious slime-formers. Aerobic bacteria grow over a wide range of pH, i.e., 0.5 ~ 10. Most have an optimum pH that is relatively neutral, i.e., 6.5 to 7.5, although many fall outside this range. The optimum temperature for growth of aerobic bacteria varies considerably, i.e., 28 °F to 149 °F, depending upon the organism.

Aerobic bacteria are also very versatile in that there is a wide range of nutrient types that they can utilize. In addition, it is exceedingly rare for any bacterium to require only one particular nutrient [3]. Even in the presence of materials for which they have no apparent enzymes many organisms are able to develop extra-cellular enzymes which will catalyze a reaction of the substrate which was previously foreign to the organism [5]. Consequently, there is almost universal agreement that for every organic substance, there are one or more bacteria which can, at least partially, degrade the organic by the production of exoenzymes. This was first referred to as the "principle of Bacterial infallibility [6]." For example, phenol is widely used as a wood preservative but can be used as a nutrient by some bacteria. A second example involves bacteria that can accumulate large amounts of mercury (a toxic heavy metal) in their cells. A third example includes the strains of tuberculosis-causing bacteria which depend upon antibiotics for satisfactory growth [3].

Bacteria which have the ability to hydrolyze various hydrocarbons were identified at least as early as the 1950s [5]. A number of these organisms are aerobic organisms. The amount of hydrocarbon hydrolyzed by these aerobes per unit time was found to be proportional to the amount of oxygen supplied to the medium. Although the oxidation of hydrocarbons has been thought to occur primarily by aerobic bacteria [7], the range and versatility of bacteria suggests that facultative or anaerobic bacteria may also be able to degrade hydrocarbons [8]. Many hydrocarbons, being relatively non-polar, are only slightly soluble in water, whereas bacteria are only viable in water. Thus, the ability of bacteria to hydrolyze hydrocarbon is proportional to water/hydrocarbon interfacial area [3, 5].

The ability of various bacterial species to oxidize hydrocarbons increases with increasing chain length and with increasing branching, although there are many conflicting reports, depending on which functional groups are present [3]. Petroleum degradation, commonly attributed in a large part due to bacterial activity, occurs primarily in fields with temps below 80 °C [9, 10].

Facultative Bacteria: These bacteria produce ATP in the presence of oxygen. When the oxygen disappears, these bacteria produce ATP by a different mechanism. Thus they can grow both in the presence and in the absence of oxygen. Which mechanism they use depends upon the concentrations of oxygen and the type and nature of nutrients in the environment.

Fermentative Bacteria (Acid-Producing Bacteria or Acetogens): Some bacterial respiration occurs by an anaerobic process called fermentation. ATP is made in this process by the conversion of glucose without oxygen. Fermentation only allows the breakdown of larger organic compounds such as some hydrocarbons, and produces low molecular weight organic acids, (called Volatile Fatty Acids), such as formic, acetic, propionic, and butyric in particular [11, 12].

Anaerobic Organisms: Anaerobic organisms either do not survive, or become dormant, in the presence of oxygen. In surface equipment, these would be found underneath the bacterial biofilm which grows in much of the equipment and in the sludge found on the bottom of tanks. These organisms grow over a wide range of conditions of pH, temperature, pressure and salinity. The fact that anaerobic organisms are able to grow in saline environments is confirmed by the high level of anaerobic activity in produced water. Relatively little is known about the extent of anaerobic degradation of organic matter in waters with a salt content greater than that of the ocean (3.5%). Several types of anaerobic organisms are involved in oil production activities. These include organisms that can produce extra-cellular enzymes which can solubilize solid materials, fermentative bacteria, methanogens, and sulfate-reducing bacteria (henceforth SRB).

Methanogens: As their name implies, methanogens produce methane. These organisms are strictly anaerobic and are rapidly killed by the presence of oxygen. Methanogenesis is the final step in the degradation of organic matter. Methanogenesis effectively removes the semi-final products of decay, i.e., various low molecular weight organic compounds including volatile fatty acids (VFA), such as acetic acid, H_2/CO_2, formic acid, methanol, methylamines, dimethylamine, trimethylamine, methionine, dimethylsulfide, methanethiol, and CO [13 - 18]. During advanced stages of organic degradation, all electron acceptors become depleted except carbon dioxide. Carbon dioxide is a product of most of the processes by which bacteria get energy, so it is not depleted like other potential electron acceptors. Without methanogenesis, a great deal of carbon (in the form of fermentation products) would accumulate in anaerobic environments. Methanogens are more sensitive to changes in temperature than most other organisms. This is due to the faster growth rate of the other groups, such as the fermentative bacteria, which can obtain energy from nutrients even at low temperatures [19]. Studies of anaerobic digestion showed that, in most ecosystems 70% or more of the methane formed is derived from acetate, depending of the type of starting organic carbon. Thus, acetate is the key intermediate in the overall fermentation of these ecosystems. Sodium can inhibit the growth of methanogens and when compared to other metal cations was found to be the strongest inhibitor on a molar basis. Sodium showed moderate inhibition at 3.5 - 5.5 g/l and strong inhibition at 8 g/l.

Sulfate-Reducing Bacteria (SRB): Of the microorganisms present in oil production, the most troublesome are those which produce H2S [20]. The most common of these belong to Desulfovibrio sp., although more recent evidence suggests that the Clostridia class are widely distributed in produced water and may be missed by the common culture-based techniques [21 - 23]. Sulfate reduction by organisms is the only known process by which H2S is formed from sulfate, in aquatic environments of moderate temperatures (0 – 75 °C) [24].

Temperature plays an important role in growth of SRB, as it does with all bacteria. In general, SRB can grow at temperatures which range from about 20 – 70 °C (68 – 158 °F), although some strains grow outside this range [12, 25, 26].

Other strains of SRB have been identified which can withstand heat sterilization used in food canning, i.e., they can survive for 30 minutes in water at 98 – 100 °C (208 – 212 °F) [27]. Desulfotomaculum, a thermophilic SRB spore-former, was found to grow at a temp of 80 °C (176 °F) and pressure of 4500 psi [28].

SRB can be separated into two general categories, assimilatory and dissimilatory. Assimilatory organisms metabolize sulfate ion and incorporate the sulfur atom into its structure. Dissimilatory SRBs metabolize an organic and reduce the sulfate ion to H2S in the process. It is the dissimilatory SRBs which are problematic [27]. SRB can be strictly anaerobic or facultative. Consequently, oxygen can either kill SRB or inhibit their growth [7]. These organisms are very hardy; they have a remarkable capacity for survival in terrestrial and aquatic environments, are widely distributed, ready to become active whenever local conditions become anaerobic [27]. One of the prerequisites for growth of SRB is that the oxidation-reduction potential (ORP) must be < ~ – 100 mV. This means the mere exclusion of air is not sufficient to ensure growth: the presence of a redox-poising agent is necessary [7, 27]. For SRB growth, pH is not as critical a factor as ORP. SRB activity occurs in pH range ~5 to ~9 [7, 20, 27, 29] although growth of some strains can occur outside of this range.

Most strains of Desulfovibrio and some other SRB can use gaseous hydrogen for the reduction of sulfate [27]. SRB can utilize simple organic acids and other low molecular weight fermentation products. The presence of VFA in formations where bacterial sulfate reduction occurs is widespread [7]. However VFA are thought by many to be the single most important carbon source for SRB [20]. Lactate is a widely employed carbon energy source, while acetate, propionate [30], n-butyrate, pyruvate, malate, and even higher chain length acids up to palmitate can also be utilized [31]. Other carbon sources include low molecular weight organics such as simple alcohols [34] and glycerol [7]. Saturated hydrocarbons have also been identified as a nutrient for some strains [32]. Dicarboxylic acids [27], aromatic compounds [33], and saturated cyclic organic acids can also be used [31].

Thermodynamically, fermentation and sulfate reduction yield much less energy than aerobic respiration and so these bacteria grow more slowly than aerobic bacteria such as Pseudomonas [27]. For example, most incompletely oxidizing SRB may double in 3 hours under ideal conditions [24] while the completely oxidizing SRB grow even more slowly, doubling in 15 hours minimum. This is contrasted with aerobic bacteria, which can double in about 20 minutes. End-products of carbon oxidation coupled to sulfate reduction by most Desulfovibrio species are acetate, water, H2S, and usually CO2 [27].

Over the past 50 years, a great many SRB have been identified. This group of organisms collectively can utilize a number of compounds as terminal electron acceptors, including sulfate, sulfite, thio-sulfate, and tetrathionate [28, 31]. Some SRB have the ability to reduce both sulfate and nitrate [31]. SRB contribute substantially to the sulfur cycle (Figure 1-5). Their main role is to bypass assimilatory sulfate reduction and generate H2S. They do this in sufficient amounts to support growth of the sulfide and sulfur oxidizing bacteria and thus they can generate a microbial ecosystem consisting of interdependent sulfur oxidizing and sulfate reducing bacteria which is called a "sulfuretum" [27].

6.1.3 Archaea (Extremeophiles):

A number of microorganisms have been identified, most of which were anaerobic, and that seemed to thrive in extreme environments that would kill most other organisms. For example, microorganisms have been identified that thrive at high temperatures (hot springs or thermal vents on the ocean floor), at very high pressures, at high brine concentrations, at very low and very high pHs. Although few of these organisms have been studied in any kind of detail, one that has been studied is Sulfolobus. Its temperature of optimum growth is about 80 °C (176 °F) and the pH of optimum growth is 2 – 3 [35]. Since their discovery, archaea have been found in a great many normal habitats common to all other microorganisms, including the soil, the ocean, and even the human colon. They are particularly numerous in the oceans and have been recognized as a major part of Earth's life and may play roles in both the carbon cycle and the nitrogen cycle. Up to 20% of microbial biomass may be due to archaea [36]. The fact that many archaea are extremeophiles means that some of these will find a hospitable environment in petroleum reservoirs. However, for the purposes of this text, archaea and bacteria will be referred to collectively as bacteria.

6.1.4 Origin of Petroleum:

How petroleum came to be in the earth's crust in the first place is a matter of some speculation. Theories vary. The most predominant theory suggests that petroleum came from life being buried in the long distant past under layers of strata, and that bacteria had some involvement in its formation:

> "..it is almost axiomatic that bacteria have contributed to the origin of oil, because all types of organic matter are susceptible to bacterial modification." [37]

Treibs found derivatives of plant pigments and other biological material present in crude oil [38 - 41]. Since then, there have been few who seriously challenged that bacteria were involved in the origin of oil [5]. Petroleum-bearing rock formations are believed to be the home to extensive communities of heterotrophic and chemoautotrophic microorganisms thriving on the immense energy reserves contained in crude oil [42, 43]. The source materials for petroleum are certainly organic in nature. That bacteria play a role in the formation of hydrocarbon is almost certain but the mechanism is not known.

6.2 Bacterial Digestion in a Petroleum Reservoir:

Bacterial Communities: In the anaerobic degradation of organic material (shown in Figure 1-6), a community (or consortia) of the bacteria described above work together to convert water-insoluble and particulate organic and inorganic materials ultimately to gases and inorganic constituents [44]. The first step in the process is the hydrolysis of particulate and water-insoluble matter to soluble organic constituents that can be processed through the bacterial cell wall. This is the rate-limiting step in the anaerobic process [45]. This step is carried out by a variety of bacteria through the release of exoenzymes that reside in close proximity to the bacteria. The soluble organic materials that are hydrolyzed by these exoenzymes consist of sugars, fatty acids, amino acids and other complex, water soluble organics.

These complex organics are then converted to CO_2 and a variety of short chain organic acids by fermentative bacteria. Other microbial groups that may be active, including sulfate-reducers, methanogens, and others may produce ammonia, hydrogen sulfide, CO_2, and methane. In the process, the bacterial consortia catalyze a wide variety of physical, chemical and biological reactions. Consequently, the most important factor in degrading waste material is the bacterial consortia [46, 47].

The interplay of organisms in petroleum reservoirs (Figure 1-7) is very similar to the interplay of organisms in anaerobic digesters. In the anaerobic digestion process, the goal is to create conditions which promote the growth of and balance between anaerobic bacteria for the purposes of degrading an organically laden waste stream into environmentally friendly and usable products, specifically a solid material that can be used as fertilizer and a biogas which can be used for fuel. A major difference between anaerobic bacterial activity in petroleum reservoirs and that of the anaerobic digester is that in a controlled anaerobic digester, an equilibrium has been achieved under the limitations of local nutrient concentrations and environmental conditions. For example, once specific organics are degraded in a region of a stagnant reservoir, the population of bacteria which utilize that specific organic will decline and bacteria which are able to utilize the organic or inorganic metabolic byproducts of those bacteria will dominate. Consequently, organisms, which may be native to the formation, may become dormant until conditions change [48, 49].

In petroleum reservoirs, anaerobic microbes tend to grow in isolated communities. These organisms rely on one another, because a metabolic byproduct of one might be a suitable food source for another. Organisms that utilize the same nutrients will compete with one another, producing antibiotics that are toxic to their competitors. Thus the biological community becomes more dynamic, once the equilibrium is shifted, e.g., bacteria-laden water introduced via drilling, fracking or other oil-field activity.

The picture is further complicated because environmental factors, such as pH, salinity, temperature, oxidation potential, and other factors may select for or against a particular organism or group of organisms. These environmental factors, especially temperature, can change from one part of a reservoir to another (Figure 1-8). Injected water, which almost always contains some level of sulfate, will cool the near wellbore, and thus create an area where SRB can grow [20]. With the drilling and completion of wells, downhole conditions change. As gas/oil/water mix moves toward the well opening, the water will move through localized areas where a different microbiological consortium has reached equilibrium and the equilibrium will shift again.

For an active oil or gas well, with gas/oil/water flowing out of the well, slow declines in pressure may require injection of water or some other fluid to keep the pressure up. Differences in the composition of the fluid being used to repressurize the reservoir will also shift the equilibrium. The net result is that there will be localized areas in the formation where the anaerobic degradation process stops.

6.3 Bacterial growth in the Subsurface:

<u>Where Bacteria Grow in the Subsurface:</u> A discussion of where bacteria grow in petroleum reservoirs requires some discussion of the nature of where petroleum is found and some discussion of the nature of the types of formations which exist. According to the most prevalent theory, the structure of the earth can be represented by a peach. The diameter of the earth is approximately 8000 miles. If we were to cut the earth in half, we would see four general layers; the inner core and the outer core (represented by the

peach pit), the mantle (represented by the fruit) and the crust (represented by the skin of the peach). The inner core (peach pit) is thought to be solid and composed mainly of iron and an outer fluid core with a radius of about 2000 miles [50]. The mantle (fruit) has a thickness of about 2000 miles.

The crust (or skin) is a relatively thin layer, thought to be about 3-30 miles (5-50 km) thick, with the thinnest parts underlying the oceans. All oilfield activities are thought to take place in the crust.

6.3.1 Conditions in the Crust:

Conditions in hydrocarbon reservoirs vary widely. There are wells that are quite deep. The deepest well to date lies under 4000 feet of water in the Gulf of Mexico, and is 35,350 feet deep (mount Everest is 29,035 feet). Conversely, in some instances, pools of oil have been found at the surface. In his writings, Marco Polo described shallow oil seeps near Baku, Azerbaijan [51]. Also note, for example, "The Beverly Hillbillies," where "up from the ground comes a bubbling crude."

Temperatures vary considerably in the earth's crust, including in oil reservoirs. There is a temperature gradient in the earth, so that, in general, as the depth increases, there is a corresponding increase in temperature. In the deeper formations, the temperature may increase by an average of 4 °C per 100 m of increasing depth [52]. In addition to temperatures, pressures, porosity, water composition, and type of rock or mineral vary considerably in oil and gas reservoirs [52].

6.3.2 The Impact of Reservoir Conditions on Growth of Microbes:

Reservoir conditions play a huge role in determining what organism can grow. These conditions include temperature, pH, pressure, salt content, and available nutrients.

pH: Bacterial action which produce fatty acids during digestion tend to lower the local pH. However, the CO_2 produced in the local area exerts substantial resistance to pH change. Oilfield brines brought to the surface may be slightly basic [54]. However, the release of CO_2 during depressurization of brines from pressures as high as 500 bar at 6,000 m [1984], has been thought to raise the pH by as much as 2 pH units [55] and so the pH of the reservoir can deviate from this. Some extremeophiles thrive at very low pHs, such as in acid mine drainage. One of these acid-loving (acidophile) organisms is Picrophilus torridus, which grows at a pH of 0 [56].

Pressure: Pressures are known to affect bacterial metabolism, including their shape and the amount of sulfide produced [7]. The overall effect of pressure on bacteria is almost always inhibition of growth though some individual process may be stimulated. One pressure-loving (barophile) organism found to date was isolated from the Mariana Trench. Shewanella benthica had a pressure of optimum growth of 15,215 psi, its ambient pressure [57].

Brine Concentration: Typical dissolved solids in flowback or produced water can range from < 10 to > 600 g/L (<10,000 to > 600,000 mg/L) but seldom exceed 350 g/L (350,000 mg/L). For comparison, seawater has a TDS of about 35,000 mg/L. The salt tolerance of most oilfield organisms can vary significantly, from 0.1 g/L (100 mg/L) to 100 g/L (100,000 mg/L) [58]. Many marine SRB species are moderately halophilic, i.e., they require 10 - 30 g/L (10,000 – 30,000 mg/L) NaCl for optimum growth. And while the activity of most SRB are inhibited if the NaCl conc exceeds 50 – 100 g/L (50,000 – 100,000 mg/L) [7, 27, 31], some organisms thrive in briny produced water. For example, some halophilic bacteria with salinity optima of up to 1000 g/L have been isolated from oilfield brines [24, 71]. After Hurricane Ike, an undersea pipeline broke and 75,000 bbl. of seawater entered the truck line, and approximately 40,000 bbl. entered the lateral. After repairs, 90 ppm H2S was detected ahead of the pig in the trunk line. When the lateral was dewatered 2 months later, H2S had spiked to 10,000 ppm [61].

Temperature: Of the factors which influence the growth of reservoir bacteria, the single largest factor which determines whether bacteria may exist in a given reservoir is temperature [52]. Reservoir temperatures range from 20 – 100 °C (68 – 212 °F), although some fall outside this range. The vast majority of bacteria are unable to grow at temperatures greater than 45 °C. The metabolic and growth rates of chemical and biochemical reactions tend to increase with temperature, within the temperature tolerances of the microorganisms. Too high a temperature, however, will cause the metabolic rate to decline, due to degradation (denaturation) of enzymes which are critical to the life of the cell. Microorganisms exhibit optimal growth and metabolic rates within a well defined range of temperatures, which is specific to each species, particularly at the upper limit which, is defined by the thermal stability of the protein molecules synthesized by each particular type of organism [53].

Microorganisms can be classified by their tolerance for temperature (Figure 1-12). For example, bacteria classified as psychrophiles (cold-loving), mesophiles (moderate temperature loving), thermophiles (warmer temperature loving), and hyperthermophiles (hot temperature loving) have been identified in various habitats. Psychrophiles (cold-loving), are those organisms which grow well at temperatures which vary from -10 °C to 25 °C. Mesophiles are those organisms which grow in the temperature range of from 20 °C to 50 °C. Thermophiles are organisms which grow at temperatures which vary from 45 °C to 80 °C. Hyperthermophiles are those organisms that grow at temperatures above about 80 °C. These temperature divisions are fairly arbitrary and show overlap, but they provide general guidelines. Temperature - pressure profiles of 2 SRB isolated from the North Sea are shown in Figure 1-4 [7]. This graph shows that there is a definite "window" within which SRB activity is possible. Under the proper conditions, Desulfovibrio desulfuricans, isolated from 700 - 900 feed drill cores from sulfur bearing Texas salt domes, was able to produce concentrations of 2000 - 2500 ppm of H2S, and do this at a rate of 1000 ppm in 24 hours [5]. In North Sea reservoirs, pressures commonly range from 3000 – 7500 psi. Temperatures vary from 140 – 212 °F (60 – 100 °C). Archaeoglobus fulgidus was found to produce H2S at a temperature of 93 °C (199 °F) [60] Stetter et al., 1987]. Other thermophilic SRB have been found, including Thermodesulfobacterium mobile [62] and Thermodesulfobacterium commune [63], which grow at a temperature of 85 °C (180 °F), and Desulfotomaculum nigrificans which grows with a max temp of 70 °C [28]. The microorganism that has been identified which can thrive at the most extreme temperature, presently, is Methanopyrus kandleeri Strain 116. This organism grows at 122 °C (252 °F) [64].

Nutrients: In addition to an organic carbon energy source, anaerobic bacteria appear to have relatively simple nutrient requirements, which include nitrogen, phosphorus, magnesium, sodium, manganese, calcium, and cobalt. Nutrient levels should be at least in excess of the optimal concentrations needed by the methanogenic bacteria, since these are the most severely inhibited by slight nutrient deficiencies. One review of biodegradation of petroleum in the deep subsurface, suggested that access to nitrogen and phosphorous could be the rate-limiting factor in the activity of microorganisms [42]. Ammonium has been measured in eastern Alberta oilfield produced waters at concentrations which vary from 1 to 1000 mg/L. The ammonium is a source of nitrogen and which can come from minerals such as feldspar and illite [65]. Phosphorus is reactive and can form insoluble complexes with Mg, Fe, and Ca. Therefore P moves very slowly through reservoir. Estimates have been made that application of oilfield chemicals may increase the concentration of nitrogen and phosphorous by 19 and 3-fold respectively [66]. Sulfate ion, used by SRB, has been found to vary from 0 to 3000 mg/L [67 - 69]

6.4 Problems Casued by Microbiological Growth:

A number of problems are caused by bacterial growth. These are broadly identified below and are discussed more completely in the designated chapter.

6.4.1 Deposits and Biofilm:

Formation Plugging by Biofilm: Bacteria can cause formation plugging, primarily by the production of biofilm. Biofilm is a biopolymer which is secreted by many types of bacteria. Biofilm is comprised mostly of polysaccharides. Biofilms are ~ 98% water, but are stabilized as gel [70 - 72]. Biofilms in nature are generally formed by a community of microorganisms [75]. This biofilm ultimately blocks paths of flow of fluid, primarily in the near well-bore region of water injection wells or in producing formations and can result in the production of substantial amounts of H2S [76]. Formation of biofilm has been harnessed for microbially enhanced oil recovery (MEOR) where the production of biofilm causes the flow routes of hydrocarbon to be effectively rerouted [72, 74]. It is important to note that control of bacteria that are free floating (and can therefore be measured in produced water) may not be an indication that bacteria attached to surfaces have been controlled. Free-floating (planktonic) bacteria are physiologically different from those attached to surfaces [73]. This is confirmed by work in potable water systems that has shown that sessile bacteria, i.e., those attached to surfaces, are 150—3000 times more resistant to hypochlorous acid than bacteria in the bulk water, while they are only 10—100 times more resistant to chloramine than bacteria in the bulk water [77]. In addition, others have shown that the resistance of a planktonic population to a non-oxidizing biocide is generally quite different from that of the sessile population [73, 78 - 80]. In general it is much easier to kill bacteria floating in the water than it would be if it were attached to a surface.

Scale and Deposit Formation: Several decades ago, bacteria were identified which form calcareous deposits by precipitation of calcium salts from water [5]. In addition to promoting the formation of calcium deposits, some bacteria and their associated biofilms provide a tacky surface where microscopic particulates of fines, marginally soluble mineral scale, corrosion products corrosion products such as iron sulfide [76], or silt can accumulate.

6.4.2 Reservoir Souring:

Reservoir Souring: Some organisms have the ability to reduce sulfate and produce H2S as a metabolic byproduct. The most troublesome sulfate reducing bacteria was identified as Desulfovibrio desulfuricans. These are strict anaerobes, that is, they do not thrive in an oxygenated environment, but in environments where oxygen is absent. They obtain their energy from the sulfate ion, SO_4^{-2}. These organisms have historically been referred to as sulfate-reducing bacteria (SRB) [27] although both archaea and bacterial sulfate reduction are known, as well as Bacilus and Clostridium [81]. H2S produced by SRB and its associated biofilm can cause souring of a formation.

The temperatures at which microbial sulfate reduction occurs are not clearly understood. Upper temperature limits have been suggested between 75 °C to 100 °C as that where reservoir souring results from SRB. Others have questioned the presence of microbiological organisms in reservoirs with temperatures much above 80 °C [9, 10]. In one report of testing 100 water samples from reservoirs with temperatures above 82 °C, no hyperthermophilic organisms could be isolated [82].

It should be pointed out that the presence of H2S does not necessarily point to a biological problem. H2S can also be produced by the thermochemical processes although a discussion of this topic is beyond the scope of this chapter.

6.4.3 Iron Sulfide:

At saturation, the solubility product of ferric iron and sulfide is very low. In other words, the solubility of iron sulfide is only a few mg/L at room temperature. Thus, whenever dissolved iron and dissolved sulfide occur in a produced water sample, iron sulfide particles are likely to form. This is a common occurrence in produced water where sulfide is generated by bacteria in produced water that already contains iron from the formation or from corrosion. There are several detrimental consequences associated with the presence of iron sulfide as discussed in Sections 2.7.4 (Iron Compounds), and Section 18.1.3 (Iron Sulfide). Iron sulfide compounds often precipitate as very fine particles (smaller than a few micron), and they form oil-wet solids which are difficult to remove (See Section 18.4 Solids Separation from Fluids).

6.4.4 Corrosion:

Corrosion: Another problem caused either directly or indirectly by bacterial biofilm is corrosion. The impact of corrosion caused by bacteria has been grossly underestimated, as work has shown that corrosion caused by microbes can be at least as significant as that caused by galvanic action [82].

Corrosion can be caused by a number of different mechanisms.

- Corrosion can be caused by direct bacterial action [82 - 84].

- Corrosion by bacterial enzyme action [85] has also been documented, and can occur even after cell death [110]. It has been found that in order to control microbially induced corrosion (MIC), it may be necessary to remove the biofilm completely, even though no viable bacteria remain within [86, 87].

- Corrosion through differential aeration cells [88] can also occur as a result of the patchy nature of biofilm [89]. In metal corrosion, the most insidious characteristic of bacterial consortia may not result from the local production and patently corrosive molecular and ionic species, but from the fact that bacterial consortia produce, in their microniches, a local difference in the concentration of protons and other cations that is substantially different from that of the general biofilm [82]. This is referred to as a "focused consortial bacterial attack." The focused attack, postulated by these authors, would explain the pits, or tunnels formed by bacteria on some metal surfaces.

- In addition, the anaerobic conditions created under a healthy biofilm promote the growth and development of the highly corrosive sulfate-reducing bacteria, Desulfovibrio desulfuricans (SRB) [90, 91]. In fact, a biofilm layer only 10—12 organisms thick can result in anaerobic conditions sufficient to promote the growth of SRB [92, 93].

- In nature, the SRB are often the dominant organism under the conditions present underneath biofilm or in sediments. SRB predominate in these circumstances because their hydrogenase enzyme has a greater affinity for hydrogen than most bacteria. The SRB are able to reduce the hydrogen concentration to such a low level that other organisms cannot compete. This ability probably accounts for the prevalence of the SRB in cases of microbioally induced corrosion.

- Some of the metabolic byproducts, other than enzymes, produced by bacterial action, including specifically sulfides, polysulfides, and elemental sulfur, are also very corrosive [24]

6.5 Bacteria Sampling and Analysis:

The commonly used method in upstream oil and gas fields for determining bioactivity is serial dilution. The most commonly used method is referred to as the "Most Probable Number" or MPN method. The method is illustrated below. One mL of a water sample is placed in a prepared container with 9 mL of a nutrient solution. The sample vial is mixed by shaking and 1 mL of the mix removed and placed in another vial with 9 mL of nutrient solution. The serial dilution is repeated as many times as is deemed necessary (typically 3 to 6 dilutions). Bottles which turn dark during the observation period are determined to be biologically active and the number of bottles which darken is reported. Since each bottle represents a factor of 10 dilution, the more bottles which darken correlates with increasing bioactivity.

Several alternative methods are available for determinging bioactivity in a water sample and these, along with MPN, are briefly summarized below. The in-field measurement of Adenosine TriPhosphate, ATP, provides a faster (10 – 20 minutes) and more quantitative measure of bioactivity compared to MPN determination. Recent advances in the reliability of ATP analysis and in the ease of use instrumentation has allowed operators to replace MPN testing with ATP analyses. In the future, additional tests are expected to become available which will not only quantify bioactivity, but also identify the genre of active bacteria. ATP testing and bio-identification should allow operators to make more informed decisions regarding the need for and the type of biocidal treatments which may or may not be required.

Bacterial Test Methods

- Serial Dilution Method

  - NACE Standard: TMO194-2014 Standard Test Method "Field Monitoring of Bacterial Growth in Oil and Gas Systems"

  - Official standard method to obtain a MPN (Most Probable Number) of bacteria in a media that encourage growth

  - Test results in 7 – 28 days

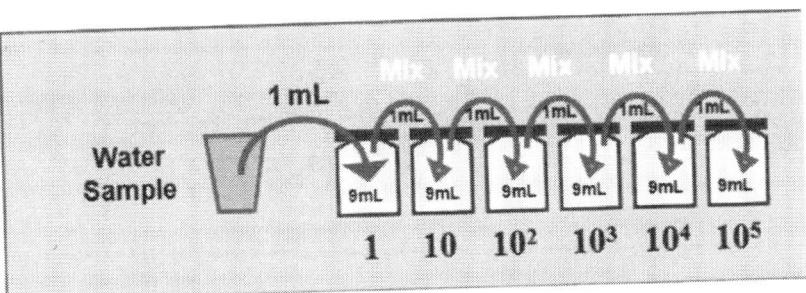

Figure 6.2

Alternative Methods for Assessing Bacterial Populations

- **DAPI (4′,6-diamidino-2-phenylindole)** is a fluorescent stain that binds strongly to A-T rich regions in DNA. It is used extensively in fluorescence microscopy. As DAPI can pass through an intact cell membrane, it can be used to stain both live and fixed cells, though it passes through the membrane less efficiently in live cells and therefore the effectiveness of the stain is lower. Note, in this microscopic method living and dead cells are counted.

Alternative Methods for Assessing Bacterial Populations

- **FISH (fluorescence in situ hybridization)** is a cytogenetic technique used to detect and localize the presence or absence of specific DNA sequences on chromosomes.

- FISH uses fluorescent probes that bind to only those parts of the chromosome with which they show a high degree of sequence complementarity.

- Fluorescence microscopy can be used to find out where the fluorescent probe is bound to the chromosomes.

- FISH is often used for finding specific features in DNA for use in genetic counseling, medicine, and species identification. Note, in this microscopic method only living and active cells are counted

Alternative Methods for Assessing Bacterial Populations

- **TGGE** (Temperature Gradient Gel Electrophoresis) and **DGGE** (Denaturing Gradient Gel Electrophoresis) are forms of electrophoresis which use either a temperature or chemical gradient to denature the sample as it moves across an acrylamide gel.

- TGGE and DGGE can be applied to nucleic acids such as DNA and RNA, and (less commonly) proteins.

- TGGE relies on temperature dependent changes in structure to separate nucleic acids.

- DGGE was the original technique, and TGGE a refinement of it.

Alternative Methods for Assessing Bacterial Populations

- **qPCR** (quantative Polymerase Chain Reaction) A real-time polymerase chain reaction is a laboratory technique of molecular biology based on the polymerase chain reaction (PCR), which is used to amplify and simultaneously detect or quantify a targeted DNA molecule. Note, this PCR method is a quantitative method.

Table 6.1

Method (MMM)	Method Based On	Living Cells Counted?	Dead Cells Counted?	Quantitative Method?	Information Yielded
DAPI	Microscopy	Yes	Yes	Yes	Total cell counts (live and dead)
FISH	Microscopy	Yes	No	Yes	Total number of live bacteria Total number of live *Archaea* Total numbers of live SRB Total numbers of live SRA
DGGE	PCR	Yes	Yes	No	Comparison of populations Identification of abundant microorganisms
qPCR	PCR	Yes	Yes	Yes	Number of total bacteria Total number of live *Archaea* Total numbers of live SRB Total numbers of live SRA Numbers of three groups of methanogens

Alternative Methods for Assessing Bacterial Populations

- **2ⁿᵈ Generation ATP** (Adenosine Triphosphate Photometry) Test Method

ATP or *Adenosine Triphosphate*, is a molecule or an energy carrier located within living biological cells that manage all biological functions, such as food consumption, maintenance, and reproduction. Like any living creature on Earth, microorganisms require ATP to survive – without it, there would be no life! **In all cases, if there are microorganisms present, there is ATP.**

- Rapid assessment in less 10 min per sample

- No underestimation of unculturable organisms

- Techniques can used in any sample type including produced fluids, oil/emulsions and solids

- Primarily a screening tool to find microbial hot spots

- Should be backed up with at least one additional method for more specific quantification

6.6 Control of Bacteria in the Oil Field:

Bacterially related problems in the oil patch was recognized in the early 1900s. As time has progressed, more and more emphasis has been placed on the importance of bacteriological control. Biocides have been employed more than any other mechanism for control of bacteria in oil production. Questions have been raised as to whether, after 50 years of biocide use, current application methodologies work. It has been suggested that biocides may not be killing, but only injuring bacterial cells within biofilm. Numerous MIC failures reported in the literature may have historically been inaccurately linked to planktonic bacteria, when the problem is biofilm bacteria [94]. There are three general types of chemistries or technologies for achieving bacterial control:

- Non-oxidizing Biocides

- Oxidizing Biocides

- Unconventional Technologies

6.6.1 Nonoxidizing Biocides:

This group of chemicals includes but is not limited to glutaraldehyde (Glut), quaternary amine compounds (Quats), acrolein, Bronopol, DBNPA, THPS, TTPS, DMO, and Dazomet. Some newer biocides have been introduced.

In general, the following criteria are some of those used for the selection of a biocide.

1. Is the biocide effective on sessile as well as free floating bacteria?

2. Is the biocide cost effective?

3. Is the biocide safe to use and environmentally friendly?

4. Is the biocide compatible with other chemicals that might be used?

5. Is the biocide compatible with materials of construction?

The answer to these questions can vary depending upon the specific application. Each biocide will be discussed briefly.

Glutaraldehyde: Glutaraldehyde, pentane-1,5-dial, is arguably the most commonly used biocide in the oil patch and is one of several that have environmental/performance profiles that merit approval for use in the North Sea, one of the most difficult to achieve processes. In one early paper, bacterial inactivation was suggested to involve a cross-linking mechanism [95]. This raises the question as to its possible impact on some of the cross-linking polymers used in hydraulic fracturing.

Researchers tested the impact of glutaraldehyde on biofouling [96]. They found that contact time and biocide concentration are not the only variables to be considered in selecting a biocide and that glutaraldehyde was less effective on sessile bacteria than on free floating bacteria. The effectiveness of glutaraldehyde on biofilm was found to be substrate dependent.

One advantage of glutaraldehyde is that a test exists for it [97]. In a laboratory study, glutaraldehyde was compared to acrolein, stabilized ClO2 (chlorite ion), unstabilized ClO2, and quat. Of these unstabilized ClO2 was found to be effective at 10 mg/L, while acrolein, a quat-MBT (methylene-bis-thiocyanate) blend, and glutaraldehyde were found to be effective at 30 mg/L. The authors also noted that much of black biofilm had been removed by the ClO2 [98].

In another study, eight biocides were tested for their biocidal efficacy towards SRB and the impact sulfide had on their biocidal efficacy. Of the eight biocides tested, glutaraldehyde and a glutaraldehyde/quat blend were superior [104]. In a follow up, this group also measured the effect of a process leak on biocidal efficacy of these [99]. A high temperature oil, diesel, hydraulic fluid, ammonia, ethylene glycol, H2S and some gases. Glutaraldehyde was adversely affected only by ammonia.

In one laboratory study the control of souring by nitrate, nitrite, or glutaraldehyde was undertaken. Their results showed that the addition of nitrate or nitrite inhibited bacterial production of H2S better than glutaraldehyde, although nitrate or nitrite was not found to be biocidal, whereas glutaraldehyde was [100]. In one test on the compatibility of glutaraldehyde and a glutaraldehyde/quat blend with friction reducer, the presence of the ammonium ion from the quat resulted in a significant degradation of the slickwater polymer, and required an increase in dosage to achieve the desired results [101]. Glutaraldehyde was observed to degrade to some extent in some produced waters. An investigation led to the root cause of this degradation being the ammonium ion, although the authors noted that it is not the only cause [105]

Quat: There is a wide range of quaternary ammonium compounds used in antimicrobial applications [102]. Each has its rather unique properties, so that some are used as antimicrobials, some are used as surfactants, some are foaming agents, and some are used in cleaner formulations, etc. However, there are some commonalities. For example, all quats are positively charged. This says something about their compatibility with anionic compounds, although some have been developed that are more compatible [103]. All have more or less good surface-active properties. In a laboratory study, quat was compared to glutaraldehyde, acrolein, stabilized ClO2 (chlorite ion), and unstabilized ClO2. The quat-MBT (methylene-bis-thiocyanate) blend was found to be effective at 30 mg/L [98].

Bronopol: Bronopol, 2-bromo-2-nitropropane-1,3-diol is a broad spectrum biocide/preservative that is used in the oil patch. Advantages include: broad spectrum activity, compatible with other biocides, low dosage, stable at acid pHs, breakdown products are also micro biologically active.

DBNPA: DBNPA, (2-2-dibromo-3-nitrilo propionamide) is an organo-bromine compound. It reacts with the protein in the cell membrane to interfere with respiration. Advantages include: broad-spectrum activity, fast acting for a non-oxidizing biocide, unaffected by hydrocarbons or solids, degrades fairly rapidly, forms non-toxic byproducts, and requires a low dosage. Disadvantages include: it is expensive, it is negatively affected by sulfides, it degrades at pH > 8.0, it is inactivated by the presence of reducing agents, it is inactivated by temperatures > 100 °F.

THPS: THPS, tetrakishydroxymethyl phosphonium sulfate, was introduced into the oil patch in the late 1980s. In addition to being an effective biocide, it also showed the ability to dissolve iron sulfide [106]. THPS has been shown to be rather effective against a wide range of bacteria, particularly SRB. A second generation THPS formulation was stated to incorporate ammonium chloride or DETPMPA phosphonate (Diethylene triamine pentamethylene phosphonic acid) and was found to be better at FeS scale removal than THPS alone [107]. Then, the second generation THPS formulation was stated to involve polyquaternary compounds [108]. Third generation formulations incorporate THPS and an anionic polymer [108 - 110]. In addition, other formulations have been proposed that increase effectiveness [111].

Dazomet: Dazomet, or Tetrahydro-3,5-dimethyl-2H-1,3,5-thiadiazine-2-thione, is a rather new biocide which has shown much promise in case histories. It has been approved for use in the North Sea. It is compatible with oxygen scavengers and friction reducers, and its effects appear to be long term. In addition to controlling acid-producing bacteria and SRB, it apparently consumes H2S in produced water. In one case history, the high H2S produced in a well was reduced to very low levels after a single treatment at ~350 – 500 ppm, where it remained for a period of about 2 years.

TTPS: Tributyl tetradecyl phosphonium chloride (TTPC) is a biocide that shows good biocidal properties against bacteria, fungi, and algae, and shows good synergy with halogens. Claims are that it controls biofilm.

DMO: Dimethyl Oxazolidine (DMO) is a colorless liquid with a pungent odor. It is a cyclic amine compound that has a relatively high vapor pressure. It is highly toxic by inhalation and contact. It is flammable but it is not an explosive or oxidizing agent. It is effective over the pH range of 7 – 11. It shows good synergy with glutaraldehyde, and at least one oil field service company has developed a blend.

Acrolein: Acrolein has been used as a bactericide and sulfide scavenger in oil field operations since the early 1960s [119 - 121]. Its use and the factors that contribute to its successful use have been described carefully [112].

Acrolein, 2-propenal, is an extremely reactive, extremely toxic chemical that is a very effective biocide at low dosages. If sterilization of the bulk water is the goal, acrolein can be cost effective. Acrolein reacts rapidly with reduced sulfur compounds, and, as a result, sulfite must be removed prior to the addition of acrolein. Acrolein does not appear to be very effective on sessile bacteria (ref here), and it is not compatible with many chemicals used in the oil patch, although an analytical method has been developed for its measurement [113, 114].

Acrolein was used to scavenge approximately 65% of H2S in less than two minutes in North Slope operations at a 1:1 molar ratio [118]. It was used in another application as a three-fold approach to address problems caused by SRB. It is an effective biocide, it scavenges H2S, and it dissolves iron sulfide [115, 116]. In one test, one group tested 16 commercial biocides and found that acrolein was effective at 30 mg/L [98].

In two field trials, extensive laboratory testing and field trial data and a performance rating based on lowest cost, acrolein provided the most economical control of FeS and H2S [117]. In an off-shore field test, injection water was treated with THPS, glutaraldehyde and later acrolein [122]. Relative efficacy of biocides tested were found to be: acrolein = acrolein with anthraquinone >> glutaraldehyde with anthraquinone > glutaraldehyde alone >> THPS alone. Acrolein has successfully been used as an adjunct with nitrate to control the growth of nitrate-fed biofilm [123]

6.6.2 Oxidizing Biocides:

Oxidizing Biocides: Oxidizing biocides are used relatively infrequently in the oil patch for a number of reasons. Of the oxidizing biocides, bleach has been used to some extent.

Bleach: Bleach has been used in some applications [124]. The advantage of bleach is that it is cheap and it is effective in that bacteria have not been identified that can develop a resistance to it. The disadvantages are that it degrades with increasing temperature, it is so reactive that it reacts with many components in the produced water, it oxidizes H2S or FeS to elemental sulfur, which is known to be highly corrosive.

Control of Bacteria in the Oil Patch with ClO2: An extensive overview of the bacterial inactivation properties of ClO2 can be found elsewhere [125]. In general, ClO2 has the advantages of a very rapid kill, second only to ozone in clean, uncontaminated systems, and unmatched in contaminated systems [126]. Bacteria are unable to develop any kind of resistance to this disinfectant, unlike most of the non-oxidizing biocides.

ClO2 is very selective, in that there are many organics that it reacts with either slowly or not at all [127]. Consequently the 'demand' for ClO2 is generally much less in contaminated systems than other oxidants, and certainly more than non-oxidizing biocides.

In addition, the chlorite ion (ClO2-) alone, is known to have biocidal properties at concentrations approaching those used in many flowback and produced waters. For example, chlorite ion exhibits a strong lethal action at a concentration of about 130 mg/l [128]. In certain applications where the pH of the water to be treated is on the acidic side, the chlorite ion shows excellent biocidal properties. Note, for example, that chlorite ion applied alone controls slime effectively on the paper machine in acidic paper making environments [129]. In addition, chlorite ion can oxidize H2S [130]. ClO2 has been found to have excellent biocidal properties in a number of oilfield applications [98, 125, 130, 131]. ClO2 can also penetrate and remove biofilm and dissolve FeS. If fed at the sufficient dosage, elemental sulfur will not be produced.

Unconventional Technologies: Several unconventional technologies have been evaluated for use as a bacterial control technology. One such technology was ultraviolet light [132]. The use of UV light was found to be effective at reducing bacterial levels by 2 or 3 logs. While a 2 or 3 log reduction may sound sufficient, consider a water which has 106 to 1010 SRB. In a case where the water contains 109 SRB, a 2 or 3 log reduction leaves 106 SRB, which is still very problematic. The single biggest drawback for the use of UV light is that UV at 254 nm, which is very close to the wavelength required to inactivate microorganisms, does not penetrate a cloudy sample. In addition, most organic components absorb UV at this wavelength.

References for Chapter 6

1. Woese, C., Kandler, O., Wheelism M., "Towards a Natural System of Organisms: Proposal for the Domains Archaea, Bacteria, and Euacrya," Proceedings of the National Academy of Science, 87(12), 4576(1990).

2. Woese, C and Fox, G., "Phylogenetic Structure of the Prokaryotic Domain: The Primary Kingdoms," Proceedings of the National Academy of Science, 74(11), 5088(1977)

3. Sharpley, J. M., Applied Petroleum Microbiology, Buckman Labs, 1961.

4. Nihalani, M., Verma, S., Kumar, J., Dubey, H., Bharali, N., Mandal, A., Lai, B., "Managing Sulphate Reducing Bacteria (SRB) Problem Associated with Produced Water in One of the Oil Fields of Oil India Limited – A Case Study," SPE 127418, WPE Oil and Gas India Conference and Exhibition, Mumbai, India, 20-22, January 2010.

5. Beerstecher, E., Petroleum Microbiology, Elsevier Press, Houston, 1954.

6. Gale, E., The Chemical Activities of Bacteria, Academic Press, New York, 1952.

7. Herbert, B., Gilbert, P., Stockdale, H., Watkinson, R., "Factors Controlling The Activity of Sulphate Reducing Bacteria in Reservoirs During Water Injection," SPE 13978/1, SPE Meeting, Richardson, TX, 1985

8. Chikere, C., Okpokwasili, G. and Ichiakor, O., "Characterization of Hydrocarbon Utilizing Bacteria in Tropical Marine Sediments," African Journal of Biotechnology, 8(11), 2541(2009)

9. Philippi, G., "Depth, Time and Mechanism of Origin of Heavy to Medium Gravity Napthenic Crude Oil," Geochimica Cosmohimica Acta, 41, 33(1977).

10. Connan, J., "Biodegradation of Crude Oils in Reservoirs," in Advances in Petroleum Geochemistry, Vol I, Brooks, J. and Welte, D., Eds., Academic Press, London, 1984.

11. Sorensen, J., Christensen, D., and Jorgensen, B., "Volatile Fatty Acids and Hydrogen as Substrates for Sulfate-Reducing Bacteria in Anaerobic Marine Sediment," Applied and Environmental Microbiology, 42(1), 5(1981).

12. Daumas, S., Cord-Ruwisch, R., and Garcia, J., "Desulfotomaculum geothermicum sp.nov., a Thermophilic, Fatty-Acid-Degrading, Sulfate-Reducing Bacterium Isolated with H2 from Geothermal Ground Water," Antonie van Leeuwenhoek, 54, 165(1988).

13. Sowers, K. and Ferry, J., "Isolation and Characterization of a Methylotrophic Marine Methanogen, Methanococcoides methylutens, gen. Nov., sp. Nov." Applied and Environmental Microbiology, 45, 684(1983).

14. Huser, B., Wuhrmann, K., and Zehnder, A., "Methanothrix soehngenii gen. nov. spec. nov., a New Acetotrophic Non-Hydrogen Oxidizing Methane Bacterium," Archives of Microbiology, 132, 1(1982).

15. Zinder, S. and Mah, R., "Isolation and Characterization of a Thermophilic Strain of Methanosarcina Unable to Use H2-CO2 for Methanogenesis," Applied and Environmental Microbiology, 38(5), 996(1979).

16. Konig, H. and Stetter, K., "Isolation and Characterization of Methanolobus tindarius, sp. Nov., a Coccoid Methanogen Growing Only on Methanol and Methylamines," Zentralbl. Bakteriol. Mikrobiol. Hyg. 1 Abt. Orig C3, 478(1982).

17. Zhilina, T., "New Obligate Halophilic Methane-Producing Bacterium," Mikrobiologiya, English translation, 52, 290(1983).

18. Zhilina, T., "Methanogenic Bacteria from Hypersaline Environments," Systematic Applied Microbiology, 7(2-3), 216(1986).

19. Schmid, L. and Lipper, R., "Swine Wastes, Characterization and Anaerobic Digestion," Pro. Of Animal Waste Management Conference, Cornell University, Ithaca, NY, 79, (1969).

20. Evans, P. and Dunsmore B., "Reservoir Simulation of Sulfate-Reducing Bacteria Activity in the Deep Sub-Surface," Paper no 06664, Corrosion 2006.

21. Rengpipat, S., Langworthy, T., and Zeikus, J., "Halobacteriodes acetoethylicus sp. Nov., a New Obligately Anaerobic Halophile Isolated from Deep Surface Hypersale Environment," Systematic and Applied Microbiology, 11, 28(1988).

22. Bhupathiraju, V., Oren, A., Sharma, P., Tanner, R., Woese, C., McInerney, M., "Haloanaerobium salsugo sp.nov., A Moderately Halophillic, Anaerobic Bacterium from a Subterranean Brine," International Journal of Systematic Bacteriology, 44, 565(1994).

23. van der Krann, G., Bruining, J., Lomans, B., van Loosdrecht, M., and Muyzer, G., "Microbial Diversity of an Oil/Water Processing Site and Its Associated Oil Field: The Possible Role of Microorganisms as Information Carriers from Oil-Associated Environments," FEMS Microbiological Ecology, 71, 428(2010).

24. Cord-Ruwisch, R., Kleinitz, W., Widdel, F., "Sulfate-Reducing Bacteria and Their Activities in Oil Production," SPE 13554, Journal of Petroleum Technology, 97, January 1987.

25. Cochrane, W., Jones, P., Sanders, P., Holt, D., and Mosley, M., "Studies on the Thermophilic Sulfate-Reducing Bacteria from a Souring North Sea Oil Field," SPE 18368, SPE European Petroleum Conference, London, UK, Oct 16-19, 1988.

26. Goorissen, H., Boschker, H., Stams, A., and Hansen, T., "Isolation of Thermophilic Desulfotomaculum Strains with Methanol and Sulfite from Solfataric Mud Pools, and Characterization of Desulfotomaculum solfataricum sp. Nov.," International Journal of Systematic and Evolutionary Microbiology, 53, 1223(2003).

27. Postgate, J. R., The Sulfate-Reducing Bacteria, Cambridge University Press, Second Edition, 1984.

28. Rosnes, J., Graue, A., and Lien, T., "Activity of Sulfate Reducing Bacteria Under Simulated Reservoir Conditions," SPE Prod. Eng., 6, 217(1991).

29. Franz, R., "Microbial Dynamics in Souring Oil Reservoirs," PhD Dissertation, Montana State University, Bozeman, MT, 1994.

30. Widdel, F. and Pfennig, N., "Studies on Dissimilatory Sulfate Reducing Bacteria that Decompose Fatty Acids. II. Incomplete oxidation of Propionate by Desulfobulbus propionicus gdn. Nov., sp. Nov.," Achives of Microbiology, 131, 360(1982).

31. Widdel, F., "Microbiology and Ecology of Sulphate and Sulphur Reducing Bacteria," in Biology of Anaerobic Microorganisms, Zehndr AJB (Editor), Wiley Interscience, USA, 469(1988).

32. Aeckersberg, F., Bak, F., and Widdel, F., "Anaerobic Oxidation of Saturated Hydrocarbons to CO_2 by a New Type of Sulfate Reducing Bacterium," Archives of Microbiology, 156, 5(1991);

33. Klemps, R., Cypionka, H., Widdel, F., and Pfennig, N., "Growth with Hydrogen and Further Physiological Characteristics of Desulfotomaculum species," Archives of Microbiology, 143, 203(1985).

34. Braun, M. and Stolp, H., "Degradation of Methanol by a Sulfate Reducing Bacterium," Archives of Microbiology, 142(1), 77(1985).

35. Brock, T., Brock, K., Belly, R., and Weiss, R., "Sulfolobus: A New Genus of Sulfur-Oxidizing Bacteria Living at Low pH and High Temperature," Archiv fur Mikrobiologie, 84(1), 54(1972).

36. DeLong, E. and Pace, N., "Environmental Diversity of Bacteria and Archaea," Systematic Biology, 50(4), 470(2001).

37. Zobell, C., "Contributions of Bacteria to the Origin of Oil," Proceedings Third World Petroleum Congress, Section 1, 414, 1950.

38. Treibs, A., "Organic Mineral Substances. II. Occurrence of Chlorophyll Derivatives in an Oil Shale of the Upper Triassic," Ann der Chem., Justus Liebigs, 509, 103(1934).

39. Treibs, A., "Organic Mineral Substances. III. Chlorophyll and Hemin Derivatives in Bituminous Rocks, Petroleums, Mineral Waxes, and Asphalts. Origin of Petroleum," Ann der Chem., Justus Liebigs, 510, 42(1934).

40. Treibs, A., "Organic Mineral Substances. IV. Chlorohyll and Hemin Derivatives in Bituminous Rocks, Petroleums, Coals, and Phosphorites," Ann der Chem., Justus Liebigs, 517, 172(1935).

41. Treibs, A., "Chlorohyll and Hemin Derivatives in Organic Mineral Substances," Angew. Chem. 49, 682(1936).

42. Head, I, Jones, D., and Larter, S., "Biological Activity in the Deep Subsurface and the Origin of Heavy Oil," Nature, 426, 344(2003).

43. Larter, S., Wilhelins, A., Head, I., Koopmans, M., Aplin, A., Di Primo, R., Zwach, C., Erdmann, M., and Telnaes, N., "The Controls on the Composition of Biodegraded Oils in the Deep Subsurface, Part 1: Biodegradation Rates in Petroleum Reservoirs," Organic Geochemistry, 34, 601(2003).

44. Winter, J., "Anaerobic Waste Stabilization," Biotechnology Advances, 2, 75 (1984).

45. Eastman, J., "Solubilisation of Organic Carbon During The Acid Phase of Anaerobic Digestion," Ph.D Thesis, University of Washington, (1977).

46. Stams, A., "Metabolic Interactions between Anaerobic Bacteria in Methanogenic Environments," Antonie van Leeuwenhoek, 66(1-3), 271(1994).

47. Burke, D., "Options for Recovering Beneficial Products from Dairy Manure," Environmental Energy Company, June 2001.

48. Onstott, T., Phelps, T., Colwell, F., Ringelberg, D., White, D., Boonje, D., McKinley, J., Stevens, T., Balkwill, D., Griffin, W., and Kieft, T., "Observations Pertaining to the Origin and Ecology of Microorganisms Recovered from the Deep Subsurface of Taylorsville Basin, Virginia," Geomicrobiology Journal, 15(4), 353(1998)

49. McGovern-Traa, C., Jyh-Yih, L., Hamilton, W., Spark, I., and Patey, I., "The Presence of Sulphate-Reducing Bacteria in Live Drilling Muds, Core Materials and Reservoir Formation Brine from New Oilfields," Petroleum Geology of the Irish Sea and Adjacent Areas, Meadows, Trueblood, Hardman and Cowen (eds.), Geological Society Special Publication No. 124, 229(1997).

50. Anon, "Discovery of Earth's Inner, Innermost Core Confirmed," Science Daily, March 10, 2008.

51. Forbes, R., Studies in Ancient Technology, E. J. Brill Publishers, Leiden, Netherlands, 1955.

52. Hulecki, J., "Chemical and Microbiological Transformations Induced in Oilfield Produced Water by Amendment with Nitrate as a Souring Control Measure: Investigation of Sulfur Cycling, Effects of Co-Amendment with Acetate and Phosphate, and Evaluation of Storage Protocols," MS Thesis, University of Alberta, Spring, 2009.

53. Wood, J., Spark, I., "Observed Variations in Hydrocarbon Reservoir Bacterial Populations with Temperature: A First Step in Modeling the Bacterial Populations of Hydrocarbon Reservoirs," SPE 54766, 1999 SPE European Formation Damage Conference, The Hague, The Netherlands, May 31 – June 1, 1999.

54. Selley, R., Elements of Petroleum Geology, W. H. Freeman and Co, NY 1985.

55. Chaston, S., Grassia, G., and Sheehy, A., "The Next Iteration Factor (NIF) Method for Calculating Equilibrium Conditions in Subsurface Waters, for the Isolation of Microorganisms from Petroleum Reservoirs," Aust. J Chem., 49, 943(1996).

56. Ciaramella, M., Napoli, A., Rossi, M., "Another Extreme Genome: How to Live at pH 0," Trends in Microbiology, 13(2), 49(2005).

57. Kaito, C., Lina, L., Nogi, Y., Nakamura, Y., Tamoka, J., and Horikoshi, K., "Extremely Borophilic Bacteria Isolated from the Mariana Trench, Challenger Deep, at a Depth of 11,000 Meters," Applied and Environmental Microbiology, 64, 1510(1998).

58. Magot, M., Ollivier, B. and Patel, B., "Microbiology of Petroleum Reservoirs," Antonie van Leeuwenhoek 77, 103(2000).

59. Ravot, G., Magot, M., Ollivier, B., Patel, B., Ageron, E., Grimont, P., Thomas, A., and Garcia, J., "Haloanaereobium congolense sp. Nov., an Anaerobic, Moderately Halophilic, Thiosulfate- ad Sulfur-Reducing Bacterium from an African Oilfield," FEMS Microbiol Lett., 147, 81(1997).

60. Stetter, K., Lauerer, G., Thomm, M., and Neuner, A,., "Isolation of Extremely Thermophilic Sulfate Reducers: Evidence for a Novel Branch of Archaebacteria," Science, 236(4803), 822(1987).

61. Powell, D., Einer, R., Check, J., Fincher, D., Gonzales, R., "Practical Experience in Re-Establishing Control over SRBs and Bacteria Generated H2S in Subsea Pipelines Following Hurricane Ike," Paper No 10064, CORROSION 2010.

62. Rozanova and Pirovarova 1988] Rozanova, E. and Pirovarova, T., "Reclassification of D. Thermophilus (Rozanoa and Khudyakova 1974)," Microbiology, 57, 85(1988).

63. Zeikus, J., Dawson, M., Thompson, T., Ingvorsen, K., and Hatchikian, E., "Microbial Ecology of Volcanic Sulphidogenesis: Isolation and Characterization of Thermodesulfobacterium commune gen. nov. and sp. Nov.," Microbiology, 129, 1159 (1983).

64. Takai K, Nakamura K, Toki T, Tsunogai U, Miyazaki M, Miyazaki J, Hirayama H, Nakagawa S, Nunoura T, Horikoshi K., "Cell Proliferation at 122°C and Isotopically Heavy CH4 Production by a Hyperthermophilic Methanogen under High-Pressure Cultivation," Proceedings of the National Academy of Science USA, 105(31), 10949(2008).

65. Manning, D. and Hutcheon, I., "Distribution and Mineralogical Controls on Ammonium in Deep Groundwaters," Appied Geochmistry, 19, 1495(2004).

66. Sunde, E., Thorstenson, T., and Torsvik, T., "Growth of Bacteria on Water Injection Additives," SPE 20690, SPE International, Richardson, TX, 1990.

67. Amajor, L. and Gbadebo, A., "Oilfield Brines of Meteoric and Connate Origin in the Eastern Niger Delta," Journal of Petroleum Geology, 15, 481(1992).

68. Barth, T., "Organic Acids and Inorganic Ions in Waters from Petroleum Reservoirs, Norwegian Continental Shelf: A Multivariate Statistical Analysis and Comparison with American Reservoir Formation Waters," Applied Geochemistry, 6, 1(1991).

69. Kharaka, Y., Callender, E., and Wallace, R., "Geochemistry of Geopressured Geothermal Waters from Frio Clay in Gulf Coast Region of Texas," Geology, 5, 241(1977).

70. Klapper I, Rupp C, Cargo R, Purevdorj B, and Stoodley P, "A Viscoelastic Fluid Description of Bacterial Biofilm Material Properties," Biotechnology and Bioengineering, 80(3):289 (2002).

71. Stoodley, P., Cargo, R., Rupp, C., Wilson, S. and Klapper, I., "Biofilm Material Properties as Related to Shear-Induced Deformation and Detachment Phenomena," Journal of Industrial Microbiology and Biotechnology, 29(6):361(2002).

72. Bottero, S., Picioreanu, C., Delft, T.,Enzien, M., van Loosdrecht, M., Bruining, H., Heimovaara, T., "Formation Damage and Impact on Gas Flow Caused by Biofilms Growing within Proppant Packing Used in Hydraulic Fracturing," SPE 128066, 2010 SPE International Symposium and Exhibition on Formation Damage Control, Lafayette, LA, Feb 10-12, 2010.

73. Cochran, W., Physiological Basis for Biofilm Resistance to Antimicrobial Agents, Doctoral Dissertation, Montana State University, Bozeman, MT, (November 1998).

74. Pintelon, T., von derSchulenburg, D., and Johns, M., "Towards Optimum Permeability Reduction in Porous Media using Biofilm Growth Simulations, Biotechnology and Bioengineering, 3(4), 767(2009).

75. Stoodley, P., Sauer, K. Davies, D. and Costerton, J., "Biofilms as Complex Differentiated Communities," Annual Reviews of Microbiology, 56(1), 2002

76. Al-Sulaiman, S., Murray, G., Islam, M., Al-Mithin, A, and Biedermann, A., "Advantages and Limitations of Using Field Test Kits for Determining Bacterial Proliferation in Oil Field Waters," Paper 08655, NACE 2008.

77. LeChevallier, M., Cawthon, C., and Lee, R., "Inactivation of Biofilm Bacteria," Applied and Environmental Microbiology, 54(10), 2492(October 1988).

78. Kajdasz, R., Young-Bandala, L., and Einstman, R., Paper No. IWC-84-60, International Water Conference, Pittsburgh, Penn., "Biocidal Efficacy with Respect to Sessile and Planktonic Organisms," October 22-24, 1984.

79. Eyc Eycott, R., Water and Wastewater Treatment, "Controlling Biofilms in Cooling Water Systems, Industrial and Process Water Treatment," 40(1990).

80. Kajdasz, R., Industrial Water Treatment, "Measuring the Biocidal Efficacy against Sessile Sulfate-Reducing Bacteria," 39(March/April 1996).

81. Al-Human, A., Rizk, T., Sunner, J., and Beech, I., "Effects of Nitrate on Bacterial Communities in an Oil Field Environment," Paper No 10249, CORROSION 2010.

82. Costerton, J. and Geesey, G., "The Microbial Ecology of Surface Colonization and of Consequent Corrosion," Biologically Induced Corrosion, NACE, Houston, 223(1986).

83. Costello, J., "The Corrosion of Metals by Micro-organisms, A Literature Survey," International Biodeterioration Bulletin, 5(3), 101(1969).

84. Tatnall, R. E., "Case Histories: Bacteria Induced Corrosion," Paper No 130, Corrosion 81, Toronto, Ontario, Canada, April, 1981.

85. Boivin, J., Laishley, E., Bryant, R., and Costerton, J., "The Influence of Enzyme Systems on MIC," Paper No. 128, Corrosion 90, Las Vegas, NV, April 23-27, 1990.

86. Characklis, W., Little, B., Stoodley, P., and McCaughey, M., "Microbial Fouling and Corrosion in Nuclear Power Plant Service Water Systems," Paper No. 281, Corrosion 91, Cincinnati, OH, March 11-15, 1991.

87. Roe Roe, F., Lewandowski, Z., and Runk, T., "Simulating Microbiologically Influenced Corrosion by Depositing Extracellular Biopolymers on Mild Steel Surfaces," Corrosion, 52(10), 744(October 1996).

88. Updegraff, D., "Microbiological Corrosion of Iron and Steel," Corrosion, 11, 44(October 1955).

89. Characklis, W. and Marshall, K., Biofilms, Wiley Interscience Publications, John Wiley & Sons, Inc., New York, 1990.

90. McKoy, J. W., Microbiology of Cooling Water, Chemical Publishing Company, New York, New York, 1980.

91. Hamilton, W. A., and Maxwell, S., "Biological and Corrosion Activities of Sulfate Reducing Bacteria Within Natural Biofilms," in Biologically Induced Corrosion, NACE, Houston, 131(1986).

92. Sanders, P. and Hamilton, W., "Biological and Corrosion Activities of Sulfate Reducing Bacteria in Industrial Process Plant," Biologically Induced Corrosion, NACE, Houston, 47(1986).

93. Pacheco, A., Dishinger, T., and Tomlin, J., "Experiences in Controlling Microbiologically-Induced Corrosion," Paper No. 376, Corrosion 87, San Francisco, CA, March 9-13, 1987.

94. Campbell, S., Duggleby, A., Johnson, A., "Conventional Application of Biocides May Lead to Bacterial Cell Injury Rather than Bacterial Kill Within a Biofilm," Paper 11234, CORROSION, 2011.

95. Eagar, R., Leder, J., and Theis, A., "Glutaraldehyde: Factors Important for Microbiocidal Efficacy," presented at the Third Conference on Progress in Chemical Disinfection, Binghamton, N. Y., April 3-5, 1986.

96. Videla, H., Satu, A., Saravia, S., Guiamet, P., Mele, M., Gaylarde, C., Beech, I, "Impact of Glutaraldehyde on Biofouling and MIC of Different Steels. A Laboratory Assessment," Paper No 105, CORROSION 91.

97. Freid, M. G., Leder, J., and Theis, A. B., "Control of Biocide Applications," Paper No. 202, Corrosion 91, Cincinnati, OH, March 11-15, 1991.

98. Farquhar, G. B., Lacey, C. A., and Deans, S. D., "Laboratory Screening of Commercial Biocides for Use in Oilfield Production," Materials Performance, 49, October, 1993.

99. Grab, L. A., Diemer, J. A., and Freid, M. G., "The Effect of Process Leak Contaminants on Biocidal Efficacy," Cooling Tower Institute Annual Meeting, Paper No. TP94-12, Houston, Texas, February 13-16, 1994.

100. Reinsel, M., Sears, J., Stewart, P., and McInerney, M., "Control of Microbial Souring by Nitrate, Nitrite or Glutaraldehyde Injection in a Sandstone Column," Journal of Industrial Microbiology, 17, 128(1996).

101. Rimassa, S., Howard, P., and Arnold, M., "Are You Buying Too Much Friction Reducer Because of Your Biocide?" SPE 119569, 2009 SPE Hydraulic Fracturing Technology Conference, The Woodlands, TX Jan 19-21, 2009.

102. Merianos, J., "Quaternary Ammonium Antimicrobial Compounds," Chapter 13, Disinfection, Sterilization and Preservation, S. Block, ed., 1991.

103. Sweeny, P. and Murray, D., "Anionic Compatible Quat?" Cooling Technology Institute, Paper No. TP 07-17, 2007 Cooling Technology Institute Annual Conference, Corpus Christi, TX, Feb 4-7, 2007.

104. Grab, L. A., and Theis, A. B., "Comparative Biocidal Efficacy vs. Sulfate-Reducing Bacteria," Materials Performance, 59, June, 1993.

105. McGinley, H., Enzien, M., Jenneman, G., Harris, J., "Studies on the Chemical Stability of Glutaraldehyde in Produced Water," SPE 141449, SPE International Symposium on Oilfield Chemistry, The Woodlands, TX April 11-13, 2011.

106. Larsen, J., Sanders, P., Talbot, R., "Experience with the Use of THPS for the Control of Downhole Hydrogen Sulfide," Paper No 00123, CORROSION 2000.

107. Gilbert, P., Grech, J., Talbot, R., Veale, M., "Tetrakishydroxymethylphosphonium Sulfate (THPS) for Dissolving Iron Sulfides Downhole and Topside – A Study of the Chemistry Influencing Dissolution," Paper 02030, Corrosion 2002.

108. Jones, C., Collins, G., Downward, B., Hernandez, K., "THPS: A Holistic Approach to Treating Sour Systems," Paper No 08659, NACE 2008.

109. Jones, C., Downward, B., Hernandez, K., Curtis, T., and Smith, F., "Extending Performance Boundaries with Third Generation THPS Formulations," Paper No 10257, NACE 2010.

110. Jones, C., Downward, B., Edmunds, S., Hernandez, K., Curtis, T., and Smith, F., "A Novel Approach to Using THPS for Controlling Reservoir Souring," Paper No 11219, NACE 2011.

111. Enzien, M., Yin, B., "New Biocide Formulations for Oil and Gas Injection Waters with Improved Environmental Footprint," OTC 21794, Offshore Technology Conference, Houston, TX, May 2-5, 2011.

112. Hess, L., Kurtz, A., Stanton, D., "Acrolein and Derivatives," Kirk Othmer Encyclopedia of Chemical Technology, 2nd Edition, 1963.

113. Brady, J., Erben A., Kissel, C., Pau, J., and Caserio, F., "Determination of Acrolein in Aqueous Systems," Oil Field Subsurface Injection of Water, ASTM STP 641, Wright, Cross, Ostroff, and Stanford, Eds, 89(1977).

114. Kissel, C., Brady, J., Guerra, A., Meshishnek, M. Rockie, B., Caserio, F., "Monitoring Acrolein in Naturally Occurring Systems," Water for Subsurface Injection, ASTM STP 735, Johnson, Stanford, Wright, and Ostroff, Eds., 102(1981).

115. Reed, C., Foshee, J., Penkala, J., Roberson, M., "Acrolein Application to Mitigate Biogenic Sulfides and Remediate Injection Well Damage in a Gas Plant Water Disposal System," SPE 93602, SPE International Symposium on Oilfield Chemistry, Houston, TX Feb 2-4, 2005.

116. Penkal Penkala, J., Reed, C., and Foshee, J., "Acrolein Application to Mitigate Biogenic Sulfides and Remediate Injection-Well Damage in a Gas-Plant Water Disposal System," SPE 98067, SPE International Symposium and Exhibition on Formation Damage Control, Lafayette, LA, Feb 15-17, 2006.

117. Salma, T., "Cost Effective Removal of Iron Sulfide and Hydrogen Sulfide from Water Using Acrolein," SPE 59708, 2000 SPE Permian Basin Oil and Gas Recovery Conference, Midland, TX, March 21-23, 2000.

118. Howell, J., and Ward, M., "The Use of Acrolein as a Hydrogen Sulfide Scavenger in Multiphase Production," SPE 21712, Production Operations Symposium, Oklahoma City, OK, April 7-9, 1991.

119. Smith, C., "Acrolein," John Wiley and Sons, New York, 1962.

120. Kissel, C., "Acrolein as a Biocide and Sulfide Inhibitor," NACE, Bossier City, LA, 1984.

121. Kissel, C., Brady, J., Gottry, H., Meshishnek, M., Preus, M., "Factors Contributing to the Ability of Acrolein to Scavenge Corrosive Hydrogen Sulfide," SPE Journal, 647(October 1985).

122. Oates, S., Gregg, M., Mulak, K., Walsh, G., Dickinson, A., "A Novel Approach to Managing a Seawater Injection Biocide Program Reduces Risk, Improves Biological Control, and Reduces Capital and Opex Costs on an Offshore Platform," Paper No 06665, CORROSION 2006.

123. Arensdorf, J., Miner, K., Ertmoed, R., Clay, B., Stadnicki, P., Voordouw, G., "Mitigation of Reservoir Souring by Nitrate in a Produced Water Re-Injection System in Alberta," SPE 121731, 2009 SPE International Symposium on Oilfield Chemistry, The Woodlands, TX April 20-22, 2009.

124. Tischler, A., Woodworth, T., Burton, S., Richards, R., "Controlling Bacteria in Recycled Production Water for Completion and Workover Operations," SPE 123450, 2009 SPE Rocky Mountain Petroleum Conference, Denver, CO, April 14-16, 2009.

125. Simpson, G., Practical Chlorine Dioxide: Volume I & II, Lightning Source Publishers, 2005.

126. Simpson, G., Laxton, G., Miller, R., and Clements, W., "A Focus on Chlorine Dioxide: The 'Ideal' Biocide," Paper No. 472, Corrosion 93, New Orleans, LA, March 8-12, 1993.

127. Simpson, G. and Phillips, E., "The Environmental Impact of Chlorine Dioxide," Chemical Oxidation: Technologies for the Nineties, Eighth International Symposium, Nashville, TN, April 1-3, 1998.

128. Pichinoty, F., Puig, J., Chippaux, M., Bigliardi-Rouvier, J., and Gender, J., "Recherches sur des mutants bactériens ayant perdu les activités catalytiques liées à la nitrate-réductase A. II. Comportement envers le chlorate et le chlorite," Ann. Inst. Pasteur, 116(4, 406(1969).

129. Carney, J. F. and Sieckhaus, J., "Efficacy of Sodium Chlorite as a Paper Mill Slimicide," unpublished report, 1978.

130. Sacco, F., "The Use of Chlorine Dioxide in a Late Life Waterflood," Proceedings Symposium on Advances in Oilfield Chemicals and Chemistry, Presented before the Division of Petroleum Chemistry, Inc., American Chemical Society, St. Louis Meeting, 605(April 8-13, 1984).

131. Knickrehm, M., Caballero, E., Romualdo, P., and Sandidge, J., "Use of Chlorine Dioxide in a Secondary Recovery Process to Inhibit Bacterial Fouling and Corrosion," Corrosion 87, Paper No. 383, March 9-13, 1987, San Francisco, CA 1987.

132. Rodvelt, G., Yeager, V., Hyatt, M., "Case History: Challenges Using Ultraviolet to Control Bacteria in Marcellus Completions," SPE 149445, SPE Eastern Regional Meeting, Columbus, OH, Aug 17-19, 2011.

CHAPTER SEVEN

Sampling and Analysis

Chapter Seven Table of Contents

7.0. Introduction:

The importance of using good sampling and analysis practices cannot be overstated. In most cases, there is a large gap between good practice and what is actually carried out in the field. Poor sampling and analysis can lead to inadequacies in design which can be costly to overcome. Some of these mistakes have cost the operators upwards of 50 million dollars, and months of downtime. Such high dollar mistakes are clearly the exception in the industry, but they do happen. More common are mistakes that create bottlenecks in an oil/water/gas separation facility which causes curtailed production and associated costs. Most sampling and analysis mistakes are simple to avoid. The perspective of this chapter is consistent with the rest of the book. The scientific details are discussed together with practical applications. In this case, practical applications include a discussion of the regulatory methods.

Regulatory Sampling and Analysis: Although regulatory requirements vary significantly from one region of the world to another, for most oil and gas operations, there is a scientific and engineering basis to most regulations. Those aspects are covered here. Important regulatory characteristics are the concentration of the dispersed oil, oily solids, and, in some cases, dissolved organics. Methods for capturing those components accurately is discussed. Some regulations specify sampling and analysis procedures. More likely though, the regulatory procedures do not adequately specify important details of sampling and analysis. Even some of the well developed regulations have multiple options for analytical procedures. This can result in variation from one lab to another. The details of the fluid mechanics, phase equilibrium thermodynamics, and analytical methods are discussed in relation to these regulations in order to make clear exactly what is being measured.

Diagnostic Sampling and Analysis: In addition to the regulatory requirements, there are many other properties and characteristics of produced water that are not measured for regulatory purposes but which have important effects on the performance of water treating equipment. As an example, for troubleshooting and diagnosing water treating problems, it is necessary to sample and determine an accurate drop/particle diameter distribution, as well as other properties of the produced water.

On-Site Sampling: Two general sets of methods are discussed in this chapter. The first set is based on grab sampling and on-site or offsite laboratory analysis. The second set is based on on-line monitoring. On-site monitoring is challenging, to say the least. Nevertheless, it is important in any monitoring or troubleshooting activity to have a combination of on-site and offsite tests performed. This will help to verify (or refute) the quality of the data. It is discussed extensively in this chapter.

On-Line Monitoring: On-line monitoring is discussed in this Chapter. On-line monitoring of oil in water should be thought of in the same way that on-line pressure, temperature, fluid levels in vessels, flow rate, or other on-line measurements are thought of. The whole point of a water treatment system is to remove oil from water. Given the dynamics of oil and gas production from wells, on-line monitoring of oil concentration can significantly help in process operation and optimization. When on-line monitoring of oil concentration is installed and used properly, it become one more source of data to add to the other process monitoring equipment such as pressure, temperature, level, flow rate.

Third-Party Labs and Standard Methods: It is common practice in the oil and gas industry to rely on third party labs to carry out most produced water analyses. For this reason, a number of references are made to US-EPA and ASTM International standard methods (originally ASTM meant American Society for Testing and Materials; in 2001 the organization changed its name to ASTM International where ASTM is not an abbreviation for anything). These standards are available through most technical library services. The advantage of identifying these methods is that they provide a very convenient way to establish an agreed protocol with third party laboratories. The important aspects of these methods are described here.

Sampling and Analysis for Bacteria: Chapter 7 does not cover the subject of sampling and analysis for bacteria. That subject is covered in Chapter 6.

7.1. Design of Sample Points:

Typical maximum oil concentration limits for produced water discharge into the sea are in the range of 30 to 40 mg/L. The concentration of other contaminants such as heavy metals is typically 10 to a thousand times lower, on the order of parts per billion. Due to these low concentrations, careful sampling practices are required to ensure samples of produced water are representative of the actual process fluids.

7.1.1. Location of Sample Points:

Effective monitoring and trouble-shooting of produced water requires good quality sampling points both upstream and downstream of all stages of separation equipment. The cost of a one inch nozzle for sampling is relatively small compared to the problems that can occur without adequate knowledge of process performance.

In selecting the location of sample points, the following guidelines should be followed, i.e. the sample point should:

- be located in a vertical section of piping where the fluid flow is upwards. The fluids in vertical piping are not subject to phase separation and stratification due to gravity.

- be at least 5 pipe diameters downstream and 3 pipe diameters upstream of any flow disturbance such as a bend, pump, manifold etc. (ASTM D-41 77-82). Bends tend to make the fluid swirl. Such a swirl motion applies centripetal and centrifugal forces which segregate the suspended material. Forces exerted on the fluid by flow disturbances (particularly bends) will result in a separation of phases of different densities.

- be located both upstream and downstream of deoiling equipment, so that separation efficiency of the equipment can be determined.

- be located both upstream and downstream of control valves and pumps so that the change in drop and particle diameter distribution caused by such equipment can be measured. Both valves and pumps generate turbulence that requires at least 10 pipe diameters to dissipate. The sample location should be at least 20 pipe diameters downstream of such equipment. The sample location upstream of this equipment can be as close as a few pipe diameters.

- be located in a region of steady, uniform turbulent flow. The Reynolds Number (Section 3.2.2) for pipe flow can be calculated and it should be upwards of 10,000. This helps to ensure that suspended material is uniformly distributed in the flow, and that the sample will be representative.

- provide an effective and safe means for the collection and disposal of the fluid flow required to flush the sample point and sample line.

For sampling points that are used for regulatory compliance monitoring purposes, the following guidance, which is extracted from a UK government (Department of Energy and Climate Change - DECC) guidance, may be considered:

- *The sampling point should be immediately after the last item of treatment equipment in, or downstream of, a turbulent region, and in any case before any subsequent dilution. The sample point should also be downstream of any installed produced water volume meter.*

- *The location of the produced water sample point must be agreed with DECC and must not be changed without written permission. Failure to request permission to change the location of a sample point is a breach of the permit conditions and will require that an OPPC non-compliance be submitted.*

7.1.2. Sample Probe:

Most sample points for water involve a simple nozzle mounted on the side of a pipe (often referred to as "side wall taping"), without any protrusion into the fluid flow. This is confirmed by a survey carried out by Shell UK looking into how produced water had been sampled. Details of the types of sample points found from the survey are given in Table 7.1.

Side wall taping, shown as a schematic diagram in Figure 7.1 (a), is simple and inexpensive. However it has a high potential for being unable to extract a representative sample from the flow line since the fluid velocity and turbulence at the pipe wall are at a minimum. When a sample is taken from such a location, some of the sample is composed of fluid from the boundary layer and some of the sample is composed of fluid from the region outside the boundary layer but still close to the wall. In general, fluid vorticity tends to grade (or segregate according to particle size) suspended material. Also, the intense shearing that occurs just outside of the boundary layer can reduce the oil drop diameter in that region. All of these effects are referred to as "wall effects". Suffice to say that wall effects can and often do bias the sample. In addition, sampling from the wall requires the flow profile to divert through 90 degrees to enter the sample piping. Due to these changes in the flow profile, the concentration and size distribution of any dispersed solid, liquid or gas bubbles in the sample is unlikely to be representative of the bulk fluid.

Table 7.1 Summary of sample points found in Shell installations in the North Sea.

Sampling	Type	Number	Comments and other issues
Manual sample	Probe, vertical pipe, flowing up (as preferred by DECC)	0	• Samples taken from part filled pipes
	Probe, horizontal pipe	1	• Sampling from a multi-function pipe
	Probe, vertical (down flow)	2	
	Side wall, vertical pipe	1	
	Top wall tapping	3	
	Side wall tapping	14	
	Bottom water tapping	8	
	Instrument tapping	3	
Online sampling	Not clear	9	• Insufficient flow

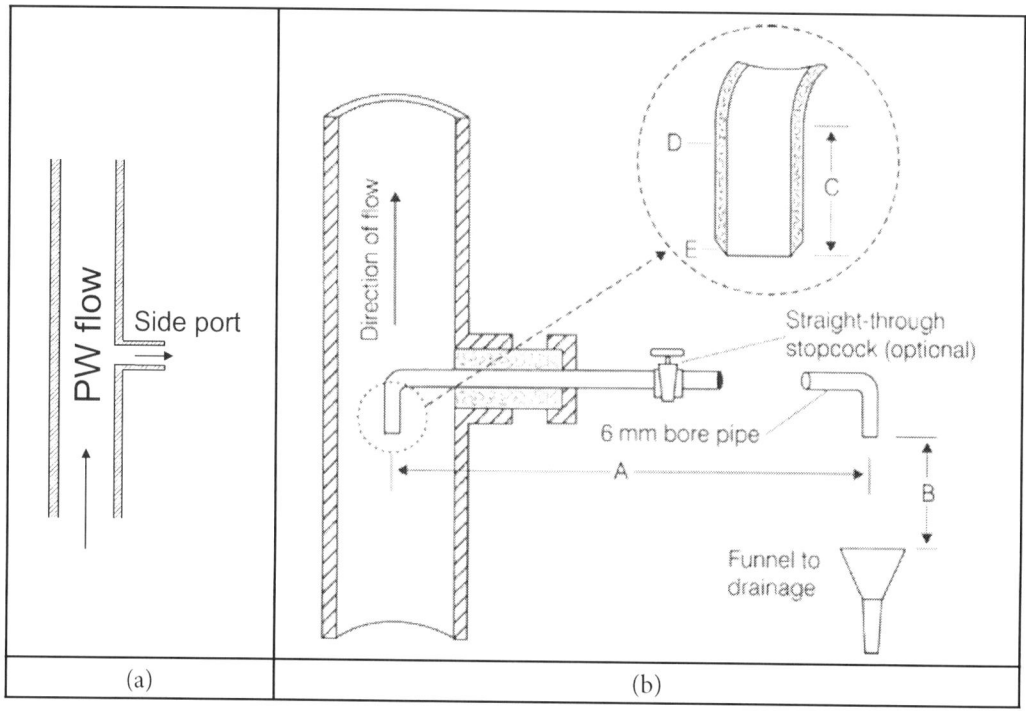

(a)   (b)

Figure 7.1 Examples of oil in water sampling probes

A better way of taking a representative sample is by using a sample probe as an example shown in Figure 7.1 (b). However sampling probes come in a variety of designs including Pitot, "NEL Pitot", circular port and 45° opening types as shown in Figure 2.

The first design has the end of the sampling probe bevelled at 45 degrees (45° opening probe). This is an inexpensive and relatively simple design. The second design (circular port probe) uses a curved end to smooth the flow path into the sample probe. Both designs have the advantage of being straight which allows the probe to be inserted in a slim fitting similar to a thermowell and allows the probe to be easily withdrawn as required for maintenance.

The Pitot types have a long radius bend designed to minimize any disturbance to the flow pattern at the tip of the probe. A larger fitting is required to allow the curved probe to be inserted. The probe is offset to ensure the tip of the probe is as far away as possible from any flow disturbances around the fitting.

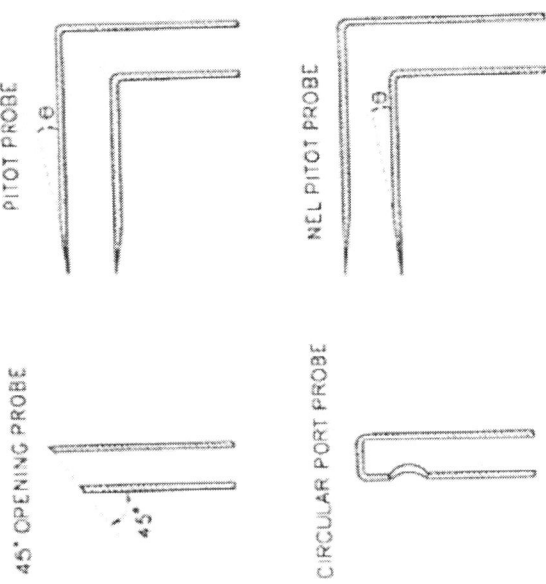

Figure 7.2 Different Designs of the Sampler Probe

The following guidelines should be followed to ensure the correct orientation of a sample point. The objective of sample probe selection and installation is to ensure a representative sampling. To an extent, the probe can overcome some inhomogeneity of the flow itself. However, the use of a sample probe usually does not overcome a poor selection of sample location, or an inadequate flow rate.

Sample points should:

- Always have a sample probe installed in a manner which ensures a representative sample will be taken. The sample probe should preferably extend to the center of the pipe or at least a distance of D/3 away from the pipe wall.

- If sampling from horizontal piping cannot be avoided, samples should not be taken from the top or bottom of pipe. The sample probe should be installed either to allow sampling from the center of the pipe or at a point along the line of the "nine o'clock" or "three o'clock" positions.

- The probe walls are generally designed to be thin or tapered to avoid creating disturbances in the flow stream and to minimize surface areas where liquid droplets can coalesce or collide and lose momentum, such that they are preferentially expelled from or drawn into the probe.

The following guidelines should be applied to all sample probes.

- The wall thickness of the probe tips should be as thin as possible and the tip of the probe should be beveled to a sharp edge. With a blunt edged probe a damming effect can occur upstream of the tip, deflecting droplets away from the probe.

- As previously mentioned, the probe should preferably extend to the center of the pipe or at least a distance of D/3 away from the pipe wall. However the length of the probe may be restricted by the need to withstand the mechanical forces exerted by the flow in the pipe. A procedure for determining the maximum length of a sample quill is given in "Standard

Practice for Automatic Sampling of Petroleum and Petroleum Products", ASTM D 4177-82 (Ref.l.07.1.c).

- The internal diameter of the sample probe must be sized to allow an iso-kinetic fluid velocity through the probe while ensuring the volumetric flow rate is within the capacity of the sampling equipment. Section 7.2.1 (Isokinetic Sampling) below discusses iso-kinetic sampling in more detail.

7.2. Sampling and Sample Handling:

This section discusses the operation of sampling and sample handling. It is assumed at this point that the facility is in operation and sample points are already available.

Bullets below provide some general guidance on routine sampling of low pressure produced water streams.

- Care must be taken to ensure that the inside of the sample container or container closure is not contaminated. Contamination could result from thin hydrocarbon films deposited from the production environment or from hands or gloves.

- The sample point should be opened and allowed to run to flush the sampling line. The volume of water required to flush a line should be calculated based on the length of the line and its internal diameter. As a general rule, five flush volumes should be used as a minimum. Ten volumes is a more robust flush volume. During the flush, the valve should be wide open in order to maximum the flow rate through the sample line. The sampling system should be designed for the safe containment and disposal of this flushing flow.

- Once the sample point has been flushed, the flow rate from the sample line should be adjusted to give the required iso-kinetic sampling flow. Once the flow has stabilized the sample may be taken.

- The sample should be taken without rinsing the sample container. Rinsing will result in higher hydrocarbon content by leaving a film of hydrocarbons on the internal surfaces of the container.

- When filling, the sample bottle should never be allowed to overflow as this can result in an inaccurate sample. When filling a bottle to a specific line on the bottle, the sample should be held at eye level. Also, when filling a sample to the top, if acid or other preservative is to be added, it must be added prior to sampling to prevent overflow.

- Samples should be clearly labeled with date, time, sample location, and a unique sample identification number. These should be recorded in a tally book. Any abnormal conditions such as process upsets, operational problems or difficulties in obtaining samples should be noted. If a sample is to be retaken, the initial sample should be discarded and a new sample taken using a fresh, clean sample bottle.

- Repeat samples should be taken to confirm unusual results. Unusual results which are not confirmed are often simply disregarded as "spurious" results which may conceal an actual operating problem.

- Samples containing dispersed hydrocarbons should not be subdivided as phase separation, coalescence of the hydrocarbons, and sticking of the hydrocarbon to the sample bottle will always occur. When determining the hydrocarbon concentration the entire sample should be analyzed and the size of the sample collected should take this into consideration. It should be noted that some regulatory procedures provide guidelines for the required sample sizes.

- If it is desired to have multiple similar samples, then filling all of the bottles to 1/3 initially, followed by filling all bottles to 2/3, followed by filling all bottle to the specified level can be carried out. In many cases, if done carefully, this results in samples that are essentially the same.

7.2.1. Isokinetic Sampling:

The objective of iso-kinetic sampling is to extract a sample from the process at a rate that does not segregate or alter the concentration or size distribution of dispersed oil droplets and solids particles. Iso-kinetic sampling is simply achieved by ensuring that the linear velocity of fluid entering the sampling quill is equal to the velocity of the fluid in the process line being sampled. In order to do this, the quill and the valve must be sized such that a sample can be withdrawn at a volumetric flow rate that allows matching of the linear velocity in the pipe. This is important when sampling streams containing a dispersed phase that has a density significantly different from the continuous phase.

When not sampling at the correct iso-kinetic velocity, the disturbance in the flow profile can result in a bias in the sampled size distribution due to the different inertial properties of the various sized elements of the dispersed phase. However, iso- kinetic sampling is only effective when a properly designed sampling probe is used. Sampling from a side wall tap type of sample point will introduce large changes in the flow profile which will tend to bias the sample whether or not iso-kinetic conditions are utilized.

To determine the iso-kinetic sampling rate, the velocity of the fluid in the main process line should be calculated. Then, knowing the internal diameter of the sample quill, the appropriate volumetric sampling flow rate can be calculated which will result in the same fluid velocity in the sample quill as in the main process line. The sampling flow rate for a non- hazardous water stream could simply be measured by timing the filling of a known volume (bucket and stop-watch technique).

For sampling of oil in produced water for concentration monitoring, while iso-kinetic sampling is a good practice and preferred, it may not be as critical. This is because the density difference between oil and water is not as large compared to that between solid and liquid, solid and gas, or liquid / solids in gas. Iso-kinetic sampling is clearly more important if the sample taken is for the analysis of solids and / or oil droplet size, or it is for solid in water concentration analysis.

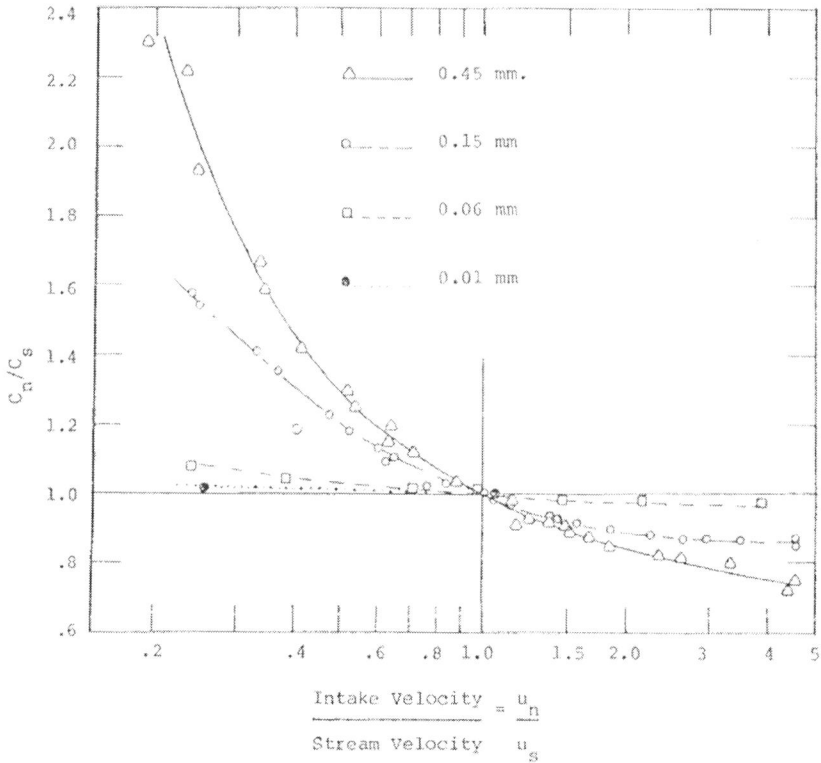

Measured relationship between concentration ratio C_n/C_s and velocity ratio u_n/u_s . (after FIASP, 1941)

Figure 7.3 Impact of non-isokentic sampling on solid in water measurement

While some experimental data, are available, see Figure 7.3 above, in the literature showing the effect of non-isokinetic sampling on the measurement of the concentration of solids in water, there is little information in the literature and / or in the established standards indicating how non-isokinetic sampling would affect water in oil or oil in water sampling. In the ISO 3171 standard, it is suggested that the velocity in the sample extractor should match the velocity in the main pipeline. But in the API 8.2 standard, a ratio of 50% to 200% of the mainline velocity seems to be acceptable.

7.2.2. Sample Containers:

Oil from produced water will typically stick to the side of the sample container. This is why almost all lab reference analysis methods involve a solvent extraction of the entire sample while the sample is in the original container into which it was taken. The solvent is intended to disssolve all of the oil in the sample, including the oil that has attached to or coated the walls of the container. In the case where an alkane is used as a solvent, it is recognized that the asphaltene fraction of the dispersed oil will not dissolve.

As discussed previously, often there is no opportunity to influence the selection, location or the design of a sample point. Selection of sample containers, however, is a different matter. More often than not, the sample container is entirely at the discretion of the field technician. In those cases where a sample container has been selected and/or is stipulated in environmental regulations, usually a correct choice has been made.

Glass: Glass sample containers are suitable for most water analyses including oil in water analysis. Glass sample bottles are preferred for non-routine samples, especially for the determination of heavy metals or when the sample will not be analyzed immediately.

The use of dark glass bottles will reduce photo-degradation of the sample. However, glass sample containers should not be used for samples from which small quantities of hardness, silica, sodium or potassium are to be determined since some produced water, such as condensed water from a gas or gas condensate well has moderately low pH which will leach these elements from the glass.

Plastic: High density linear polyethylene, polypropylene and Teflon containers are suitable for most water analysis duties. However, users should be aware that these containers are slightly permeable to light volatile hydrocarbons and gases such as carbon dioxide. Mercury will also be lost from the sample through the walls of the sample container.

Plastic containers may be acceptable for routine determination of hydrocarbon content of water samples, as errors introduced by sample acquisition considerations will far outweigh potential errors from sample container materials. Glass sample containers should preferably be used for non-routine analyses, heavy metal analyses, or when samples will not be analyzed immediately.

Closures, Tops, Lids: In addition to the sample containers themselves, some consideration should also be given to the sample container lids and stoppers. These should also be made from glass or suitable inert plastics which will not absorb hydrocarbons from the sample. Aluminum foil lined closures should not be used when analyzing for aluminum in samples that are strongly alkaline or acidic. Absorbent materials such as cork should not be used for lids or stoppers. Greases should not be used to assist the sealing of ground glass fittings. Teflon lined caps are preferred.

Cleaning Sample Containers: All sample containers, new or used, should be thoroughly cleaned before use. Sample bottles and lids should be scrupulously clean of all hydrocarbon residues. No traces of cleaning detergents should be left in the bottles as these may influence the coalescing and settling of the dispersed phase. The following typical glassware cleaning procedure has been extracted from ASTM D 3325-90, "Standard practice for preservation of water-borne oil samples" (Ref.l.07.2.a).

- *The cleaning steps consist of an initial wash with a warm aqueous detergent mixture followed by six hot tap water rinses, two rinses with reagent water, a rinse with reagent grade acetone and a final rinse with solvent such as pentane, hexane, cyclohexane, dichloromethane, or chloroform followed by drying in a clean oven at 105°C or hotter for 30 minutes.*

- *If the glassware requires cleaning under field conditions, it should be washed with warm aqueous detergent followed by extensive water rinsing. A solvent rinse with acetone should be made, if possible, followed by a lengthy air drying to remove residual solvent*

More stringent cleaning procedures for both plastic and glass sample containers can be found in ASTM D 3694-92, "Standard Practice for Preparation of Sample Containers and for Preservation of Organic Constituents". Sample containers for determination of trace substances such as heavy metals will require an even more thorough cleaning procedure, for example:

- *Sample containers are cleaned by standing overnight at room temperature containing 1:1 hydrochloric acid. The containers are then rinsed clean with de-ionized water and allowed to stand overnight at room temperature containing 20% v/v nitric acid. The containers are finally thoroughly rinsed clean with de-ionized water, then stored in a particle and fume free environment prior to dispatch. Acids used should be analytical reagent grade.*

7.2.3. Sample Preservation and Storage:

After sampling, a produced water sample will tend to "age" through the action of chemical and biological reactions. This ageing will tend to alter the composition and characteristics of the sample and thus should be prevented or minimized. Samples should preferably be analyzed or utilized immediately after sampling to minimize the effect of ageing. However when this is not possible then appropriate storage and preservation techniques should be used to extend the representative life of the sample.

Brief summaries on sample preservation and storage are given here. Additional information can be obtained from the following references:

- "Standard Practice for the Preservation of Waterborne Oil Samples", ASTM D 3325-90 (Ref.l.07.2.a).

- "Standard Practice for Preparation of Sample Containers and for Preservation of Organic Constituents", ASTM D 3694-92 (Ref.1.07.2.b).

- "Standard Practice for Estimation of Holding Time for Water Samples Containing Organic Constituents", ASTM D 4515-85 (Ref.l.07.2.c).

The life of a sample can be extended by the addition of preservative chemicals to the sample. These chemicals inhibit the chemical and biological changes that may alter the composition of the sample over time. The most common preservation technique used is the acidification of samples which are to be analyzed for hydrocarbon content. Acidification (typically using hydrochloric acid) to a pH < 2 will inhibit the biological degradation of organic materials. For a 500 ml sample, a 2.5 ml 1:1 HCl solution is usually sufficient for lowering the pH of the sample to less than 2.

Different preservation chemicals may be required for the preservation of other components of the sample. In some cases separate samples using different preservation chemicals may be required. To maximize the storage life the sample should be stored in dark glass bottles in a dark area to minimize photo-degradation Storage at a temperature of approximately 5°C will inhibit biological and chemical reactions. Freezing should be avoided, especially for emulsions where the interfacial film may be destroyed by the formation of ice crystals. Air, or more specifically oxygen, should be excluded from the sample to prevent oxidation reactions. Air may be displaced from sample containers using an inert gas such as nitrogen.

7.2.4. Taking Duplicate Samples:

Taking duplicate samples may become necessary, for example, if one intends to compare two different oil-in-water analysis methods using field collected samples. There are a number of ways that one may possibly use to take duplicate samples. One is the use of so called "Y" shaped or "T" shaped sampling devices. Here a "Y" shaped or "T" shaped piece is attached to the sample point allowing produced water stream to be sampled from both the left and right hand side of the device. Providing such a device is well manufactured, the samples taken from the left hand side of the devices should have the same composition to the one taken from the right hand side.

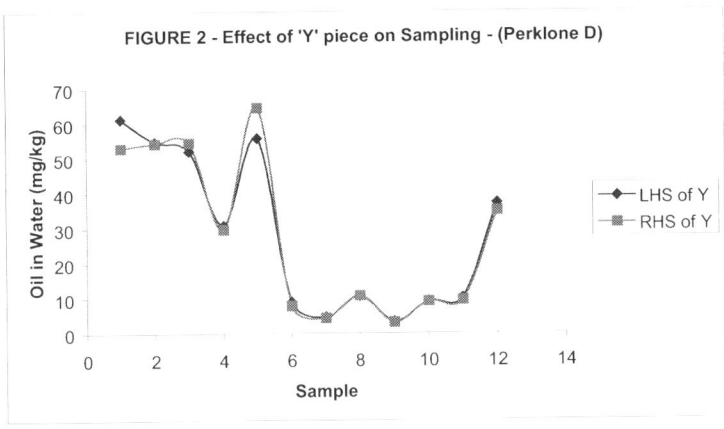

Figure 7.4 Comparison of oil-in-water results of duplicate samples taken using a "Y" shaped piece [1]

A study was carried out by Shell UK in the North Sea [1], in which the viability of obtaining duplicate samples using a "Y" shaped piece was investigated. In the study, 12 sets of samples were collected over a period of twelve hours and the oil-in-water content of the left and right hand samples of the "Y" shaped piece was determined for each set using solvent extraction and Infrared analysis. Figure 7.4 shows the results obtained. The fact that the two sets of results track each other closely demonstrates the viability of the use of such a device for taking duplicate samples, in particular when the sample concentration is less 40 mg/l.

Another way of taking duplicate samples, which may be particularly useful in the absence of a "Y" or "T" shaped sampling device, is to mark sample bottles at 1/3, 2/3, and full sample volume, then fill each sample bottle 1/3 at a time in turn until two duplicate samples are obtained. It should be emphasized that when duplicate samples are taken in this way, it should be done when the produced water treatment process is reasonably steady and is expected to remain so for the duration in which the samples are taken. This particular sampling procedure was used in the North Sea where duplicate samples were collected from a substantial number of offshore installations and then sent onshore for analysis by the OSAPR GC-FID method and also by the old OSPAR Infrared method so that the two methods could be compared. Results showed good agreement between samples.

7.3. Dissolved Mineral Content:

Water is unique in many respects. For the purpose of this discussion, its important properties include a high dielectric permittivity, strong electron donor capacity, strong proton donor capacity, and very strong hydrogen bonding tendency. These properties result in a fluid that has a very high capacity to dissolve polar molecules, and allow ionization of salts, ionization of acid, certain hydroxyl, and nitrogen groups. From a practical standpoint, the dissolved components of produced water are mostly the organic acids, alcohols, BTEX, and salts of dissolved minerals.

The concentration of dissolved minerals, organic acids, and other compounds depends very much on the history of the water. Suffice to say here that produced water composition depends on the ancient origin of the water, its temperature history, what minerals the water came in contact with, whether or not mixing occurred with meteorological or ground water, and the composition of the crude oil from which polar compounds would migrate and dissolve. A wide range of chemistry and composition have been observed for produced water around the globe.

Precipitation of sparingly soluble salts, metals, particularly metal sulfides, presents risk to many oilfield operations. Therefore knowing the chemical composition of the water is critical in the water treatment process design phase, monitoring process performance and in troubleshooting. Understanding what constituents to monitor or of which to know the concentrations in designing or operating a water process is critical. Below is a table which shows what constituents' concentrations should be known and their relative importance depending on what is happening to the water. Note that many constituents/contaminants have "Total" and "Filtered" values. In order to develop a complete Analytical Profile, each water sample should be divided into two parts. One part is filtered through 0.45 micron filter paper – delivering the "Filtered" values. Filtered can be interpreted as "reactive" or "dissolved." Total minus Filtered = Suspended; the concentration of the constituent present as a suspended particle. For example, Filtered Iron will react with polymers used in Polymer Floods or Fracking. Total Fe – Filtered Fe yields the concentration of iron that is present as particulate. Particulate iron is much less likely to cause trouble by reacting with polymers so is not considered a real problem. In chemical treatment, iron chelating agents would be added to control the dissolved iron while the particulate iron would be removed through clarification/filtration.

Note that some parameters are best determined in the field, including pH, alkalinity and turbidity. These parameters are quite likely to change substantially in a relatively short amount of time. Determining pH on a Produced Water sample that is a week old will yield a value that is highly suspect and therefore not a good value to include in any analysis, chemical program development or water process design.

Table 7.2 Water Profile Analytical Data – Relative Importance by Application 1 to 5, with 5 = High

Constituent or Contaminant	PW Re-injection	PW Offshore Discharge	PW Onshore Discharge	Frac Fluid	SWD	CEOR	TEOR
pH*	5	5	5	5	5	5	5
Specific Gravity	2	1	1	3	1	2	2
P Alkalinity*	4	1	2	2	1	2	4
M Alkalinity*	4	1	2	2	1	2	4
Total Alkalinity, (P + M)	4	1	2	2	1	2	4
TSS	5	2	5	5	5	5	5
TDS	3	3	3	3	3	3	3
Turbidity*	5	3	5	5	5	5	5
Conductivity	5	5	5	5	5	5	5
Viscosity	1	1	1	3	4	5	3
Al, Filtered	5	1	3	5	5	5	5
Al, Total	5	1	5	5	3	3	5
Arsenic, Total	1	1	5	1	1	1	1
Arsenic Filtered	1	1	3	1	1	1	1
B, Filtered	1	1	1	5	1	1	1
B, Total	1	1	1	5	1	1	1
Ba	3	1	1	5	1	3	3
Cadmium, Filtered	1	1	3	3	1	3	3
Cadmium, Total	1	1	5	3	1	1	3
Calcium, Filtered	5	1	3	3	4	4	5
Calcium, Total	5	1	3	3	3	3	5
Chromium, Filtered	1	1	3	3	1	1	1

Constituent or Contaminant	PW Re-injec-tion	PW Offshore Discharge	PW Onshore Discharge	Frac Fluid	SWD	CEOR	TEOR
Chromium Total	1	1	5	3	1	1	1
Co, Filtered	1	1	3	3	1	1	1
Co, Total	1	1	5	3	1	1	1
Cu	1	1	5	1	1	1	3
F	1	1	3	3	1	1	1
Fe, Filtered	5	1	3	5	5	5	5
Fe, Total	5	1	5	5	3	3	5
Hg	3	1	5	1	1	1	1
Mg, Filtered	3	1	3	3	3	3	5
Mg, Total	3	1	3	3	3	3	5
Mn	5	1	3	5	5	5	5
Mo	1	1	5	1	1	1	1
Nitrogen, Kirkendall	1	1	5	1	1	1	1
Ni, Filtered	1	1	3	1	1	1	1
Ni, Total	1	1	5	1	1	1	1
Phosphate, Filtered	3	1	3	1	1	1	5
TIP, Filtered	3	1	3	1	1	1	5
Organic Phosphate	1	1	3	1	1	1	5
Phosphate, Total	1	1	5	1	3	3	5
Pb, Filtered	1	1	3	1	1	1	1
Pb, Total	1	1	5	1	1	1	1
Se, Filtered	1	1	5	3	1	1	1
Se, Total	1	1	5	3	1	1	1
Silica, Filtered	4	1	3	3	4	4	5
Silica, Total	4	1	3	3	3	3	5
Sn	1	1	5	1	1	1	1
Sulfate	5	1	3	5	5	5	5
Ti, Filtered	1	1	3	1	1	1	1
Ti, Total	1	1	5	1	1	1	1
V. Filtered	1	1	3	1	1	1	1
V, Total	1	1	5	1	1	1	1
Zn, Filtered	1	1	3	1	1	1	1
Zn, Total	1	1	5	1	1	1	1
Zr, Total	1	1	5	1	1	1	1
OIW/TPH	5	5	5	5	5	5	5
BOD$_5$	1	1	5	1	1	1	1
COD	1	1	5	1	1	1	1

* Indicates determination must be made at time of sample collection – not laboratory.

PW – Produced Water

Relative importance will vary within certain identified categories. For example, for CEOR, if a Polymer Flood or ASP, then viscosity is important. If a surfactant program then viscosity is not important.

Onshore discharge assumes discharge 'onto a receiving stream' or discharge into the environment.

Offshore discharge assumes water going overboard from a platform.

Relative importance for 'Discharge' categories will vary with local regulations which requires a thorough knowledge of local regulations. Determining analytical parameters to be monitored should be accomplished in conjunction with local environmental authorities. Analytical requirement will be determine by regulation. Onshore – metals will be key focus area as will BOD$_5$ and COD.

Determining concentration of constituents of High (5) relative importance should be part of routine monitoring.

In separating oil from water and other liquid/liquid or liquid/solids separation activities the nature and condition of the interface is a critical determine factor in the difficulty of affecting the separation. One of the most important features of produced water is the fact that the dissolved ions (both the minerals and organic acids) can migrate to the oil / water or solid / water interface. This, as discussed below, has a strong effect to neutralize the repulsive electrostatic energy barrier between droplets and surfaces that would otherwise occur due to the presence of interfacially active components.

7.3.1. Analytical Equipment:

Atomic absorption spectrophotometry (AAS): is now the most widely used tool for determining dissolved metal ions in oilfield waters. Due to its low cost, high sensitivity and accuracy, atomic absorption spectrophotometry has made the determination of many trace metals in oilfield waters feasible.

For waters with salinities greater than seawater and brackish waters, the *additions method* is used in the atomic absorption spectrophotometric determination of lithium, potassium, sodium, magnesium, calcium and strontium ions. The trace metals chromium, cobalt, cadmium, copper, manganese, lead, nickel and zinc in saline waters are chelated and extracted before being determined by AAS. This removes interferences and concentrates the ions.

The Inductively coupled plasma (ICP) spectrometer is becoming popular for analysis of elements in waters with low solids content. Although more expensive, it is more convenient, sensitive and faster. It allows for many elements to be analyzed simultaneously. Dissolved metals are determined in filtered and acidified samples. There are interferences in samples with high dissolved solids (<1,500 mg/L), which suggest, samples may have to be diluted.

Other instrumental techniques are also used. **Atomic emission spectroscopy (AES)** using a dc argon plasma jet as an excitation source and an echelle grating are also being applied to water analyses. Nonmetals such as boron, silicon and phosphorous that are difficult to measure using AAS may be analyzed using AES.

X-ray spectrometric: analysis is sometimes used for analysis of some elements in brines. The instrument cost limits the use of this technique.

Mobile laboratories include equipment for many routine determinations in the field, such as **membrane test rigs** for measurement of TDS; equipment and kits for measurement of dissolved oxygen, H2S, CO2 and other gases; spectrometers for determining various ions, chlorine and the level of oil in water; and **coulter counters** for particle size analysis. A popular portable spectrophotometer made by Hach® is now used in oil fields all around the world. Other kits and equipment allow biological determination in the field.

Test methods and Checks

There are several analytical methods to determine Ions dissolved in water. These methods have been developed specifically for water analysis. Organizations such as NACE, American Petroleum Institute (API), The American Society for Testing and Materials (ASTM) have established standards for oil and gas water testing.

Approved water analysis procedures include methods for both saline as well as fresh waters – The *Annual Book of ASTM Standards, Water*. And the *Standard Methods for the Examination of Water and Waste Water*, which is updated periodically.

Several variations on the AAS method exists.

The flame AAS method – where a sample is aspirated into a flame and atomized. A light beam is directed through the flame into a monochromator and onto a detector that measures the amount of light absorbed by the atomized element in the flame. Because each metal has its own characteristic absorption wavelength, a source lamp composed of that element is used. This makes this method relatively free from spectral or radiation interferences. In the air-acetylene flame AAS method, metals are aspirated directly into an air-acetylene flame.

For trace elements, electrothermal AAS permits determination of most metallic elements with sensitivities and detection limits of 20-1,000 times lower than conventional AAS, with many as low as 1.0μg/L. An electrical heated atomizer or graphite furnace replaces the standard burner head. Arsenic and selenium are determined by hydride-generation AAS

Inductively Coupled Plazma (ICP): The ICP method is fundamentally an atomic or elemental analysis technique. It provides determination of elemental composition. Samples are typically acidified before measurement. Acidification and 24-hour settling is referred to as digestion in some analytical laboratories. The purpose of digestion is to dissolve the mineral components such as the carbonate, and the iron compounds. The ICP source consists of a flowing stream of argon gas ionized by an applied radio frequency field typically oscillating at 27.1 MHz. This field is inductively coupled to the ionized gas by a water-cooled coil surrounding a quartz torch that supports and confines the plasma. A sample aerosol is generated in an appropriated nebulizer and spray chamber and injected into the ICP, heating it to temperatures that result in complete dissociation of molecules.

Nonmetallic anions may be measured individually using electrometric, colorimetric, or titrimetric methods, or measured rapidly and sequentially using ion chromatography. A water sample is injected into a stream of carbonate-bicarbonate eluent and passed through a series of ion exchangers. Anions are separated and measured by conductivity.

Routine checks should be made to validate water analyses. Firstly check that the molar-equivalent sum (in meq-L) of the cations equal the sum of the anions. An accuracy of 1% is acceptable. A discrepancy of more than 5% indicates problems with the overall procedure.

Another useful validation is to compare the measured TDS, or the measured specific gravity to the sum of the anions and cations as measured by ICP. Calculating a specific gravity using the ionic constituents and comparing this with the measured value or comparing the TDS with that of a sodium chloride solution of the same specific gravity as that measure in the analysis are useful comparisons. Comparing the reported resistivity with that of a sodium chloride solution of the same TDS should result in comparable values. Some deviations will occur, but gross differences should make one suspicious of the analysis. Another check is that the pH of an unacidified, depressurized sample should be between 5-9. Above 9 indicates acidization, and abnormal samples.

An indirect check is to establish that the water analysis is consistent with the reservoir conditions at its source. With a few exceptions of where the water is formed, such as geological salt domes – water can be saturated in $CaCO_3$. Hence if $Ca^{2+}/HCO_3^- >> 1$, the degree of saturation should be close to unity.

7.4. Oil in Produced Water:

7.4.1. What is Oil in Water?

Before the discussion of monitoring and analysis methods, the term "oil-in-water" must be first defined. As previously discussed in Chapter 1, it is not straightforward to define what is meant by oil-in-water. Oil is composed of thousands, if not millions, of different chemical components. The components differ in their vapor pressure, partitioning into extraction solvents, acidity, response to oxygen exposure, response to adsorbent media (such as silica gel), and response to spectroscopic measurements such as IR and UV light, and chromatographic response. Given this, the quantity that is measured as oil-in-water cannot be universally defined. Instead, the measured oil-in-water concentration is dependent on the measurement method used.

Nevertheless, it is useful to discuss three general categories for oil-in-water. There is no precise definition for these categories, for the reasons just given, results depend on measurement method. Thus, these classifications are very general:

Free oil usually refers to oil floating on the surface of water or very large oil droplets that would rise to the surface very quickly, within one or two minutes. Free oil has essentially the same composition as dispersed oil (below) but is formed of larger droplets that quickly rise to the top of the water phase.

Dispersed oil refers to oil in produced water in the form of smaller droplets, which may have diameters that range from sub-microns to hundreds of microns. Dispersed oil may contain aliphatic and aromatic hydrocarbons and other organics, e.g. organic acids, phenols, and other polar organic entities such as resins and asphaltenes.

Dissolved oil refers to oil in produced water in a soluble form. Dissolved oil contains very little aliphatic hydrocarbon since aliphatic hydrocarbons in general have very low solubility in water, typically single digit or low double digit mg/liter. Dissolved oil contains mostly the aromatic hydrocarbons, in particular the single ring BTEXs (Benzene, Toluene, Ethyl-Benzene and Xylenes), and two ring NPDs (Naphthalene, Phenanthrene and Dibenzothiophene), together with organic acids (e.g. fatty acids and naphthenic acids), and phenols.

There is considerable overlap in the chemistry of these groups. But in general, the dissolved components are polar and have a low hydrocarbon portion. The dispersed components may also have polar groups, but the size of the hydrocarbon portion of the molecules is larger and therefore causes the molecule to have low solubility in water.

Each analytical method will measure a portion of dispersed and a portion of the dissolved oil in produced water depending on the details of the methodology, the measurement principle, degree of acidification of the sample before extraction, type and quantity of solvent used for the extraction, how the sample is extracted, the use of florisil or silica gel for removing polar compounds, how the calibration is established, what calibration oil is used etc. For the discharge of produced water in the OSPAR region and USA Gulf of Mexico (GoM), oil in produced water is well defined. OSPAR and U.S. GoM are significantly different from each other, but they are both well-defined.

Oil-in-Water Monitoring Club: Over the last two decades or so, there have been attempts to provide relatively standard methods of analysis. In the North Sea this was driven by the differing cultures of the surrounding countries, and their differing attitudes toward health and the environment. In the late 1990's a collaborative organization was formed known as the Oil-in-Water Monitoring Club. It was organized by the National Engineering Lab (NEL) of Glasgow, Scotland. Initially it was comprised mostly of operating companies. The main agenda was to test monitoring methods and provide input to regulators. Over the years, they have become a focal point for information exchange, through annual workshops, for not only the monitoring methods but also for the treatment of produced water.

7.4.2. EPA-1664 and other Gravimetric Methods:

The most widely used gravimetric method is the USA EPA Method 1664A [2]. Outside the US, the EPA-1664 method is often cited as Standard Method 5520B. The method was developed following the phase-out of Freon as mandated by the Montreal Protocol on Substances that Deplete the Ozone Layer. In the method, a 1-litre sample of oily water is acidified to pH less than 2 and then extracted using three aliquots of n-hexane whose total volume is 100 ml. After separating the solvent (now containing oil) from the water sample, it is transferred into a flask, which has been weighed beforehand. The flask is placed into a temperature controlled water bath and the solvent is evaporated at a specified temperature, condensed and collected. After the solvent is evaporated, the flask, now containing the residue oil, is dried and weighed. Knowing the original clean weight of the empty flask, the amount of residue oil can be calculated. The oil content is referred to as HEM (Hexane Extractable Material).

Depending upon whether the hexane extract undergoes a silica gel treatment (cleaning) process, the residue oil obtained is either called Hexane Extractable Material (HEM) or Silica Gel Treated - Hexane Extractable Material (SGT-HEM).

For clarity, it should be noted that the evaporation step for EPA 1664 stops when the condenser head temperature reaches 70°C. However, the introductory notes for the EPA 1664 documentation clearly state that compounds which are volatile below 85°C are generally not detected by the method.

Due to the evaporation procedure at elevated temperature light constituents will be lost. In fact, the text of the EPA 1664 method states specifically that hydrocarbons boiling below 85°C are generally not detected by the method. However, compared to the Infrared based method in which polar components such as the fatty acids and phenols are removed by a cleaning step using silica gel or Florisil, these constituents are included as "oil and grease" in the gravimetric method such as the EPA 1664 A.

7.4.3 Relation between Dispersed, Dissolved, TPH, TOG and WSO:

In the USA, HEM (Hexane Extractable Material) is synonymous with "Oil and Grease" or TOG. For regulatory compliance monitoring for the offshore oil and gas industry, it is this "Oil and Grease" as determined by the EPA Method 1664 A that is regulated at a monthly average of < 29 mg/l and daily maximum of 42 mg/l. WSO (Water Soluble Organics) and TPH (Total Petroleum Hydrocarbons) are not regulated. The EPA-1664 analytical procedure generates values for these quantities which can be used as a guide for the chemical type of material that contributes to TOG.

In the USA, TOG, TPH (Total Petroleum Hydrocarbon) and WSO are defined as:

TOG = HEM (mg/L n-Hexane Extractable Material)

TPH = SGT-HEM (mg/L Silica Gel Treated, n-Hexane Extractable Material)

WSO = TOG – SGT-HEM = (calculated value, mg/L Silica Gel Adsorbed, n-Hexane Extractable Material)

Based on these definitions, the following relationship is evident:

$$TOG = TPH + WSO$$

It is important to note that neither TOG, WSO, nor TPH is strictly equivalent to the dispersed oil content of an oily water sample since some dissolved organics are included, due to the analytical procedure, and other classes of organics are excluded by the analytical procedure. See Chapter 21 (Applications – Water Soluble Organics) for discussion of this point. Further, WSO is not a direct or even an indirect measure of dissolved organics in produced water. Most of the compounds that are dissolved in the produced water will not extract into hexane and are therefore not measured as WSO. This is discussed further below and in the chapter on Applications.

Dissolved/Dispersed versus WSO/TOG: The name "water soluble organics" conjures the idea of organic compounds dissolved in produced water. However, given the analytical procedure [2, 3], there is little reason why WSO should be equivalent to dissolved organics. Keep in mind that the measured value of WSO are those compounds that are extracted from the water into the hexane, that do not vaporize when the hexane is evaporated, and that stick to the silica gel. To understand what WSO are and how they differ from dissolved organics, it is helpful to discuss two ways of characterizing produced water.

Total Oil & Grease (TOG) and Water Soluble Organics (WSO): There are two frameworks for (ways of thinking about) organic compounds in produced water. One framework is based on the physical picture of dissolved versus dispersed organics. This is the phase partitioning framework. The other framework is the analytical framework. This framework distinguishes between TPH and WSO on the basis of the EPA 1664 analytical method.

The difference between these two frameworks can be described in the following figure.

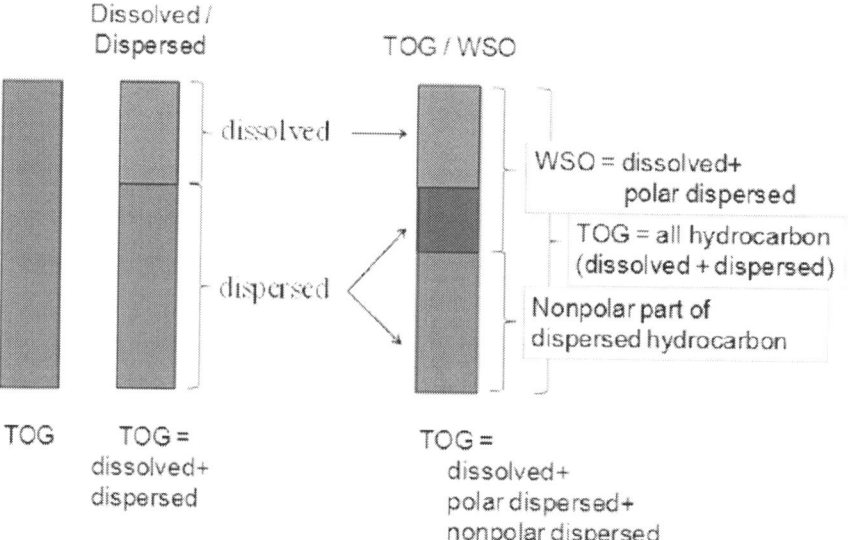

Figure 7.5 Schematic diagram of the two different frameworks for reporting of concentrations of organic compounds in produced water. On the far left, the total oil and grease (TOG) is shown having a certain value represented by the height of the bar graph. This TOG can be characterized as a combination of dissolved and dispersed organics, as shown by the middle bar graph. As shown on the right hand side, this same TOG value can also be characterized by WSO (Water Soluble Organics) plus TPH (Total Petroleum Hydrocarbons - the nonpolar part of the dispersed organics).

The term, WSO is misleading. As shown in the figure, the WSO are a combination of two groups of compounds. One group are the truly dissolved components. The other group are dispersed (not dissolved) compounds that have at least one polar group such that they adhere to the silica gel and do not elute during the EPA-1664 test. The fundamental definition of WSO is that fraction of the dissolved and dispersed hexane extractable material that does not flash off at 85 °C and which sticks to silica gel (EPA-1664 method). Thus, WSO contain both dissolved and dispersed compounds.

The term, Total Petroleum Hydrocarbon (TPH), is also misleading since, strictly speaking, a hydrocarbon compound is composed only of hydrogen and carbon. According to IUPAC there are four classifications of hydrocarbons (alkanes, unsaturated alkanes, cycloalkanes, and aromatics). In the EPA-1664 method, the reported TPH portion of a produced water sample generally contains a wide variety of compounds, only a fraction of which are hydrocarbons. Discussions of these concepts are available in the literature [4 - 9]. In this book, a discussion of the topic is given in Chapter 21, Applications.

7.4.4. Gas Chromatography and Flame Ionization Detection (GC-FID):

In the OSPAR region, there is a distinction between total oil and dispersed oil. Total oil means total hydrocarbons. Dispersed oil means the hydrocarbons as determined according to the OSPAR defined reference method, which currently is a modified version of the ISO 9377-2 method. Thus dispersed oil is defined as:

"The sum of the concentrations of compounds extractable with n-pentane, not adsorbed on florisil and which may be chromatographed with retention times between those of n-heptane (C_7H_{16}) and n-tetracontane ($C_{40}H_{82}$) excluding the concentrations of the aromatic hydrocarbons toluene, ethyl benzene and the three isomers of xylene (TEX)."

It is worth noting that in the above definition, oil in produced water will essentially be the hydrocarbons with carbon number between C7-C40 excluding the TEX, acids, and phenols. Those hydrocarbons with carbon number above C40 and below C7 will not be included.

In a typical GC-FID method, an oily water sample is acidified and extracted by a solvent just like gravimetric reference methods. The extract is dried and purified before a small amount of the extract is injected into a Gas Chromatography (GC) instrument that contains a chromatographic column which separates the different hydrocarbon constituents. With the help of a carrier gas, different groups of hydrocarbons leave the column at different times and can be quantitatively detected. As hydrocarbon components leave the column, they are burned and detected by a Flame Ionization Detector (FID).

For oil in water analysis, it is the sum of all the responses within a specific carbon range or retention time that is related to the oil concentration by reference to standards of known concentrations.

Two examples of GC-FID based reference methods are given in Table 7.

Table 7.5 Examples of reference GC-FID methods

Reference Method	Country	Solvent Used	Hydrocarbon Index	Status
ISO 9377-2	International	Solvents within the boiling point range (36 °C to 69 °C)	C_{10}-C_{40}	In use
OSPAR GC-FID	OSPAR countries	n-pentane	C_7-C_{40} minus TEX	In use

Notes: TEX stands for Toluene, Ethyl-benzene and Xylene.

The OSPAR GC-FID method, which is currently used as the reference method for measuring oil in produced water in the OSPAR countries, is a modified version of the ISO 9377-2. Compared to the ISO 9377-2, the OSPAR GC-FID method requires a higher resolution GC so that the TEX can be separated and excluded from the area integration between C_7-C_{40}.

The TEX constituents are relatively soluble in water, and were historically discounted in the North Sea when comparing oil in water results to the 40 mg/l performance standard. The TEXs were not considered as a significant component of the "dispersed" oil upon which the 40 mg/l performance standard was based. Thus when the new oil in water method based on GC-FID was made available; OSPAR felt the need to remove the TEX from the oil and grease measurement.

It should be pointed out that although the OSPAR GC-FID is the official reference method for the measurement of oil in produced water in the North Sea region, the use of GC-FID instruments off-shore for produced water oil in water analysis has been limited. In the UK sector, implementation of the OSPAR GC-FID has been based on using an Infrared method offshore that is correlated to the OSPAR method onshore.

7.4.5. Reference (Regulatory) Methods:

Reference methods are important. Without them, discharge limits or performance standards become meaningless. With the use of a reference method, consistent oil-in-water measurement data can be obtained and used to evaluate current discharge water quality on a uniform basis as well as formulate future legislation based on sound information.

For historical reasons, and also because produced water is viewed as a waste stream, there has never been one oil in water measurement reference method that is universally accepted and adopted by regulators and operators across the world. It must be emphasized again that different methods will produce different measured results. Therefore, one should not compare results if they are obtained from different methods. It should also be said that different methods will require different instruments, procedures and personnel skill sets, which has implications on cost (capital and operational); training, health and safety etc.

Generally speaking, Infrared methods are very well established, commonly used and easy to deploy with portable fixed wavelength instruments. However infrared methods require either an infrared transparent solvent (such as a CFC – chlorofluorocarbon), or they require evaporation of the solvent. Both options have drawbacks. Also, infrared methods do not give compositional data. Gravimetric methods are simple and relatively inexpensive to utilize, but again they too do not provide details of compositions. Also due to the evaporation procedures involved in the gravimetric methods, there will be some loss of volatile components. For GC-FID based methods, there is no need to use CFC solvents. Although the GC-FID methods do not use an evaporation step in the method, they still do not detect C_7 and lighter hydrocarbons because of the large solvent peak in the GC trace which initially comes off of the GC column. The one advantage which the GC-FID methods provide is that they have the potential to provide detailed information on compositions. But they need more sophisticated instruments which require also more skilled personnel to operate.

7.4.6. Field Measurement Methods:

While reference / regulatory methods are essential for the definition of oil in water values and for compliance monitoring, they are not always user-friendly, and in some cases they may even be impractical for making offshore oil in produced water measurements.

From an operational and process optimization stand point, time, effort and training required are all factors that must be weighed alongside of accuracy in selecting a suitable monitoring method for field application. Analysis of oil in water using some of the reference methods, e.g. GC-FID and gravimetric based methods, can be time consuming. Therefore field measurement methods and devices that are easy to use, inexpensive and quick to run, offer advantages.

Field measurement instruments and methods can be divided into two main categories – laboratory bench top and online monitors. Both are discussed below. Bench top instruments and methods are used for routine oil in water analyses. If the methods are properly correlated to the reference methods, then results from these methods allow operators to evaluate their system's performance on a spot basis. Online monitors provide continuous data for process trending and the detection of process upsets. Online methods can be very useful in process optimization and trouble shooting.

7.4.7. Bench-Top Oil-in-Water Analytical Methods:

There are several bench-top oil in water measurement methods (and instruments). The commonly used methods include:

- Colorimetric;

- Infrared based Horizontal Attenuated Total Reflection (HATR);

- Infrared based Horiba instrument using S-316 as a solvent;

- Infrared based solventless method;

- Infrared based using a non-conventional wavelength for detection;

- UV fluorescence;

- Other new developments.

(a) **Colorimetric:**

For a typical colorimetric based method, oil in water is determined by extracting an oily water sample with a solvent and then directly measuring the color of the sample extract using a visible spectrophotometer at a wavelength, for example, of 450 nm. In order for the colorimetric method to work, the oil in question must show color. Oil that possesses little color will not work well with the method. Thus for water samples collected from gas and gas condensate installations, the method may not work properly. The Hach OiW analytical procedure is one example of an established colorimetric method. It has a stated measurement range of 0 to 80 ppm.

(b) **Infrared Based Horizontal Attenuated Total Reflection (HATR):**

In a typical Infrared absorption based method, an oily water sample is first acidified, and then extracted by a chlorofluorocarbon (CFC) solvent. Following the separation of the extract from water sample, the extract is then removed, dried and purified with the removal of polar compounds. A portion of the extract is then placed into an infrared instrument, where the absorbance is measured. By comparing the absorbance obtained from a sample extract to those of calibration standards with known concentrations, one can calculate the oil concentration of the sample extract and the original oil in water sample. The fundamental principle of an Infrared based measurement is that of the Beer-Lambert law, which is given by the equation:

$$A = \log I_o/I = ELc \qquad \text{Eqn (7.1)}$$

Where

      A is the absorbance;

      I_o is the incident light intensity;

      I is the transmitted light intensity;

      E is a constant;

      L is the path length;

      c is the hydrocarbon concentration in the sample extract.

Depending upon the number of wavelengths used for measurement and quantification, there are two types of Infrared based reference methods:

- Single wavelength Infrared methods;

- Triple peak or three wavelength methods.

In a single wavelength method, measurement of the infrared absorbance is conducted by using a single wavelength, usually at around 2930 cm^{-1}, which corresponds to the methylene C-H stretch vibration frequency. Two examples of single wavelength based Infrared reference methods are given in Table 7.6.

Table 7.6 Examples of single wavelength based reference methods

Reference Method	Wavelength cm^{-1}	Solvent Used	Calibration	Status
DECC IR Method	2930	Tetrachloroethylene or S-316	Field Specific oil	Still in use in the UK
ASTM D 7066-04	2930	S-316	n-hexadecane + isooctane or oil in question	Still in use

In theory, a single wavelength method quantifies all of the -CH$_2$ contained in a sample extract. These include all of the aliphatic hydrocarbons and those that are contained in aromatic hydrocarbons such as that in ethyl-benzene. Also if production chemicals and other process chemicals that contain -CH$_2$ are present in the produced water sample and extracted by the solvent, then they will also contribute to the measurement. For a single wavelength Infrared method, both fixed wavelength and scanning Infrared instruments can be used. Many of the fixed wavelength instruments available on the market are portable, and easy to use, which is advantageous for use in field analyses.

In a triple peak or three wavelength Infrared method, instead of measuring the absorbance at one fixed wavelength, Infrared absorbance at three different wavelengths is recorded. The three wavelengths are respectively related to the stretch vibration frequency of aromatic C-H at 3030 cm^{-1}, methylene CH-H at 2930 cm^{-1} and methyl CH$_2$-H at 2960 cm^{-1}. Oil content is quantified by using an equation such as one shown in equation (2) below, which takes into account of the absorbance obtained at the three wavelengths. The equation has three coefficients (X, Y, Z) that will have been determined by measuring the absorbance at the same three wavelengths of known concentration calibration standards, which are often prepared using n-hexadecane, pristine and toluene respectively.

$$C_{total} = \{[X(A_{2930})] + [Y(A_{2960})] + [Z(A_{3030} - A_{2930}/F)]\}10vD/VL \qquad \text{Eqn (7.2)}$$

Where

C$_{Total}$ is the oil in water concentration;

X, Y and Z are factors calculated using absorbance obtained with known concentration calibration standards;

v is the volume of extraction solvent;

D is the dilution factor (if the sample is not diluted, D=1);

V is the volume of produced water sample;

A is the absorbance at the specified wavelengths;

F is A$_{2930}$/A$_{3030}$ for hexadecane standard;

L is the cell path length.

With a three wavelength method, the aromatic part of hydrocarbons in the sample is more properly quantified. Also such a method allows for the calculation of aliphatic and aromatic hydrocarbons as given in the equation (7.3) and (7.4).

$$C_{aliphatic} = [X(A_{2930})] + [Y(A_{2960})]\,10vD/VL \qquad \text{Eqn (7.3)}$$

$$C_{aromatic} = C_{total} - C_{aliphatic} \qquad \text{Eqn (7.4)}$$

For a triple peak method, while a fixed wavelength instrument may be used, often a scanning Infrared instrument is employed. Two examples of three wavelength Infrared methods are given in Table 7.7

Table 7.7 Examples of three wavelength based Infrared methods

Reference Method	Country	Solvent Used	Calibration	Status
IP 426/98 (Energy Institute)	UK	Tetrachloroethylene	n-hexadecane, pristane and toluene	Still in use
GB/T 17923-1999	China	Carbon Tetrachloride	n-hexadecane, pristane and toluene	Still in use
Dutch Standard NEN-6675	Netherlands	freon	synthetic oil	Still in use

In both of the IP 426/98 and GB/T 17923 -1999 methods, individual calibration standards of n-hexadecane, pristane and toluene were prepared by respectively dissolving a certain amount of each synthetic oil into a fixed volume of solvent. It is important to point out that although three wavelength infrared methods quantify both aliphatic and aromatic hydrocarbons and allow the calculation of aliphatic, aromatic and total hydrocarbons, it does not mean that oil in water results obtained from a three wavelength based method will be higher than those obtained from single wavelength based methods.

The Agilent Cary 630 FTIR (Fourier Transform Infrared) instrument is an example of a bench top unit that measures absorption at stretch vibration frequencies corresponding to C-H stretch of aromatic, methylene and methyl groups. Actually, since it is a Fourier Transform method it measures all wavelengths simultaneously. Thus, it is possible to use this instrument for determination of alkyl and aromatic constituents. Meijer and Kuijvenhoven [10] discuss the use of the triple peak Infrared method according to NEN-6675 [11, 12]. For aromatics, they compared the results to those obtained by gas chromatography. The former method being more suitable for offshore application, and the later method more suited for onshore application. Their interest in these methods stemmed from the deployment of the Macro Porous Poly Extraction technology which is used for removal of aromatics and PAH compounds. They found good agreement between the two methods.

Extraction Solvents play an important role in any of the Infrared based reference methods discussed above. A good extraction solvent should possess following properties:

- Good extraction ability of oil in water;

- Do not absorb Infrared at the measurement wavelength or wavelengths;

- Environmentally friendly;

- Safe to use;

- Heavier than water;

- Affordable and readily available;

Over the years, different solvents have been used, these include:

- Carbon tetrachloride (CCl_4);

- Freon 113 ($C_2F_3Cl_3$);

- Tetrachloroethylene (C_2Cl_4);

- S-316 ($CClF_2CClFCClFCClF_2$).

All of the above solvents are Infrared transparent and known to be a good extractants for oil in water measurement. Due to the fact that carbon tetrachloride is known to be carcinogenic and Freon 113 was found to be an ozone depleting substance, both of these chemicals have being phased out and gradually replaced by tetrachloroethylene and S-316. It should be pointed out that tetrachloroethylene is suspected to be carcinogenic and that S-316 is expensive to purchase. The use of all halogenated solvents on platforms in US waters was banned by the EPA in _19xx_. It was this regulation which spurred the development of EPA 1664 using hexane as the extraction solvent.

In summary Infrared methods have been widely used until recently. They are excellent methods for the measurement of oil in water. They are less being used now mainly due to environmental and health relate issues with the solvents.

With HATR, oil is extracted from a produced water sample using hexane. A pipette is used to take and transfer 50 microliter sample of the hexane / hydrocarbon mixture into the ATR crystal measurement plate (on the top of the unit). A timer provided by the instrument is used to allow evaporation of the solvent for a set amount of time. The evaporation occurs at room temperature. Infrared light is then used to measure the total absorption and hence amount of hydrocarbon on the plate. The calibration can be programmed into the instrument such that a direct output of the oil-in-water concentration is reported.

The measurement principle can be explained using the schematic diagram in Figure 10a. At each reflection, Infrared light is absorbed in a similar way as in the conventional (single wavelength) Infrared analysis. Therefore by calibration and comparing absorbance obtained from a real sample extract to that obtained from calibration standards, one can calculate the oil concentration in a real sample.

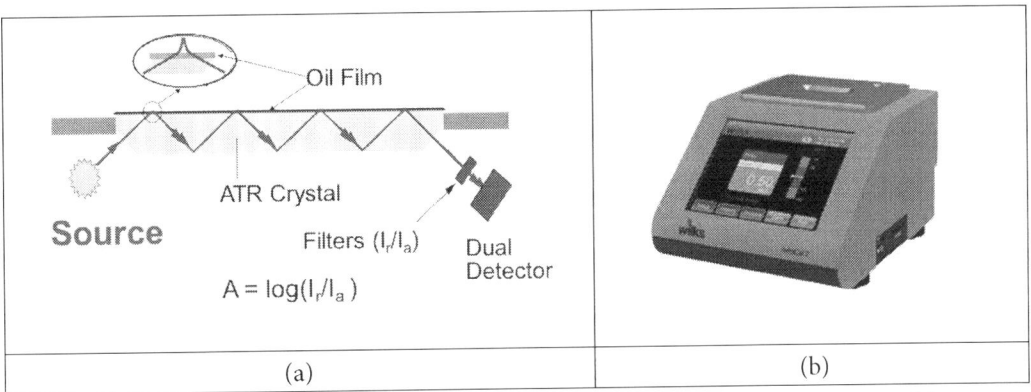

Figure 7.9 A schematic diagram of the HATR technique and a picture of the Wilks ATR-SP model.

HATR is a well established OIW measurement technology. Portable instruments as an example shown in Figure 10b are available from Wilks Enterprise Inc. (now part of Spectro Scientific). Early versions of the technology had a measurement range of 4 to 1000 ppm While the newer model, the ATR-SP, can measure oil in water concentration less than 1 ppm.

For the purpose of correlating to the EPA-1664 method, this procedure can be carried out using acidized samples, as required by the EPA-1664 test. Also, the hexane extracted sample can be contacted with silica gel which would distinguish between TOG and TPH, as in the EPA-1664 test. When this is done, the only remaining difference between the IR method and the EPA method is that the EPA method involves temperature elevation to evaporate the solvent. Thus, the IR technique measures more of the light components than does the EPA method. The difference between TOG and TPH is the WSO, discussed above. The practice of running acidized versus unacidized samples in order to understand WSO concentration gives misleading results because it preferentially measures the partitioning of the small molecular weight acids that would normally evaporate in the elevated temperature used to evaporate the solvent in the EPA-1664 test.

It should be pointed out that the HATR method is not suited for measuring oil in water when the oil contains volatile components such as those from gas and gas condensate fields. This is because the sample extract undergoes an evaporation procedure before the measurement takes place. As with the EPA 1664 gravimetric procedure, it is inevitable that the volatile components in the sample extract will be lost during the evaporation process and not measured. For light condensates, a non-evaporation method should be used such as a UV method, the solvent-less method discussed below, or a method that uses a transparent solvent.

(c) **Infrared Based Methods using IR-Transparent Solvents:**

In this method, hydrocarbon is extracted from the produced water sample using a solvent that is transparent in the IR spectrum. Two such solvents are Vertrel and S-316. Since these solvents are transparent, they do not need to be evaporated. Keathley and Konrad [13] discuss the method, and compare results to the HATR method above.

As an example, the Horiba OCMA-310 instrument is shown in Figure 11. Using the Horiba instrument, a small volume of oil in water sample and extraction solvent are injected using a syringe into the instrument. In the instrument, extraction takes place automatically. After the extraction and separation of the solvent from water, a portion of the extraction solvent is diverted into the instrument measurement cell. Oil concentration is obtained by measuring the Infrared absorbance. The measurement method is essentially a semi-automatic version of a conventional (single wavelength) Infrared analysis method as described above. But it is not a reference method even though it is widely used in some areas of the world. According to the instrument supplier, Horiba, it has a measurement range of 0 to 200 ppm.

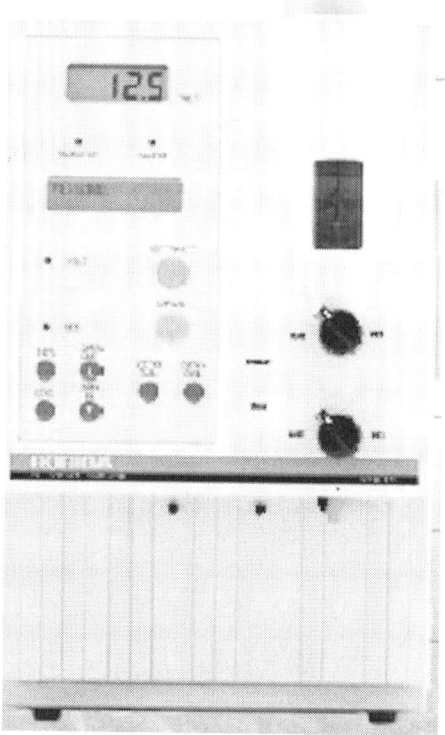

Figure 7.10 A picture the Horiba OCMA-310 instrument

(d) Infrared Based Solventless Approach:

One of the key issues related to the use of an Infrared based oil in water measurement method is the solvent utilized for the extraction. Therefore, any approach that allows for the continued use of Infrared quantification without the need of using a solvent is of great interest. A solventless approach has been developed in the recent years. In this method, a volume of oil in water sample is filtered through a proprietary membrane which can retain oil (dispersed). The membrane is then dried before it is placed into an Infrared instrument, and Infrared quantification is made similarly to the conventional Infrared reference method at a fixed wavelength. The steps involved are illustrated in Figure 12. The Membrane is the key to this analytical method. It acts like a cuvette containing pure CFC solvent in a conventional Infrared method. It does not absorb Infrared light itself. According to the ASTM method, it has a measurement range of 3 to 200 ppm.

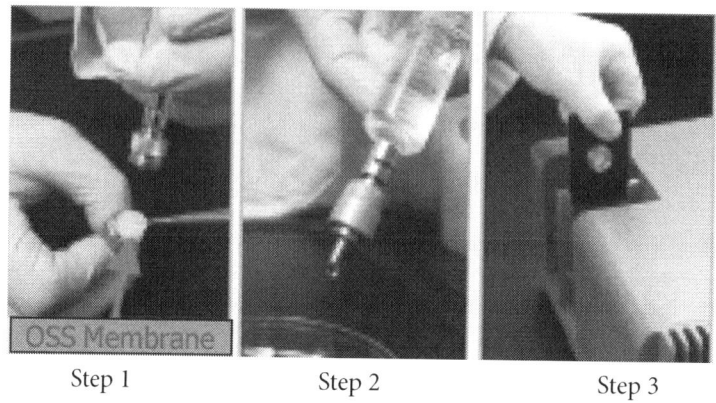

Step 1   Step 2   Step 3

Figure 7.11 Pictures showing the steps involved in the OSS solventless Infrared technology

Conventional Infrared methods are based on measuring the absorbance of sample extract contained in a cuvette cell.

(e)   Infrared Based Using a Non-conventional Wavelength for Detection:

This is a relatively new development based on using Quantum Cascade Laser Infrared (QCL-IR) technology. The method involves in an extraction of oil using a cyclic hydrocarbon, such as cyclohexane or cyclopentane (compared to the use of CFC in a conventional Infrared method). Quantification is based on measuring the absorbance of infra-red light at a wavelength in the region of 1350 – 1500 cm^{-1} using mid-infrared spectroscopy that employs a Quantum Cascade Laser as a light source.

The principle of measurement is straightforward. As illustrated in Figure 13, in which attenuated total reflection (ATR) absorbance spectra for samples collected in the North Sea is compared to that of the extraction solvent cyclohexane. For crude oil in water samples, absorption in the region 1350 – 1400 cm^{-1} takes place, while for the solvent cyclohexane, no such absorption occurs. Therefore by measuring the absorbance of a sample extract at a wavelength around 1450 cm-1, and comparing that to those from calibration standards, one can calculate the oil concentration in a water sample. It has a measurement range of 0.5 to 1000 ppm.

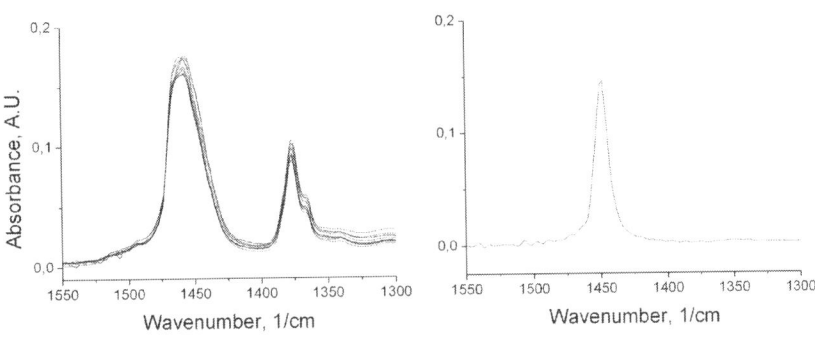

Fig. 2a– FTIR absorbance spectra of different samples from the North Sea measured with the ATR technique.

Fig. 2b– FTIR absorbance spectrum of cyclohexane measured with the ATR technique.

Figure 7.12 FTIR Absorbance spectra of different samples from the North Sea measured with the ATR technique (left) and that from the solvent cyclohexane (right)

An instrument named "Eracheck" is marketed by Quantared and the 2015 version of the instrument is shown in Figure 14.

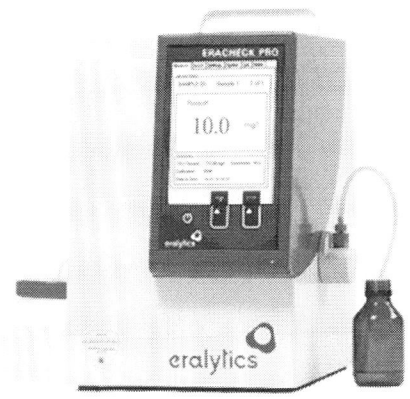

Figure 7.13 A picture of Eralylics from Quantared

(f) UV Fluorescence:

When aromatic hydrocarbons are present in a crude oil, they absorb ultra-violet (UV) light and emit fluorescent light at a longer wavelength. The phenomenon is called UV fluorescence. Therefore by measuring the intensity of the UV fluorescence light, one can calculate the amount of aromatic hydrocarbons in the sample, which can in turn be related to the total hydrocarbons present in the sample as long as the ratio of aromatic to the total hydrocarbons remains reasonably constant. UV fluorescence based devices have been in widespread use by the oil and gas industry since the early 2000s for the measurement of oil in produced water. As an example, portable instruments like the TD-500 from Turner Designs Hydrocarbon Instruments and Fluorocheck from Arjay Engineering are shown in Figure 16.

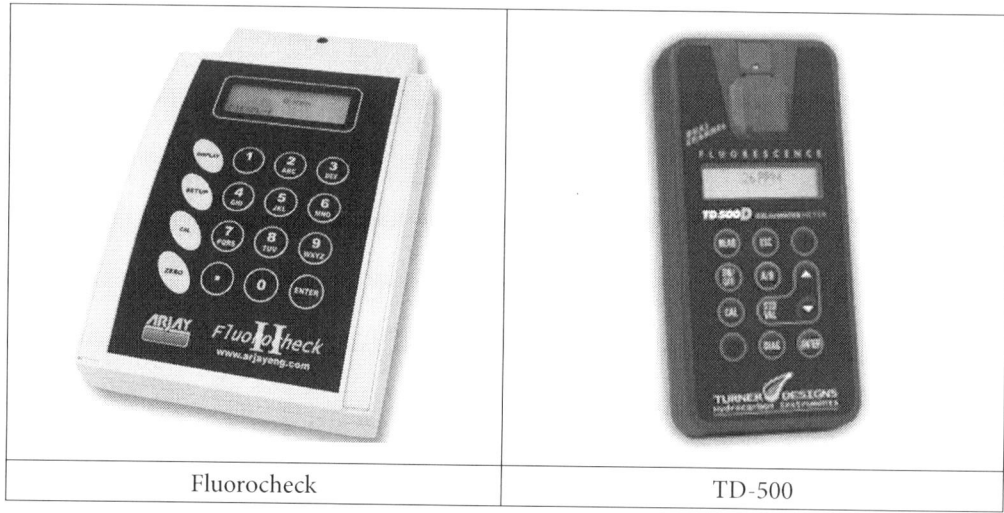

| Fluorocheck | TD-500 |

Figure 7.14 Portable / handheld UV fluorescence based devices from Arjay Engineering from Turner Designs Hydrocarbon Instruments

The UV fluorescence method is sensitive and does not need to use a CFC solvent. However, the method relies on a stable ratio of aromatic to total hydrocarbons. If this ratio changes, for example, when a new reservoir fluid is brought into the process, then it is important to have a new calibration developed, otherwise it can lead to erroneous oil in water results.

(g) Other New Developments:

Two additional OIW analytical methods have been developed which are worth mentioning. One is based on using Laser Induced Fluorescence (LIF) and one is based on using light scattering.

The Advanced Sensors' LIF based HD-1000 system, as shown in Figure 16(a), requires no solvent extraction. A probe is inserted directly into an oily water sample, fluorescence light is then detected via fiber optics, which is related to the total oil concentration through calibration. The method is simple and produces results very quickly since there is no extraction involved. Obtaining an accurate OIW measurement with this type of instrument requires that oil droplet creaming or loss to the container walls does not take place during the time required for the measurement All UV fluorescence measurements exhibit some sensitivity to the oil droplet size distributions in the sample. As droplet size increases, the penetration of UV light into the droplet diminishes and UV fluorescence emitted deep inside the droplet can be attenuated by re-adsorption. Both effects reduce the intensity of the observed UV fluorescence and result in an artificially low oil concentration being reported.

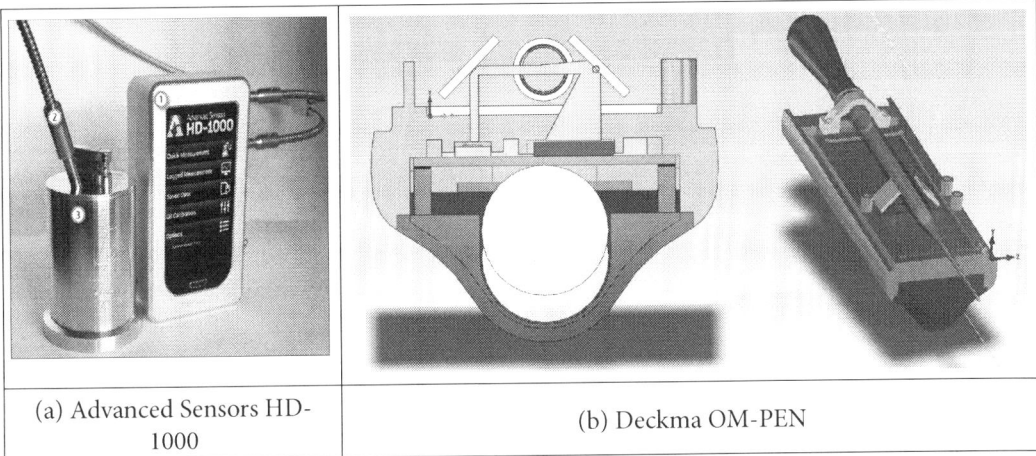

(a) Advanced Sensors HD-1000	(b) Deckma OM-PEN

Figure 7.15 Newly developed bench top methods from Advanced Sensors and Deckma

Light scattering is another established technique for online oil in water monitoring. This technology will be discussed in more detail in the next section. A bench top Oil-Measurement PEN (OM-PEN) is available which uses the light scattering technique. As shown in Figure 16 (b), an oily water sample is withdrawn using a commonly available glass pipette. The pipette is inserted into a system which includes a mirrored holder and visible light is shined onto the pipette containing the oily water sample. The intensity of scattered light is then detected. Again through calibration, oil in water can be measured. As of 2015 the technology is new and remains under development. However, the concept is simple and the device is easy to use.

7.4.8. Online Oil in Water Monitors:

There are a number of technologies utilized by online oil in water monitors. Some of the most popular ones are listed and described below.

- Focused ultrasonic acoustics;

- Microscopy based image analysis;

- Light scattering;

- UV fluorescence (including Laser-Induced Fluorescence (LIF)).

(a) **Focused Ultrasonic Acoustics:**

In this technique, a highly focused acoustic transducer is inserted directly into a produced water stream. The focal area of the transducer and a time interval window determines the volume element to be used for the measurement. Any particles, such as oil droplets, solid particles and gas bubbles that pass through the measurement volume, will produce acoustic echoes which are utilized for the measurement. These acoustic echoes are detected, classified and used to determine particle sizes and size distribution. The number of detected particles along with their size distribution is used to calculate oil concentration. The technology was initially developed by TNO TDP and then licensed to Roxar. It is currently (2016) owned by Mirmorax in Norway.

It should be pointed out that the technology is non-optical based. The use of ultrasonic transducers and sensors is less affected by fouling than optical based technologies. According to Mirmorax, particles with size less than 2-3 microns will not be measured. The technology may also struggle with particles larger than 70 microns. However according to Mirmorax, it can measure oil up to 2500 ppm. The company is also developing the technology for subsea produced water quality measurement applications.

(b) Microscopy Based Image Analysis:

A typical microscopy based image analysis system utilizes a high resolution video microscope and a light source to examine the content of a sample stream as illustrated in Figure 17(a). A sequence of video images as shown in Figure 17(b) are captured. Image analysis software analyzes the size and the shape of particles in the images. In focus particles are counted, their sizes measured and volumes calculated.

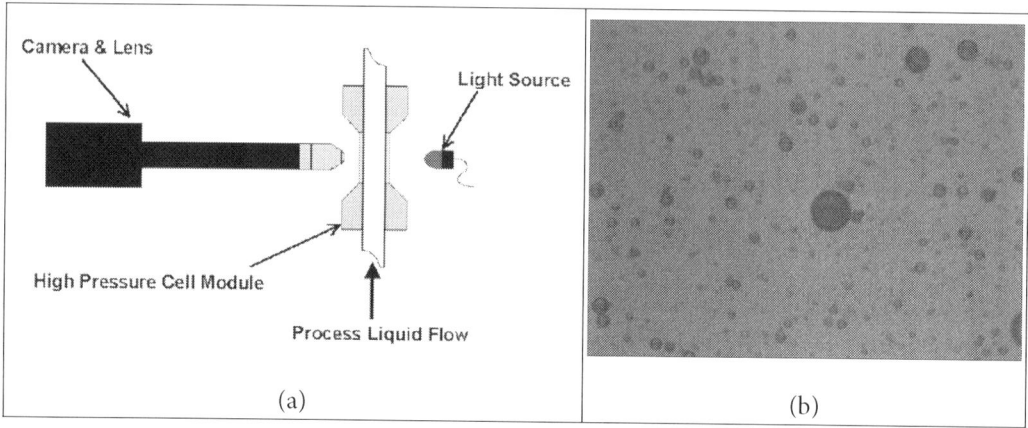

Figure 7.16 Schematic of a microscopy based oil in water analyzer (a) and an image of oil in water dispersion (b)

To distinguish oil droplets from solid particles, a shape factor is used. As an example, for the ViPA (Visual Process Analyzer) from Jorin, this shape factor is defined by: $(4\pi \times Area) / Perimeter^2$. For a perfect circle (sphere), the shape factor is always 1. As the length of perimeter increases compared to the area enclosed, the shape factor value decreases very quickly. For a particle to be classified as an oil droplet, the shape factor has to be very close to 1.

Gas bubbles may be present in a produced water stream. They are spherical and will also be captured by the microscopy image system due to their optical properties, the gas bubbles are recognized and excluded from oil droplet size and concentration calculations.

Several microscopy based image analysis systems are available for produced water quality measurement. Suppliers include Jorin, J M Canty, Fluid Image and Advanced Sensors. Microscopy image based analysis systems have been installed by a number of operators for produced water re-injection and top side process optimization applications.

One of the issues encountered in the past by these microscopy based image systems has been the short life expectance of their light sources. In the recent years, LED has been increasingly used as the light source by the instrument manufacturers, which significantly improves the reliability of the instruments. To mitigate fouling of optical windows, ultra-sonic window cleaning technology has been incorporated by some suppliers, e.g. J M Canty and Advanced Sensors. Both Jorin and J M Canty have also been developing a system suitable for subsea produced water quality measurement.

(c) Light Scattering:

Light scattering as a measurement technique is shown in Figure 18. The technique involves passing a visible light through an oily water sample. Particles (oil droplets, solid particles and gas bubbles) present in the sample will absorb as well as scatter the light. By measuring the amount of transmitted light together with the amount of light that is scattered at different angles, it is possible to distinguish the oil droplets from solid particles and gas bubbles, and to calculate the oil in water concentration.

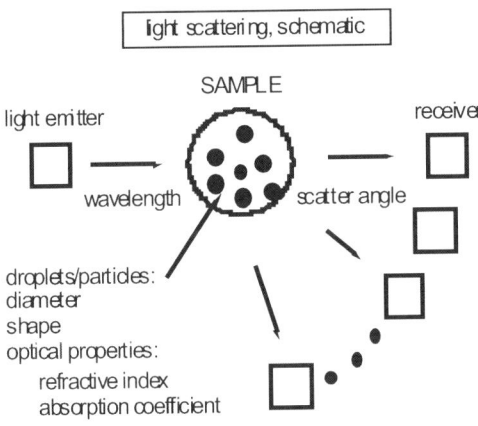

Figure 7.17 Schematic showing light scattering oil in water measurement principle

Light scattering is probably the most popular technique for online oil in water measurement. Most of ships above certain size (400 gross tonnage) are fitted with an oil content meter (OCM) that uses the light scattering technique. The application of this technology to produced oil in water measurement has had only limited success. Some of the main problems are related to fouling and also the non-ideal light scattering properties of solid particles and gas bubbles. Also the light scattering properties of oil droplets are affected by their size and this complicates the calculation of oil and solids concentrations.

There are several manufacturers who supply online light scattering based oil in water monitors. Well known suppliers for the oil and gas industry include Deckma and Optek.

(d) UV Fluorescence:

The use of UV fluorescence for oil in water measurement is described in the previous sections of this chapter. The key difference between an online monitor and a bench top analyzer is that no solvent extraction is required for online monitors. UV fluorescence is now the most widely used technique for online oil in produced water measurement applications.

There are two groups of fluorescence based online monitors, one is based on using a UV lamp and another one based on Laser Induced Fluorescence (LIF) as schematic diagrams show in Figure 19. With LIF, a probe type of device is made possible. Examples of suppliers for UV lamp based online monitors include Arjay Engineering, Sigrist, and Turner Designs Hydrocarbon Instruments. All three have a significant number of instruments installed at both onshore and offshore facilities.

Advanced Sensors and ProAnalysis are examples of suppliers of LIF instruments. Since about 2010 the development and commercialization of these monitors has been rapid. The incorporation of an ultrasonic cleaning technology by both Advanced Sensors and ProAnalysis has helped to reduce the amount of instrument maintenance and increased the reliability of these instruments. In addition, the availability of probe based LIF instruments means that such a device can be directly inserted into a produced water stream without the need of having to analyze a by-pass stream. Both Advanced

Sensors and ProAnalysis have developed the capability of a central processing unit being able to handle multiple probes, which is useful in terms of monitoring and assessing the efficiency of produced water treatment equipment.

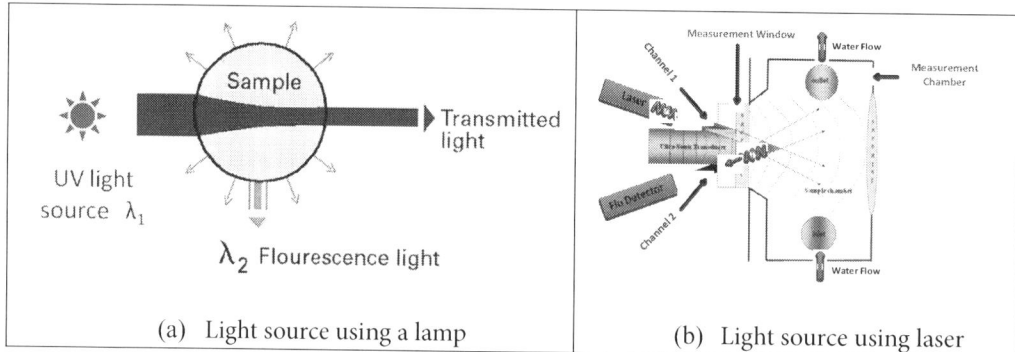

Figure 7.18 Schematic showing UV fluorescence based online monitoring technique

7.4.4. Calibration:

Most of oil in water measurement requires calibration. Calibration is a set of operations that establish, under specific conditions, the relationship between the output of a measurement system (i.e. the response of an instrument) and the accepted values of the calibration standards. Best practice guidance on calibration in general can be found. The brief discussion here will focus on calibration issues that are specifically related to oil in water measurement.

For a laboratory based oil in water measurement method, it is important to consider the following two issues when conducting a calibration:

- What calibration oil to use;

- How the calibration standards are prepared.

What calibration oil is used and how the calibration standards are prepared could have a significant impact on oil in water measurement results.

A Norwegian study carried out in 2000 showed that using a synthetic oil (a 50:50 mixture of isooctane and cetane) as a calibration oil for a single wavelength based Infrared method produced an oil in water result that is on average 19% less than that from using crude oil as a calibration standard. The significant difference is thought to have resulted from the different responses respectively given by the synthetic oil and crude oil to the Infrared quantification at the measurement wavelength of 2930 cm^{-1}.

Also calibration standards may be prepared in different ways. For example:

- Direct dissolving the calibration oil in a solvent to form a stock solution;

- Spiking the calibration oil into a water and then back extracting the oil to form a stock solution.

When a stock solution is prepared by direct dissolving, a high instrument signal is likely to be obtained compared to the back extraction approach. This is because with a back extraction, not all of the calibration oil will be extracted by the solvent. An example of how the calibration curves from the two techniques for preparing calibration standards may differ (not real data) is shown in Figure 20. When an actual sample is analyzed, using the calibration curve obtained by measurements on standard solutions prepared using the direct dissolving method would result in a lower oil in water value being reported. In the example illustrated in Figure 20, the Direct Dissolving calibration curve would report an oil in water value of about 56 mg/liter while the Back Extraction calibration curve would report an oil in water value of about 67 mg/liter.

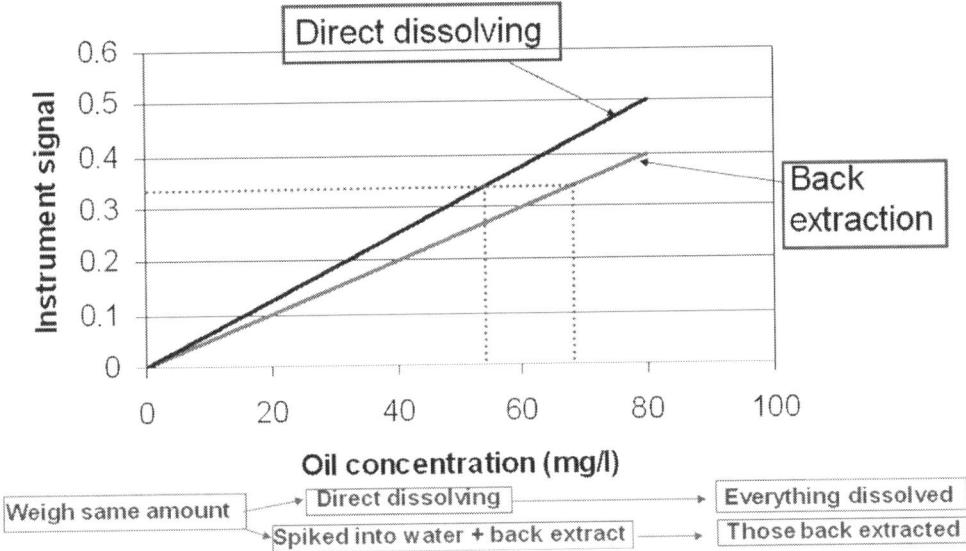

Figure 7.19 A schematic showing the effect of calibration standard preparation on oil in water results

Direct dissolving methods producing higher instrument signals have also been linked to solvent containing hydrocarbons. Tetrachloroethylene has been used as a solvent for an Infrared based single wavelength method. However the solvent supplied often included stabilizers that contain $-CH_2$, which absorbs Infrared at 2930 cm^{-1}. At the same time it has been found that some of these stabilizers could actually leach into the water being extracted. Thus, when using a direct dissolving method to establish a calibration curve, a higher infrared absorbance could be obtained compared to the back extraction approach. With a higher absorbance calibration curve, when a real sample is analyzed, a lower oil in water figure would result.

Compared to calibrating a laboratory bench top or a reference method, properly calibration of an online oil in water monitor is even more difficult. One of the key issues in calibrating an online oil in water monitor is related to the fact that produced water is extremely difficult to mimic in a laboratory. Thus, in the authors' view, proper calibration can only be realistically achieved in a field after the instrument is installed and only if a reference oil in water analysis method is available and representative samples parallel to online monitoring can be obtained. There is also the challenge with the availability of online samples with a wide range of oil in water concentrations. Thus actual field calibration curves are often based on a limited range of oil in water calibration samples with the curves extrapolated without validation to higher and lower values.

7.5. Organic Content:

While oil in water or TOG is measured for compliance monitoring of produced water discharges offshore, for onshore discharges of produced water, the regulation framework is often more stringent and involves the measurement of not only oil in water or TOG but also parameters such as COD, BOD and TOC. This section therefore discussed the parameters of COD, BOD, TOC and their relationships.

7.5.1. Chemical Oxygen Demand (COD):

The chemical oxygen demand measures the amount of oxygen that is consumed by a sample when all of the organic carbon is oxidized to CO_2. The measurement is carried out using hot chromic acid. Interference from a number of commonly found contaminants is possible. Steps must be taken to suppress those interferences.

7.5.2. Total Inorganic and Organic Carbon (TIC TOC):

As practiced in the oilfield, TOC is intended to measure the carbon content of dissolved organic material. For this reason it is also sometimes referred to as Total Oxidizable Carbon. TOC is reported as mg/liter carbon. However, it must be kept in mind that the mass of carbon per liter is not the same thing as mass of organic material per liter. Carbon is only one element in the organic material. Other elements include oxygen, hydrogen, nitrogen, and sulfur. The percent carbon in the dissolved organic material would have to be determined by other tests such as an elemental analysis. By multiplying the results by 12 (molecular weight of carbon), the moles C/L (moles of carbon per liter of sample) can be calculated. However, again, the moles of carbon per liter is not the same thing as moles of organic molecules per liter.

In order to measure only the dissolved material, the sample should be filtered first in order to remove any dispersed material. This results in a somewhat ambiguous result because droplets of light oil can pass through a typical 0.45 micron Millipore filter. Finer filtration can be used but the ambiguity will always remain due to the difficulty of deciding at what micron size are the contaminants dissolved or dispersed. Worse yet, some protocols for the measurement of TOC do not indicate whether the sample should be filtered or not.

As indicated, TOC is a measure of organic carbon. Compounds such as CO_2, H_2CO_3, Na_2CO_3, for example, contain carbon but are not organic and therefore are not measured by the test. In order to exclude these compounds from the test, the sample is acidified and purged with nitrogen. This drives the carbonate equilibria to CO_2 which is then removed by the nitrogen.

TIC (Total Inorganic carbon) is measured by adding an oxidizer and an acid to the sample. The oxidizer reacts with all inorganic carbon (bicarbonate carbonate ions) to release CO2. A nitrogen purge is used to remove the CO2. The mass of CO2 is measured and reported as TIC. The remaining inorganic carbon-free sample is then oxidized and subjected to UV radiation which works with the oxidant to break down all remaining carbon bonds in the sample to release a second quantity of CO2. The mass of carbon dioxide generated from the oxidation process is measured and reported as the TOC.

There are a number of TOC analyzers available on the market. They use a variety of different analytical methods. These methods include combustion, UV persulfate oxidation, or ozone promoted oxidation. With the combustion method, analysis is determined when carbon compounds are combusted in an oxygen-rich environment, resulting in the complete conversion of carbon-to-carbon dioxide. In UV persulfate oxidation, the carbon dioxide is purged from the sample and measured by a detector calibrated to directly display the mass of carbon dioxide. This mass is proportional to the mass of analyte in the sample. Persulfate reacts with organic carbon in the sample at 100°C to form carbon dioxide that is purged from the sample and detected. The ozone promoted method oxidizes the carbon by exposing it to ozone. One of the most important specifications for total organic carbon analyzers is the measuring range. Carrier gas flow rate, average analysis time and process temperature are important as well. Accuracy and resolution are also important to consider. An example of a carbon analyzer is the Sievers Model 800 Carbon Analyzer.

One method for introducing the sample to the analyzer is via syringe. Another is loop sampling, in which the sample loop introduction system allows repeatable analysis over a wide range of concentrations while avoiding the inherent dead volumes of syringe-based systems. On-line total organic carbon analyzer systems have an analyzer that is mounted in a process line and the sample is introduced via a connection to the process. Vial auto samplers are another way to introduce the sample. The liquid-sample transfer auto-sampler removes specific sample volumes from a standard vial and transfers the sample to the common analysis vessel. A sample carousel is loaded with up to fifty vials and placed in the auto-sampler for unattended analysis. In addition to measuring total organic carbon, total organic carbon analyzers may sometimes be used to detect total carbon, total inorganic carbon, and purgeable and nonpurgeable organic carbon.

The amount of TOC in a sample depends on the concentration of organic acids, esters, alcohol, phenols, BTEX (benzene, toluene, ethyl benzene, xylene). Many of these compounds are dissolved in the water and do not completely extract into hydrocarbon solvents and so do not fully contribute to typical measurements of TOG (Total Oil & Grease as measured by the EPA-1664 test). Thus, TOG together with TOC provides a more comprehensive understanding of the total amount of organic contamination in a sample. However, for oilfield purposes, TOC is usually not required from a regulatory standpoint.

Theoretical Calculation of TOC: In the case where a known compound is added to a sample, the TOC can be calculated theoretically. In that case, a pure sample of a known compound would have been added to the sample at a known amount, and known concentration. This is useful in establishing accuracy and precision of sample analysis. The calculation is carried out as follows.

The stoichiometric formula for an organic compound can be written as: $C_xH_yO_z$

In the explanation given here, the basis for the calculation is 1 mol of the known compound in a liter of waste water. The TOC test will result in the compound being oxidized to CO_2 and H_2O. Thus, x moles of CO_2 will be generated and measured due to the presence of 1 mol of the compound in the water.

$$C_xH_yO_z + (x + y/4 - z/2)\ O_2 \rightarrow x\ CO_2 + (y/2)\ H_2O$$

As a check of this formula, consider the following numerical example calculation:

$$C_3H_6O_5 + 2\ O_2 \rightarrow 3\ CO_2 + 3\ H_2O$$

Using the formula for the coefficient of oxygen gives:

$$x + y/4 - z/2 = 3 + 1.5 - 2.5 = 2$$

This is the correct answer.

Upon oxidation of 1 mol of the contaminant, x = 3 moles of CO_2 will be formed. The carbon content of 3 moles of CO_2 is 3 x 12 grams. Since the volume of waste water is 1 liter, the TOC is expressed as 36 gr C / L water. In terms of the general formula, the TOC is 12 x g C / L water, where x is the stoichiometric coefficient of carbon in the compound of interest. Note that the basis for this calculation was 1 mol of the contaminant. Thus the theoretical TOC for this example is really 12 x g C / mol of contaminant / L water sample.

If the basis had been 1 gram of contaminant, then the theoretical TOC would be calculated as:

$$\text{Theoretical TOC} = 12\ x\ /\ (12\ x + y + 16\ z)$$

The denominator is just the molecular weight of the contaminant. In this example the molecular weight is 122 gr / mol. Thus, the value of TOC is 36 / 122 = grams of TOC / gram of contaminant / L of waste water.

Theoretical Calculation of COD: The theoretical COD can also be calculated in much the same way. In fact, the stoichiometric equation for COD oxidation is the same as for TOC:

$$CxHyOz + (x + y/4 - z/2) \ O_2 \ \rightarrow \ x \ CO_2 + (y/2) \ H_2O$$

As before, we consider a molar basis and a gram mass basis. If we consider 1 mol of the known compound in a liter of waste water, the COD test will result in the consumption of $(x + y/4 – z/2)$ moles of oxygen. As a check of this formula, consider the following numerical example calculation:

$$C_3H_6O_5 + 2 \ O_2 \ \rightarrow \ 3 \ CO_2 + 3 \ H_2O$$

Upon oxidation of 1 mol of the contaminant, 2 moles of O_2 will be consumed. The mass of 2 moles of 2 is 2 x 32 grams, or 64 grams. Since the volume of waste water is 1 liter, the COD is expressed as 64 gr COD / mol contaminant / L water. In terms of the general formula, the COD is 32 $(x + y/4 – z/2)$ g COD / mol contaminant / L water.

If the basis had been 1 gram of contaminant, then the theoretical TOC would be calculated as:

$$\text{Theoretical COD} = 32 \ (x + y/4 – z/2) \ / \ (12 \ x + y + 16 \ z)$$

The denominator is just the molecular weight of the contaminant. In this example the molecular weight is 122 gr / mol. Thus, the value of TOC is 36 / 122 = grams of TOC / gram of contaminant / L of waste water.

7.5.3. The Ratio COD / TOC:

COD and TOC are obviously related chemically. For many sources of produced water there will be a correlation between measured TOC and measured COD. The ratio of TOC and COD indicates the molar ratio of oxygen in the organic material. During partial oxidation, the COD will decrease immediately. This is due to the incorporation of oxygen into the organic molecules forming acid, ether, ester and alcohol groups from CH_2 groups. The TOC will only decrease once the organic material has been converted to CO_2 and H_2O. Like TOC, a COD sample must be pre-filtered in order to remove TOG. The COD value represents both biodegradable material and non-biodegradable material.

Theoretical Ratio of COD/TOD: The ratio of COD / TOC can be calculated from the formulas already given:

$$\text{COD} / \text{TOC} = 32 \ (x + y/4 – z/2) \ / \ (12 \ x) = 8/3 + (2y – 4z)/3x$$

Table 7.8

Component	X	Y	Z	I mg COD mg^{-1} $C_xH_yO_z$	II mg TOC mg^{-1} $C_xH_yO_z$	III mg COD mg^{-1} TOC
Oxalic acid	2	2	4	0.18	0.27	0.67
Formic acid	1	2	2	0.35	0.26	1.33
Citric acid	2	4	3	0.64	0.32	2.00
Glucose	6	12	6	1.07	0.40	2.67
Lactic acid	3	6	3	1.07	0.40	2.67
Acetic acid	2	4	2	1.07	0.40	2.67
Glycerine	3	8	3	1.22	0.39	3.11
Phenol	6	6	1	2.38	0.77	3.11
Ethyl glycol	2	6	2	1.29	0.39	3.33
Benzene	6	6	0	3.08	0.92	3.33
Acetone	3	6	1	2.21	0.62	3.56
Palmitic acid	16	32	2	3.43	0.75	3.83
Cyclohexane	6	12	0	3.43	0.86	4.00
Ethylene	2	4	0	3.43	0.86	4.00
Ethanol	2	6	1	2.09	0.52	4.00
Methanol	1	4	1	1.50	0.38	4.00
Ethane	2	6	0	3.73	0.80	4.67
Methane	1	4	0	4.00	0.75	5.33

A useful check of TOC measurement is to calculate the COD / TOC ratio. In fact, this ratio is a test of both COD and TOC. It is possible to calculate a theoretical ratio for many compound classes. For example, the COD / TOC ratio for methane can be calculated from the following chemical formula:

$$CH_4 + 2\,O_2 = CO_2 + 2\,H_2O$$

The ratio of COD/TOC is given by:

2 moles oxygen x molecular weight of oxygen / (1 mol carbon x molecular weight of carbon) = 2 x 32 / 12 = 5.33

According to the above formulas, the COD/TOC ratio can also be calculated by:

$$COD / TOC = 8/3 + (2y - 4z)/3x = 2.67 + 2.67 = 5.33$$

The general formula given above can also be used to calculate the maximum value of COD / TOC that can be found by simple organic molecules. To maximize the COD / TOC value, the value of z would have to be a minimum such as zero (no oxygen in the molecule). The value of y/x would need to be a maximum which can only be achieved by the methane molecule which has 4 hydrogens, one for each valence electron of carbon. Thus, methane represents the highest achievable value of COD / TOC. This is also verified in the table above.

This value of 5.33 is useful to keep in mind. Whenever TOC and COD results are reported, a useful check of their validity is to calculate the COD / TOC ratio and confirm that it is less than 5.33.

7.5.4. Biological Oxygen Demand (BOD):

Biological Oxygen Demand. This parameter is intended to measure the mass of organic material that can be readily biodegraded. The test measures the amount of oxygen that is consumed by common bacteria as they digest the organic material in a water sample. The test is carried out over a five day period at 20 °C. The BOD value is reported as ppm of oxygen. It is a measure of whether or not a bioreactor could be used to meaningfully decrease the organic carbon content.

7.6. Solids Content (TSS – Total Suspended Solids):

Measurement of suspended solid particles is important when it comes to produced water re-injection. Supended solid particles may include sand, clay, precipitated salts, scales such as calcium carbonate, barium and strontium sulphate and corrosion products such as iron oxides and iron sufidea. Measurement of suspendeds solid is usually termed as measurement of Total Suspended Solids (TSS).

Total Suspended Solids (TSS) is commonly defined as those solids which will not pass through a standard 0.45 micron filter. These include both those solids that will settle or float and the lighter nonsettleable (colloidal) solids.

Total Suspended Solids should be differentiated from Total Solids. The term Total Solids is applied to the material left in a dish after the evaporation of a sample and its subsequent drying in an oven at a defined temperature. Total Solids will include both Total Suspended Solids and Total Dissolved Solids. Total Dissolved Solids usually refers to those solids that will pass through a standard 0.45 micron filter.

7.6.1. Laboratory Methods:

For a typical TSS analysis, a well-mixed measured sample is poured into a filtration apparatus and, with the aid of a vacuum pump or aspirator, drawn through a pre weighed standard laboratory glass fiber filter. After filtration, the glass fibre filter is dried at a certain temperature, typically 103105°C, cooled, and reweighed. The increase in weight of the filter and solids compared to the filter alone represents the Total Suspended Solids (TSS).

There are a number of standard methods available for measuring TSS. Some of these are listed in Table 9.

Table 7.9 commonly used TSS analysis methods

Reference Method	Principle	Measurement range	Precision Info	Status
2540 D	Glass fiber filter, e.g. Whatman grade 934AH with particle retention above 1.5 μm.	2.5 mg/l to 20,000 mg/l	The standard deviation was 5.2 mg/L (coefficient of variation 33%) at 15 mg/L, 24 mg/L (10%) at 242 mg/L, and 13 mg/L (0.76%) at 1707 mg/L in studies by two analysts of four sets of 10 determinations each. Single-laboratory duplicate analyses of 50 samples of water and wastewater were made with a standard deviation of differences of 2.8 mg/L.	In use
NS-EN 872	Glass fiber filter (1.6 μm)	2 mg/l		In use
ASTM D5907-10	Glass fiber filter	4 to 20,000 mg/l		In Use

The TSS methods listed in Table 9 are commonly used for water, waste water and effluent water quality assessment. However from the table it is clear that these methods are suited for total suspended solid contents more than 2 mg/l. For produced water total suspended solids measurement, there is no universally agreed standard method. Often a similar method to those given in the Table 9 is adopted. However the glass fiber filter would normally be replaced by 0.45-micron cellulose acetate Millipore filter.

Also it should be pointed out that produced water samples will almost always contain oil droplets that will be filtered out with the suspended solid particles. Furthermore solid particles will likely to be coated with oil. Therefore to obtain the correct TSS in produced waster, the content collected on the filter will need to be washed with solvent to remove both the dispersed oil as well as oil coated on the surface of the solid particles before the removing the filters for drying and weighing. Upon request, most labs will report TSS that includes oil on the solids, and the amount of oil on the solids as determined after rinsing the filtered solids with warm toluene. The weight of oil-free solids is then reported separately.

It should also be noted that to obtain a true TSS value, the produced water should be filtered online. This avoids capturing solids formed by scale mineral precipitation due to changes in the samples temperature, pressure, and pH. TSS determined by laboratory filtration of aged samples is always at risk of error due to scale mineral precipitation and the oxidation of soluble ferrous iron to insoluble ferric iron.

7.6.2. Online Methods:

Laboratory methods as described above are time consuming. Online methods provide a big advantage in that they will give a result much more quickly.

For the measurement of total suspended solids in runoff and river discharges, there are technologies on the market which have been reviewed by Jessica Brnigan. These are shown in Table 10.

Table 7.10 Online TSS Sensors and Probes

Product	Method of Measurement	Range (mg/L)	Accuracy	Repeatability
Paab SS Probe[1]	90° Scattering/Light Absorption	0-30,000	± 3% of reading	98%
Galvanic Monitek Acoustic SS Probe[2]	Ultrasonic Reflection	0-10,000	± 5% of reading	± 4% of reading
Hach TSS Sensor[3]	Modified Absorption Measurement	1-500,000	Based on sampling technique	< 4% of reading
Insite IG Portable SS Analyzer[4]	Single Gap Optical	0-30,000	± 3% of reading or ± 20 mg/L	± 0.5% of reading
Insite IG SS Analyzer[5]	Single Gap Optical	0-30,000	3% ± of reading	± 0.5% of reading
Royce Water Process Analyzer + Sensor[6]	Single Gap Optical	10-80,000	± 5% of reading or ± 5mg/L	± 1% of reading or ± 2mg/L
Royce TSS Analyzer + Sensor[7]	Single Gap Optical	10-80,000	± 5% of reading or ± 5mg/L	± 1% of reading or ± 2mg/L
Royce Portable TSS Analyzer[8]	Single Gap Optical	10-10,000	± 5% of reading or ± 100 mg/L	± 1% of reading or ± 20mg/L

While the technologies listed in Table 10 may work well for the measurement of TSS in runoff water, most of them are not designed to differentiate solids particle from oil droplets. As a result these methods are not particulary suitable for the measurement of TSS in produced water.

For the measurement of TSS in produced water online, the technology must be able to differentiate solids from oil droplets. The only method that is currently on the market is those using microscopy based image analysis as described above.

7.7. Oil Droplet and Solid Particle Size:

The size and size distribution of the dispersed hydrocarbon phase is one of the most important parameters governing the performance of most deoiling equipment.

Droplet size analysis is a useful tool in the characterization of oil/water dispersions. When done properly, the results allow the selection of separation equipment to occur with a high degree of confidence that the equipment will perform as intended. When used for troubleshooting, it often is a strong indicator of the source of produced water treating problems. Performance of deoiling equipment is often defined in terms of a percentage removal efficiency for a given range of drop sizes. Pumping, transport in pipelines, and change in pressure (especially across valves) can all modify the properties of an oil/water mixture, especially the droplet size distributions. Monitoring of these changes can give insight into these effects and can help in optimization of separation plant operation at e.g. oil/gas production platforms.

For produced water reinjection, knowing the size and size distribution of solid particles is also extremely important. Solid particles are well known to impair injectivity and different sized solid particles will have different impacts on the well's injection characteristics.

7.7.1. Sampling for Droplet Size and Solid Particle Size Measurement:

Sampling for accurate droplet diameter and solid particle size and size distribution measurements requires representative sampling. In the case for the measurement of drop diameter, additional shear must not be introduced during the sampling process.

Representative sampling is discussed earlier in this Chapter under the topic of iso-kinetic sampling. Most routine sampling does not involve measurement of oil drop diameters and / or solid particle sizes in the produced water. Thus, it is unlikley that sample locations and sample point design will have the features required to ensure reliable sampling and measurement.

- For sampling for oil droplet diameter determination and solid particle size measurement, the following guidance should be considered in addition to those that were discussed in the earlier sections of this Chapter.

- Insure that the sample is representative. Obtaining a representative sample of a two phase liquid can be very difficult and a number of samples may be required with the results averaged.

- Insure that the sample point is correctly designed with a quill extension into the main fluid flow away from the piping wall. In some cases samples may be required from different points to ensure that the droplet distribution does not alter over the cross-section of the area being sampled (e.g. stratified flow).

- The distance between the sample probe and the sample valving should always be minimized, as the longer the sample line, the greater the chance of errors influencing any subsequent analysis.

- Make every effort to have iso-kinetic sampling used at the point where the sample enters the sample system. This ensures the sample is not biased by changes in the flow profile.

- Insure that iso-energetic conditions are maintained after the sample has entered the sample system. This will ensure that sampling does not alter the droplet size distribution by changing the dynamic equilibrium between coalescence and emulsification.

- Depressurization of the sample should be avoided. Depressurization is likely to alter the droplet size distribution through the imposition of shear forces on the sample and may lead to problems due to the evolution of gas bubbles. A sample cylinder should be used to avoid de-pressurizing the sample.

7.7.2. Laboratory methods:

In addition to the sampling considerations given above, the following points should also be considered when measuring particle size and size distributions using grab sample methods.

- Most equipment for the measurement of droplet size distribution operates at atmospheric pressure. If the sample is pressurized, it must be depressurized without shear.

- Some equipment for the measurement of droplet size distribution may be temperature sensitive, requiring the sample to be cooled before measurement. This cooling may also affect the droplet size distribution.

- Coalescence may rapidly alter the size distribution of dispersed droplets. Samples must be analyzed as quickly as possible.

- Solid particles and gas bubbles present in the sample may be detected as droplets, leading to a skewed size distribution or vice versa.

There are three main laboratory methods that are currently available for the measurement of oil droplet and solid particle size and size distribution.

Aperture Counting Methods: Aperture counting based particle analyzers were originally developed by Coulter Inc (now Beckman Coulter). In this method, a tube with a small aperture is immersed into a beaker containing particles suspended in a low concentration electrolyte. Two electrodes are placed each side of the aperture creating a "sensing zone". As a particle passes through the aperture, a volume of electrolyte equivalent to the immersed volume of the particle is displaced from the sensing zone, which produces a voltage pulse or current pulse. The pulse height is proportional to the volume of the particle. Thus by measuring the number and magnitude of the electrical pulses the size distribution of the particles can be determined.

The correct size of the aperture is linked to the particle size distribution being measured. Aperture size typically ranges from 20 to 2000 microns. According to Beckman Coulter, each aperture can be used to measure particles within a size range of 2 to 80% of its size, which means that in theory the instrument can cover an overall particle size range of 0.4 to 1600 microns. However, the ability of the technology to analyze particles is limited to those particles that can be successfully suspended in an electrolyte solution. In reality the upper limit is more likely to be around 500 µm for sand and perhaps smaller for oil droplets because of settling.

Selecting a correct aperture size for the right application is important. If the aperture size is too small, deformation of the oil droplet may occur. In the case of solid particles, a blockage may happen as particles pass through the aperture. According to Beckman Coulter, if the sample to be measured is composed of particles largely within a 30:1 diameter size range, then a single aperture size may be required. For example, a 30 μm aperture can measure particles from about 0.6 to 18 μm in diameter. A 140 μm aperture can measure particles from about 2.8 to 84 μm. otherwise two or more apertures will have to be used and the measured particle results be overlapped to provide a complete particle size distribution. In some cases the sample will need to be diluted with an isotonic solution to achieve a suitable droplet population density, however this dilution may influence the droplet size distribution. It is also worth mentioning that the aperture counting method requires the continuous phase to be electrically conductive. Fortunately this is normally not an issue with produced water.

Laser Diffraction Methods: Laser diffraction measurement is based on the fact that laser light scatters when it passes through a produced water sample containing oil droplets and solid particles. It is understood that large particles scatter light at small angles relative to the incident laser beam and small particles scatter light at large angles. Thus by measuring the angular scattering intensity data, particles size and size distribution can be calculated using available theory.

Diffraction based methods are sensitive to the number of particles in the sample. Too few particles may not give a statistically significant result while too many droplets may result in optical interference with the measurement. Not all equipment warns the user that these errors may be occurring.

One of the best established laser diffraction based instrument is the Mastersizer from Malvern[ ]. A schematic of such an instrument is shown in Figure 21. According to Malvern, the instrument can provide a particle size measurement range from 0.01 micron to 3500 micron using a single optical measurement path. To assist the analysis, different sample handling systems are also available from Malvern. These sample handling systems will help ensure that particles delivered to the instrument for measurement are stable, representative of the samples.

Laser Obscuration Time: Laser Obscuration Time (formerly called Time of Transition) methods measure the interaction pulses generated as particles intersect a rotating laser beam. As a particle intersects with the rotating laser beam as depicted in Figure 22, the instrument detects the exact time the particle obscures the laser beam. With the speed of the rotation laser beam known and the time of the obscuration of the laser beam with the particle, the particle size can be determined.

The technology was originally developed in Israel in the 1980's. Instruments are now available from Ankersmid (EyeTech) and Donner Technology (CIS 100). The newer versions of these instruments have also incorporated both LOT and image analysis to allow for not only particle size measurement but also shape analysis. According to the suppliers, the instruments have a measurement range of 0.1 to 3600 microns.

The interaction pulses are affected by the shape of the particle allowing differentiation between particles of different shape. Thus solid particles such as sand or corrosion products, which tend to have asymmetrical shapes can be differentiated from symmetrical hydrocarbon droplets.

7.7.3. Online Methods:

While laboratory methods described above are useful, they are not always helpful when it comes to measuring oil droplet and solid particle sizes and size distributions. Also one of the difficulties with produced water is that it contains dispersed oil which can coalesce resulting in the droplet size and size distribution changes during the sample handling and laboratory analysis process. Online methods help minimize the risk.

There are a number of techniques that have been developed for online partile size and size distribution measurement. These may include:

- Laser difraction;

- Laser reflection based;

- Microscopy image analysis based;

- Sound based;

Laser Diffraction: The principle of laser diffraction is described above. The Malvern Mastersizer is an example of an instrument for laboratory based particle size analysis. Over the years, Malvern has been trying to develop a laser diffraction based online instrument for the measurement of oil in produced water, which has resulted in the adoption of their Insitec device for this application. A picture of the instrument is shown in Figure 23. A wet flow cell is required. Laboratory trials have been carried out at NEL. It was found that as long as the flow cell can be kept clean, the instrument performed well.

The Malvern Insitec can cover a measurement range of 0.1 to 1000 microns.

Figure 7.20 A picture showing the Malvern Insitec instrument

Laser Reflection: A well established laser based system is the Focused Beam Reflectance Measurement (FBRM) instrument from Mettler-Toledo. For a typical FBRM, laser light is transmitted through fiber-optics to a probe tip. A rotating optical lens at the probe tip deflects the laser. When the probe is inserted into a dispersion system that contains particles, e.g. solids and droplets, the laser emitted light is reflected if it scans across the surface of a particle. By measuring the reflection time, and with the known speed of the rotating lens, FBRM can calculate the chord length. The principle of FBRM is shown in the figure below.

The technology and has found successful applications in the droplet size measurement of emulsions encountered in oil and gas multiphase separators. It is a probe type instrument and therefore can be easily inserted into a pipeline. It has a wide size measurement range of 0.5 to 2,500 μm. It is one of few instruments that can measure very high concentration dispersions as well as low concentration dispersions.

FBRM probes are commonly rated to 10 bar as standard, with options up to 250 bar depending on the type of mounting used. One of the issues with the FBRM is that it measures the chord length, not the particle diameter. Experiment results show that FBRMs in general undersize particles in comparison to those determined by for an example microscopy based image analysis systems. Also the system does not provide good information on concentration. A FBRM system has been tested at NEL for water-in-oil droplet size measurement in a pipeline.

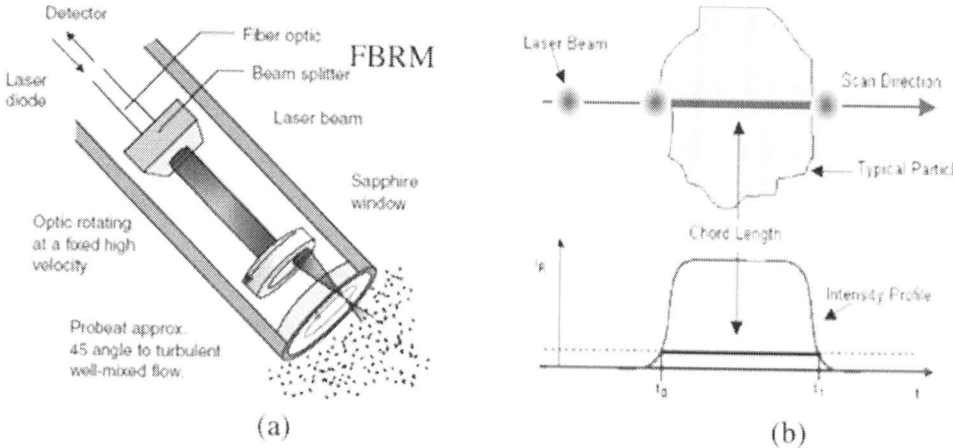

Figure 7.21 (a) Schematic of the FBRM probe technique; (b) Chord length determination

Similar technologies are available from HEL Group and Sequip. The LaserTRACK probe from HEL Group is constructed in a similar fashion as the Mettler-Toledo's FBRM. A laser beam is projected through the sapphire window of the probe and focused into a solution. The focused beam is then moved at a velocity of around 2 m/s so it follows an elliptical path within the solution. As particles pass by the window surface, the focused beam will intersect the edge of a particle, which begins to backscatter laser light until the beam reaches the particles opposite edge. Particle cord length is then calculated in the same way as the FBRM system. A schematic showing the measurement principle is given in Figure 25.

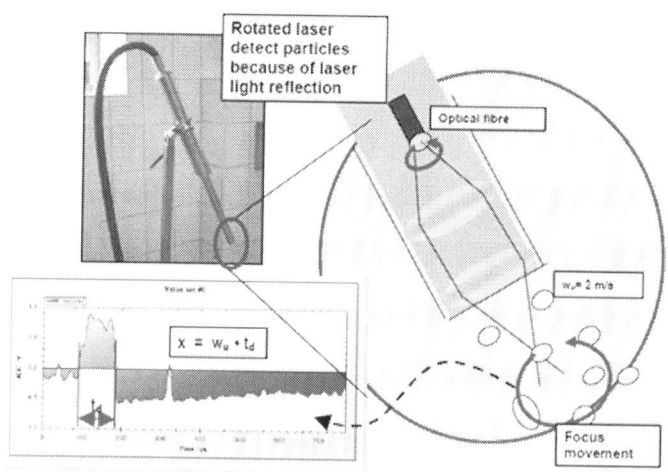

Figure 7.22 Schematic showing measurement principle of the HEL Group's 2D-ORM system

The key difference between the LaserTRACK and FBRM seems to be related to the arrangement of where and how the laser light is focused and moved. According to HEL's presentation, LaserTRACK works better than the FBRM in terms of true representation of particles, trend monitoring and accuracy and reproducibility. However, an independent study shows that the LaserTRACK system under-sizes particles even more than the FBRM.

Acoustic Based Systems: Two well-known sound systems exist. One is based on using focused ultrasonic signals into a produced water stream. Particles (solids, oil droplets and gas bubbles) that pass through a small defined focused ultrasonic beam area will generate reflected waves or acoustic echoes that are detected and linked to particle sizes. The principle of the technology was summarized above.

The second sound system is constructed based so called ultrasonic extinction. A schematic of the principle for determination of frequency dependent ultrasonic extinction is shown in the Figure 26 (a). An electrical RF generator is connected to a piezoelectric ultrasonic transducer, which generates ultrasonic waves that pass through a solution media. Particles present in the solution media will interact with the ultrasonic wave and reduce the intensity, (hence called attenuation / extinction). The extinction of the ultrasonic waves is calculated from the intensity of the ultrasonic waves at the transmitter end and that of the received end, which is then converted to the particle size and size distribution through complex mathematical algorithms.

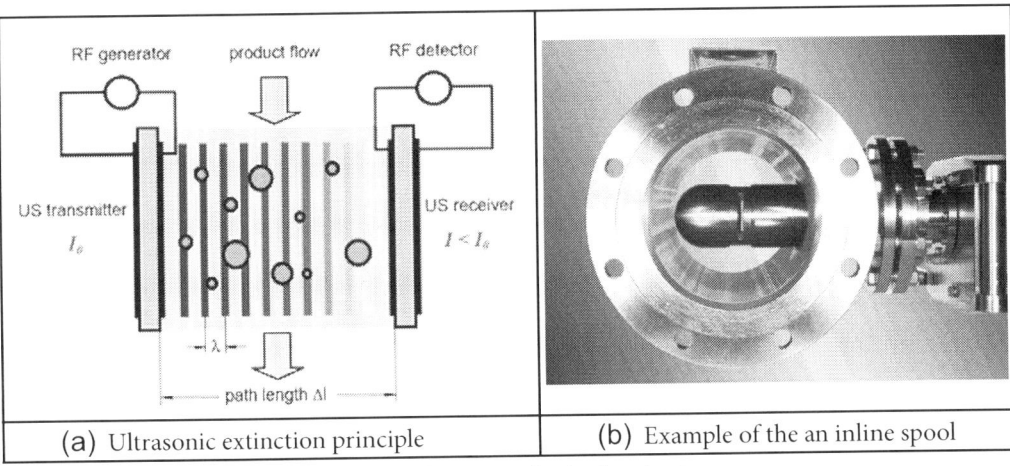

| (a) Ultrasonic extinction principle | (b) Example of the an inline spool |

Figure 7.23 Ultrasonic extinction based system

The technique has been developed by a German company called Sympatec, who now supplies its inline Opus instrument as shown in Figure 26 (b). The Opus instrument is designed as a finger probe which can be adapted to different process pipes. Its standard version provides real time particle size analysis with following process conditions:

- Temperature: 0-120 °C;
- Pressure: 0 - 40 bar;
- Particle size: 0.001 to 3000 µm;
- Concentration: 1-70%

The instrument has been used to characterize water-in-oil dispersions at various locations. Most recently it has been tested at NEL for a similar application.

Image Analysis Systems: Microscopy based image analysis technique has already been discussed for the measurement of oil in produced water, and there are a number of suppliers with significant oilfield presence. More systems that use a similar approach are discussed here. These include the EyeTech Probe from Ankersmid. The EyeTech probe, as schematic diagram shown in Figure 27, has the following specifications:

- Size range: customizable (1-300 µm, 2-1300 µm; 0.75 – 500 µm; 10-5000 µm)
- Concentration: 1-15 volume %;
- Reproducibility: 2%;
- Pressure: 1-6 bar customizable;
- Temperature: to 160 °C

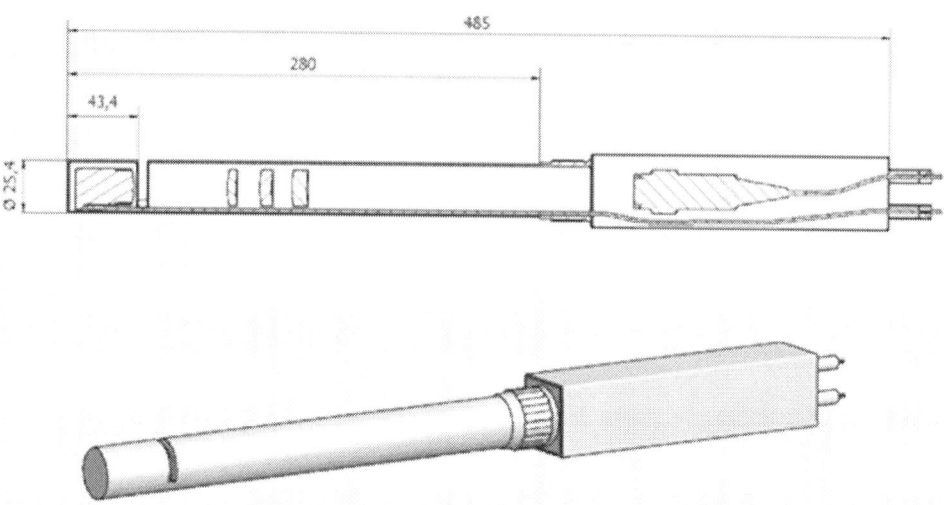

Figure 7.24 A schematic diagram of the EyeTech Probe

Instead of having fluids pass through a gap between a light source and a camera, there is also a different approach where both light source and camera stay at the same side. The PVM (Particle Vision Microscope) probe from Mettler-Toledo uses such an approach. The probe consists of six near-IR lasers which illuminate a small area in front of the probe face as a schematic diagram shown in the Figure 28 (a). It then records digital images, an example of which is shown in Figure 28 (b), with the illuminated area having a field of view of 1075 μm x 825 μm for the model V825 Ex. The instrument has a resolution of 2 μm. It can handle process condition as follows:

- Temperature range: to 120 °C

- Pressure: to 10 bar but customizable to 150 bar

- Power: 230 VAC, 50-60 Hz, 0.2 , 21 W

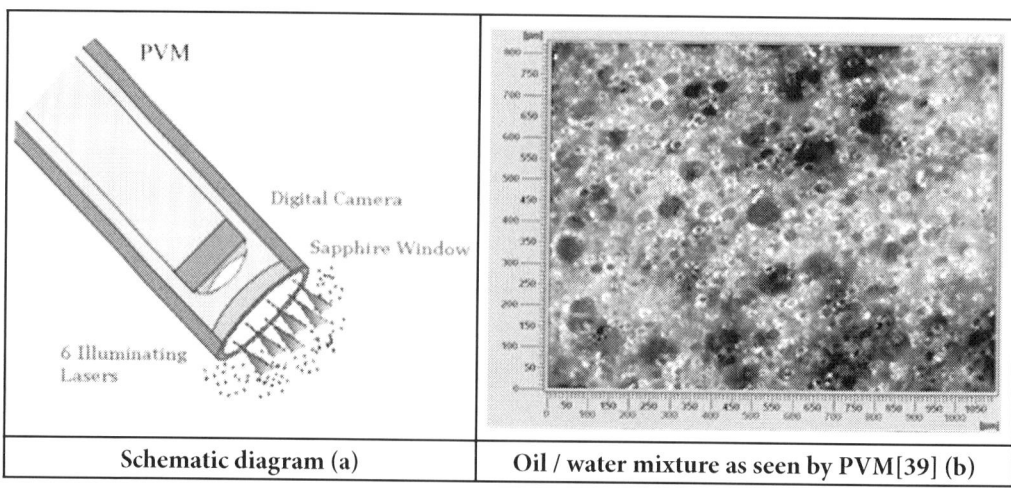

| Schematic diagram (a) | Oil / water mixture as seen by PVM[39] (b) |

Figure 7.25 A schematic diagram showing how the PVM works

Despite the availability of these online monitoring methods, only the microscopy based image systems likes those from J M Canty, Jorin, Fluid Image and Advanced Sensors have found applications in produced water oil droplet and solid particle size and size distribution measurement.

7.8. Precision of Analysis Methods:

Precision is a measure of the range of results obtained when a number of analyses are performed by the same analyst using the same procedures. Precision is not the same as accuracy. A series of analytical measurements may be very repeatable (and thus precise), but this does not mean that the results are accurate. When it comes to a discussion of the precision of an analytical method, two parameters are often referred to:

- Repeatability

- Reproducibility

Repeatability is the closeness of the agreement between repeated measurements of the same property using the same method on identical test material in the same laboratory by the same operator with the same equipment within short intervals of time.

Reproducibility is the closeness of the agreement between measurements of the same property obtained with the same method on identical test material in different laboratories with different operators using different equipment.

Precision is important as it forms part of the information required to assess the uncertainty associated with a measurement result, in this case, the oil in water or solids in water concentration or particle size.

Also the performance of an online method is often judged by comparing the result from an online method to that from a laboratory based reference method. If the reference method has a poor precision, i.e. poor repeatability and poor reproducibility, such comparison may be unreliable for the assessment of the performance of an online method. In addition, measurement precision may affect how produced water equipment performance guarantees can be specified. The authors have seen examples of performance guarantees that cannot be realistically implemented due to the level of precision reported for some of the reference methods. This section looks into the precision data available from the various analytical methods.

7.8.1. Reference Oil in Water Analysis Methods:

As discussed in Section 7.4 (Oil in Produced Water), there are several oil in water analysis methods based on three main measurement principles: Infrared absorption, ultraviolet absorption, gravimetric and GC-FID.

For each of the measurement principles, well known reference methods with available precision data have been tabulated. These are respectively shown in Table 12 to Table 15. Some of the methods may have already been superseded or withdrawn, but it is still useful to include them here.

Table 7.11 Examples of using single wavelength based IR oil in water analysis reference methods

Reference Method	Principle	Measurement range	Precision Information	Status
USA EPA 413.2	Freon extraction & IR	0.2-1000 mg/l	14 mg/l oil and grease, recovery was 99% with a standard deviation of ±1.4 mg/l	Superseded
ASTM 3921-85	Freon extraction & IR	0.5–100 mg/l	Overall precision in the concentration range of 0.6 to 66 mg/l may be expressed: $S_t=0.167\ x + 0.333$ where S_t is the overall precision and x is the oil and grease concentration, mg/l	Superseded
ASTM D 7066-04[6]	S-316 extraction & IR	5 – 100 mg/l, could be extended with a larger or smaller sample	Precision of 47.1%, 49.9%, 66.3%, 50.3%, 37.2% and 24.7% were found for mean values respectively at 30.5 mg/l, 21.2 mg/l, 6.6 mg/l., 6.4 mg/l, 429.9 mg/l, 551.2 mg/l	In use, but not commonly used
ASTM D 7575-10	membrane extraction and IR	5 – 200 ml/l	Precision of 16.8%, 19.6%, 11.6%, 10.5%, 10.2% were respectively found at oil-in-water concentration of 10.8 mg/l, 9.3 mg/l, 58.8 mg/l, 91.8 mg/l, 86.6 mg/l	In use, but not commonly used

Table 7.12 Examples of using three wavelength based IR reference oil in water analysis methods

Reference Method	Principle	Measurement range	Precision Information	Status
IP 426/98 (Energy Institute)	Tetrachloro-ethylene, and IR, using Florisil	0.5 – 150 mg/l	repeatability = 1 mg/kg over 0.5 mg/kg – 5 mg/kg range: Repeatability = 0.2127 (x+5) mg/kg where x is the mean of the oil-in-water concentration over 5 mg/kg – 150 mg/kg range	Still in use, mainly by the refineries and the offshore industry for bi-annual produced water analysis in the UK

Table 7.13 Examples of gravimetric based oil in water reference methods

Reference Method	Principle	Measurement range	Precision Information	Status
EPA 1664 B[3]	N-Hexane extraction & gravimetric	5-1000 mg/l (oil and grease) with a minimum detection limit of 0.91mg/l	Initial precision of 11% is acceptable.	In use, official oil and grease measurement method in the USA

Table 7.14 Examples of GC-FID based reference methods

Reference Method	Principle	Measurement range	Precision Information	Status
ISO 9377-2:2000	Solvent extraction and GC-FID C10 to C40 hydrocarbon	0.1 mg/l – 150 mg/l	Respective precision of 9.6%, 33.5%, 21.1% and 40.5% were observed at oil concentration of 2.99 mg/l, 0.7 mg/l, 3.61 mg/l and 1.04 mg/l	In use
OSPAR GC-FID	Solvent extraction and GC-FID C7 to C40 hydrocarbon	0.1 mg/l – 150 mg/l	Respective precision of 11.4%, 20.1%, 21.7%, 21.4% and 36.0% were respectively found at oil concentration of 6 mg/l, 18 mg/l, 25 mg/l, 43 mg/l and 100 mg/l.	In use, reference method for oil-in-produced water analysis in the North Sea

From the tables above, it is clear that even for the best established reference oil in water analysis methods, the repeatability could be well above 10%. The Reproducibility range will normally be higher than the repeatability range.

Take as an example of the IP 426/98 method. Within the concentration range of 0.5 mg/kg to 5 mg/kg, repeatability is 1 mg/l. Therefore, if someone takes a sample and gets an oil-in-water figure of 5 mg/kg, for the same person to repeat the measurement, the result (with 68% confidence) could be anything between 4 mg/kg and 6 mg/kg.

Take as another example, the OSPAR GC-FID method. NEL conducted an inter-laboratory study and found that the average reproducibility coefficient is about 20% over the concentration range of 6 to 100 mg/l. This means that for an oil-in-water value of 6 mg/l from one person, another person from a different lab could get a value between 4.8 mg/l or 7.2 mg/l, and be within the method's precision (with a 68% confidence).

Note that the 68% confidence interval is basically ± 1 standard deviation around the mean value of the results. For a 96% level of confidence in the measurement, the confidence interval will be ± 2 standard deviations.

7.8.2. Online Oil in Water Measurement Methods:

Compared to the reference methods discussed above, few online monitors have gone through the rigorous process that is required to establish the precision information for the instrument. However, most online monitor suppliers do provide a data sheet with information on "accuracy". Table 16 provides details on accuracy information from some of the well-established online oil in water monitors.

Table 7.15 Examples of online oil-in-water monitors and suppliers

Technique	Suppliers	Model	Claimed "accuracy"
Image analysis	Advanced Sensors	Ex-400m	±4% over 0-1000 ppm
	J M Canty	Inflow	±1%
Light scattering	Deckma	OMD-17	±5 ppm over 0-99 ppm range
	Inventive Systems	Oil Sentry OS-100m	±2 ppm at 15 ppm; ±10 at 100 ppm
UV Fluorescence	Arjay Engineering	HydroSense 2410	±1ppm over ranges either 0-10 ppm or 0 -5000ppm
	Turner Design	TD-4100XD	Range: 1 ppb to 1000 ppm
	Advanced Sensors	EX-100P	1% over 0 to 20,000ppm
	ProAnalysis	Argus Environment	less than 10% over 0 – 1000 ppm

Accuracy is defined as the closeness of the agreement between a measurement result and the "true" value. The problem with oil in water analyses is that there is no absolutely "true" value for the several reasons previously discussed.

It is not always clear how each of the manufacturers generated their accuracy data and what was considered as the "true" value in calculating the accuracy figures.

7.8.3. TSS Analysis Methods:

Precision information of the commonly used reference methods is shown in Table 9. Compared to oil in produced water measurements, there have been much fewer analytical methods available and much less information on precision associated with TSS in produced water measurements. In the authors' view, precision of TSS analysis methods is likely to be much worse than a typical oil in water analysis method.

7.8.4. Particle Size Measurements:

Particle size is a notion introduced for comparing the dimensions of particles. Particles are three dimensional objects and unless they are perfect spheres, they cannot be fully described by a single dimension such as diameter. Thus for particle size measurement, it is often convenient to define the particle size using the concept of equivalent spheres. In this case, the particle size is defined by the diameter of an equivalent sphere having the same property as the actual particle such as, for example, volume, mass, or surface area.

For the Malvern Mastersizer, the following specifications have been given. However Malvern does indicate that the figures may be dependent on samples and how the samples are prepared.

- "Accuracy"- Better than 1%

- "Repeatability" - Better than 0.5%

- "Reproducibility" - Better than 1%

Laser diffraction does not require calibration but can be verified by using certified particle standards. It should be noted that certified particle standards are generally spherical or close to spherical in shape – which may not be the case for TSS recovered from produced water.

7.9. Subsea Produced Water Quality Measurement:

Subsea separation and produced water re-injection and / or discharge is an integral part of the subsea processing strategy. Subsea processing is increasingly considered and accepted as a way forward for maximizing the oil recovery of offshore oil fields. While subsea separation and produced water re-injection is already happening, e.g. the Troll pilot and Tordis in the North Sea by Statoil, and Marlim in Brazil, there are still technology gaps regarding subsea water quality measurement. Without closing these gaps, monitoring of water quality subsea can only be achieved by taking subsea samples using ROVs (Remotely Operated Vehicles) and bringing them up to the surface for analysis, which is not only extremely expensive and time consuming, but more importantly it does little to help oil production that requires continuous and timely data for control and operations.

This section discusses the technical requirements of subsea water quality measurement devices, potential technologies, and research and development efforts that have been carried out and on-going.

7.9.1. Technical Specifications:

In order to develop a subsea water quality measurement instrument, one needs to define the technical requirements specifically for such a device. Through survey and review, two sets of technical specifications for subsea water quality measurement devices – one for subsea separation / processing and one for subsea separation and produced water reinjection (PWRI) were developed through a JIP. These are shown in Table 17.

Table 7.16 Technical specifications for subsea water quality measurement devices

Parameter	Devices for PW Re-injection	Devices for Separation / Processing
Solid concentration (mg/l)	0-300	0-1000
Solid particle size (μm)	0-200	0-200
Oil concentration (mg/l)	0-5000	0-20,000
Oil droplet size (μm)	1-100	10-300
PW temperature (°C)	4 to 175	4 to 175
PW pressure (barg)	220 upstream of the injection pump, up to 690 downstream of the pump. Maximum design pressure should be 690.	220
Sea water temperature (°C)	4	4
Sea water pressure (barg)	300	300
Water depth (m)	3000	3000
Maximum flow velocity (m/s)	4.6	4.6
Device accuracy (%)	15	15
Response time	2 minutes for oil content 30 minutes for solid particles	
Mean Time Between Failures (MTBF) (Year)	5 (minimum)	

In comparison to surface operations, a subsea device will have to operate reliably and accurately at a water depth up to 3000 m. It will also need to withstand a much higher operating temperature and pressure. In addition, these devices need to have the capability of measuring oils as well as solids both in terms of concentration and particle sizes. In the case of subsea separation and produced water re-injection, measurement of solids (both in terms of concentration and particle size and size distribution) is generally considered to be more important than the measurement of oil.

While the above specifications were developed for subsea separation processing and PWRI applications, a recent RPSEA project has developed technical specifications for a subsea produced water discharge sensor. Some of the key specifications for such a sensor are given in Table 18.

Table 7.17 Technical specifications for subsea produced water discharge sensors

Parameter	Requirements for a subsea discharge sensor	Comments
Oil concentration (mg/l)	0-100	Typical 15
Solid concentration (mg/l)	0-100	Measurement of solids not required. But the presence of solids may affect oil and grease measurement
Accuracy	10-15% or statistically equivalent	Comparison to the EPA Method 1664
Water depth (ft)	10,000	
Sea water temperature (F)	33	28 as the next step
Design temperature (F)	33 – 300	350 as the next step
Operating / Service temperature (F)	20 - 200	
Design pressure (psig)	10,000	15,000 as the next step
Operating / service (psig)	0-5000	220
Flow velocity (ft/s)	Max up to 15 ft/s	desired
Oil density coverage (API)	20 – min 35, maybe 60	
Salinity (ppm)	0-250,000	
Response time	Hourly or better	More from a regulatory compliance monitoring stand point.
Design service life (years)	25	
Mean Time Between Failures (MTBF) (Year)	5	

It should be pointed out that the technical specifications for subsea produced water quality measurement sensors are still evolving. Those given in Table 17 and Table 18 should provide a useful starting point if one is to consider the development of a subsea water quality measurement instrument.

7.9.2. Potential Technologies:

Content of solid particles and oil droplets as well as their size and size distributions. While for subsea produced water discharge, it is most likely that the main parameter required to be monitored will be the content of oil. To measure the water quality, there are a number of potential technologies and instruments available on the market that may be considered. Most of these technologies have already been discussed above. Their comparative capabilities are summarized in Table 19. Extensive reviews of the various technologies have been carried out by NEL through its JIPs and involvement with the RPSEA project.

For subsea oil concentration measurement, Laser Induced Fluorescence (LIF), microscopy based image analysis systems, and light scattering based technologies currently appear to be potentially be suitable. In fact vendors such as Advanced Sensors, ProAnalysis, J M Canty, Jorin and Digitrol have all been involved in developing a subsea oil in water measurement device.

For the monitoring of both oil and solids content, as well as size and size distribution, which is important for subsea produced water re-injection applications, those that are based on microscopy image analysis have the best potential. Ultrasonic based technology may also have some potential, but in comparison, this technologie is less established than the use of microscopy and image analysis. However, since ultrasonic based technology uses no optical window, it would be less susceptible to the issue of fouling.

Table 7.18 Potential techniques for subsea water quality measurement

Technique	Oil content	Solid content	Particle size & distribution	Topside applications	Subsea applications
Ultrasonic	√	√	√	√	X
Microscopy	√	√	√	√	X
LIF	√	X	X	√	X
Light scattering	√	X	X	√	√

7.9.3. Research and Development Activities:

In the recent years, a significant amount of Research and Development effort have been put into the development of a subsea water quality measurement device. Among the operators, Statoil, Exxon-Mobil and Petrobras have been most active.

Statoil started an R&D project in 2012 aimed at developing and qualifying an online subsea oil in water monitor in use for a subsea production facility (Subsea Factory). The technical specifications developed by Statoil are given in Table 20.

Table 7.19 – Statoil Subsea OIW Monitor Specifications

Parameters	Range
Oil concentration (ppm)	10 – 3000
Solid concentration (ppm)	0 – 100 (optional)
Solid particle size (micron)	1 – 200 (optional)
Process temperature (C)	4 – 100
Process pressure (bar)	20 – 100
Flow velocity (m/s)	1 – 10
Water depth (m)	3000
Measurement accuracy (%)	20 (relative to conc.)
Response time	< 1 minutes
Mean time between failure	5 year

Through laboratory testing and evaluation, Statoil have now selected three technologies that will undergo offshore field trial. These technologies include:

- Microscopy based image analysis system from Aker Solutions (utilizing Jorin's ViPA as the core technology);
- LIF based system from ProAnalysis;
- Ultrasonic based from Mirmorax.

Following the field trial, further development may follow in terms of marinizing and integration.

ExxonMobil have recently co-developed and tested two PWQM (Produced Water Quality Monitoring) prototypes based on using J M Canty's microscopy based image analysis technique. ExxonMobil think that the key technical gaps for subsea water quality monitoring include: inadequate oil in water measurement range, lack of subsea experience and limitation in detecting both oil and solids in water simultaneously. Also fouling of the optical sensors must be addressed.

The developed prototypes have incorporated water jetting to mitigate fouling. With a microscopy based system, it is able to measure both solid and oil droplets in terms concentration as well as size and size distribution. The two prototypes have a measurement range of 0 - 2,500 ppm and 0 – 50,000 ppm OIW respectively.

Flow loop tests conducted have demonstrated that the prototypes can measure OIW concentration in the designed range and that they can also provide size and size distribution measurements for oil droplets and solid particles. The prototypes were also found to be not affected by operating conditions such as temperature, pressure, salinity and, to an extent, flow velocity.

Petrabras has gone a step further. It has actually qualified and installed one subsea oil in water sensor at their Marlim field where a subsea separation and produced water re-injection system was implemented in 2012. Few details are available from Petorbras, but the authors understand that the subsea oil in water monitor was based on using a light scattering technology with the core technology provided by Optek. To prevent fouling, a hydrodynamic cleaning mechanism has been incorporated into the subsea measurement device.

In addition to the work that has been carried out by the individual operators, collaborative JIPs have been conducted.

In the past 5 years, NEL has conducted two Joint Industry Projects (JIPs) aimed at accelerating the development of these devices. While the first JIP focused on reviewing potential technologies and establishing functional specifications, the second concentrated on performance testing of a number of selected technologies and identifying technology gaps. With the success of these two projects, a third JIP was initiated in 2014. The third JIP, supported by three operators and one subsea separation systems provider, is aimed at developing subsea water quality measurement technology to TRL (Technology Readiness Level) 5.

In the USA, a government supported project entitled "Subsea Produced Water Sensor Development Project" was initiated in 2014. The primary goal of the project is to develop sensors that can be used for regulatory compliance monitoring purposes for the discharge of produced water separated subsea. The project, which aims at bringing potential sensors to a TRL 3 or 4 as defined in API RP17N, consists of two phases. Phase 1 (9 months) is about concept proof of a new and emerging oil-in-water measurement technique based on using Confocal Laser Fluorescence Microscopy (CLFM); developing technical requirements for subsea produced water discharge sensors and conducting a gap analysis on some of the existing oil-in-water measurement sensors. Phase 2 (15 months) is about the design, development and performance testing of a number of sensors that will have been selected from Phase 1. The twenty four month project is expected to be completed by September 2016.

References to Chapter 7

1. M. Yang, "Challenges for water quality measurement for produced water handling subsea," OTC – 23099, paper presented at the Offshore Technology Conference, Houston, TX (2012).

2. U.S. EPA, "Method 1664 Revision A: n-hexane Extractable Material and Silica Gel Treated n-Hexane Extractable Material by Extraction and Gravimetry," National Service Center for Environmental Publications EPA-821-R-98-002 (1999).

3. P.R. Hart, "Removal of Water Soluble Organics from Produced Brine without Formation of Scale," SPE 80250 (2003).

4. L.-G. Faksness, P.G. Grini, P.S. Daling, "Recent results from correlating dispersed oil and PAH content in produced water, and its impact on the selection of treatment technologies," TUVNEL Produced Water Workshop, Aberdeen (March 2003).

5. P.R. Hart, "Progress Made in Removing Water Soluble Organics From GOM Produced Water," Sept (2006).

6. L.G. Faksness, G. Grini, S. Daling, "Partitioning of Semi-Soluble Organic Compounds Between the Water Phase and Oil Droplets in Produced Water," Marine Pollution Bulletin 48, pp. 731-742 (2004).

7. S.A. Ali, L.R. Henry, J.W. Darlington, J. Occhipinti, "Novel Filtration Process Removes Dissolved Organics from Produced Water and Meets Federal Oil and Grease Guidelines," PWS Meeting (1999).

8. Paul R. Hart, "Removal of Water Soluble Organics from Produced Water in the Gulf of Mexico," Produced Water Society meeting, Houston (2005).

9. A. Descousse, K. Monig, K. Voldum, "Evaluation Study of Various Produced-Water Treatment Technologies to Remove Dissolved Aromatic Components," SPE 90103 (2004).

10. D. Th. Meijer, C.A.T. Kuijvenhoven, "Field-proven removal of dissolved hydrocarbons from offshore produced water by the Macro Porous Polymer-Extraction technology," OTC – 13217, paper presented at the Offshore Technology Conference, Houston (2001).

11. Dutch Standard NEN 6675, "Determination of mineral oil content by infrared spectroscopy – (three wavelength)," (1989).

12. M. Yang, "Oil in Produced Water Analysis and Monitoring in the North Sea," SPE 102991 (2006).

13. J. Keathley, K. Konrad, "Comparison of hexane and DuPont Vertrel MCA as extraction solvents in the analysis of oil in produced water using EPA method 1664 and the Wilks Model HATR Infracal analyzer," paper presented at the 10th Annual Produced Water Society Seminar, Houston (2000).

Index

The section number is given here for the major subjects covered in this book.

Made in the USA
Columbia, SC
07 July 2024

38184119R00274